STUDENT'S SOLUTIONS MANUAL

NANCY S. BOUDREAU
Emeritus Associate Professor
Bowling Green State University

STATISTICS

THIRTEENTH EDITION

James McClave
Info Tech, Inc.
University of Florida

Terry Sincich
University of South Florida

PEARSON

Boston Columbus Indianapolis New York San Francisco
Amsterdam Cape Town Dubai London Madrid Milan Munich Paris Montréal Toronto
Delhi Mexico City São Paulo Sydney Hong Kong Seoul Singapore Taipei Tokyo

The author and publisher of this book have used their best efforts in preparing this book. These efforts include the development, research, and testing of the theories and programs to determine their effectiveness. The author and publisher make no warranty of any kind, expressed or implied, with regard to these programs or the documentation contained in this book. The author and publisher shall not be liable in any event for incidental or consequential damages in connection with, or arising out of, the furnishing, performance, or use of these programs.

Reproduced by Pearson from electronic files supplied by the author.

Publishing as Pearson, 501 Boylston Street, Boston, MA 02116.

ISBN-13: 978-0-13-408112-0
ISBN-10: 0-13-408112-9

7

www.pearsonhighered.com

PEARSON

TABLE OF CONTENTS

Preface

This solutions manual is designed to accompany the text, *Statistics*, Thirteenth Edition, by James T. McClave and Terry Sincich. It provides solutions to the odd-numbered exercises for each chapter in the text.

Other methods of solution may also be appropriate; however, the author has presented one that she believes to be the most instructive to the beginning statistics student. The student should first attempt to solve the assigned exercises without the help from this manual. Then, if unsuccessful, the solution in the manual will clarify points necessary to the solution. The student who successfully solves an exercise should still refer to the manual's solution. Many points are clarified and expanded upon to provide maximum insight into and benefit from each exercise.

Instructors will also benefit from the use of this manual. It will save time in preparing presentations of the solutions and possibly provide another point of view regarding their meaning.

Some of the exercises are subjective in nature and thus omitted from the Answer Key at the end of *Statistics*. The subjective decisions regarding these exercises have been made and are explained by the author. Solutions based on these decisions are presented; the solution to this type of exercise is often most instructive. When an alternative interpretation of an exercise may occur, the author has often addressed it and given justification for the approach taken.

Nancy S. Boudreau, Associate Professor Emeritus
Bowling Green State University
Bowling Green, Ohio

Chapter 1

Statistics, Data, and Statistical Thinking

1.1 Statistics is a science that deals with the collection, classification, analysis, and interpretation of information or data. It is a meaningful, useful science with a broad, almost limitless scope of applications to business, government, and the physical and social sciences.

1.3 The first element of inferential statistics is the population of interest. The population is a set of existing units. The second element is one or more variables that are to be investigated. A variable is a characteristic or property of an individual population unit. The third element is the sample. A sample is a subset of the units of a population. The fourth element is the inference about the population based on information contained in the sample. A statistical inference is an estimate, prediction, or generalization about a population based on information contained in a sample. The fifth and final element of inferential statistics is the measure of reliability for the inference. The reliability of an inference is how confident one is that the inference is correct.

1.5 Quantitative data are measurements that are recorded on a meaningful numerical scale. Qualitative data are measurements that are not numerical in nature; they can only be classified into one of a group of categories.

1.7 A population is a set of existing units such as people, objects, transactions, or events. A sample is a subset of the units of a population.

1.9 An inference without a measure of reliability is nothing more than a guess. A measure of reliability separates statistical inference from fortune telling or guessing. Reliability gives a measure of how confident one is that the inference is correct.

1.11 The data consisting of the classifications A, B, C, and D are qualitative. These data are nominal and thus are qualitative. After the data are input as 1, 2, 3, and 4, they are still nominal and thus qualitative. The only differences between the two data sets are the names of the categories. The numbers associated with the four groups are meaningless.

1.13 a. The population of interest to the researchers is all young women who recently participated in a STEM program.

b. The sample of interest is the 159 young women recruited to complete an on-line survey.

c. Of the 159 young women who recently participated in a STEM program, 27% felt that participation in the STEM program increased their interest in science. We can infer that 27% of all young women who recently participated in a STEM program felt that participation in the STEM program increased their interest in science.

1.15 a. The experimental units for this study are the 15 earthquake sites.

b. From 1940 to 1995, there were many more than 15 earthquakes. Thus, these 15 earthquakes represent a sample.

c. There are 3 variables in this problem. The variable 'type of ground motion' has 3 levels (short, long, or forward directive). Thus, this variable is qualitative. The variable 'earthquake magnitude' is measured on a Richter scale which results in a meaningful number. Thus this variable is quantitative. The variable peak ground acceleration is measured in feet per second and is quantitative.

1.17 a. The experimental units are the pipe locations.

b. The sample is the 19 different randomly selected pipe locations.

c. The population is the set of all possible pipe locations.

d. Since not all the pipe locations were sampled, this is an example of inferential statistics. The researchers want to predict if one test is more desirable than the other for all possible locations.

1.19 a. Species of sea buckthorn is a category. Therefore, the variable is qualitative.

b. Altitude of collection location is a number. Therefore, it is a quantitative variable.

c. Total flavonoid content is a number. Therefore, it is a quantitative variable.

1.21 a. The town where the sample was collected is a name and thus, is a qualitative variable.

b. The type of water supply is a category. This is a qualitative variable.

c. The acidic level is a meaningful number. Thus, this is a quantitative variable.

d. The turbidity level is measured on a numerical scale and is a quantitative variable.

e. The temperature is measured on a numerical scale and is a quantitative variable.

f. The number of fecal coliforms per 100 milliliters is measured on a numerical scale and is thus, a quantitative variable.

g. The free chlorine-residual (milligrams per liter) is measured on a numerical scale and is a quantitative variable.

h. The presence of hydrogen sulphide is measured on a categorical scale (yes or no). This is a qualitative variable.

1.23. This study represents a descriptive statistical study. Data were collected and used to portray or describe the condition of the U.S. Treasury in 1861. No inferences were made.

1.25 a. Length of maximum span can take on values such as 15 feet, 50 feet, 75 feet, etc. Therefore, it is quantitative.

b. The number of vehicle lanes can take on values such as 2, 4, etc. Therefore, it is quantitative.

c. The answer to this item is "yes" or "no," which are not numeric. Therefore, it is qualitative.

d. Average daily traffic could take on values such as 150 vehicles, 3,579 vehicles, 53,295 vehicles, etc. Therefore, it is quantitative.

e. Condition can take on values "good," "fair," or "poor," which are not numeric. Therefore, it is qualitative.

f. The length of the bypass or detour could take on values such as 1 mile, 4 miles, etc. Therefore, it is quantitative.

g. Route type can take on values "interstate," U.S.," "state," "county," or "city," which are not numeric. Therefore, it is qualitative.

1.27 a. The data collection method used by the cancer researchers is a designed experiment. Those participating in the study were randomly assigned to one of the two screening methods.

b. The experimental units in this study are the 50,000 smokers.

c. The variable measured in this study is the age at which the scanning method first detects a tumor. Since age is represented by a meaningful number, it is quantitative.

d. The population of interest is all smokers in the U.S. The sample of interest is the set of 50,000 smokers participating in the study.

e. The inference of interest is the difference in mean age at which each of the scanning methods first detects a tumor.

1.29 a. The data collection method used by the researchers was a designed experiment. Half of the boxers received the massage and half did not.

b. The experimental units are the amateur boxers.

c. There are two variables measured on the boxers – heart rate and blood lactate level. Both of these variables are measured on a numeric scale and thus, are quantitative.

d. The inferences drawn from the analysis are: there is no difference in the mean heart rates between the two groups of boxers (those receiving massage and those not receiving massage) and there is no difference in the mean blood lactate levels between the two groups of boxers. Thus, massage did not affect the recovery rate of the boxers.

e. No. Only amateur boxers were used in the experiment. Thus, all inferences relate only to boxers.

1.31. a. Since only students were sampled, the population of interest would be all students. The sample would be the 155 students who participated in the experiment.

b. The data were collected using a designed experiment. Each of the participants was randomly assigned to a treatment (emotional state).

c. The researchers inferred that a higher proportion of students in the guilty-state group chose not to repair the car than those in the neutral-state and anger-state groups.

d. One factor that could affect the reliability of the inference from this study is how representative the sample is. The sample was 155 volunteer students. Students in psychology classes typically have to volunteer for a certain number of experiments. These students may not take the study seriously or may not be representative of all students.

1.33. a. Data for age would be quantitative. Age is measured on a numerical scale. Data for gender would be qualitative. Gender is either 'male' or 'female'. Data for dating experience is quantitative. The number of dates is measured on a numerical scale. Data for the extent to which a student was willing to tell parents about a dating issue would be qualitative. The responses could be 'never tell', 'rarely tell', 'sometimes tell', 'almost always tell', and 'always tell'.

b. If students were recruited from a primarily European American middle-class school district, then the responses would probably not be typical of all high school students. Upper-class students and lower-class students' behaviors might be quite different from middle-class students' behaviors. In addition, the behaviors of minority students might be quite different from that of European American students. Thus, the inferences from this study may not be reflective of the entire population of high school students.

1.35. a. The data collection method for this study is observational. Specifically, a survey was conducted to collect the data.

b. The data collected for this study are qualitative. The responses were either 'Obama' or 'Palin'.

c. The results of this study are very suspect. The sample used was self-selected. Only those readers who elected to respond to the question were included. People who respond to these types of surveys typically have very strong views and may not be representative of the entire population. In addition, readers who might want to respond but who do not have access to an online computer would not be able to participate in the survey. Thus, the results would very likely not be representative of the population in general.

Chapter 2

Methods for Describing Set of Data

2.1 The class frequency is the number of observations in the data set falling in a particular class. The class relative frequency is the class frequency divided by the total number of observations in the data set. The class percentage is the class relative frequency multiplied by 100.

2.3 In a bar graph, a bar or rectangle is drawn above each class of the qualitative variable corresponding to the class frequency or class relative frequency. In a Pareto diagram, the bars of the bar graph are arranged in order from the largest to the smallest from left to right.

2.5 a. To find the frequency for each class, count the number of times each letter occurs. The frequencies for the three classes are:

Class	Frequency
X	8
Y	9
Z	3
Total	20

b. The relative frequency for each class is found by dividing the frequency by the total sample size. The relative frequency for the class X is 8/20 = .40. The relative frequency for the class Y is 9/20 = .45. The relative frequency for the class Z is 3/20 = .15.

Class	Frequency	Relative Frequency
X	8	.40
Y	9	.45
Z	3	.15
Total	20	1.00

c. The frequency bar chart is:

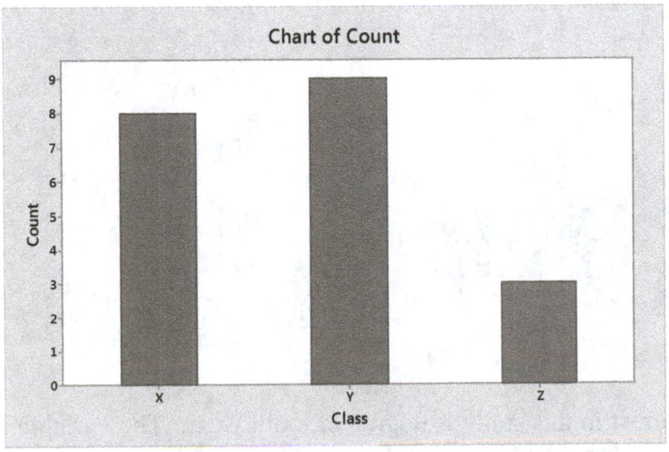

d. The pie chart for the relative frequency distribution is:

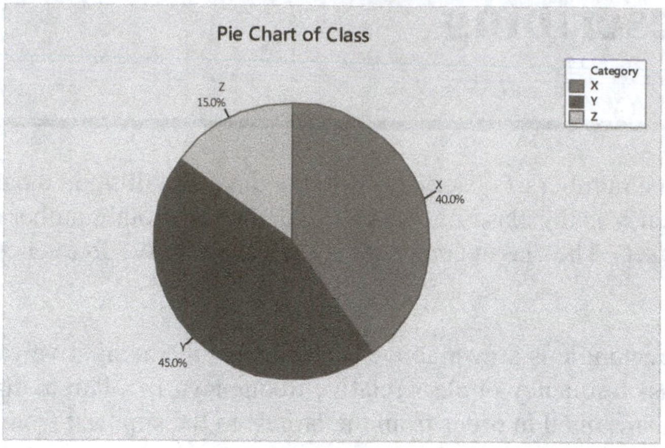

2.7 a. The type of graph is a bar graph.

b. The variable measured for each of the robots is type of robotic limbs.

c. The social robot design used the most is the one with legs only.

d. The relative frequencies are computed by dividing the frequencies by the total sample size. The total sample size is $n = 106$. The relative frequencies for each of the categories are:

Type of Limbs	Frequency	Relative Frequency
None	15	15/106 = 0.142
Both	8	8 / 106 = 0.075
Legs ONLY	63	63/106 = 0.594
Wheels ONLY	20	20/106 = 0.189
Total	106	1.000

e. Using MINITAB, the Pareto diagram is:

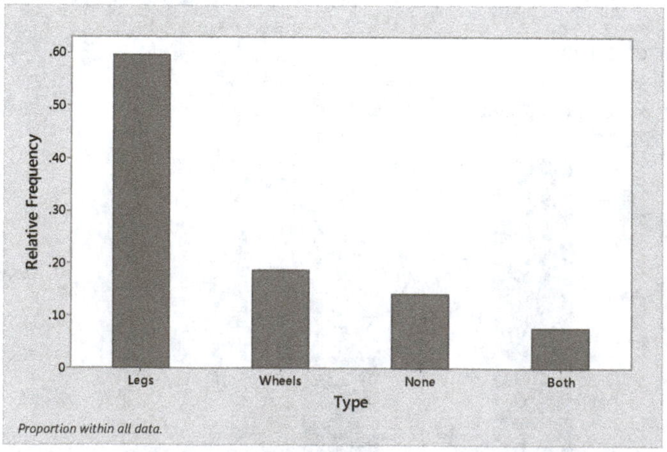

2.9 a. The variable of interest in this study is degree of tooth wear. The variable is qualitative. Possible responses include 'Unknown', 'Heavy', Moderate-heavy', etc.

b. The number of cheek teeth in each wear category is:

Degree of Wear	Frequency	Relative Frequency
Unknown	5	5/18 = 0.278
Unworn	2	2/18 = 0.111
Slight	4	4/18 = 0.222
Light-moderate	2	2/18 = 0.111
Moderate	3	3/18 = 0.167
Moderate-heavy	1	1/18 = 0.056
Heavy	1	1/18 = 0.056
Total	**18**	**18/18 = 1**

c. To compute the relative frequency, you divide the frequency by the total number of observations. For this example, the total number of observations is 18. The relative frequency for 'Unknown' is 5/18 = .278. The rest of the calculations are done in a similar manner and appear in the table in part b.

d. Using MINITAB, a relative frequency bar graph is:

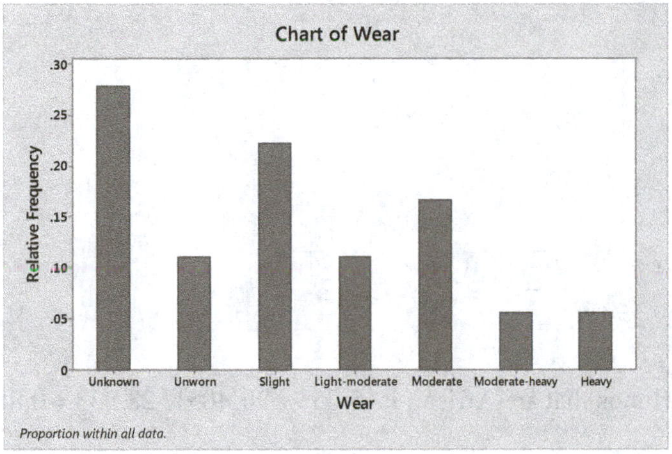

e. A Pareto diagram is a bar graph with the categories arranged in order from the largest to the smallest from left to right. The Pareto diagram for this data is:

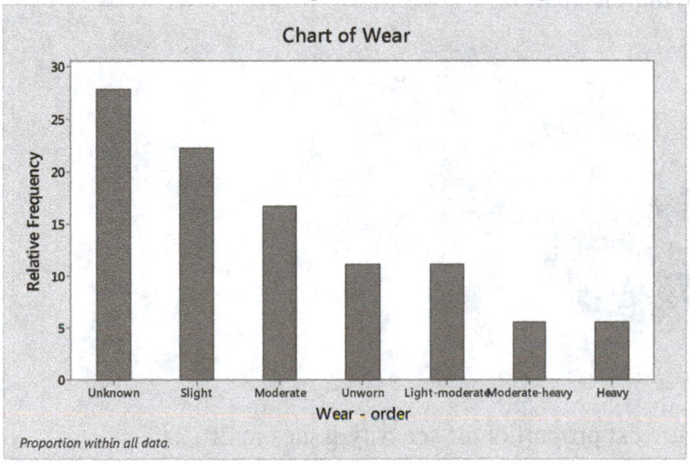

f. The degree of wear category that occurred most often is 'Unknown'.

2.11 a. To construct a relative frequency table for data, we must find the relative frequency for each species of rhinos. To find the relative frequency, divide the frequency by the total population size, 17,800. The relative frequency for African Black rhinos is 3,610/17,800 = 0.203. The rest of the relative frequencies are found in a similar manner and are reported in the table.

Rhino species	Population Estimate	Relative Frequency
African Black	5,055	5055 / 28933 = 0.175
African White	20,405	20405 / 28933 = 0.705
(Asian) Sumatran	100	100 / 28933 = 0.003
(Asian) Javan	40	40 / 28933 = 0.001
(Asian) Indian	3,333	3333 / 28933 = 0.115
Total	**28,933**	**0.999**

b. Using MINITAB, the relative frequency bar chart is:

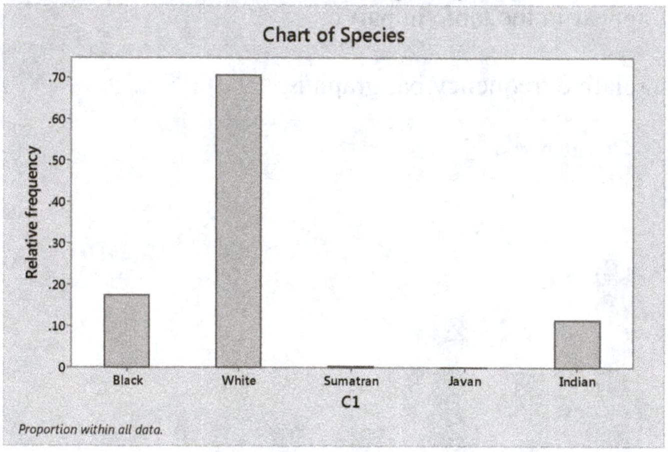

c. The proportion of rhinos that are African is $(5,055+20,405)/28,933=0.880$.

The proportion of rhinos that are Asian is $(100+40+3,333)/28,933=0.120$.

2.13 a. Using MINITAB, the pie chart is:

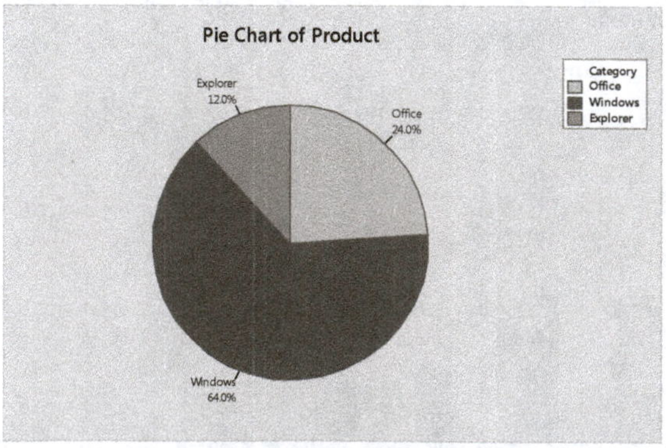

Explorer had the lowest proportion of security issues in 2012.

b. Using MINITAB, the Pareto diagram is:

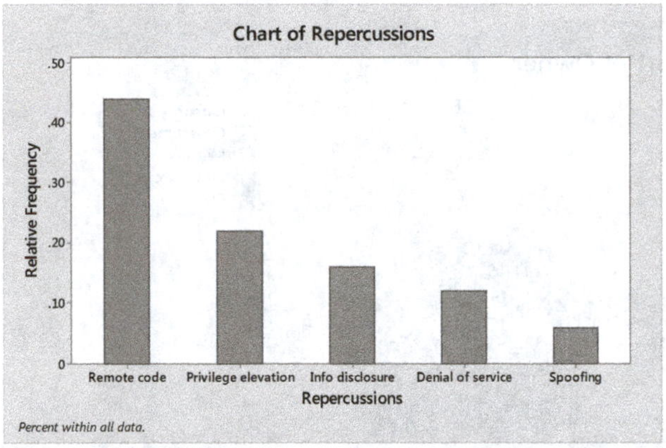

We would advise Microsoft to focus on remote code since it has the highest relative frequency.

2.15 a. From the summary table, the proportion of melt ponds that had landfast ice is 38.89 / 100 = 0.3889.

b. Yes. From the summary table 17.46% of the melt ponds had first-year ice.

c. Using MINITAB, the Pareto diagram is:

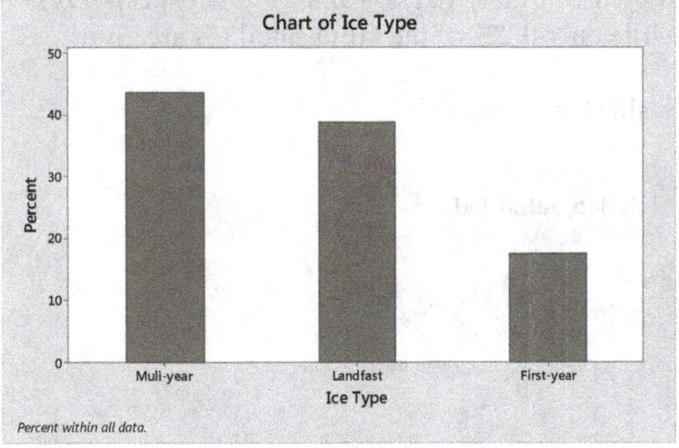

The most commonly found type of ice is multi-year ice, followed by landfast ice. The least frequently found type of ice is first-year ice.

2.17 Using MINITAB, pie charts for the two sets of data are:

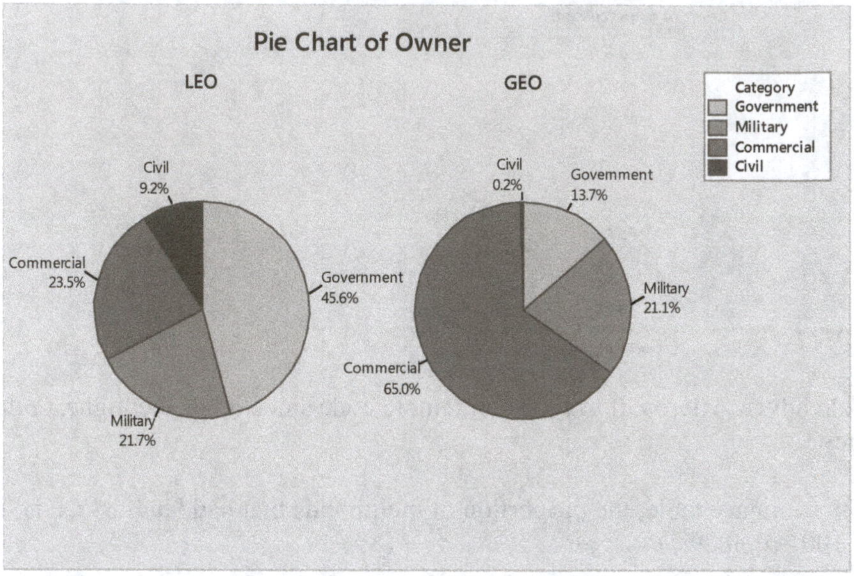

In the LEO, most of the satellites are owned by the government (45.6%), while in GEO, most of the satellites are commercially owned (5.0%). About the same percentage of both LEO and GEO satellites are owned by the military (21.7% and 21.1% respectively). About 9% of the LEO satellites are civil, while only 0.2% of the GEO satellites are civil.

2.19 Using MINITAB, the side-by-side bar charts are:

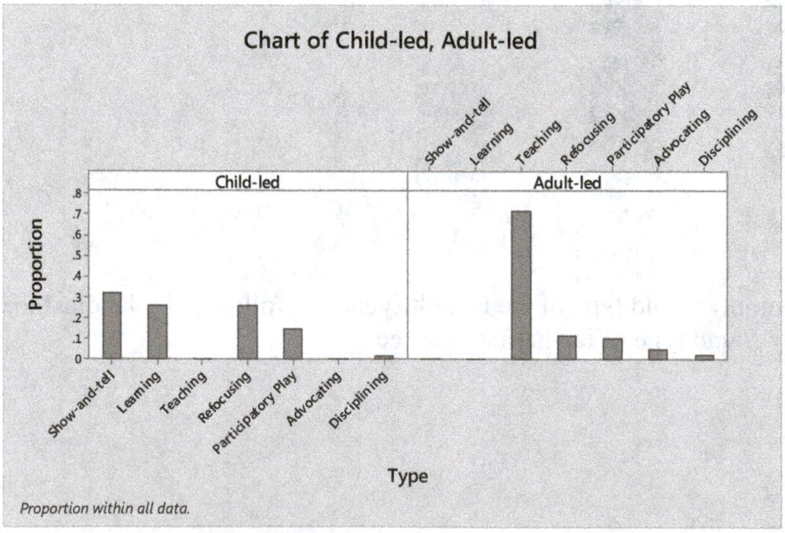

The four main categories for child-led interactions are show-and-tell, learning, refocusing, and participatory play. Adults had only one main category which was teaching.

2.21 a. The frequency table is:

Response	Frequency	Relative Frequency
1	1527	1527/4826 = 0.316
2	2231	2231/4826 = 0.462
3	996	996/4826 = 0.206
4	72	72/4826 = 0.015
Totals	**4826**	**0.999**

b. Using MINITAB, the pie chart for the data is:

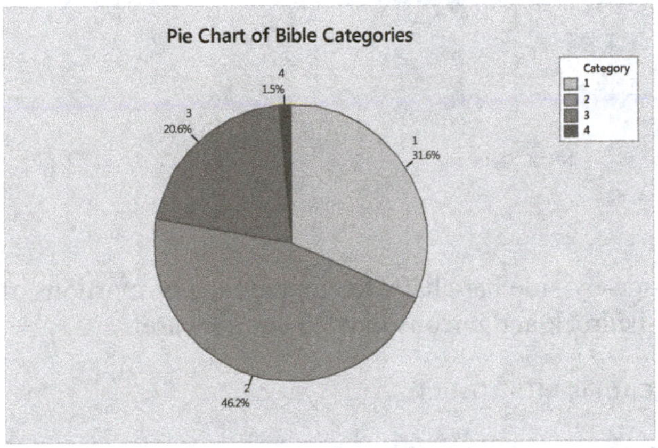

c. The response with the highest frequency is 2, 'the Bible is the inspired word of God but not everything is to be taken literally'. About 46% of the respondents selected this answer. A little less than one-third of the respondents answered 1, 'the Bible is the actual word of God and is to be taken literally'. Very few (1.5%) of the respondents chose response 4, 'the Bible has some other origin'.

2.23 a. Using MINITAB, pie charts for Well Class, Aquifer, and Detect MTBE are:

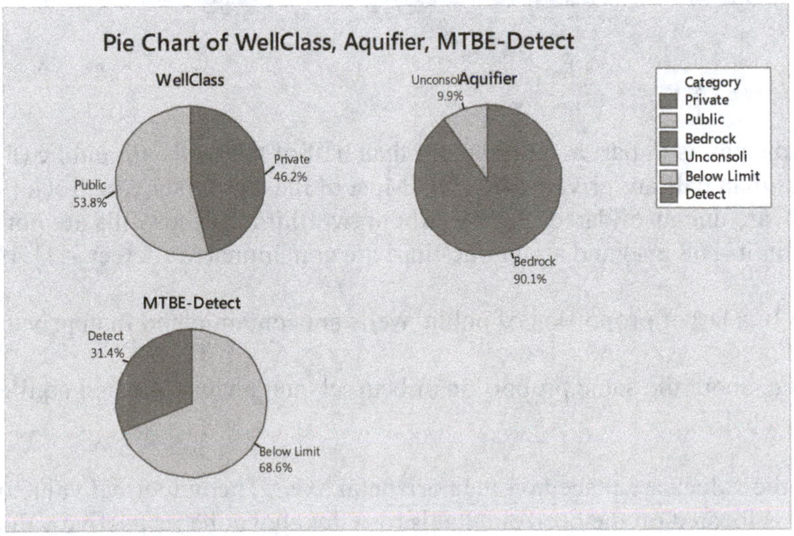

b. Using MINITAB, the side-by-side bar charts to compare the proportions of contaminated wells for private and public wells classes are:

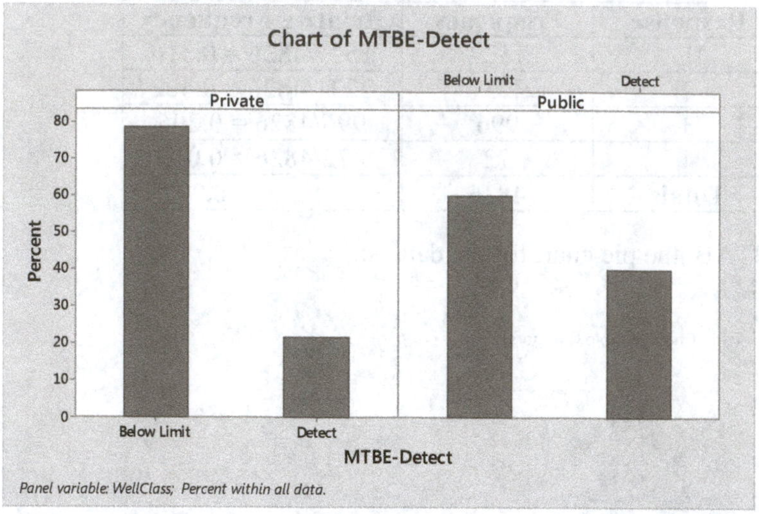

c. Using MINITAB, the side-by-side bar charts to compare the proportions of contaminated wells for bedrock and unconsolidated aquifers are:

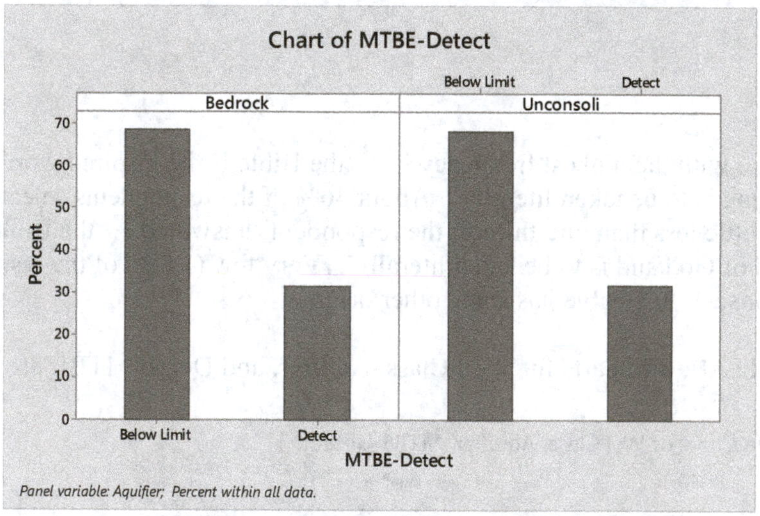

d. For the pie charts in part **a**, a little more than half of the wells are public (53.8%) and a little less than half are private (46.2%). Most of the aquifers are bedrock (90.1%) and very few are unconsolidated (9.9%). About two-thirds of the wells are not contaminated (below limit – 68.6%) and about one-third are contaminated (detect – 31.4%).

For part **b**, a larger proportion of public wells are contaminated than private wells.

For part **c**, about the same proportion of bedrock and unconsolidated aquifers are contaminated.

2.25 In a dot plot, the values are placed on the horizontal axis. The numerical value of each measurement is located on the horizontal axis by a dot above the respective value. When values repeat, the dots are placed one above the other. In a stem-and-leaf display, the stems are the left-most digits of the measurements in the data set. They are arranged vertically from the smallest to the largest. The leaves are the right-most digit of the measurements in the data set. Each leaf is placed beside the appropriate stem. For each stem, the leaves are arranged in order from the smallest to the largest.

2.27 In a stem-and-leaf display, the stem is the left-most digit of a measurement, while the leaf is the right-most digit of a measurement.

2.29 As a general rule for data sets containing between 25 and 50 observations, we would use between 7 and 14 classes. Thus, for 50 observations, we would use around 14 classes.

2.31 a. The original data set has $1+3+5+7+4+3=23$ observations.

b. For the bottom row of the stem-and-leaf display:

The stem is 0.
The leaves are 0, 1, 2.
The numbers in the original data set are 0, 1, and 2.

c. The dot plot corresponding to all the data points is:

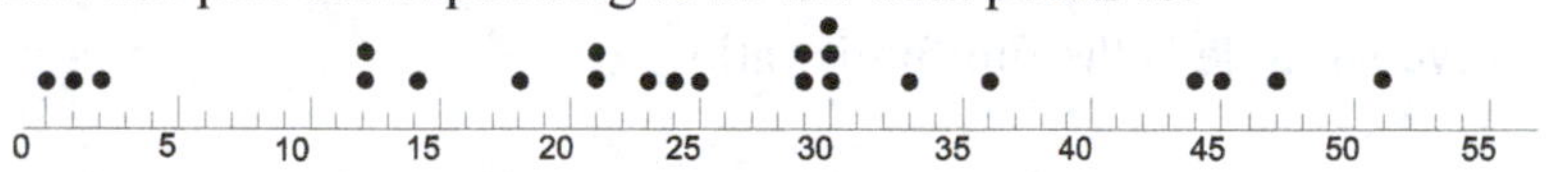

2.33 To find the number of measurements for each measurement class, multiply the relative frequency by the total number of observations, $n = 500$. The frequency table is:

Measurement Class	**Relative Frequency**	**Frequency**
0.5 – 2.5	0.10	500(0.10) = 50
2.5 – 4.5	0.15	500(0.15) = 75
4.5 – 6.5	0.25	500(0.25) = 125
6.5 – 8.5	0.20	500(0.20) = 100
8.5 – 10.5	0.05	500(0.05) = 25
10.5 – 12.5	0.10	500(0.10) = 50
12.5 – 14.5	0.10	500(0.10) = 50
14.5 – 16.5	0.05	500(0.05) = 25
		500

The frequency histogram is:

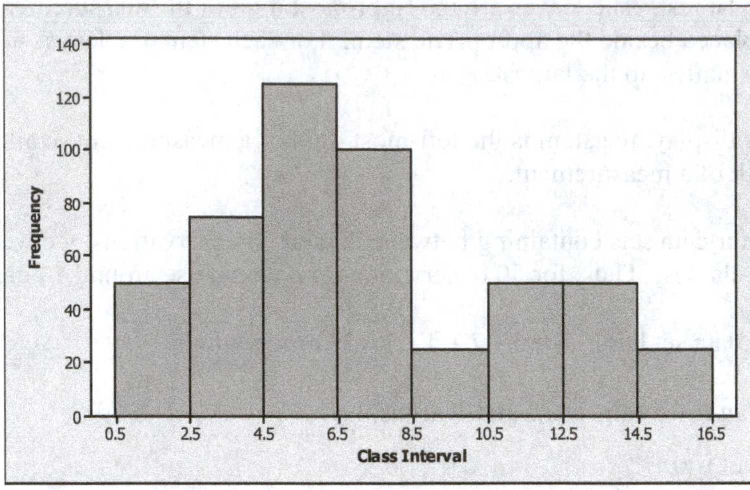

2.35 a. Even though there are gaps between the bars, this is a frequency histogram because the fup/fumic ratio is a quantitative variable.

b. The quantitative variable is the fup/fumic ratio.

c. The proportion of fup/fumic ratios that fall above 1 is $(9+4+1)/416=0.034$.

d. The proportion of fup/fumic ratios that fall below 0.4 is $(182+107)/416=0.695$.

2.37 Using MINITAB, a dot plot is:

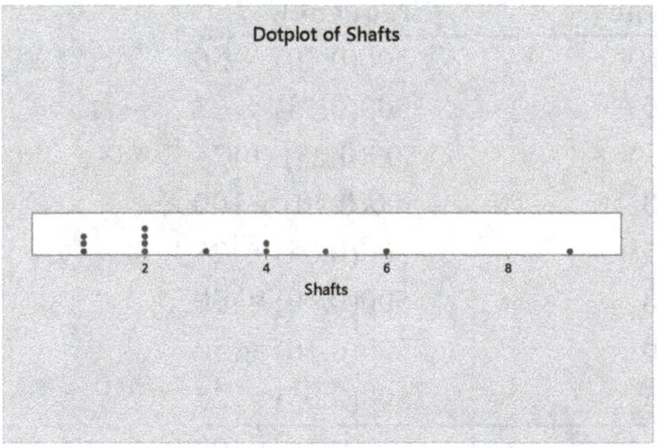

The number of sword shafts that was observed most often was 2.

2.39 a. Using MINITAB, the stem-and-leaf plot is:

Stem-and-Leaf Display: PAI

```
Stem-and-leaf of PAI  N  = 8
Leaf Unit = 1.0

  1    3  9
  3    4  01
 (2)   4  23
  3    4
  3    4
  3    4
  3    5  1
  2    5
  2    5  45
```

b. The value of 42 is highlighted in gray on the plot in part **a**.

c. The PAI range most likely to occur is 40-49. Half of the observations are in this range. No other ranges have as many observations.

2.41 a. Using MINITAB, the stem-and-leaf display of the data is:

Stem-and-Leaf Display: Score

```
Stem-and-leaf of Score  N  = 186
Leaf Unit = 1.0

  1     6    9
  1     7
  2     7    3
  3     7    4
  4     7    6
  5     7    8
  7     8    11
  8     8    3
 11     8    445
 20     8    666677777
 25     8    99999
 36     9    00001111111
 54     9    222222333333333333
 89     9    44444444444444444555555555555555555
(42)    9    666666666666666666667777777777777777777777
 55     9    888888888888888889999999999999999999
 19    10    0000000000000000000
```

The stems are 60, 70, 80, … 100. The leaves are units from 0 to 9.

b. From the stem-and-leaf display, we see that there are only 11 observations with sanitation scores less than the acceptable score of 86. The proportion of ships that have an acceptable sanitation standard would be $(186-11)/186=0.941$.

c. The sanitation score of 69 is highlighted in gray on the stem-and-leaf display in part **a**.

2.43 Using MINITAB, a histogram of the data is:

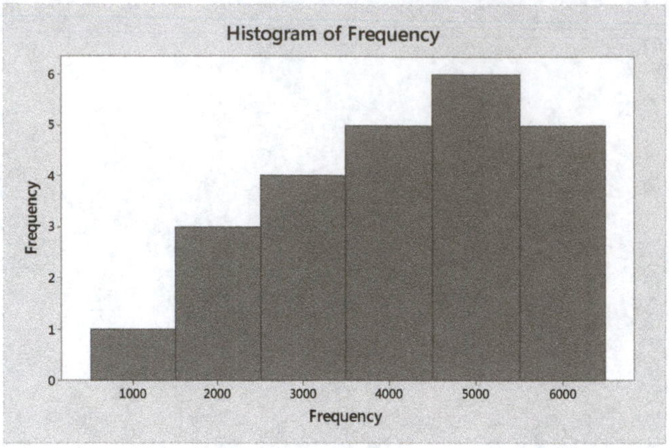

2.45 The dot plot for these data is:

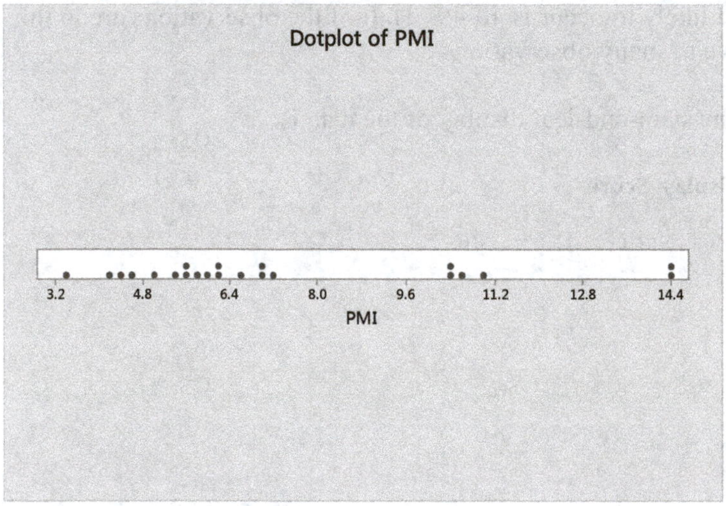

From the dot plot, most of the PMI's range from 3 to 7.5 (16 of the 22). The rest of the PMI's range from 10 to 15.

2.47 Using MINITAB, a histogram of the data is:

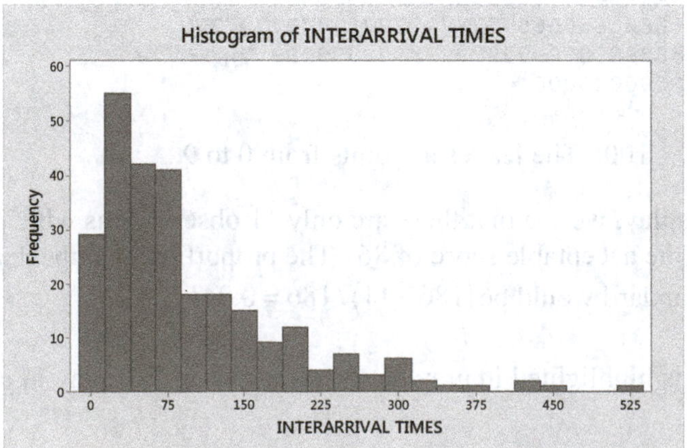

This histogram looks very similar to the one shown in the problem. Thus, it appears that there was minimal or no collaboration or collusion from within the company. We could conclude that the phishing attack against the organization was probably not an "inside job".

2.49 Three measures of central tendency are the mean, the median, and the mode.

2.51 The sample mean is represented by $\bar{x}$. The population mean is represented by μ.

2.53 The two factors that impact the accuracy of the sample mean as an estimate of the population mean are sample size and variability or spread of the data. For the sample size, the larger the sample size, the more accurate the estimate. For variability, all other factors remaining constant, the more variable the data the less accurate the estimate of the mean.

2.55 Assume the data are a sample. The sample mean is:

$$\bar{x} = \frac{\sum x}{n} = \frac{3.2 + 2.5 + 2.1 + 3.7 + 2.8 + 2.0}{6} = \frac{16.3}{6} = 2.717$$

The median is the average of the middle two numbers when the data are arranged in order (since $n = 6$ is even). The data arranged in order are: 2.0, 2.1, 2.5, 2.8, 3.2, 3.7. The middle two numbers are 2.5 and 2.8. The median is: $\frac{2.5 + 2.8}{2} = \frac{5.3}{2} = 2.65$

2.57 a. $\bar{x} = \frac{\sum x}{n} = \frac{85}{10} = 8.5$

b. $\bar{x} = \frac{400}{16} = 25$

c. $\bar{x} = \frac{35}{45} = .78$

d. $\bar{x} = \frac{242}{18} = 13.44$

2.59 a. $\bar{x} = \frac{\sum x}{n} = \frac{7 + \cdots + 4}{6} = \frac{15}{6} = 2.5$

Median $= \frac{3+3}{2} = 3$ (mean of 3rd and 4th numbers, after ordering)

Mode = 3

b. $\bar{x} = \frac{\sum x}{n} = \frac{2 + \cdots + 4}{13} = \frac{40}{13} = 3.08$

Median = 3 (7th number, after ordering)

Mode = 3

c. $\bar{x} = \frac{\sum x}{n} = \frac{51 + \cdots + 37}{10} = \frac{496}{10} = 49.6$

Median $= \frac{48 + 50}{2} = 49$ (mean of 5th and 6th numbers, after ordering)

Mode = 50

2.61 The sample mean is: $\bar{x} = \frac{\sum_{i=1}^{13} x_i}{n} = \frac{10.94 + 13.71 + 11.38 + ... + 6.77}{13} = \frac{126.32}{13} = 9.72$

The average rebound length of these 13 observations is 9.72.
The sample median is found by finding the middle observation once the data are arranged in order. The middle number is 10.94. Thus, the sample median is 10.94. Half of all rebound lengths are longer than 10.94 and half are shorter.

2.63 a. From the printout, the mean is 67.755. The statement is accurate.

b. From the printout, the median is 68.000. The statement is accurate.

c. From the printout, the mode is 64. The statement is not accurate. A better statement would be: "The most common reported level of support for corporate sustainability for the 992 senior managers was 64."

d. Because the values of the mean and median are quite close, the data are fairly symmetric. The histogram in Exercise 2.36 looks to be fairly symmetric.

2.65 a. The sample mean is $\bar{x} = \frac{\sum_{i=1}^{n} x_i}{n} = \frac{18.12 + 19.48 + 19.36 + ... + 16.20}{18} = \frac{296.99}{18} = 16.499$.
The average depth of all the teeth collected is 16.499 mm. If the largest depth measurement were doubled, the mean would increase.

b. There are 18 measurements in the data set so the median is the average of the middle 2 numbers once the data have been arranged in order. The 9th and 10th observations when arranged in order are 16.12 and 16.20. The average is $\frac{16.12 + 16.20}{2} = \frac{32.32}{2} = 16.16$.
Thus, the median is 16.16. Half of the depths are larger than 16.16 and half are smaller. If the largest depth measurement were doubled, the median would not change. It would stay the same.

c. Since no observation occurs more than once, then we say there is either no mode or we say that all of the observations are modes.

2.67 a. The salaries of all persons employed by a large university are probably skewed to the right. There will be a few individuals with very large salaries (i.e. president, football coach, Dean of the Medical school). However, the majority of the employees will have salaries in a rather small range.

b. The grades on an easy test will probably be skewed to the left. Most students will get very high grades on the test. Since there is an upper limit to the grades (i.e. 100%), there will likely be many grades in this upper range. However, even on an easy test, a few individuals will still not do well.

c. The grades on a difficult test will probably be skewed to the right. Most students will get fairly low grades on the test. However, even on a difficult test, a few individuals will still do quite well.

d. The amounts of time students in your class studied last week will probably be close to symmetric. Some individuals will not study very much, while others will study quite a bit. However, most students will study an average amount of time.

e. The ages of cars on a used car lot will probably be skewed to the left. Most of the cars will be fairly new. However, there will probably be a few fairly old cars.

f. The amounts of time spent by students on a difficult examination will probably be skewed to the left. If there is a maximum time limit, then most students will take that amount of time or close to it. There will probably be a few students who take less time than the maximum allowed.

2.69 a. For Group A, the mean is 73.62 and the median is 70.45. The average permeability value for Group A is 73.62. Half of the permeability values for Group A are less than or equal to 70.45 and half are greater than 70.45.

b. For Group B, the mean is 128.54 and the median is 139.30. The average permeability value for Group B is 128.54. Half of the permeability values for Group B are less than or equal to 139.30 and half are greater than 139.30.

c. For Group C, the mean is 83.07 and the median is 78.65. The average permeability value for Group C is 83.07. Half of the permeability values for Group C are less than or equal to 78.65 and half are greater than 78.65.

d. The mode for group C is 70.9. Three observations in Group C had values of 70.9.

e. Weathering type B appears to result in faster decay. The mean, median and mode for this group are greater than the corresponding values in both Groups A and C.

2.71 a. Using MINITAB, the measures of central tendency are:

Descriptive Statistics: PRDiff_outlier

```
                                            N for
Variable          N    Mean  Median  Mode    Mode
PRDiff_outlier   14  -1.091  -0.655     *       0
```

The average peripheral refraction difference is -1.091. The median peripheral refraction difference is -0.655. Half of the peripheral refraction difference values are less than or equal to -0.655 and half are greater than -0.655. No observation appeared more than once, so there is no mode.

b. The one large negative value is -8.11.

c. Using MINITAB, the measures of central tendency with -8.11 changed to -0.11 are:

Descriptive Statistics: PRDiff

```
                                          N for
Variable    N     Mean   Median   Mode    Mode
PRDiff     14   -0.519   -0.520      *       0
```

The mean increases from -1.091 to -0.519 or increases by 0.572. The median increases from -0.655 to -0.520 or increases by 0.135. The mean is much more affected by correcting the outlier than the median.

2.73 The range of a data set is the difference between the largest and smallest measurements.

2.75 The sample variance is the sum of the squared deviations from the sample mean divided by the sample size minus 1. The population variance is the sum of the squared deviations from the population mean divided by the population size.

2.77 If the standard deviation increases, this implies that the data are more variable.

2.79 a. Range $= 42 - 37 = 5$

$$s^2 = \frac{\sum x^2 - \frac{\left(\sum x\right)^2}{n}}{n-1} = \frac{7935 - \frac{199^2}{5}}{5-1} = 3.7 \qquad s = \sqrt{3.7} = 1.92$$

b. Range $= 100 - 1 = 99$

$$s^2 = \frac{\sum x^2 - \frac{\left(\sum x\right)^2}{n}}{n-1} = \frac{25{,}795 - \frac{303^2}{9}}{9-1} = 1{,}949.25 \qquad s = \sqrt{1{,}949.25} = 44.15$$

c. Range $= 100 - 2 = 98$

$$s^2 = \frac{\sum x^2 - \frac{\left(\sum x\right)^2}{n}}{n-1} = \frac{20{,}033 - \frac{295^2}{8}}{8-1} = 1{,}307.84 \qquad s = \sqrt{1{,}307.84} = 36.16$$

2.81 This is one possibility for the two data sets.

Data Set 1: 1, 1, 2, 2, 3, 3, 4, 4, 5, 5
Data Set 2: 1, 1, 1, 1, 1, 5, 5, 5, 5, 5

$$\bar{x}_1 = \frac{\sum x}{n} = \frac{1+1+2+2+3+3+4+4+5+5+}{10} = \frac{30}{10} = 3$$

$$\bar{x}_2 = \frac{\sum x}{n} = \frac{1+1+1+1+1+5+5+5+5+5}{10} = \frac{30}{10} = 3$$

Therefore, the two data sets have the same mean. The variances for the two data sets are:

$$s_1^2 = \frac{\sum x^2 - \frac{\left(\sum x\right)^2}{n}}{n-1} = \frac{110 - \frac{30^2}{10}}{9} = \frac{20}{9} = 2.2222$$

$$s_2^2 = \frac{\sum x^2 - \frac{\left(\sum x\right)^2}{n}}{n-1} = \frac{130 - \frac{30^2}{10}}{9} = \frac{40}{9} = 4.4444$$

The dot diagram for the two data sets is shown below.

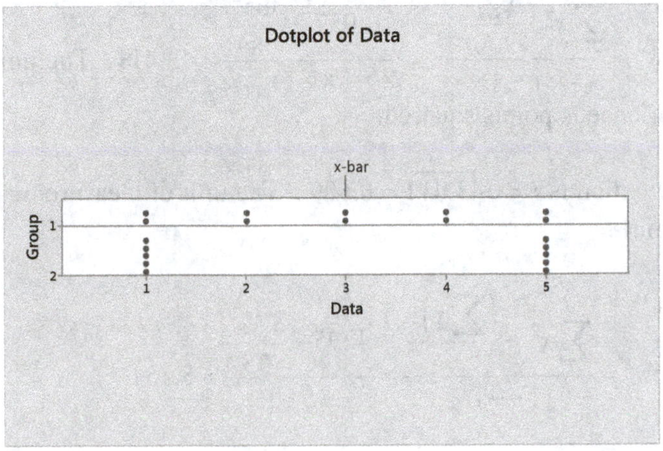

2.83 a. Range = 3 − 0 = 3

$$s^2 = \frac{\sum x^2 - \frac{\left(\sum x\right)^2}{n}}{n-1} = \frac{15 - \frac{7^2}{5}}{5-1} = 1.3 \qquad s = \sqrt{1.3} = 1.1402$$

b. After adding 3 to each of the data points,

Range = 6 − 3 = 3

$$s^2 = \frac{\sum x^2 - \frac{\left(\sum x\right)^2}{n}}{n-1} = \frac{102 - \frac{22^2}{5}}{5-1} = 1.3 \qquad s = \sqrt{1.3} = 1.1402$$

c. After subtracting 4 from each of the data points,

Range = −1 − (−4) = 3

$$s^2 = \frac{\sum x^2 - \frac{\left(\sum x\right)^2}{n}}{n-1} = \frac{39 - \frac{(-13)^2}{5}}{5-1} = 1.3 \qquad s = \sqrt{1.3} = 1.1402$$

d. The range, variance, and standard deviation remain the same when any number is added to or subtracted from each measurement in the data set.

2.85 a. From the printout, the range is 51.26.

b. From the printout, the variance is 128.57.

c. From the printout, the standard deviation is 11.34.

d. If we are interested in just these 76 captured turtles, then the data would represent a population. The variance would be represented by σ^2 and the standard deviation would be represented by σ .

2.87 a. The sample variance is $s^2 = \dfrac{\sum x^2 - \dfrac{\left(\sum x\right)^2}{n}}{n-1} = \dfrac{16{,}957 - \dfrac{365^2}{8}}{8-1} = 43.411$. The unit of measurement for the variance is points squared.

b. The sample standard deviation is $s = \sqrt{43.411} = 6.589$. The unit of measurement for the standard deviation is points.

2.89 a. The sample variance is: $s^2 = \dfrac{\sum x^2 - \dfrac{\left(\sum x\right)^2}{n}}{n-1} = \dfrac{4295 - \dfrac{375^2}{35}}{35-1} = 8.151$

The standard deviation is $s = \sqrt{8.151} = 2.855$

b. The sample variance is: $s^2 = \dfrac{\sum x^2 - \dfrac{\left(\sum x\right)^2}{n}}{n-1} = \dfrac{2631 - \dfrac{275^2}{33}}{33-1} = 10.604$

The standard deviation is $s = \sqrt{10.604} = 3.256$

c. The sample variance is: $s^2 = \dfrac{\sum x^2 - \dfrac{\left(\sum x\right)^2}{n}}{n-1} = \dfrac{1881 - \dfrac{241^2}{37}}{37-1} = 8.646$

The standard deviation is $s = \sqrt{8.646} = 2.940$

d. The DM dosage group appears to have the most variability since the variance of the DM dosage group is larger than the variances of the other 2 groups. The honey dosage group appears to have the least variability since the variance of the honey dosage group is smaller than the variances of the other 2 groups. However, the variance of the honey dosage group and the control group are very similar.

2.91 a. The range is the difference between the largest and smallest observations and is $19.70 - 13.25 = 6.45$ mm. If the largest depth measurement were doubled, the range would increase. The new range would be $39.40 - 13.25 = 26.15$.

b. The variance is: $s^2 = \frac{\sum x^2 - \frac{\left(\sum x\right)^2}{n}}{n-1} = \frac{4966.173 - \frac{296.99^2}{18}}{18-1} = 3.883$

If the largest depth measurement were doubled, the variance would increase. The new variance would be: $s^2 = \frac{\sum x^2 - \frac{\left(\sum x\right)^2}{n}}{n-1} = \frac{6130.443 - \frac{316.69^2}{18}}{18-1} = 32.861$

d. The standard deviation is $s = \sqrt{3.883} = 1.970$. If the largest depth measurement were doubled, the standard deviation would increase. The new standard deviation would be $s = \sqrt{32.861} = 5.732$.

2.93 Chebyshev's rule can be applied to any data set.

2.95 a. The unit of measurement of the variable of interest is dollars (the same as the mean and standard deviation). Based on this, the data are quantitative.

b. Since no information is given about the shape of the data set, we can only use Chebyshev's rule.

\$900 is 2 standard deviations below the mean, and \$2100 is 2 standard deviations above the mean. Using Chebyshev's rule, at least 3/4 of the measurements (or $3/4 \times 200 = 150$ measurements) will fall between \$900 and \$2100.

\$600 is 3 standard deviations below the mean and \$2400 is 3 standard deviations above the mean. Using Chebyshev's rule, at least 8/9 of the measurements (or $8/9 \times 200 = 178$ measurements) will fall between \$600 and \$2400.

\$1200 is 1 standard deviation below the mean and \$1800 is 1 standard deviation above the mean. Using Chebyshev's rule, nothing can be said about the number of measurements that will fall between \$1200 and \$1800.

\$1500 is equal to the mean and \$2100 is 2 standard deviations above the mean. Using Chebyshev's rule, at least 3/4 of the measurements (or $3/4 \times 200 = 150$ measurements) will fall between \$900 and \$2100. It is possible that all of the 150 measurements will be between \$900 and \$1500. Thus, nothing can be said about the number of measurements between \$1500 and \$2100.

2.97 According to the Empirical Rule:

a. Approximately 68% of the measurements will be contained in the interval $\bar{x} - s$ to $\bar{x} + s$.

b. Approximately 95% of the measurements will be contained in the interval $\bar{x} - 2s$ to $\bar{x} + 2s$.

c. Essentially all the measurements will be contained in the interval $\bar{x} - 3s$ to $\bar{x} + 3s$.

2.99 Using Chebyshev's rule, at least 8/9 of the measurements will fall within 3 standard deviations of the mean. Thus, the range of the data would be around 6 standard deviations. Using the Empirical Rule, approximately 95% of the observations are within 2 standard deviations of the mean. Thus, the range of the data would be around 4 standard deviations. We would expect the standard deviation to be somewhere between Range/6 and Range/4.

For our data, the range $= 760 - 135 = 625$.

The Range $/6 = 625/6 = 104.17$ and Range $/4 = 625/4 = 156.25$.

Therefore, I would estimate that the standard deviation of the data set is between 104.17 and 156.25.

It would not be feasible to have a standard deviation of 25. If the standard deviation were 25, the data would span 625/25 = 25 standard deviations. This would be extremely unlikely.

2.101 a. $\bar{x} = \dfrac{\sum x}{n} = \dfrac{17{,}566}{186} = 94.441$

$$s^2 = \frac{\sum x^2 - \frac{\left(\sum x\right)^2}{n}}{n-1} = \frac{1{,}664{,}214 - \frac{17{,}566^2}{186}}{186-1} = 28.464$$

$s = \sqrt{s^2} = \sqrt{28.464} = 5.335$

b. $\bar{x} \pm s \Rightarrow 94.441 \pm 5.335 \Rightarrow (89.106,\ 99.776)$

$\bar{x} \pm 2s \Rightarrow 94.441 \pm 2(5.335) \Rightarrow 94.441 \pm 10.670 \Rightarrow (83.771,\ 105.111)$

$\bar{x} \pm 3s \Rightarrow 94.441 \pm 3(5.335) \Rightarrow 94.441 \pm 16.005 \Rightarrow (78.436,\ 110.446)$

c. There are 142 out of 186 observations in the first interval. This is $(142/186)100\% = 76.3\%$.
There are 175 out of 186 observations in the second interval. This is $(175/186)100\% = 94.1\%$. There are 178 out of 186 observations in the third interval. This is $(178/186)100\% = 95.7\%$.

The percentage for the first interval (74.3%) is somewhat larger than what we would expect using the Empirical Rule (approximately 68%). The percentage in the second interval (94.1%) is close to what we would expect using the Empirical Rule (approximately 95%). The percentage in the third interval (95.7%) is smaller than what we would expect using the Empirical Rule (approximately 99.7%). Chebyshev's Theorem says that at least ¾ or 75% of the observations will fall within 2 standard deviations of the mean and at least 8/9 or 88.8% of the observations will fall within 3 standard deviations of the mean. It appears that our observed percentages agree with Chebyshev's Theorem better that the Empirical Rule.

2.103 a. The mean is 79 and the standard deviation is 23. Thus, a measurement of 102 is 1 standard deviation above the mean. If nothing is known about the shape of the distribution, we must use Chebyshev's Rule to describe the distribution. Using this, we know nothing about the number of observations within 1 standard deviation of the mean, and thus, know nothing about the percentage of observations less than 102.

b. If we assume that the distribution is mound-shaped, we can use the Empirical Rule to describe the distribution. If the distribution is mound-shaped, then it is symmetric. Thus, we know that half of the observations will be below 79. We know that about 68% of the observations are within 1 standard deviation of the mean. Therefore, about 34% of the observations will be between the mean and 1 standard deviation above the mean, or between 79 and 102. Thus, the percentage of observations less than 102 is about $50\% + 34\% = 84\%$.

2.105 Because the histogram in Exercise 2.36 is approximately mound-shaped, the Empirical Rule can be used. The interval $\bar{x} \pm s$ is $67.755 \pm 26.871 \Rightarrow (40.884, 94.626)$. Thus, approximately 68% of the observations will fall between the values 41 and 95. There are a total of 992 observations. Therefore, approximately $0.68(992) = 674.56 \approx 675$ observations will fall in this interval.

2.107 For parts a and b, we must use Chebyshev's Rule to describe the data sets. For both the hand rubbers and the hand washers, the standard deviation is greater than the mean. Since no observations can be less than 0, we know that the smallest observation in the data sets is less than 1 standard deviation from the mean. Thus, the distributions cannot be symmetric, but rather skewed to the right. We cannot use the Empirical Rule.

a. Using Chebyshev's Rule, we know that at least $1 - \frac{1}{2^2} = 1 - \frac{1}{4} = \frac{3}{4} = 0.75$ of the observations fall within 2 standard deviations of the mean.
Thus, the interval $\bar{x} \pm 2s$ will contain at least 0.75 of the observations.

For handrubbers: $\bar{x} \pm 2s \Rightarrow 35 \pm 2(59) \Rightarrow 35 \pm 118 \Rightarrow (0,\ 153)$

b. For handwashers: $\bar{x} \pm 2s \Rightarrow 69 \pm 2(106) \Rightarrow 69 \pm 212 \Rightarrow (0,\ 281)$

c. Since the interval for handwashers is much larger than that for handrubbers and the sample mean for handrubbers is less than that for handwashers, we can conclude that handrubbing is more effective than handwashing for reducing bacterial counts.

2.109 a. Since no information is given about the distribution of the velocities of the Winchester bullets, we can only use Chebyshev's rule to describe the data. We know that at least 3/4 of the velocities will fall within the interval:

$$\bar{x} \pm 2s \Rightarrow 936 \pm 2(10) \Rightarrow 936 \pm 20 \Rightarrow (916, 956)$$

Also, at least 8/9 of the velocities will fall within the interval:

$$\bar{x} \pm 3s \Rightarrow 936 \pm 3(10) \Rightarrow 936 \pm 30 \Rightarrow (906, 966)$$

b. Since a velocity of 1,000 is much larger than the largest value in the second interval in part **a**, it is very unlikely that the bullet was manufactured by Winchester.

2.111 a. The interval $\bar{x} \pm 3s$ for the flexed arm group is $\bar{x} \pm 3s \Rightarrow 59 \pm 3(4) \Rightarrow 59 \pm 12 \Rightarrow (47, 71)$. The interval for the extended arm group is $\bar{x} \pm 3s \Rightarrow 43 \pm 3(2) \Rightarrow 43 \pm 6 \Rightarrow (37, 49)$. We know that at least 8/9 or 88.9% of the observations will fall within 3 standard deviations of the mean using Chebyshev's Rule. Since these 2 intervals barely overlap, the information supports the researchers' theory. The shoppers from the flexed arm group are more likely to select vice options than the extended arm group.

b. The interval $\bar{x} \pm 3s$ for the flexed arm group is $\bar{x} \pm 3s \Rightarrow 59 \pm 3(10) \Rightarrow 59 \pm 30 \Rightarrow (29, 89)$. The interval for the extended are group is $\bar{x} \pm 3s \Rightarrow 43 \pm 3(15) \Rightarrow 43 \pm 45 \Rightarrow (-2, 88)$. Since these two intervals overlap almost completely, the information does not support the researcher's theory. There does not appear to be any difference between the two groups.

2.113 Since we do not know if the distribution of the heights of the trees is mound-shaped, we need to apply Chebyshev's rule. We know $\mu = 30$ and $\sigma = 3$. Therefore,

$$\mu \pm 3\sigma \Rightarrow 30 \pm 3(3) \Rightarrow 30 \pm 9 \Rightarrow (21, 39)$$

According to Chebyshev's rule, at least 8/9 or 0.89 of the tree heights on this piece of land fall within this interval and at most 1/9 or 0.11 of the tree heights will fall above the interval. However, the buyer will only purchase the land if at least $\frac{1000}{5000}$ or 0.20 of the tree heights are at least 40 feet tall. Therefore, the buyer should not buy the piece of land.

2.115 Using the definition of a percentile:

	Percentile	Percentage Above	Percentage Below
a.	75th	25%	75%
b.	50th	50%	50%
c.	20th	80%	20%
d.	84th	16%	84%

2.117 a. $z = \frac{x - \bar{x}}{s} = \frac{40 - 30}{5} = 2$ (sample) 2 standard deviations above the mean.

b. $z = \frac{x - \mu}{\sigma} = \frac{90 - 89}{2} = 0.5$ (population) .5 standard deviations above the mean.

c. $z = \frac{x - \mu}{\sigma} = \frac{50 - 50}{5} = 0$ (population) 0 standard deviations above the mean.

d. $z = \frac{x - \bar{x}}{s} = \frac{20 - 30}{4} = -2.5$ (sample) 2.5 standard deviations below the mean.

2.119 Since the element 40 has a z-score of -2 and 90 has a z-score of 3,

$$-2 = \frac{40-\mu}{\sigma} \text{ and } 3 = \frac{90-\mu}{\sigma}$$

$$\Rightarrow -2\sigma = 40-\mu \qquad \Rightarrow 3\sigma = 90-\mu$$

$$\Rightarrow \mu - 2\sigma = 40 \qquad \Rightarrow \mu + 3\sigma = 90$$

$$\Rightarrow \mu = 40 + 2\sigma$$

By substitution,

$$40 + 2\sigma + 3\sigma = 90$$

$$\Rightarrow 5\sigma = 50$$

$$\Rightarrow \sigma = 10$$

By substitution, $\mu = 40 + 2(10) = 60$

Therefore, the population mean is 60 and the standard deviation is 10.

2.121 The percentile ranking for the age of 25 years in the distribution of all ages of licensed drivers stopped by police is $100 - 79 = 21^{st}$ percentile.

2.123 From Exercises 2.41 and 2.101, $\bar{x} = 94.441$ and $s = 5.335$.

a. The z-score for the Nautilus Explorer is: $z = \frac{x-\bar{x}}{s} = \frac{74-94.441}{5.335} = -3.83$

The score for the Nautilus Explorer is 3.83 standard deviations below the mean for all the cruise ships.

b. The z-score for the Star Princess is: $z = \frac{x-\bar{x}}{s} = \frac{92-94.441}{5.335} = -0.46$

The score for the Star Princess is 0.46 standard deviations below the mean for all the cruise ships.

2.125 a. Using SAS, the output is:

```
                         EGG LENGTH

                  The UNIVARIATE Procedure
                     Variable:  length

                           Moments

N                         130    Sum Weights                130
Mean               60.6538462    Sum Observations          7885
Std Deviation      43.9861168    Variance            1934.77847
Skewness            2.1091362    Kurtosis            4.70026314
Uncorrected SS         727842    Corrected SS        249586.423
Coeff Variation    72.5199136    Std Error Mean      3.85783765
```

```
                 Basic Statistical Measures

    Location                    Variability

Mean     60.65385     Std Deviation            43.98612
Median   49.50000     Variance                     1935
Mode     35.00000     Range                   220.00000
                      Interquartile Range      32.00000


                 Tests for Location: Mu0=0

   Test               -Statistic-      -----p Value------

   Student's t      t  15.72224     Pr > |t|      <.0001
   Sign             M        65     Pr >= |M|     <.0001
   Signed Rank      S    4257.5     Pr >= |S|     <.0001


                  Quantiles (Definition 5)

                  Quantile       Estimate

                  100% Max          236.0
                  99%               218.0
                  95%               160.0
                  90%               122.0
                  75% Q3             67.0
                  50% Median         49.5
                  25% Q1             35.0
                  10%                23.0
                  5%                 19.0
                  1%                 16.0
                  0% Min             16.0


                        EGG LENGTH

                  The UNIVARIATE Procedure
                     Variable:  length

                   Extreme Observations

          ----Lowest----          ----Highest---

          Value      Obs          Value      Obs

           16.0      102            195      128
           16.0       90            205      123
           17.0      101            216      122
           18.0      103            218      129
           18.5      100            236      130


                     Missing Values

                               -----Percent Of-----
    Missing                                 Missing
     Value          Count       All Obs        Obs

       .               2          1.52      100.00
```

The 10th percentile egg length is 23. This means that 10% of all egg lengths are less than 23 and 90% are greater than 23.

b. From the printout, $\bar{x} = 60.65$ and $s = 43.99$. The z-score corresponding to the *moas* (*P. australis*) bird species' egg length is: $z = \frac{x - \bar{x}}{s} = \frac{205 - 60.65}{43.99} = 3.28$

The *z*-score for *moas* is 3.28 standard deviations above the mean. This is a fairly large value for a *z*-score. This indicates that the egg length for *moas* could be very unusual.

2.127 a. The z-score for the number of sags is: $z = \frac{x - \bar{x}}{s} = \frac{400 - 353}{30} = 1.57$. The number of sages for this transformer is 1.57 standard deviations above the mean.

b. The z-score for the number of swells is: $z = \frac{x - \bar{x}}{s} = \frac{100 - 184}{25} = -3.36$. The number of swells for this transformer is 3.36 standard deviations below the mean.

2.129 a. The *z*-score for Harvard is 5.08. This means that the productivity score for Harvard was 5.08 standard deviations above the mean. This is an unusually high *z*-score.

b. The *z*-score for Howard is -0.81. This means that the productivity score for Howard was 0.81 standard deviations below the mean.

c. Yes. If only 44 out of 129 schools had positive z-scores, this indicates that the data are skewed to the right. Other indications that the data are skewed to the right are the *z*-scores associated with the smallest and largest scores. The largest z-score is 5.08 and the smallest is -0.81. The largest observation is much further from the mean than the smallest observation.
Using MINITAB, a histogram of the data is:

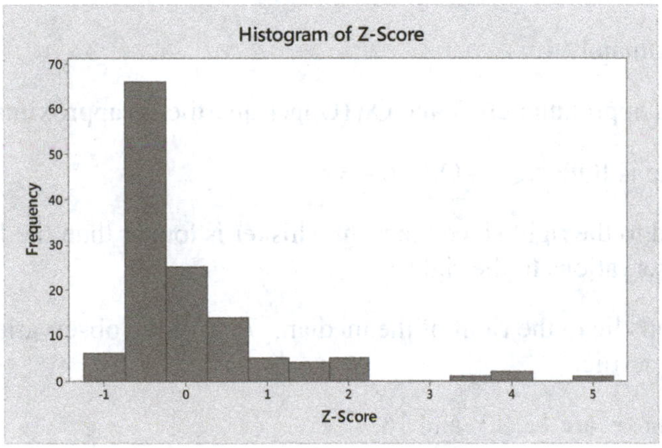

The data are, in fact, skewed to the right.

2.131 a. From the problem, $\mu = 2.7$ and $\sigma = .5$

$$z = \frac{x - \mu}{\sigma} \Rightarrow z\sigma = x - \mu \Rightarrow x = \mu + z\sigma$$

For $z = 2$, $x = 2.7 + 2(0.5) = 3.7$

For $z = -1.0$, $x = 2.7 - 1(0.5) = 2.2$

For $z = 0.5$, $x = 2.7 + 0.5(0.5) = 2.95$

For $z = -2.5$, $x = 2.7 - 2.5(0.5) = 1.45$

b. For $z = -1.6$, $x = 2.7 - 1.6(0.5) = 1.9$

c. If we assume the distribution of GPAs is approximately mound-shaped, we can use the Empirical Rule.

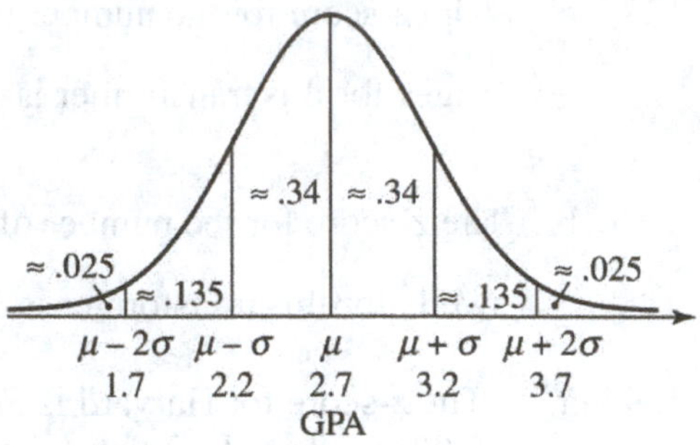

From the Empirical Rule, we know that ≈ 0.025 or $\approx 2.5\%$ of the students will have GPAs above 3.7 (with $z = 2$). Thus, the GPA corresponding to summa cum laude (top 2.5%) will be greater than 3.7 ($z > 2$).

We know that ≈ 0.16 or $\approx 16\%$ of the students will have GPAs above 3.2 ($z = 1$). Thus, the limit on GPAs for cum laude (top 16%) will be greater than 3.2 ($z > 1$).

We must assume the distribution is mound-shaped.

2.133 The hinges of a box plot are the upper quartile and the lower quartile (top and bottom of the rectangle).

2.135 An observation that is unusually large or small relative to the data values we want to describe is an outlier.

2.137 a. The median is approximately 4.

b. Q_L (Lower quartile) is approximately 3 and Q_U (Upper quartile) is approximately 6.

c. The interquartile range is $IQR = Q_U - Q_L \approx 6 - 3 = 3$.

d. The data set is skewed to the right since the right whisker is longer than the left and there are outlying observations to the right.

e. 50% of the observations lie to the right of the median. 75% of the observations lie to the left of the upper quartile.

f. The outliers in the data set are 12, 13, and 16.

2.139 Using MINITAB, the box plot for the data is given below.

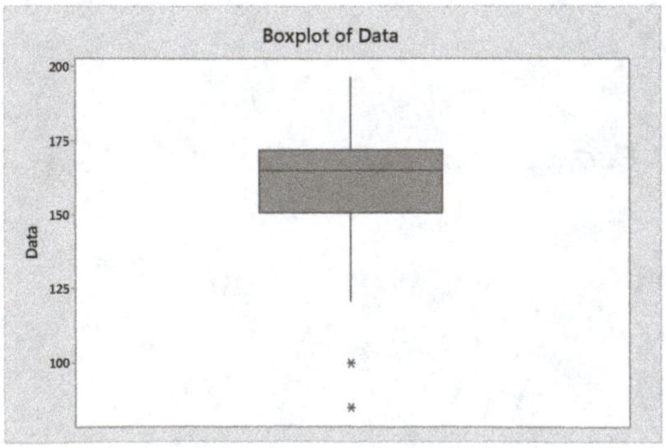

There are 2 suspect outliers. These measurements correspond to values 85 and 100.

2.141 For genotype CC, there appears to be 3 outliers. One outlier has a value larger than 3 and two outliers have values less than -2.4. For genotype CT, there appears to be 2 outliers. One outlier has a value greater than 2 and one has a value less than -2.4. For genotype TT, there are no outliers.

2.143 a. The z-score is $z = \frac{x - \overline{x}}{s} = \frac{3.3 - 7.3}{3.18} = -1.26$.

b. A PMI score of 3.3 would not be considered an outlier. The z-score is -1.26. A z-score this small is not considered unusual.

2.145 a. The z-score corresponding to 400 sags is: $z = \frac{x - \overline{x}}{s} = \frac{400 - 353}{30} = 1.57$. Because this observation is only 1.57 standard deviations from the mean, it is not considered an outlier. It would not be an unusual value.

b. The z-score corresponding to 100 swell is: $z = \frac{x - \overline{x}}{s} = \frac{100 - 184}{25} = -3.36$. Because this observation is more than 3 standard deviations from the mean, it is considered a highly suspect outlier.

2.147 a. Using MINITAB, the boxplot is:

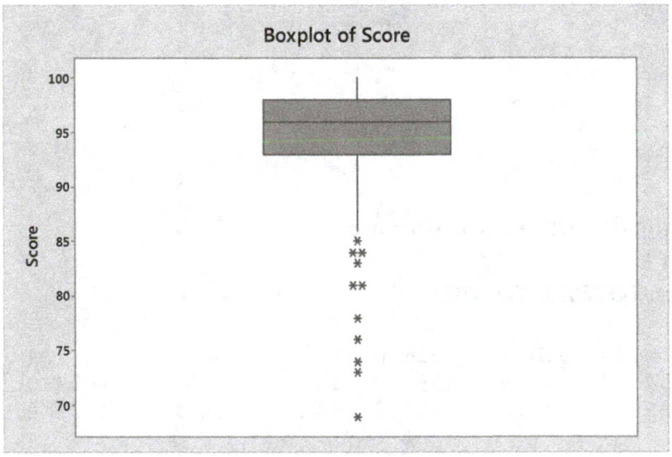

From the boxplot, there appear to be 11 outliers: 69, 73, 74, 76, 78, 81, 81, 83, 84, 84, and 85.

b. From Exercise 2.101, $\overline{x} = 94.441$ and $s = 5.335$. Since the data are skewed to the left, we will consider observations more than 3 standard deviations from the mean to be outliers. An observation with a z-score of 3 would have the value:

$$z = \frac{x - \overline{x}}{s} \Rightarrow 3 = \frac{x - 94.441}{5.335} \Rightarrow 3(5.335) = x - 94.441 \Rightarrow 16.005 = x - 94.441 \Rightarrow x = 110.446$$

An observation with a z-score of -3 would have the value:

$$z=\frac{x-\overline{x}}{s}\Rightarrow -3=\frac{x-94.441}{5.335}\Rightarrow -3(5.335)=x-94.441\Rightarrow -16.005=x-94.441\Rightarrow x=78.436$$

Observations greater than 110.446 or less than 78.436 would be considered outliers. Using this criterion, the following observations would be outliers: 69, 73, 74, 76, and 78.

c. No, these methods do not agree. Using the boxplot, 11 observations were identified as outliers. Using the z-score method, only 5 observations were identified as outliers. Since the data are very highly skewed to the left, the z-score method may not be appropriate.

2.149 a. Using MINITAB, the side-by-side box plots are:

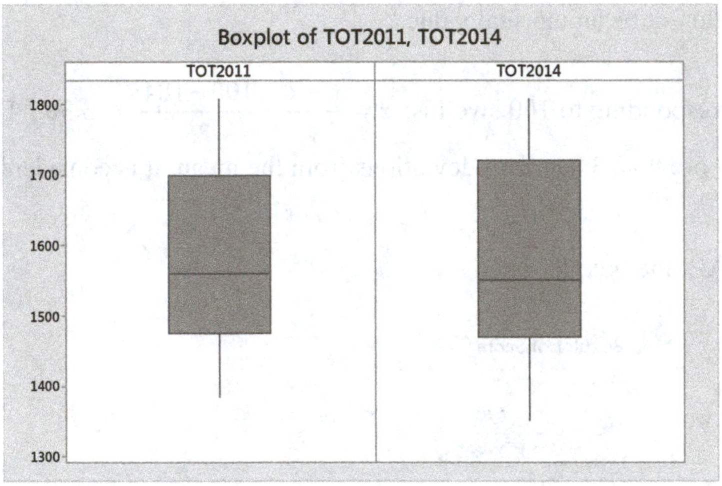

b. Using MINITAB, the descriptive statistics are:

Descriptive Statistics: TOT2011, TOT2014

Variable	N	Mean	StDev	Minimum	Q1	Median	Q3	Maximum
TOT2011	51	1587.7	121.5	1385.0	1475.0	1559.0	1699.0	1807.0
TOT2014	51	1586.6	134.0	1351.0	1470.0	1551.0	1721.0	1807.0

The standard deviation for 2011 is 121.5 and the standard deviation for 2014 is 134.0. Also, the IQR for 2011 is $Q_U-Q_L=1699-1475=224$ while the IQR for 2014 is $Q_U-Q_L=1721-1470=251$. Thus, the variability for 2014 is slightly greater than that for 2011.

c. Since there are no observations outside the inner fences for either year, there are no outliers.

2.151 a. Using MINITAB, the boxplot of measurements from Group A is:

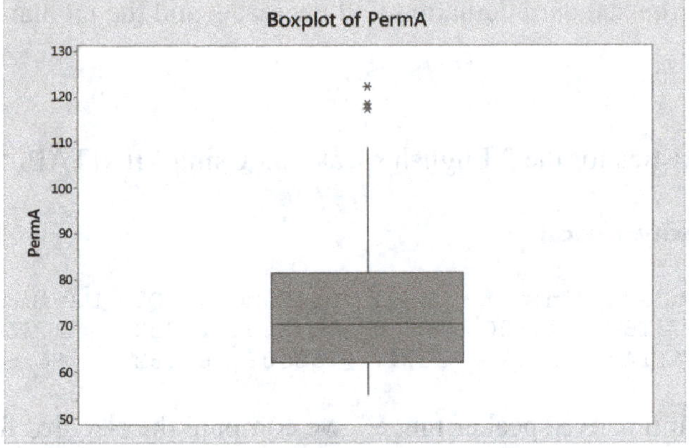

From the boxplot, there appear to be 3 outliers in the data set. These observations correspond to the values 117.3, 118.5, and 122.4.

b. Using MINITAB, the boxplot of measurements from Group B is:

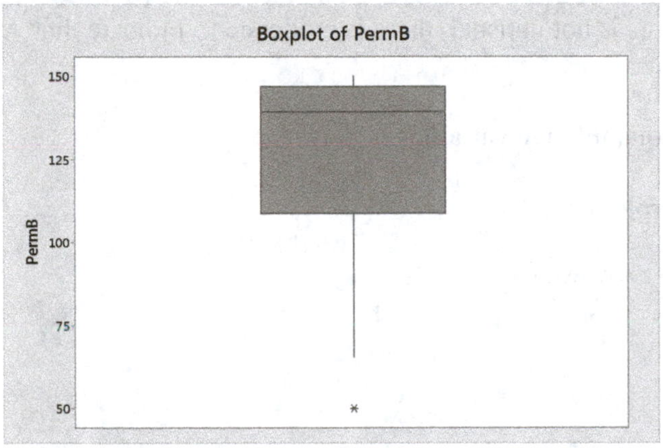

From the boxplot, there appears to be 1 outlier in the data set. This observation corresponds to the value 50.4.

c. Using MINITAB, the boxplot of measurements from Group C is:

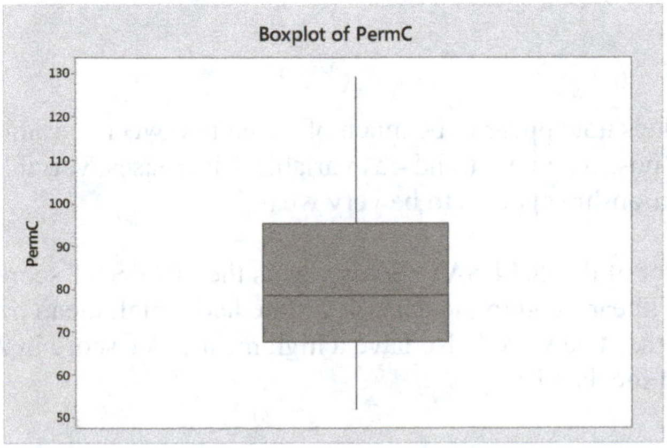

From the boxplot, there do not appear to be any outliers in the data set.

d. If the outliers are removed from the Group A data set, the mean and standard deviation will decrease and the median will slightly decrease. If the outlier in Group B is removed, the mean will increase, the standard deviation will decrease, and the median will slightly increase.

2.153 First, we find the simple statistics for the 3 English speakers. Using MINITAB, they are:

Descriptive Statistics: Complexity, Lexical

Variable	N	Mean	StDev	Minimum	Q1	Median	Q3	Maximum
Complexity	3	1.973	0.216	1.820	1.820	1.880	2.220	2.220
Lexical	3	59.87	4.14	56.97	56.97	58.03	64.62	64.62

To determine if "Alex is within native speaker range", we compute the *z*-scores for each of Alex's scores.

The *z*-score for Alex's syntactic complexity score is: $z = \dfrac{x - \overline{x}}{s} = \dfrac{1.73 - 1.973}{0.216} = -1.125$.

The *z*-score for Alex's lexical diversity score is: $z = \dfrac{x - \overline{x}}{s} = \dfrac{55.88 - 59.87}{4.14} = -0.964$.

Because each of these *z*-scores is not unusual, there is evidence to indicate that Alex is within native speaker range.

2.155 Scatterplots are useful with quantitative variables.

2.157 Using MINITAB, the scatterplot is:

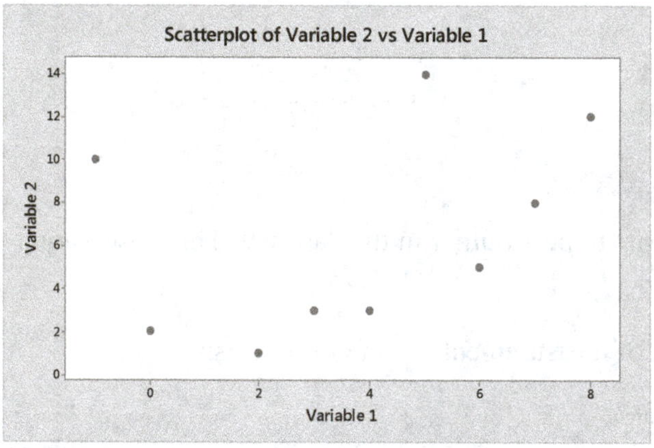

From the scatterplot, there does not appear to be much of a trend between variable 1 and variable 2. There is a slight positive linear trend - as variable 1 increases, variable 2 tends to increase. However, this relationship appears to be very weak.

2.159 If we constructed a scatterplot of the 2014 SAT scores versus the 2011 SAT scores, we would expect there to be a positive linear trend to the data. If a state had a high mean SAT score in 2011, we would expect that the state would also have a high mean SAT score in 2014. Using MINITAB, the scatterplot of the data is:

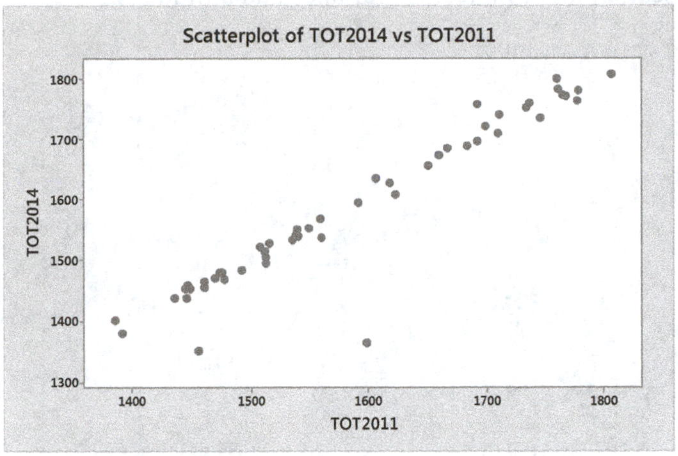

2.161 Using MINITAB, the scatterplot of the data is:

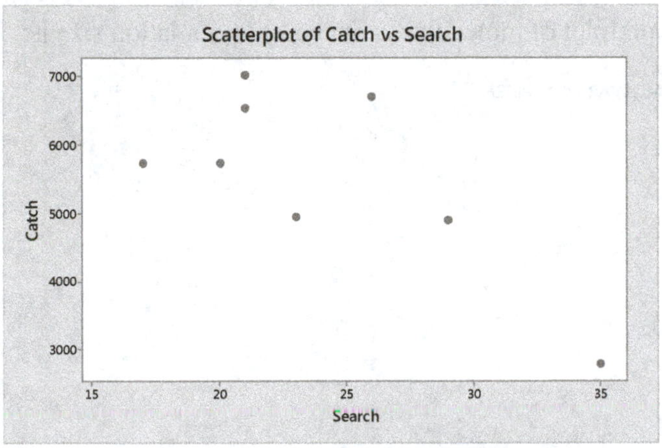

There is an apparent negative linear trend between the search frequency and the total catch. As the search frequency increases, the total catch tends to decrease.

2.163 Using MINITAB, the scatterplot of the data is:

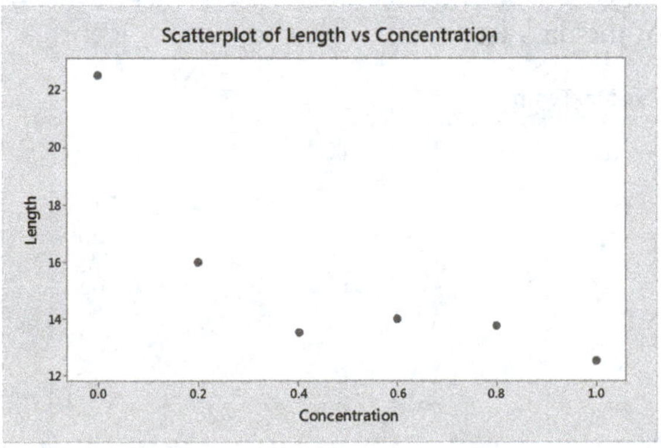

There appears to be a negative trend to the data. As the concentration increases, the wicking length tends to decrease. It appears that the relationship may not be linear, but rather curvilinear.

2.165 a. Using MINITAB, the scatterplot of molecular variance and altitude is:

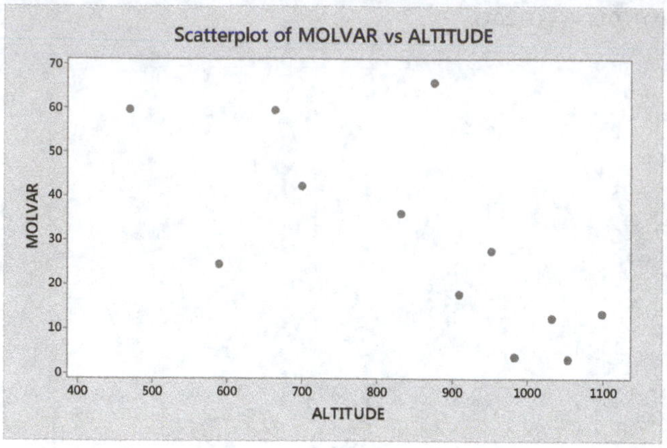

There appears to be a weak, negative linear trend to the data. As altitude increases, the molecular variance tends to decrease.

b. Using MINITAB, the scatterplot of molecular variance and population size is:

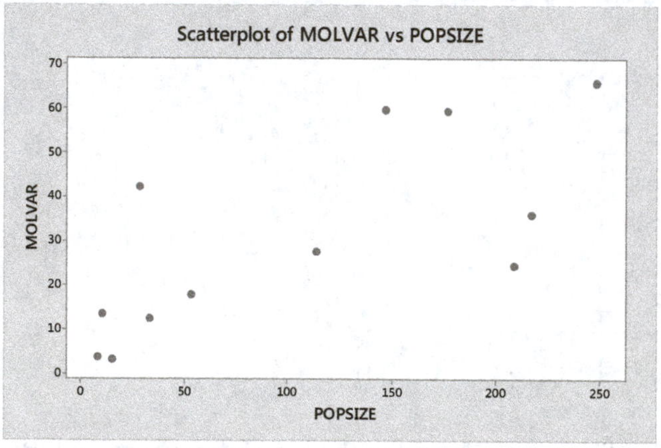

There appears to be a weak, positive linear trend to the data. As population size increases, the molecular variance tends to increase.

2.167 a. Using MINITAB, a scatterplot of the data is:

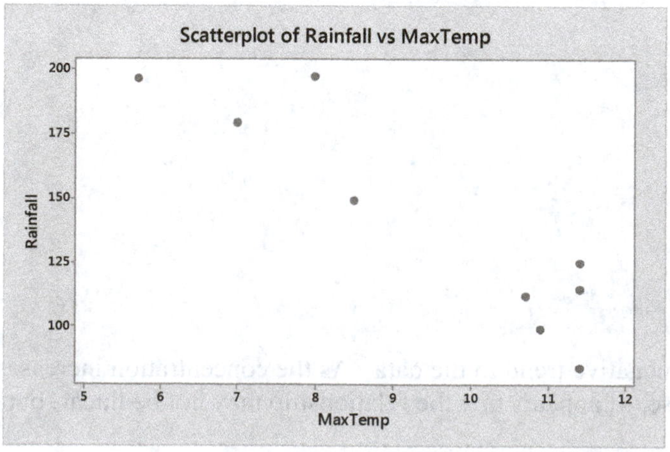

There appears to be a negative, linear relationship between the annual rainfall and the maximum daily temperature. As the maximum daily temperature increases, the amount of annual rainfall tends to decrease.

b. Using MINITAB, a scatterplot of the relationship between annual rainfall and total plant cover is:

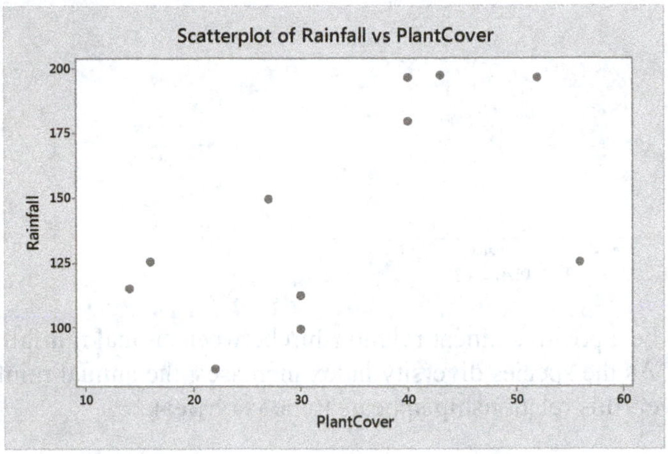

There appears to be a positive, linear relationship between annual rainfall and total plant cover. As the total plant cover increases, the annual rainfall tends to increase.

Using MINITAB, a scatterplot of the relationship between annual rainfall and number of ant species is:

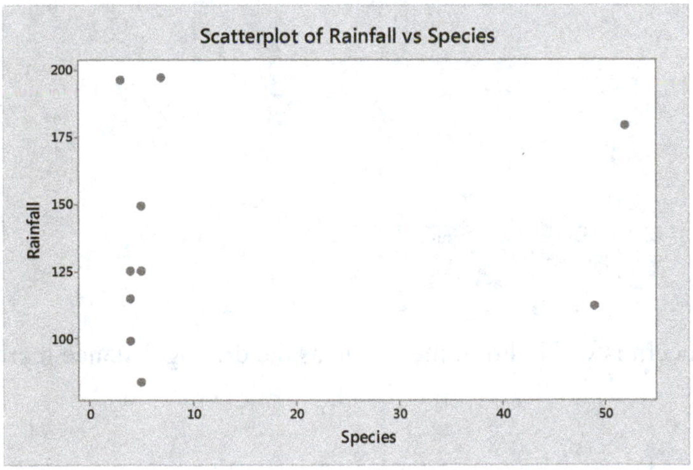

There does not appear to be a relationship between annual rainfall and number of ant species.

Using MINITAB, a scatterplot of the relationship between annual rainfall and species diversity index is:

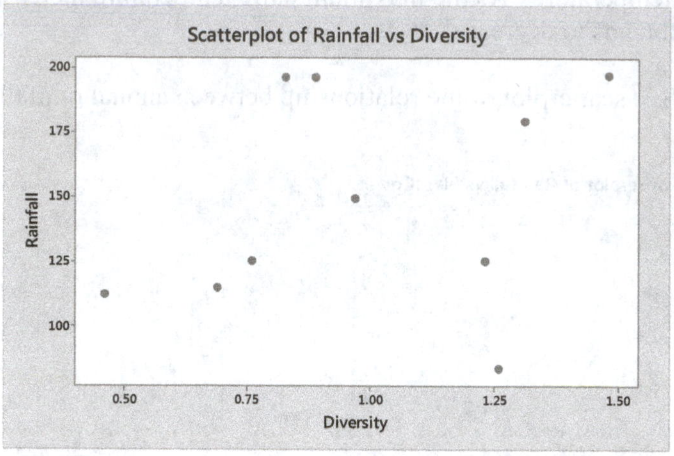

There appears to be a positive, linear relationship between annual rainfall and species diversity index. As the species diversity index increases, the annual rainfall tends to increase. However, this relationship appears to be very weak.

2.169 Using MINITAB, a graph of the accuracy scores versus the driving distance is as follows:

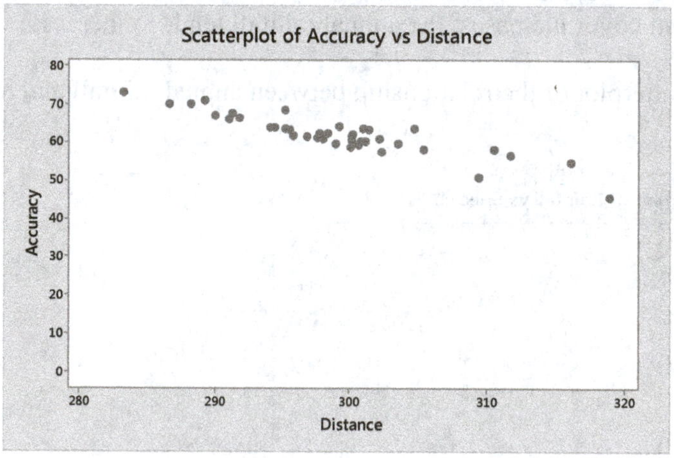

Yes, the golfer's concern is valid. From the graph, as the driving distance increases, the accuracy decreases.

2.171 a. Using MINITAB, the scatterplot is:

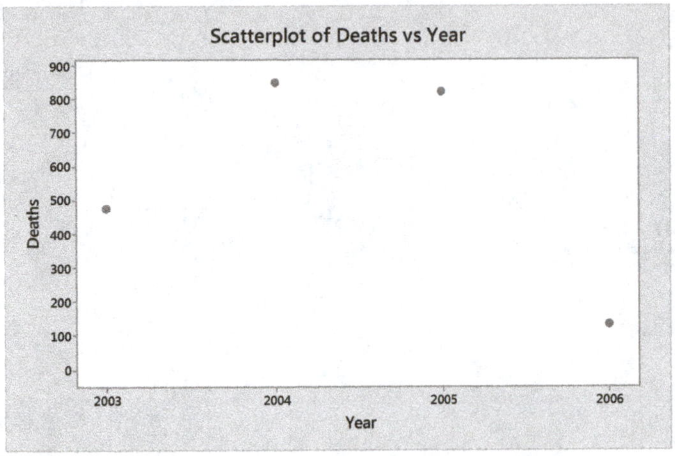

b. The scatterplot is misleading because the information for 2006 is incomplete – it is based on only 2 months while all of the rest of the years are based on 12 months.

c. In order to construct a plot that accurately reflects the trend in American casualties from the Iraq War, we would want complete data for 2006 and information for the years 2007 through 2014.

2.173 a. If one wanted to magnify the difference in error rates between the two conditions, one would use the mean because it is larger than the median.

b. If one wanted to diminish the difference in error rates between the two conditions, one would use the median because it is smaller than the mean.

c. The values of the absolute difference between error rates would have a smaller range and smaller standard deviation because all the measures would be positive. The absolute difference would measure the difference between the two error rates without taking into account which condition was higher or lower. This would more accurately reflect the difference in the two conditions.

2.175 The median is preferred to the mean as a measure of central tendency when the data set is skewed either right or left. If the data set has some extreme observations, the median better measure of central tendency than the mean.

2.177 A stem-and-leaf display is generally preferred over a histogram when the data set is relatively small.

2.179 a. $z = \dfrac{x - \mu}{\sigma} = \dfrac{50 - 60}{10} = -1$

$z = \dfrac{70 - 60}{10} = 1$

$z = \dfrac{80 - 60}{10} = 2$

b. $z = \frac{x-\mu}{\sigma} = \frac{50-60}{5} = -2$

$z = \frac{70-60}{5} = 2$

$z = \frac{80-60}{5} = 4$

c. $z = \frac{x-\mu}{\sigma} = \frac{50-40}{10} = 1$

$z = \frac{70-40}{10} = 3$

$z = \frac{80-40}{10} = 4$

d. $z = \frac{x-\mu}{\sigma} = \frac{50-40}{100} = 0.1$

$z = \frac{70-40}{100} = 0.3$

$z = \frac{80-40}{100} = 0.4$

2.181 The relative frequency histogram is:

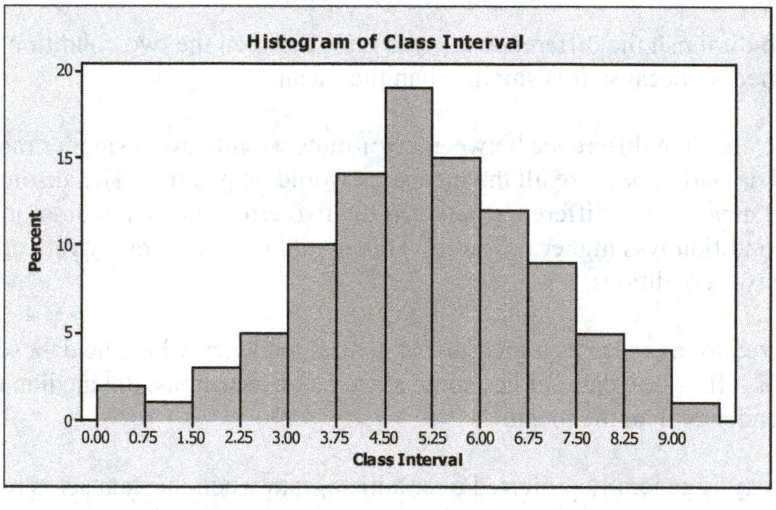

2.183 a. $\sum x = 4+6+6+5+6+7 = 34$ $\sum x^2 = 4^2+6^2+6^2+5^2+6^2+7^2 = 198$

$\bar{x} = \frac{\sum x}{n} = \frac{34}{6} = 5.67$

$$s^2 = \frac{\sum x^2 - \frac{\left(\sum x\right)^2}{n}}{n-1} = \frac{198 - \frac{34^2}{6}}{6-1} = \frac{5.3333}{5} = 1.0667$$

$s = \sqrt{1.0667} = 1.03$

b. $\sum x = -1 + 4 + (-3) + 0 + (-3) + (-6) = -9$

$$\sum x^2 = (-1)^2 + 4^2 + (-3)^2 + 0^2 + (-3)^2 + (-6)^2 = 71$$

$$\bar{x} = \frac{\sum x}{n} = \frac{-9}{6} = -\$1.5$$

$$s^2 = \frac{\sum x^2 - \frac{\left(\sum x\right)^2}{n}}{n-1} = \frac{71 - \frac{(-9)^2}{6}}{6-1} = \frac{57.5}{5} = 11.5$$

$$s = \sqrt{11.5} = \$3.39$$

c. $\sum x = \frac{3}{5} + \frac{4}{5} + \frac{2}{5} + \frac{1}{5} + \frac{1}{16} = 2.0625$

$$\sum x^2 = \left(\frac{3}{5}\right)^2 + \left(\frac{4}{5}\right)^2 + \left(\frac{2}{5}\right)^2 + \left(\frac{1}{5}\right)^2 + \left(\frac{1}{16}\right)^2 = 1.2039$$

$$\bar{x} = \frac{\sum x}{n} = \frac{2.0625}{5} = .4125\%$$

$$s^2 = \frac{\sum x^2 - \frac{\left(\sum x\right)^2}{n}}{n-1} = \frac{1.2039 - \frac{2.0625^2}{5}}{5-1} = \frac{0.3531}{4} = 0.0883\% \text{ squared}$$

$$s = \sqrt{0.0883} = 0.297\%$$

d. (a) $\text{Range} = 7 - 4 = 3$

(b) $\text{Range} = \$4 - (\$-6) = \$10$

(c) $\text{Range} = \frac{4}{5}\% - \frac{1}{16}\% = \frac{64}{80}\% - \frac{5}{80}\% = \frac{59}{80}\% = .7375\%$

2.185 Using MINITAB, the scatterplot is:

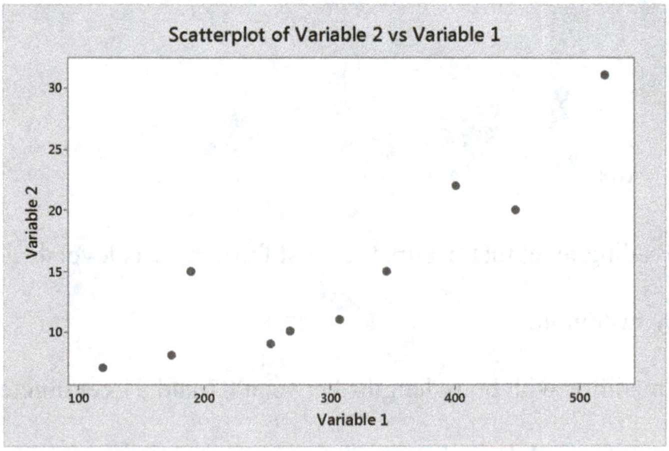

From the scatterplot, it appears that there is a positive trend. As variable 1 increases, variable 2 also tends to increase.

2.187 a. The proportion of books at reading level 1 is found by dividing the number of books at level 1 by the total number of books, or $39/266 = 0.147$.

b. For reading level 2, the proportion of books is $76/266 = 0.286$. For reading level 3, the proportion of books is $50/266 = 0.188$. For reading level 4, the proportion of books is $87/266 = 0.327$. For reading level 5, the proportion of books is $11/266 = 0.041$. For reading level 6, the proportion of books is $3/266 = 0.011$.

c. The sum of the proportions is $0.147 + 0.286 + 0.188 + 0.327 + 0.041 + 0.011 = 1.000$.

d. Using MINITAB, a bar graph of the data is:

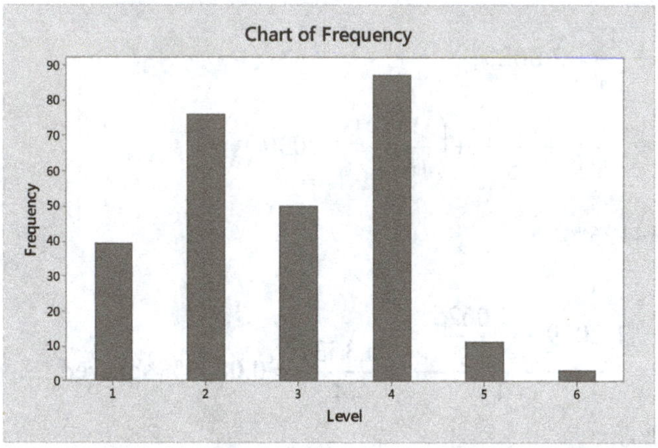

e. Using MINITAB, the Pareto diagram is:

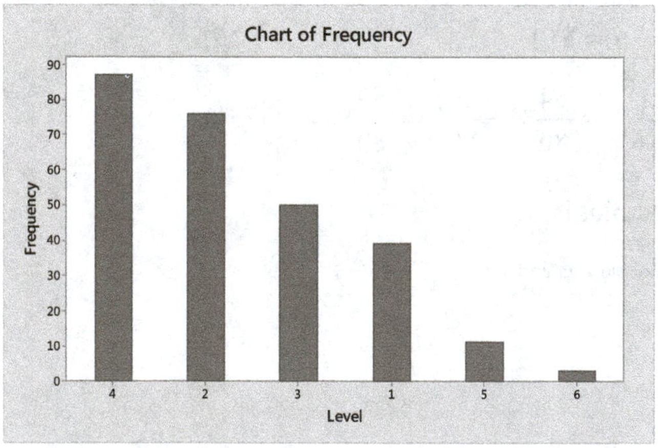

From this diagram, the reading level that occurs the most frequently is level 4.

2.189 a. This graph is a frequency histogram.

b. There are $16 + 22 = 38$ armadillos with body lengths between 87 and 91 centimeters.

c. The proportion of 38armadillos with body lengths between 87 and 91 centimeters is the number observed divided by the total number of captured turtles or $38/80 = 0.475$.

d. The number of illegal armadillos is $2+4+3+1+1=11$. The proportion is $11/80 = 0.1375$

2.191 a. The sample mean is 603.7. This is the average of the 98 observations. The standard deviation is StDev = 185.4. We would expect most of the observations to fall within 2 standard deviations of the mean. The minimum value is 216. This is the smallest observation in the data set. The Q1 value is 475. This is the 25th percentile. Twenty-five percent of all observations in the data set are less than or equal to 475. The median is 605.0. Half of the observations in the data set are above 605 and half are below. The Q3 value is 724.3. This is the 75th percentile. Seventy-five percent of all observations in the data set are less than or equal to 724.3. The maximum value is 1240.0. This is the value of the largest observation in the data set.

b. The z-score is: $z = \dfrac{x - \overline{x}}{s} = \dfrac{408 - 603.7}{185.4} = -1.06$

A head-injury rating of 408 is less than the mean head-injury rating. It is a little more than one standard deviation below the mean.

2.193 a. The mean response was 5.87. The average rating (across the 15 respondents) of the statement "The Training Game is a great way for students to understand the animal's perspective during training" was 5.87. Since the highest rating the statement could receive was 7, this indicated that on the average, the students rather strongly agreed with this statement.

The mode response was 6. More students rated the statement with a 6 than any other rating. A rating of 6 indicated a rather strong agreement to the statement.

b. Since the mode is only slightly larger than the mean, there probably is no skewness present. If any skewness exists, the data would probably be skewed to the left since the mode is larger than the mean and because the mode is very close to the largest value possible.

2.195 Using MINITAB, a bar graph is:

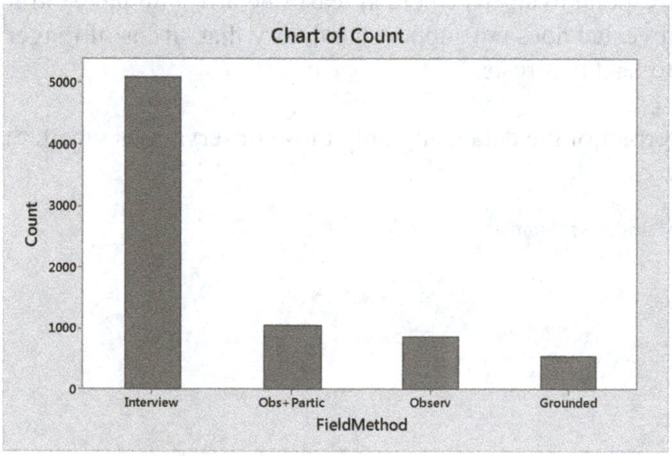

Most of the types of papers found were interviews. There were about twice as many interviews as all other types combined.

2.197 a. Regardless of the shape of the distribution, most of the observations will fall within 3 standard deviations of the mean. Thus, for the SAT-Math scores, an interval likely to contain a student's change in score is:

$$\bar{x} \pm 3s \Rightarrow 19 \pm 3(65) \Rightarrow 19 \pm 195 \Rightarrow (-176, 214)$$

b. Regardless of the shape of the distribution, most of the observations will fall within 3 standard deviations of the mean. Thus, for the SAT-Verbal scores, an interval likely to contain a student's change in score is:

$$\bar{x} \pm 3s \Rightarrow 7 \pm 3(49) \Rightarrow 7 \pm 147 \Rightarrow (-140, 154)$$

c. For the SAT-Verbal, the maximum increase in scores is about 154 points. For the SAT-Math, the maximum increase in scores is approximately 214. Thus, a student is more likely to get a 140-point increase on the SAT-Math test.

2.199 a. Using MINITAB, the scatterplot of the data is:

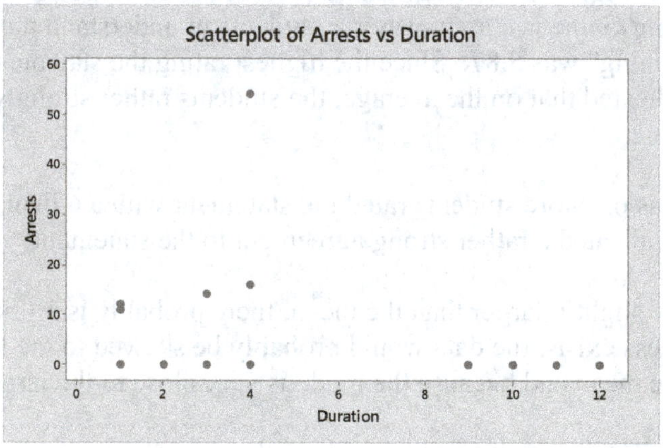

The observations that correspond to sit-ins where at least one arrest all have durations less than 5. The pattern revealed does not support the theory that sit-ins of longer duration are more likely to lead to arrests.

b. Using MINITAB, a scatterplot of the data using only those observations where there was at least one arrest is:

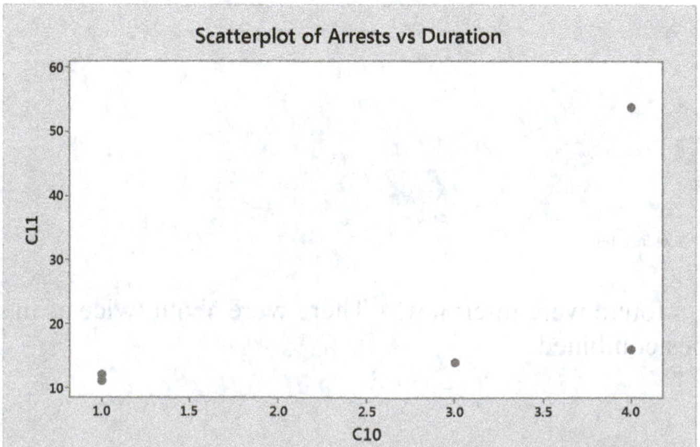

There appears to be a positive relationship between the number of arrests and the duration of the sit-in. As the duration increases, the number of arrests tends to increase.

c. Since there are only 5 observations used in the scatterplot in part **b**, the reliability is suspect.

2.201 A relative frequency bar graph is used to depict the data:

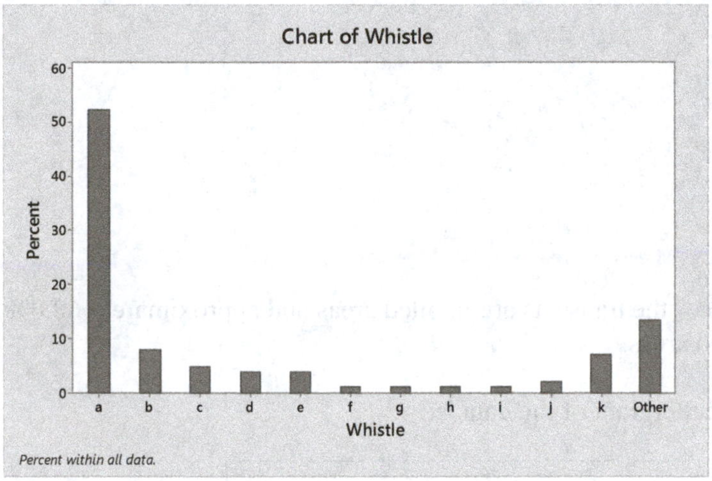

From the bar graph, over half of the whistle types were "Type a." The next most frequent category was the "Other types" with a relative frequency of about .14. Whistle types b and k were the next most frequent. None of the other whistle types had relative frequencies higher than .05.

2.203 a. Using MINITAB, the boxplot is:

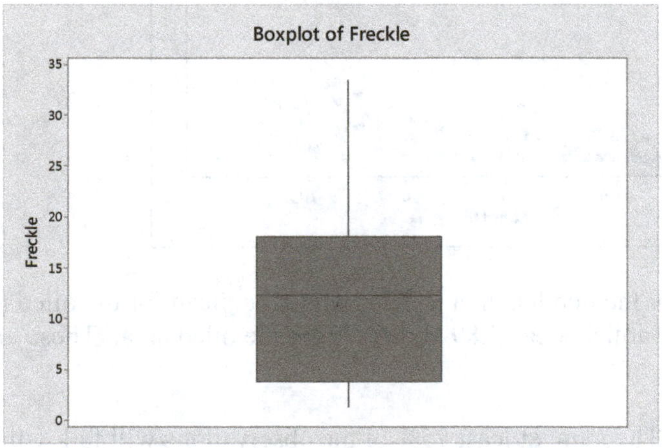

From the boxplot, there are no outliers detected.

b. Since there are no outliers, we cannot find the z-score for the points identified as outliers.

2.205 a. For each transect, three variables were measured. The number of seabirds found is quantitative. The length of the transect is also quantitative. Whether or not the transect was in an oiled area is qualitative.

b. The experimental unit is the transect.

c. A pie chart of the oiled and unoiled areas is:

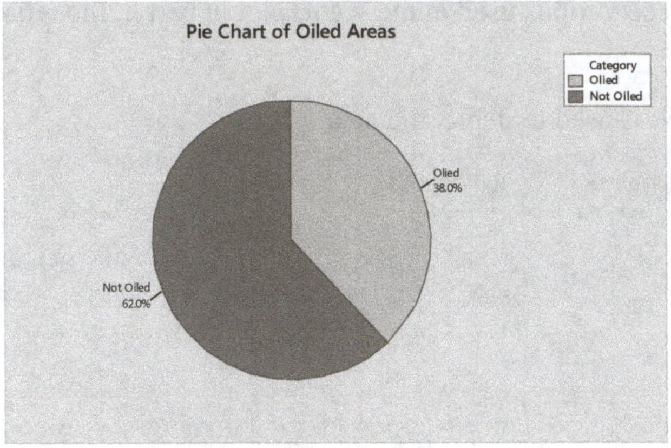

Approximately 38.0% of the transects are in oiled areas and approximately 68.0% of the transects are in unoiled areas.

d. Using MINITAB, a scattergram of the data is:

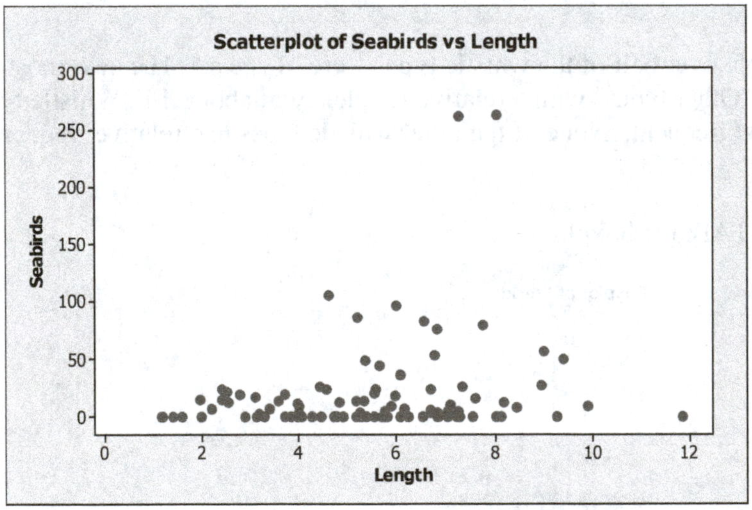

e. The mean density for the unoiled area is 3.27, while the mean for the oiled area is 3.495. The median for the unoiled area is .89 and is .70 for the oiled area. These are both fairly similar.

f. Using Chebyshev's Theorem, at least 75% of the observations will fall within 2 standard deviations of the mean. This interval for unoiled areas would be:

$\bar{x} \pm 2s \Rightarrow 3.27 \pm 2(6.7) \Rightarrow 3.27 \pm 13.4 \Rightarrow (-10.13, 16.67)$

g. Using Chebyshev's Theorem, at least 75% of the observations will fall within 2 standard deviations of the mean. This interval for oiled areas would be:

$\bar{x} \pm 2s \Rightarrow 3.495 \pm 2(5.968) \Rightarrow 3.495 \pm 11.939 \Rightarrow (-8.441, 15.431)$

h. From the above two intervals, we know that at least 75% of the observations for the unoiled area will fall between −10.31 and 16.67 and at most 25% of the observations will fall above 15.431 for the oiled areas. Thus, the unoiled areas would be more likely to have a seabird density of 16.

2.207 a. Using Minitab, a stem-and-leaf display of the data is:

Stem-and-Leaf Display: Galaxy

```
Stem-and-leaf of Galaxy  N  = 51
Leaf Unit = 100

1    18  4
3    18  79
12   19  001112444
18   19  566788
20   20  12
21   20  7
21   21
23   21  99
(5)  22  11344
23   22  5666777777889
10   23  001222344
1    23
1    24
1    24  9
```

b. From this stem-and-leaf display, it is fairly obvious that there are two different distributions since there are two groups of data.

c. Since there appears to be two distributions, we will compute two sets of numerical descriptive measures. We will call the group with the smaller velocities A1775A and the group with the larger velocities A1775B.

For A1775A:

$$\bar{x} = \frac{\sum x}{n} = \frac{408,707}{21} = 19,462.2$$

$$s^2 = \frac{\sum x^2 - \frac{\left(\sum x\right)^2}{n}}{n-1} = \frac{7,960,019,531 - \frac{408,707^2}{21}}{21-1} = 283,329.3$$

$$s = \sqrt{283,329.3} = 532.29$$

For A1775B:

$$\bar{x} = \frac{\sum x}{n} = \frac{685,154}{30} = 22,838.5$$

$$s^2 = \frac{\sum x^2 - \frac{\left(\sum x\right)^2}{n}}{n-1} = \frac{15,656,992,942 - \frac{685,154^2}{30}}{30-1} = 314,694.88$$

$$s = \sqrt{314,694.88} = 560.98$$

d. To determine which of the two clusters this observation probably belongs to, we will compute z-scores for this observation for each of the two clusters.

$$\text{For A1775A: } z = \frac{20,000 - 19,462.2}{532.29} = 1.01$$

Since this z-score is so small, it would not be unlikely that this observation came from this cluster.

$$\text{For A1775B: } z = \frac{20,000 - 22,838.5}{560.98} = -5.06$$

Since this z-score is so large (in magnitude), it would be very unlikely that this observation came from this cluster.

Thus, this observation probably came from the cluster A1775A.

2.209 First, compute the z-score for 1.80: $z = \frac{x - \mu}{\sigma} = \frac{1.80 - 2.00}{.08} = -2.5$

If the data are actually mound-shaped, it would be very unusual (less than 2.5%) to observe a batch with 1.80% zinc phosphide if the true mean is 2.0%. Thus, if we did observe 1.8%, we would conclude that the mean percent of zinc phosphide in today's production is probably less than 2.0%.

2.211 a. Due to the "elite" superstars, the salary distribution is skewed to the right. Since this implies that the median is less than the mean, the players' association would want to use the median.

b. The owners, by the logic of part **a**, would want to use the mean.

2.213 The answers to this will vary. Some things that should be included in the discussion are:

Even though 4,500 completed questionnaires were returned, the response rate was extremely small, only 4.5%. Also, it is very likely that this is not a representative sample of the population of American women. First, the questionnaires were not distributed to a random sample of women from the U.S. The questionnaires were sent specifically to certain groups and the directors of those groups were asked to distribute the surveys to members of their organizations. Thus, if a woman did not belong to one of the targeted groups, she had a very small chance of participating in the survey.

In these types of mail-in surveys, usually only those with very strong opinions respond to the surveys. Thus, those who actually responded to the survey were probably not the same as the population in general. The results are very likely not very reliable.

Chapter 3

Probability

3.1 An experiment is an act or process of observation that leads to a single outcome that cannot be predicted with certainty.

3.3 A sample space of an experiment is a collection of all its sample points.

3.5 Two probability rules for sample points are:

1. All sample point probabilities **must** lie between 0 and 1, i.e. $0 \le p_i \le 1$.

2. The probabilities of all sample points within a sample space **must** sum to 1, i.e., $\sum p_i = 1$.

3.7 The probability of an event A is calculated by summing the probabilities of the sample points in the sample space for A.

3.9 a. Since the probabilities must sum to 1,

$$P(E_3)=1-P(E_1)-P(E_2)-P(E_4)-P(E_5)=1-0.1-0.2-0.1-0.1=0.5$$

b. $$P(E_3)=1-P(E_1)-P(E_2)-P(E_4)-P(E_5)=1-P(E_3)-P(E_2)-P(E_4)-P(E_5)$$
$$\Rightarrow 2P(E_3)=1-P(E_2)-P(E_4)-P(E_5)$$
$$\Rightarrow 2P(E_3)=1-0.1-0.2-0.1 \Rightarrow 2P(E_3)=0.6 \Rightarrow P(E_3)=0.3$$

c. $$P(E_3)=1-P(E_1)-P(E_2)-P(E_4)-P(E_5)=1-0.1-0.1-0.1-0.1 =0.6$$

3.11 $$P(A)=P(1)+P(2)+P(3)=0.05+0.20+0.30=0.55$$

$$P(B)=P(1)+P(3)+P(5)=0.05+0.30+0.15=0.50$$

$$P(C)=P(1)+P(2)+P(3)+P(5)=0.05+0.20+0.30+0.15=0.70$$

3.13 a. $$\binom{5}{2}=\frac{5!}{2!(5-2)!}=\frac{5!}{2!3!}=\frac{5\cdot 4\cdot 3\cdot 2\cdot 1}{2\cdot 1\cdot 3\cdot 2\cdot 1}=10$$

b. $$\binom{6}{3}=\frac{6!}{3!(6-3)!}=\frac{6!}{3!3!}=\frac{6\cdot 5\cdot 4\cdot 3\cdot 2\cdot 1}{3\cdot 2\cdot 1\cdot 3\cdot 2\cdot 1}=20$$

c. $$\binom{20}{5}=\frac{20!}{5!(20-5)!}=\frac{20!}{5!15!}=\frac{20\cdot 19\cdot 18\cdots 2\cdot 1}{5\cdot 4\cdot 3\cdot 2\cdot 1\cdot 15\cdot 14\cdot 13\cdots 2\cdot 1}=15{,}504$$

3.15 a. If we denote the marbles as B_1, B_2, R_1, R_2, R_3, then the ten equally likely sample points in the sample space would be:

$$S: \begin{bmatrix} (B_1,B_2),(B_1,R_1),(B_1,R_2),(B_1,R_3),(B_2,R_1) \\ (B_2,R_2),(B_2,R_3),(R_1,R_2),(R_1,R_3),(R_2,R_3) \end{bmatrix}$$

Notice that order is ignored, as the only concern is whether or not a marble is selected.

b. Each of these ten would be equally likely, implying that each occurs with a probability $1/10$.

c. $P(A)=\frac{1}{10}$ $P(B)=6\left(\frac{1}{10}\right)=\frac{6}{10}=\frac{3}{5}$ $P(C)=3\left(\frac{1}{10}\right)=\frac{3}{10}$

3.17 a. The sample points of the study are: legs only, wheels only, both legs and wheels, and neither legs nor wheels.

b. Reasonable probabilities would be the relative frequencies:

$P(\text{legs only})=\frac{63}{106}=0.594$ $P(\text{wheels only})=\frac{20}{106}=0.189$

$P(\text{both})=\frac{8}{106}=0.075$ $P(\text{neither})=\frac{15}{106}=0.142$

c. $P(\text{wheels}) = P(\text{wheels only}) + P(\text{both}) = 0.189 + 0.075 = 0.264$

d. $P(\text{legs}) = P(\text{legs only}) + P(\text{both}) = 0.594 + 0.075 = 0.669$

3.19 a. There are 6 sample points for this experiment: Blue, Orange, Green, Yellow, Brown, and Red.

b. According to the Mars Corporation, reasonable probabilities would be:

$P(\text{Blue})=0.24$, $P(\text{Orange})=0.20$, $P(\text{Green})=0.16$, $P(\text{Yellow})=0.14$, $P(\text{Brown})=0.13$, and $P(\text{Red})=0.13$.

c. $P(\text{Brown})=0.13$.

d. $P(\text{Red, Green, or Yellow}) = P(\text{Red}) + P(\text{Green}) + P(\text{Yellow}) = 0.13 + 0.16 + 0.14 = 0.43$.

e. $P(\text{Not Blue}) = 1 - P(\text{Blue}) = 1 - 0.24 = 0.76$.

3.21 a. Let A = {chicken passes inspection with fecal contamination}. $P(A)=1/100=0.01$.

b. Yes. The relative frequency of passing inspection with fecal contamination is $306/32{,}075=0.0095\approx 0.01$.

3.23 $P(\text{urban})=\frac{107}{174}=0.615$; $P(\text{not rural}) = P(\text{urban}) + P(\text{suburban}) = \frac{107}{174}+\frac{57}{174}=\frac{164}{174}=0.943$

3.25 P(tooth shows slight or moderate amount of wear)

$= P$(tooth shows slight amount of wear) + P(tooth shows moderate amount of wear)

$= 4/18 + 3/18 = 7/18 = 0.389$.

3.27 a. $P(\text{track \#7}) = \frac{1}{11} = 0.091$

b. P(track #2 or #5 or #10) = P(track #2) + P(track #5) + P(track #10)

$= \frac{1}{11} + \frac{1}{11} + \frac{1}{11} = \frac{3}{11} = 0.273$

3.29 a. There are a total of $\binom{5}{3} = \frac{5!}{3!(5-3)!} = \frac{5!}{3!2!} = \frac{5\cdot4\cdot3\cdot2\cdot1}{3\cdot2\cdot1\cdot2\cdot1} = 10$ ways to get 3-grill displays.

These 10 display combinations are:

1, 2, 3	1, 3, 4	2, 3, 4	3, 4, 5
1, 2, 4	1, 3, 5	2, 3, 5	
1, 2, 5	1, 4, 5	2, 4, 5	

However, since grill #2 must be selected, there are only 6 possibilities:

1, 2, 3	2, 3, 4
1, 2, 4	2, 3, 5
1, 2, 5	2, 4, 5

b. We can estimate the probabilities by using the relative frequency for each sample point. The relative frequency is found by dividing the frequency by the total sample size of 124. These estimates are contained in the following table:

Grill Display Combination	Number of Students	Probability
1-2-3	35	0.282
1-2-4	8	0.065
1-2-5	42	0.339
2-3-4	4	0.032
2-3-5	1	0.008
2-4-5	34	0.274
TOTAL	124	1.0000

c. Of the 6 sample points, only 3 of them contain Grill #1.

P(Grill #1 chosen) = P(1-2-3 or 1-2-4 or 1-2-5) $= 0.282 + 0.065 + 0.339 = 0.686$.

3.31 a. There are a total of $\binom{8}{2} = \frac{8!}{2!(8-2)!} = \frac{8!}{2!6!} = \frac{8\cdot7\cdot6\cdot5\cdot4\cdot3\cdot2\cdot1}{2\cdot1\cdot6\cdot5\cdot4\cdot3\cdot2\cdot1} = 28$ possible Quinella bets.

b. If all the players are of equal ability, then the probability of getting any combination is $1/28$.

3.33 The total number of pairs of bullets is a combination of 1,837 bullets taken 2 at a time or:

$$\binom{1{,}837}{2} = \frac{1{,}837!}{2!1{,}835!} = \frac{1{,}837(1{,}836)}{2} = 1{,}686{,}366$$

The probability of finding a match or a false positive is $693/1{,}686{,}366 = 0.0000411$.

This probability is very small. The confidence in the FBI's forensic evidence should be very high.

3.35 a. $$\binom{6}{2} = \frac{6!}{2!(6-2)!} = \frac{6!}{2!4!} = \frac{6\cdot5\cdot4\cdot3\cdot2\cdot1}{2\cdot1\cdot4\cdot3\cdot2\cdot1} = 15$$

b. $$\binom{6}{3} = \frac{6!}{3!(6-3)!} = \frac{6!}{3!3!} = \frac{6\cdot5\cdot4\cdot3\cdot2\cdot1}{3\cdot2\cdot1\cdot3\cdot2\cdot1} = 20$$

c. $$\binom{6}{4} = \frac{6!}{4!(6-4)!} = \frac{6!}{4!2!} = \frac{6\cdot5\cdot4\cdot3\cdot2\cdot1}{4\cdot3\cdot2\cdot1\cdot2\cdot1} = 15$$

d. $$\binom{6}{5} = \frac{6!}{5!(6-5)!} = \frac{6!}{5!1!} = \frac{6\cdot5\cdot4\cdot3\cdot2\cdot1}{5\cdot4\cdot3\cdot2\cdot1\cdot1} = 6$$

e. In addition to the combinations computed in parts a – d, we need to find the number of 0-drug combinations, 1-drug combination, and 6-drug combinations.

$$\binom{6}{0} = \frac{6!}{0!(6-0)!} = \frac{6!}{0!6!} = \frac{6\cdot5\cdot4\cdot3\cdot2\cdot1}{1\cdot6\cdot5\cdot4\cdot3\cdot2\cdot1} = 1$$

$$\binom{6}{1} = \frac{6!}{1!(6-1)!} = \frac{6!}{1!5!!} = \frac{6\cdot5\cdot4\cdot3\cdot2\cdot1}{1\cdot5\cdot4\cdot3\cdot2\cdot1} = 6$$

$$\binom{6}{6} = \frac{6!}{6!(6-6)!} = \frac{6!}{6!0!} = \frac{6\cdot5\cdot4\cdot3\cdot2\cdot1}{6\cdot5\cdot4\cdot3\cdot2\cdot1\cdot1} = 1$$

The total number of ways the 6 drugs can be combined is $15+20+15+6+1+6+1=64$.

3.37 No. The probabilities will be the same. If 50% of the volunteers are randomly selected to be tested, then the probability of a volunteer being selected would be 0.50. Now, if all 980 career firefighters and 160 volunteers are combined and 50% are selected, the probability of a volunteer being selected would still be 0.50. If we let N equal the total number of firefighters, then $N = 980 + 160 = 1140$. If we let n equal the total number of firefighters selected to be tested, then $n = \frac{1140}{2} = 570$. The total number of ways to select n firefighters from N is $\binom{N}{n} = \frac{N!}{n!(N-n)!}$. Now if we want to find the probability that volunteer #1 is selected, we have to find the number of arrangements that contain volunteer #1. If volunteer #1 has to be in the arrangement, then there are only $N-1$ firefighters to choose from and only $n-1$ firefighters to choose. Thus, the total number of arrangements containing volunteer #1 is $\binom{N-1}{n-1} = \frac{(N-1)!}{(n-1)!(N-n)!}$. Finally, the probability that volunteer #1 is selected would be

$$\frac{\binom{N-1}{n-1}}{\binom{N}{n}} = \frac{\frac{(N-1)!}{(n-1)!(N-n)!}}{\frac{N!}{n!(N-n)!}} = \frac{n}{N}.$$

For this example, $\frac{n}{N} = \frac{570}{1140} = 0.50$. This is the same probability of selecting volunteer #1 if we selected 50% of all volunteers.

3.39 The union of 2 events A and B is the event that occurs if either A or B or both occur on a single performance of the experiment.

3.41 The complement of an event A is the event that A does not occur – that is, the event consisting of all sample points that are not in event A.

3.43 The Additive Rule of Probability for mutually exclusive events is: If 2 events A and B are mutually exclusive, the probability of the union of A and B equals the sum of the probabilities of A and B; that is, $P(A \cup B) = P(A) + P(B)$.

3.45 a.

A: {*HHH, HHT, HTH, THH, TTH, THT, HTT*}
B: {*HHH, TTH, THT, HTT*}
$A \cup B$: {*HHH, HHT, HTH, THH, TTH, THT, HTT*}
A^c: {*TTT*}
$A \cap B$: {*HHH, TTH, THT, HTT*}

b. If the coin is fair, then each of the 8 possible outcomes is equally likely, with probability $1/8$.

$P(A) = \frac{7}{8}$ $\quad$ $P(B) = \frac{4}{8} = \frac{1}{2}$ $\quad$ $P(A \cup B) = \frac{7}{8}$

$P(A^c) = \frac{1}{8}$ $\quad$ $P(A \cap B) = \frac{4}{8} = \frac{1}{2}$

c. $P(A \cup B) = P(A) + P(B) - P(A \cap B) = \frac{7}{8} + \frac{1}{2} - \frac{1}{2} = \frac{7}{8}$

d. No. $P(A \cap B) = \frac{1}{2}$ which is not 0.

3.47 a. $P(A) = P(E_1) + P(E_2) + P(E_3) + P(E_5) + P(E_6) = \frac{1}{5} + \frac{1}{5} + \frac{1}{5} + \frac{1}{20} + \frac{1}{10} = \frac{15}{20} = \frac{3}{4}$

b. $P(B) = P(E_2) + P(E_3) + P(E_4) + P(E_7) = \frac{1}{5} + \frac{1}{5} + \frac{1}{20} + \frac{1}{5} = \frac{13}{20}$

c. $P(A \cup B) = P(E_1) + P(E_2) + P(E_3) + P(E_4) + P(E_5) + P(E_6) + P(E_7)$
$= \frac{1}{5} + \frac{1}{5} + \frac{1}{5} + \frac{1}{20} + \frac{1}{20} + \frac{1}{10} + \frac{1}{5} = 1$

d. $P(A \cap B) = P(E_2) + P(E_3) = \frac{1}{5} + \frac{1}{5} = \frac{2}{5}$

e. $P(A^c) = 1 - P(A) = 1 - \frac{3}{4} = \frac{1}{4}$

f. $P(B^c) = 1 - P(B) = 1 - \frac{13}{20} = \frac{7}{20}$

g. $P(A \cup A^c) = P(E_1) + P(E_2) + P(E_3) + P(E_4) + P(E_5) + P(E_6) + P(E_7)$
$= \frac{1}{5} + \frac{1}{5} + \frac{1}{5} + \frac{1}{20} + \frac{1}{20} + \frac{1}{10} + \frac{1}{5} = 1$

h. $P(A^c \cap B) = P(E_4) + P(E_7) = \frac{1}{20} + \frac{1}{5} = \frac{5}{20} = \frac{1}{4}$

3.49 The experiment consists of rolling a pair of fair dice. The simple events are:

1, 1	2, 1	3, 1	4, 1	5, 1	6, 1
1, 2	2, 2	3, 2	4, 2	5, 2	6, 2
1, 3	2, 3	3, 3	4, 3	5, 3	6, 3
1, 4	2, 4	3, 4	4, 4	5, 4	6, 4
1, 5	2, 5	3, 5	4, 5	5, 5	6, 5
1, 6	2, 6	3, 6	4, 6	5, 6	6, 6

Since each die is fair, each simple event is equally likely. The probability of each simple event is 1/36.

a. A: {(1, 6), (2, 5), (3, 4), (4, 3), (5, 2), (6, 1)}

B: {(1, 4), (2, 4), (3, 4), (4, 4), (5, 4), (6, 4), (4, 1), (4, 2), (4, 3), (4, 5), (4, 6)}

$A \cap B$: {(3, 4), (4, 3)}

$A \cup B$: {(1, 4), (2, 4), (3, 4), (4, 4), (5, 4), (6, 4), (4, 1), (4, 2), (4, 3), (4, 5), (4, 6), (1, 6), (2, 5), (5, 2), (6, 1)}

A^c: {(1, 1), (1, 2), (1, 3), (1, 4), (1, 5), (2, 1), (2, 2), (2, 3), (2, 4), (2, 6), (3, 1), (3, 2), (3, 3), (3, 5), (3, 6), (4, 1), (4, 2), (4, 4), (4, 5), (4, 6), (5, 1), (5, 3), (5, 4), (5, 5), (5, 6), (6, 2), (6, 3), (6, 4), (6, 5), (6, 6)}

b. $P(A) = 6\left(\frac{1}{36}\right) = \frac{6}{36} = \frac{1}{6}$

$P(B) = 11\left(\frac{1}{36}\right) = \frac{11}{36}$

$P(A \cap B) = 2\left(\frac{1}{36}\right) = \frac{2}{36} = \frac{1}{18}$

$P(A \cup B) = 15\left(\frac{1}{36}\right) = \frac{15}{36} = \frac{5}{12}$

$P(A^c) = 30\left(\frac{1}{36}\right) = \frac{30}{36} = \frac{5}{6}$

c. $P(A \cup B) = P(A) + P(B) - P(A \cap B) = \frac{1}{6} + \frac{11}{36} - \frac{1}{18} = \frac{6+11-2}{36} = \frac{15}{36} = \frac{5}{12}$

d. *A* and *B* are not mutually exclusive. To be mutually exclusive, $P(A \cap B)$ must be 0.

Here, $P(A \cap B) = \frac{1}{18}$

3.51 a. $P(A) = 0.50 + 0.10 + 0.05 = 0.65$

b. $P(B) = 0.10 + 0.07 + 0.50 + 0.05 = 0.72$

c. $P(C) = 0.25$

d. $P(D) = 0.05 + 0.03 = 0.08$

e. $P(A^c) = 0.25 + 0.07 + 0.03 = 0.35$ (Note: $P(A^c) = 1 - P(A) = 1 - 0.65 = 0.35$)

f. $P(A \cup B) = P(B) = 0.10 + 0.07 + 0.50 + 0.05 = 0.72$

g. $P(A \cap B) = P(A) = 0.50 + 0.10 + 0.05 = 0.65$

h. Two events are mutually exclusive if they have no sample points in common or if the probability of their intersection is 0.

$P(A \cap B) = 0.50 + 0.10 + 0.05 = 0.65$. Since this is not 0, *A* and *B* are not mutually exclusive.

$P(A \cap C) = 0$. Since this is 0, *A* and *C* are mutually exclusive.

$P(A \cap D) = .05$. Since this is not 0, *A* and *D* are not mutually exclusive.

$P(B \cap C) = 0$. Since this is 0, *B* and *C* are mutually exclusive.

$P(B \cap D) = .05$. Since this is not 0, *B* and *D* are not mutually exclusive.

$P(C \cap D) = 0$. Since this is 0, C and D are mutually exclusive.

3.53 Define the following events:
LO: {robot has legs only}
WO: {robot has wheels only}
B: {robot has both legs and wheels}
N: {robot has neither legs nor wheels}

$$P(\text{either legs or wheels}) = P(N^c) = 1 - P(N) = 1 - \frac{15}{106} = 1 - 0.142 = 0.858$$

3.55 a. A Venn Diagram that illustrates the results of the gene profiling analysis is:

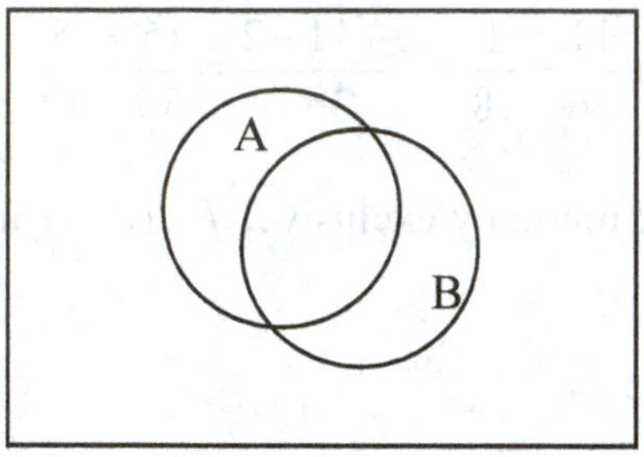

b. From the problem, we know $P(A) = 0.41$, $P(B) = 0.42$, and $P(A \cap B) = 0.40$
$P(A \cup B) = P(A) + P(B) - P(A \cap B) = 0.41 + 0.42 - 0.40 = 0.43$.

c. $P(\text{Neither}) = P(A \cup B)^c = 1 - P(A \cup B) = 1 - 0.43 = 0.57$

3.57 a. There are 9 sample points for this experiment. They consist of all possible combinations for previous sleep stage and current stage. If we let the first event be the level of previous sleep stage and the second event be the level of current stage, the sample points are:

Non-REM, Non-REM
Non-REM, REM
Non-REM, Wake
REM, Non-REM
REM, REM
REM, Wake
Wake, Non-REM
Wake, REM
Wake, Wake

b. Reasonable probabilities would be the relative frequencies for each of the combinations. The total number of observations is $33{,}814 + 8{,}127 + 7{,}987 = 49{,}928$. Dividing each frequency by the total gives the following probabilities:

	Previous Sleep Stage			
Current Stage	Non-REM	REM	Wake	**Totals**
Non-REM	0.6385	0.0032	0.0347	0.6764
REM	0.0069	0.1524	0.0035	0.1628
Wake	0.0318	0.0072	0.1218	0.1607
Totals	0.6772	0.1628	0.1600	

c. $P(\text{current REM}) = P(\text{Non-REM, REM}) + P(\text{REM, REM}) + P(\text{Wake, REM})$
$= 0.0069 + 0.1524 + 0.0035 = 0.1628$

d. $P(\text{previous Wake}) = P(\text{Wake, Non-REM}) + P(\text{Wake, REM}) + P(\textit{Wake}, \text{Wake})$
$= 0.0347 + 0.0035 + 0.1218 = 0.160$

e. $P(\text{REM, Non-REM}) = 0.0032$

3.59 a. Define the following events:
A: {Student choose stated option}
B: {Student did not choose stated option}
C: {Emotion state is Guilt}
D: {Emotion state is Anger}
E: {Emotion state is Neutral}

$$P(C) = \frac{45+12}{171} = \frac{57}{171} = 0.333$$

b. $$P(A) = \frac{45+8+7}{171} = \frac{60}{171} = 0.351$$

c. $$P(C \cap A) = \frac{45}{171} = 0.263$$

d. $P(C \cup A) = P(C) + P(A) - P(C \cap A) = 0.333 + 0.351 - 0.263 = 0.421$

3.61 a. Define the following events:

F: {Fight}
N: {No fight}
W: {Initiator Wins}
T: {No Clear Winner}
L: {Initiator Loses}

$P(F \cap W) = 26/167 = 0.156$

b. $P(N) = 103/167 = 0.617$

c. $P(T) = 35/167 = 0.210$

d. $P(F \cup L) = (26+23+15+11)/167 = 75/167 = 0.449$

e. Yes. $P(T \cap L) = 0$.

3.63 a. $P(\text{responds to all 3 source odors}) = 0.19$

b. $P(\text{responds to kairomone}) = 0.19 + 0.025 + 0.025 + 0.165 = 0.405$

c. P(responds to Mups A and Mups B but not kairomone) = 0.275

3.65 There are a total of $\binom{10}{2} = \frac{10!}{2!(10-2)!} = 45$ ways to choose 2 balls from the 10. There are $\binom{3}{2} = \frac{3!}{2!(3-2)!} = 3$ ways to get 2 yellow balls, $\binom{2}{2} = \frac{2!}{2!(2-2)!} = 1$ way to get 2 blue balls, and $\binom{3}{2} = \frac{3!}{2!(3-2)!} = 3$ ways to get 2 purple balls. Thus, the total number of ways to get 2 balls of the same color is $3+1+3=7$. Finally,

$P(\text{detecting clone attack}) = \frac{7}{45} = 0.156$.

3.67 In the Exercise, it says that an expression represents a sample point. There are 3 types of expressions, so there are 3 sample points. Let D = dimension declarations, P = plain expressions, and λ=lambda abstractions. From the description of the problem, ACC contains P and D, CFA contains P and λ, and CCC contains D, P, and λ. A possible Venn diagram is:

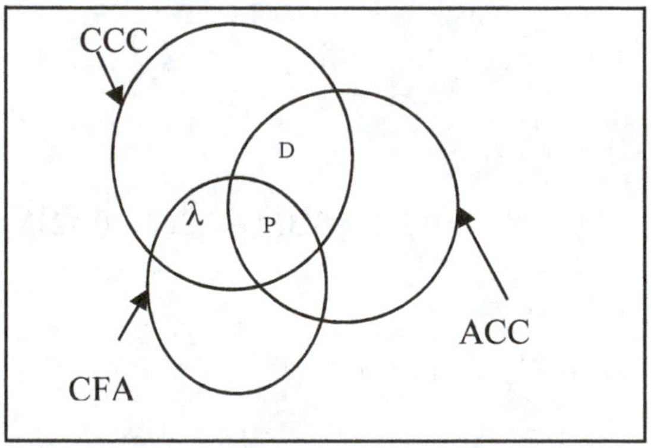

3.69 a. The Multiplicative Rule of Probability for two independent events is:

$P(A \cap B) = P(A)P(B)$

b. The Multiplicative Rule of Probability for any two events is:

$P(A \cap B) = P(B \mid A)P(A) = P(A \mid B)P(B)$

3.71 a. "Dependent events are always mutually exclusive" is not true. Suppose we define an experiment as follows. Two fair coins are tossed and the results are recorded. The possible outcomes are (H,H), (H, T), (T, H), (T, T). If the coins are fair, then these outcomes are equally likely with probabilities ¼. Define the following events:

A: {Both outcomes are H}
B: {Both outcomes are the same}

Then $P(A)=1/4$ and $P(B)=1/2$. The event $A \cap B$ is the event that both outcomes are H or event A. Thus, $P(A \cap B)=1/4$. Since this probability is not 0, events A and B are not mutually exclusive. However, events A and B are dependent:

$P(A \mid B)=\frac{P(A \cap B)}{P(B)}=\frac{1/4}{1/2}=\frac{1}{2}$ and this does not equal $P(A)=1/4$, so events A and B are dependent.

Thus, dependent events are **not** always mutually exclusive.

b. "Mutually exclusive events are always dependent" is true. Suppose we define a third event:

C: {Both outcomes are T}

Now, events A and C are mutually exclusive because $P(A \cap C)=0$. Also,

$P(C \mid A)=\frac{P(C \cap A)}{P(A)}=\frac{0}{1/4}=0$ and this does not equal $P(C)=1/4$. Thus, events A and C are dependent.

For any two mutually exclusive events A and C, $P(A \cap C)=0$. Thus, $P(C \mid A)=P(A \mid C)=0$. As long as $P(A)>0$ and $P(C)>0$, then $P(C \mid A) \neq P(C)$ and $P(A \mid C) \neq P(A)$.

c. "Independent events are always mutually exclusive" is not true. Suppose we define 2 additional events:

D: {Both outcomes are different}
E: {The first coin is a T}

Now, $P(D)=1/4=1/2$, $P(E)=2/4=1/2$, and $P(D \cap E)=1/4$. Thus, since $P(D \cap E)=1/4$ which is not 0, D and E are not mutually exclusive. However, D and E are independent:

$P(D \mid E)=\frac{P(D \cap E)}{P(E)}=\frac{1/4}{1/2}=\frac{1}{2}$ and this does equal $P(D)=1/2$. Thus, events D and E are independent. Therefore, events D and E are independent but not mutually exclusive.

3.73 a. $P(A \mid B)=\frac{P(A \cap B)}{P(B)}=\frac{0.1}{0.2}=0.5$

b. $P(B \mid A)=\frac{P(A \cap B)}{P(A)}=\frac{0.1}{0.4}=0.25$

c. No, events A and B are not independent. If A and B were independent, then $P(A|B)=P(A)$. For this problem, $P(A|B)=0.5$ and $P(A)=0.4$.

3.75 a. $P(A\cap B)=P(A)P(B)=0.4(0.2)=0.08$ (Since A and B are independent)

b. $P(A|B)=\dfrac{P(A\cap B)}{P(B)}=\dfrac{0.08}{0.2}=0.4$

c. $P(A\cup B)=P(A)+P(B)-P(A\cap B)=0.4+0.2-0.08=0.52$

3.77 a. $P(A)=P(E_1)+P(E_2)+P(E_3)=0.1+0.1+0.2=0.4$

$P(B)=P(E_2)+P(E_3)+P(E_5)=0.1+0.2+0.1=0.4$

$P(A\cap B)=P(E_2)+P(E_3)=0.1+0.2=0.3$

b. $P(E_1|A)=\dfrac{P(E_1\cap A)}{P(A)}=\dfrac{0.1}{0.4}=0.25$

$P(E_2|A)=\dfrac{P(E_2\cap A)}{P(A)}=\dfrac{0.1}{0.4}=0.25$

$P(E_3|A)=\dfrac{P(E_3\cap A)}{P(A)}=\dfrac{0.2}{0.4}=0.5$

We are given that $P(E_1)=0.1$, $P(E_2)=0.1$, and $P(E_3)=0.2$

Thus, $P(E_1)=P(E_2)$ and $P(E_3)=2P(E_1)$

From above, $P(E_1|A)=P(E_2|A)$ and $P(E_3|A)=2P(E_1|A)$

$P(E_1|A)+P(E_2|A)+P(E_3|A)=0.25+0.25+0.50=1.00$

c. Using the sum of the conditional probabilities,

$$P(B|A)=P(E_2|A)+P(E_3|A)=0.25+0.50=0.75$$

Using the formula,

$$P(B|A)=\frac{P(A\cap B)}{P(A)}=\frac{0.3}{0.4}=0.75$$

3.79 The 36 possible outcomes obtained when tossing two dice are listed below:

(1, 1) (1, 2) (1, 3) (1, 4) (1, 5) (1, 6)
(2, 1) (2, 2) (2, 3) (2, 4) (2, 5) (2, 6)
(3, 1) (3, 2) (3, 3) (3, 4) (3, 5) (3, 6)
(4, 1) (4, 2) (4, 3) (4, 4) (4, 5) (4, 6)
(5, 1) (5, 2) (5, 3) (5, 4) (5, 5) (5, 6)
(6, 1) (6, 2) (6, 3) (6, 4) (6, 5) (6, 6)

A: {(1, 2), (1, 4), (1, 6), (2, 1), (2, 3), (2, 5), (3, 2), (3, 4), (3, 6), (4, 1), (4, 3), (4, 5), (5, 2), (5, 4), (5, 6), (6, 1), (6, 3), (6, 5)}

B: {(3, 6), (4, 5), (5, 4), (5, 6), (6, 3), (6, 5), (6, 6)}

$A \cap B$: {(3, 6), (4, 5), (5, 4), (5, 6), (6, 3), (6, 5)}

If A and B are independent, then $P(A)P(B) = P(A \cap B)$.

$P(A) = \frac{18}{36} = \frac{1}{2}$ $\quad P(B) = \frac{7}{36}$ $\quad P(A \cap B) = \frac{6}{36} = \frac{1}{6}$

$P(A)P(B) = \frac{1}{2} \cdot \frac{7}{36} = \frac{7}{72} \neq \frac{1}{6} = P(A \cap B)$. Thus, A and B are not independent.

3.81 Define the following events:
B: {rough diamond is a blood diamond}
I: {rough diamond is processed in Surat, India}

From the exercise, we know $P(B) = \frac{1}{4} = 0.25$, $P(I) = 0.90$, and $P(B \mid I) = \frac{1}{3}$.

a. $P(B^c) = 1 - P(B) = 1 - 0.25 = 0.75$

b. $P(B \cap I) = P(B \mid I)P(I) = \frac{1}{3}(0.90) = 0.30$

3.83 Define the following events:
C: {Cereal crop damaged}
WBO: {Wheat, barley or oats cereal crop damaged}

From the exercise, we know $P(C) = 0.45$ and $P(WBO \mid C) = 0.71$.

$P(WBO) = P(WBO \mid C)P(C) = 0.71(0.45) = 0.3195$

3.85 a. Define the following events:
A: {adult owns at least one gun}
B: {adult owns a hand gun}

From the exercise, we know that $P(A) = 0.26$ and $P(B \mid A) = 0.05$. Thus, $P(A) = 0.26$.

b. $P(A \cap B) = P(B \mid A)P(A) = 0.26(0.05) = 0.013$

3.87 Define the following events:

C: {Speeding is cause of fatal crash}
M: {Missing curve is cause of fatal crash}

From the problem, $P(C) = 0.3$ and $P(C \cap M) = 0.12$

$$P(M|C)=\frac{P(C\cap M)}{P(C)}=\frac{0.12}{0.30}=0.40$$

3.89 The probability that at least one acquires gastroenteritis is equal to 1 minus the probability that neither acquires gastroenteritis. We will assume that the events that you acquire gastroenteritis and the event that your friend acquires gastroenteritis are independent.

$P(\text{at least one acquires gastroenteritis}) = 1 - P(\text{neither acquires gastroenteritis})$

$= 1 - P(\text{you do not acquire gastroenteritis})P(\text{friend does not acquire gastroenteritis})$

$$=1-\left(\frac{994}{1000}\right)\left(\frac{993}{1000}\right)=1-0.987=0.013$$

3.91 Define the following events:
H: {no SOP for detecting hydrogen cyanide}
C: {no SOP for detecting carbon monoxide}

From the exercise, we know $P(H)=0.80$, $P(C)=0.49$, and $P(H\cup C)=0.94$.

$P(H\cup C)=P(H)+P(C)-P(H\cap C)$
Thus, $P(H\cap C)=P(H)+P(C)-P(H\cup C)=0.80+0.49-0.94=0.35$

3.93 Define the following event:

R: {Fish is red snapper}

From the problem, $P(R^c)=0.77$

a. $P(R)=1-P(R^c)=1-0.77=0.23$

b. P(At least one customer is served red snapper) = 1 – P(No customers are served red snapper)
$=1-P(R^c\cap R^c\cap R^c\cap R^c\cap R^c)=1-0.77(0.77)(0.77)(0.77)(0.77)=1-0.271=0.729$

3.95 a. $P(\text{Species extinct})=38/132=0.2879$

b. If 9 species are chosen and all are extinct, then for the 10th pick, there are only $132-9=123$ species to pick from, of which $38-9=29$ are extinct. P(Species extinct on 10th selection | first 9 species are extinct) $=29/123=0.236$.

3.97 a. Define the following events:
A: {ambulance travels to location A under 8 minutes}
B: {ambulance travels to location B under 8 minutes}
C: {ambulance is busy}

From the exercise, we know that $P(A|C^c)=0.58$, $P(B|C^c)=0.42$, and $P(C)=0.3$.

$P(A) = P(A \mid C^c)P(C^c) = 0.58(1-0.3) = 0.58(0.7) = 0.406$

b. $P(B) = P(B \mid C^c)P(C^c) = 0.42(1-0.3) = 0.42(0.7) = 0.294$

3.99 $P(A \mid B) = 0$

3.101 a. Define the following event:

A: {Psychic picks box with crystal}
If psychic is just guessing, $P(A) = 1/10 = 0.1$

b. P(Psychic guesses correct at least once in 7 trials)
= 1 − P(Psychic does not guess correct in 7 trials)

$= 1 - P(A^c \cap A^c \cap A^c \cap A^c \cap A^c \cap A^c \cap A^c) = 1 - 0.9(0.9)(0.9)(0.9)(0.9)(0.9)(0.9)$
$= 1 - 0.478 = 0.522$
(This assumes that the trials are independent)

c. If the psychic is just guessing, then P(Psychic does not guess correct in 7 trials) $= 0.478$.

Thus, if the person was not a psychic (and was merely guessing) we would expect the person to guess wrong in all seven trials about half the time. This would not be a rare event. Thus, this outcome would support the notion that the person is guessing.

3.103 Define the following events:
E_1: {erroneous ciphertext}
E_2: {error in restoring plaintext}

From the exercise, we know $P(E_1) = \beta$, $P(E_2 \mid E_1) = 0.5$, and $P(E_2 \mid E_1^c) = \alpha\beta$.

Then $P(E_2) = P(E_2 \mid E_1)P(E_1) + P(E_2 \mid E_1^c)P(E_1^c) = 0.5\beta + \alpha\beta(1-\beta) = 0.5\beta + \alpha\beta - \alpha\beta^2$

3.105 Suppose a basketball team has a total of 10 players. There are 5 distinct positions on the team: center, forward 1, forward 2, shooting guard, and point guard. How many different ways could a coach arrange the starting lineup, assuming that all players could play all positions?

We can apply the Permutations Rule to find the number of combinations.

$$P_n^N = N(N-1)(N-2)\cdots(N-n+1) = \frac{N!}{(N-n)!} = \frac{10!}{(10-5)!}$$

$$= \frac{10!}{5!} = \frac{10 \cdot 9 \cdots 1}{5 \cdot 4 \cdot 3 \cdot 2 \cdot 1} = 10 \cdot 9 \cdot 8 \cdot 7 \cdot 6 = 30{,}240.$$

3.107 The Combinations Rule does not take order into account, while the Permutations Rule does take order into account.

3.109 Since the number of possible outcomes from tossing a coin is always two on each flip, the multiplicative rule would yield a product of a series of 2's.

a. $(2)(2) = 2^2 = 4$

b. $(2)(2)(2) = 2^3 = 8$

c. $(2)(2)(2)(2)(2) = 2^5 = 32$

d. 2^n

3.111 a. $\binom{7}{3} = \frac{7!}{(3!)(4!)} = \frac{7 \cdot 6 \cdot 5}{3 \cdot 2 \cdot 1} = 35$

b. $\binom{6}{2} = \frac{6!}{(2!)(4!)} = \frac{6 \cdot 5}{2 \cdot 1} = 15$

c. $\binom{30}{2} = \frac{30!}{(2!)(28!)} = \frac{30 \cdot 29}{2 \cdot 1} = 435$

d. $\binom{10}{8} = \frac{10!}{(8!)(2!)} = \frac{10 \cdot 9}{2 \cdot 1} = 45$

e. $\binom{q}{r} = \frac{q!}{r!(q-r!)}$

3.113 a. We want to select 3 grills from the 5 possible grills. The total number of possible three-grill-displays is a combination of 5 things taken 3 at a time or $\binom{5}{3} = \frac{5!}{3!(5-3)!} = \frac{5(4)(3)(2)(1)}{3(2)(1)(2)(1)} = 10$.

b. If Grill #2 must be in the display, then we have to choose 2 grills from the remaining 4. Thus, the number of different displays with Grill #2 included is $\binom{4}{2} = \frac{4!}{2!(4-2)!} = \frac{(4)(3)(2)(1)}{(2)(1)(2)(1)} = 6$

c. Now order matters and Grill #2 must be in the display. If Grill #2 is in the first position, then there are $1(4)(3) = 12$ ways to order the three grills. If Grill #2 is in the second position, then there are $4(1)(3) = 12$ ways to order the 3 grills. If Grill #2 is in the third position, then there are $4(3)(1) = 12$ ways to order the 3 grills. Thus, there are a total of $12 + 12 + 12 = 36$ ways to view the grills.

3.115 a. Since order is not important, the number of choices is:

$$\binom{8}{5} = \frac{8!}{5!3!} = 56$$

b. Order is important because it is necessary to distinguish between the positions (guard, forward, and center). However, the two players chosen as guards and the two forwards are indistinguishable once they are chosen. That is, choosing player #1 as the first guard and player #3 as the second guard results in the same team as choosing player #3 for the first guard and player #1 for the second guard. Denote the five positions to be filled as G_1, G_2, F_1, F_2, and C. The number of distinct teams, assuming each position was distinct is:

$$P_5^8 = \frac{8!}{3!} = 8(7)(6)(5)(4) = 6720$$

However, since G_1 and G_2 are indistinguishable, this number must be divided by 2, or $6720/2 = 3360$ possible teams. Similarly, F_1 and F_2 are indistinguishable, so that there are in fact only $3360/2 = 1680$ possible teams.

c. In this situation, all five positions are distinct. Thus, there are $P_5^8 = 6720$ possible teams.

3.117 Using the counting rule, there would be $2(2)(3) = 12$ different purchase order cards. This is the same number of possible options found in Exercise 3.32.

3.119 a. To visit all 4 cities and then return home, the order of the visits is important. The number of different orders is a permutation of 4 things taken 4 at a time or

$$P_4^4 = \frac{4!}{(4-4)!} = \frac{4 \cdot 3 \cdot 2 \cdot 1}{1} = 24$$

b. If the salesperson visits B first, then he only has 2 choices for the next city (D or E), since B and C are not connected. Once in D or E, he has 2 choices for the next city (if in D, can choose C or E; if in E, can choose D or C). Once in the third city, there is only one choice for the fourth city. The possible trips with B first are $2 \times 2 \times 1 = 4$: BDEC, BDCE, BEDC, BECD.

Similarly, if city C is the first city, there are 2 choices for the second city (D or E), 2 choices for the third, and one for the fourth. Thus, there are $2 \times 2 \times 1 = 4$ choices: CDEB, CDBE, CEDB, CEBD.

If city D is the first chosen, then to visit all four cities exactly once, city B or C must be chosen next. The third city must be E, and the fourth city the only one left. Thus, there are $2 \times 1 \times 1 = 2$ choices: DBEC, DCEB.

Similarly, if city E is the first chosen, then B or C must be the second city, D must be the third, and the fourth city must be the one remaining. Thus, there are $2 \times 1 \times 1 = 2$ choices: EBDC, ECDB.

The total number of routes is $4 + 4 + 2 + 2 = 12$.

3.121 We will select 2 women from the 50 who were assigned to the kiwifruit breakfast. There will be a combination of 50 things taken 2 at a time or

$$\binom{50}{2} = \frac{50!}{2!(50-2)!} = \frac{50(49)(48)\cdots(1)}{(2)(1)(48)(47)(46)\cdots(1)} = 1225$$ ways to select 2 women from this group.

There will also be 1225 ways to select 2 women from the 50 who were assigned to the banana breakfast. The total number of ways to select 2 women from each group is $1225(1225) = 1{,}500{,}625$.

3.123 a. The total number of distributors that make up Elaine's group is $6(5) = 30$.

b. The total number of distributors that make up Elaine's group now is $6(5)(7)(5) = 1{,}050$.

3.125 There are a total of $3(2)(1) = 6$ to assign the 3 professors to the 3 students. The arrangements are:

aA bB cC, aA bC cB, aB bA cC, aB,bC cA, aC bA cB, aC bB cA

a. $P(aA\ bB\ cC) = \frac{1}{6}$

b. $P(aA) = \frac{2}{6} = \frac{1}{3}$

c. There are only 2 outcomes where no preferred matches exist: aB,bC cA and aC bA cB.

$P(\text{no preferred matches}) = \frac{2}{6} = \frac{1}{3}$

3.127 a. There are a total of $2 \times 3 \times 3 = 18$ maintenance organization alternatives.

b. If all alternatives are equally likely, then each has a probability of $1/18$ of being selected. If 4 of these alternatives are feasible, then P(feasible alternative) $= 4/18 = 0.222$.

3.129 The professor will choose 5 of 10 questions. Of these 10 questions, the student has prepared answers for 7, and has not prepared the other three. The 5 questions on the exam can be chosen in

$$\binom{10}{5} = \frac{10!}{5!5!} = 252 \text{ ways}$$

a. If the student has prepared all 5 exam questions, the professor has chosen 5 from the 7 prepared questions. This can be done in

$$\binom{7}{5} = 21 \text{ ways}$$

Hence, the desired probability is $21/252$.

b. Define the following events.

A: {The student is prepared for no questions}
B: {The student is prepared for 1 question}
C: {The student is prepared for 2 questions}

A will occur if the professor picks 0 questions from the 7 prepared and 5 from the 3 unprepared. This is impossible, as is *B*. *C* will occur if the professor picks 2 questions from the 7 prepared and 3 questions from the 3 unprepared. This can happen in

$$\binom{7}{2}\binom{3}{3} = 21(1) = 21 \text{ ways}$$

Hence, $P(\text{prepared for less than } 3) = P(A) + P(B) + P(C) = 0 + 0 + 21/252 = 21/252$

c. The professor must pick 4 from the 7 prepared questions and 1 from the remaining 3 questions. This can happen in

$$\binom{7}{4}\binom{3}{1} = 35(3) = 105 \text{ ways}$$

Hence, $P(\text{prepared for exactly } 4) = 105/252$

3.131 a. The number of hands possible is:

$$\binom{52}{5} = \frac{52!}{5!47!} = \frac{52 \cdot 51 \cdot 50 \cdot 49 \cdot 48}{5 \cdot 4 \cdot 3 \cdot 2 \cdot 1} = 2{,}598{,}690$$

b. Every set of five cards would be equally likely. Therefore, the probability of any one type of hand can be calculated by finding the number of ways of obtaining that particular type of hand and dividing by the total number of hands possible. For example,

$$P(A) = \frac{\text{Number of ways of obtaining a flush}}{\text{Number of hands possible}}$$

To determine the number of flushes possible, it is necessary to recognize that there would be the same number of flushes in spades as hearts, diamond, or clubs. If we can determine the number of ways of obtaining a flush in any particular suit, then four times that amount would be the number of flushes possible. However, since there are 13 cards of each suit, it follows that the number of flushes for any particular suit is:

$$\binom{13}{5} = \frac{13!}{5!8!} = 1{,}287$$

Then the total number of flushes possible would be $4(1{,}287) = 5{,}148$

Then $P(A) = \dfrac{5{,}148}{2{,}598{,}960} \approx 0.002$

c. To compute $P(B)$, it is necessary to recognize that there are 10 different sets of cards' values which result in straights:

{A, 2, 3, 4, 5}
{2, 3, 4, 5, 6}
{3, 4, 5, 6, 7}
{4, 5, 6, 7, 8}
{5, 6, 7, 8, 9}
{6, 7, 8, 9, 10}
{7, 8, 9, 10, J}
{8, 9, 10, J, Q}
{9, 10, J, Q, K}
{10, J, Q, K, A}

We can determine the total number of possible straights by determining the number of ways that any one particular type of straight could occur and then multiply this amount by 10. But the number of ways of obtaining any one straight, say {10, J, Q, K, A}, is done easily by the multiplicative rule. Since there would be one 10 from four to be selected and one jack from four to be selected, etc., the number of ways of obtaining a straight of the form {10, J, Q, K, A} is:

$$4(4)(4)(4)(4)=1024$$

There would then be $10(1024)=10,240$ possible straights, so that

$$P(B)=\frac{10,240}{2,598,960}\approx 0.00394$$

d. There are only four straight flushes of each type. For example, a straight flush of the type {10, J, Q, K, A} can occur in spades, hearts, diamonds, and clubs. Since there are 10 types of straights, there are $4(10)=40$ possible straight flushes.

Thus, $P(A\cap B)=\dfrac{40}{2,598,960}=0.0000154$

3.133 If events A and B are independent, then the probability that event B occurs does not depend on whether event A has occurred or not. If events A and B are independent, then $P(B\mid A)=P(B)$.

3.135 a. $P(B_1\cap A)=P(A\mid B_1)P(B_1)=0.3(0.75)=0.225$

b. $P(B_2\cap A)=P(A\mid B_2)P(B_2)=0.5(0.25)=0.125$

c. $P(A)=P(B_1\cap A)+P(B_2\cap A)=0.225+0.125=0.35$

d. $P(B_1\mid A)=\dfrac{P(B_1\cap A)}{P(A)}=\dfrac{0.225}{0.35}=0.643$

e. $P(B_2\mid A)=\dfrac{P(B_2\cap A)}{P(A)}=\dfrac{0.125}{0.35}=0.357$

3.137 If A is independent of B_1, B_2, and B_3, then $P(A \mid B_1) = P(A) = 0.4$.

Then $P(B_1 \mid A) = \frac{P(A \mid B_1)P(B_1)}{P(A)} = \frac{0.4(0.2)}{0.4} = 0.2$

3.139 Define the following events:

T: {Athlete illegally uses testosterone}
P: {Test for testosterone is positive}
N: {Athlete does not illegally use testosterone}

a. $P(P \mid T) = \frac{50}{100} = 0.50$

b. $P(P^c \mid N) = 1 - P(P \mid N) = 1 - \frac{9}{900} = 1 - 0.01 = 0.99$

c. $P(T \mid P) = \frac{P(P \mid T)P(T)}{P(P)}$.

Now, $P(P) = P(P \mid N)P(N) + P(P \mid T)P(T) = 0.01(0.9) + 0.5(0.1) = 0.009 + 0.05 = 0.059$

$P(T \mid P) = \frac{P(P \mid T)P(T)}{P(P)} = \frac{0.5(0.1)}{0.059} = \frac{0.05}{0.059} = 0.847$

3.141 Using Bayes' Rule, the probability is:

$$P(D \mid N) = \frac{P(D \cap N)}{P(N)} = \frac{P(D)P(N \mid D)}{P(D)P(N \mid D) + P(D^c)P(N \mid D^c)}$$

$$= \frac{\frac{1}{80}\left(\frac{1}{2}\right)}{\frac{1}{80}\left(\frac{1}{2}\right) + \frac{79}{80}(1)} = \frac{\frac{1}{160}}{\frac{1}{160} + \frac{79}{80}} = \frac{\frac{1}{160}}{\frac{1}{160} + \frac{158}{160}} = \frac{\frac{1}{160}}{\frac{1}{160} + \frac{158}{160}} = \frac{\frac{1}{160}}{\frac{159}{160}} = \frac{1}{159}$$

3.143 Define the following events:
D: {sample is dolomite}
S: {sample is shale}
G: {Gamma ray greater than 60}

From the exercise
$P(D) = \frac{476}{771} = 0.61738$, $P(S) = \frac{295}{771} = 0.38262$, $P(G \mid D) = \frac{34}{476} = 0.07143$, and $P(G \mid S) = \frac{280}{295} = 0.94915$.

$$P(D|G)=\frac{P(D\cap G)}{P(G)}=\frac{P(D)P(G|D)}{P(D)P(G|D)+P(D^c)P(G|D^c)}$$

$$=\frac{0.6174(0.0714)}{0.6174(0.0714)+0.3826(0.9492)}=\frac{0.04408}{0.04408+0.36316}=\frac{0.04408}{0.40724}=0.1082$$

No, the probability that the area should be mined is 0.1082.

3.145 From Exercise 3.98, we defined the following events:

I: {Intruder}
N: {No intruder}
A: {System A sounds alarm}
B: {System B sounds alarm}

Also, from Exercise 3.98, $P(A\cap B|I)=0.855$ and $P(A\cap B|N)=0.02$.
Thus,
$P(A\cap B)=P(A\cap B|I)P(I)+P(A\cap B|N)P(N)=0.855(0.4)+0.02(0.6)=0.342+0.012=0.354$

$P(I\cap A\cap B)=P(A\cap B|I)P(I)=0.855(0.4)=0.342$

$$P(I|A\cap B)=\frac{P(I\cap A\cap B)}{P(A\cap B)}=\frac{0.342}{0.354}=0.966$$

3.147 a. If $\frac{P(T|E)}{P(T^c|E)}<1$, then $P(T^c|E)>P(T|E)$. Thus, T^c is more likely to occur than T.
This supports the theory that more than 2 bullets were used.

b. Using Bayes' Rule, we know:

$$P(T|E)=\frac{P(T)P(E|T)}{P(T)P(E|T)+P(T^c)P(E|T^c)} \text{ and } P(T^c|E)=\frac{P(T^c)P(E|T^c)}{P(T^c)P(E|T^c)+P(T)P(E|T)}$$

If we take the ratio of these 2 equations, we get:

$$\frac{P(T|E)}{P(T^c|E)}=\frac{\dfrac{P(T)P(E|T)}{P(T)P(E|T)+P(T^c)P(E|T^c)}}{\dfrac{P(T^c)P(E|T^c)}{P(T^c)P(E|T^c)+P(T)P(E|T)}}=\frac{P(T)P(E|T)}{P(T^c)P(E|T^c)}$$

3.149 a. $A\cup B$

b. B^c

c. $A\cap B$

d. $A^c|B$

3.151 a. Since events A and B are mutually exclusive, $P(A \cap B) = 0$.

$$P(A \mid B) = \frac{P(A \cap B)}{P(B)} = \frac{0}{0.3} = 0$$

b. If Events A and B are independent, then $P(A \mid B) = P(A)$. From part **a**, we know $P(A \mid B) = 0$. However, we also know $P(A) = 0.2$. Thus, events A and B are not independent.

3.153 We know $P(A \mid B) = \frac{P(A \cap B)}{P(B)}$. Thus, $P(B) = \frac{P(A \cap B)}{P(A \mid B)} = \frac{0.4}{0.8} = 0.5$.

3.155 a.

(1, H)	(2, 1)	(3, H)	(4, 1)	(5, H)	(6, 1)
(1, T)	(2, 2)	(3, T)	(4, 2)	(5, T)	(6, 2)
	(2, 3)		(4, 3)		(6, 3)
	(2, 4)		(4, 4)		(6, 4)
	(2, 5)		(4, 5)		(6, 5)
	(2, 6)		(4, 6)		(6, 6)

b. Each simple event is an intersection of two independent events. Each simple event whose first element is 1, 3, or 5 has probability

$$\left(\frac{1}{6}\right)\left(\frac{1}{2}\right) = \frac{1}{12}$$

while each simple event whose first element is 2, 4, or 6 has probability

$$\left(\frac{1}{6}\right)\left(\frac{1}{6}\right) = \frac{1}{36}$$

c. $P(A) = P\{(1,H),(3,H),(5,H)\} = \frac{1}{12} + \frac{1}{12} + \frac{1}{12} = \frac{3}{12} = \frac{1}{4}$

$P(B) = P\{(1,H),(1,T),(3,H),(3,T),(5,H),(5,T)\} = \frac{6}{12} = \frac{1}{2}$

d. A^c: {all except (1, H), (3, H), and (5, H)}

B^c: {(2, 1), (2, 2), (2, 3), (2, 4), (2, 5), (2, 6), (4, 1), (4, 2), (4, 3), (4, 4), (4, 5), (4, 6), (6, 1), (6, 2), (6, 3), (6, 4), (6, 5), (6, 6)}

$A \cap B$: {(1, H), (3, H), (5, H)}

$A \cup B$: {(1, H), (1, T), (3, H), (3, T), (5, H), (5, T)}

e. $P(A^c) = 1 - P(A) = 1 - \frac{1}{4} = \frac{3}{4}$

$P(B^c) = 1 - P(B) = 1 - \frac{1}{2} = \frac{1}{2}$

$P(A \cap B) = P(A) = \frac{1}{4}$

$$P(A \cup B) = P(A) + P(B) - P(A \cap B) = \frac{1}{4} + \frac{1}{2} - \frac{1}{4} = \frac{1}{2}$$

$$P(A \mid B) = \frac{P(A \cap B)}{P(B)} = \frac{1/4}{1/2} = \frac{1}{2}$$

$$P(B \mid A) = \frac{P(A \cap B)}{P(A)} = \frac{1/4}{1/4} = 1$$

f. $P(A \cap B) \neq 0$, so that A and B are not mutually exclusive.
$P(A \mid B) \neq P(A)$, so that A and B are not independent.

3.157 a. Because events A and B are independent, we have:

$$P(A \cap B) = P(A)P(B) = 0.3(0.1) = 0.03$$

Thus, $P(A \cap B) \neq 0$, and the two events cannot be mutually exclusive.

b. $P(A \mid B) = \frac{P(A \cap B)}{P(B)} = \frac{0.03}{0.1} = 0.3$ $\quad P(B \mid A) = \frac{P(A \cap B)}{P(A)} = \frac{0.03}{0.3} = 0.1$

c. $P(A \cup B) = P(A) + P(B) - P(A \cap B) = 0.3 + 0.1 - 0.03 = 0.37$

3.159 a. There are 5 sample points for this experiment: None, 1 or 2, 3-5, 6-9, and 10 or more.

b. We can use the percentages given as reasonable estimates of the probabilities:
$P(\text{none}) = 0.25$, $P(1 \text{ or } 2) = 0.31$, $P(3-5) = 0.25$, $P(6-9) = 0.05$, $P(10 \text{ or more}) = 0.14$

c. $P(\text{more than } 2) = P(3-5) + P(6-9) + P(10 \text{ or more}) = 0.25 + 0.05 + 0.14 = 0.44$.

3.161 Use the relative frequency as an estimate for the probability. Thus,
$P(\text{physically assaulted}) = 600/12{,}000 = 0.05$.

3.163 From the problem, to find probabilities, we must convert the percents to proportions by dividing by 100. Thus, $P(1 \text{ star}) = 0$, $P(2 \text{ stars}) = 0.0408$, $P(3 \text{ stars}) = 0.1735$, $P(4 \text{ stars}) = 0.6020$, and $P(5 \text{ stars}) = 0.1837$.

a. False. The probability of an event cannot be 4. The probability of an event must be between 0 and 1.

b. True. $P(4 \text{ or } 5 \text{ stars}) = P(4 \text{ stars}) + P(5 \text{ stars}) = 0.6020 + 0.1837 = 0.7857$.

c. True. No cars have a rating of 1 star, thus $P(1 \text{ star}) = 0$.

d. False. $P(2 \text{ stars}) = 0.0408$ and $P(5 \text{ stars}) = 0.1837$. Since 0.0408 us smaller than 0.1837, the car has a better chance of having a 5-star rating than a 2-star rating.

3.165 a. Define the following event:

A: {Beech tree is damaged by fungi}

$$P(A) = \frac{49}{188} = 0.261$$

b. There would be 3 sample points for this experiment: trunk, leaves, and branch. We can convert the percentages to proportions and use these to estimate the probabilities.

Sample Points	**Probabilities**
Trunk	0.85
Leaves	0.10
Branch	0.05
Total	**1.00**

3.167 a. The sample space is the listing of all possible nearshore bar conditions. The sample space is {Single, shore parallel; Other; Planar}.

b. Define the following events:

SP: {Single, shore parallel}
P: {Planar}
O: {Other}

Assuming that all the sample points are equally likely, $P(SP) = 2/6 = 0.333$. $P(O) = 2/6 = 0.333$. $P(P) = 2/6 = 0.333$

c. $P(P \cup SP) = 4/6 = 0.667$

d. The sample space is the listing of all possible beach conditions. The sample space is {No dunes/flat; Bluff/scarp; Single dune; Not observed}.

e. Define the following events:

N: {No dunes/flat}
B: {Bluff/scarp}
D: {Single dune}
NO: {Not observed}

Assuming that all the sample points are equally likely, $P(N) = 2/6 = 0.333$. $P(B) = 1/6 = 0.167$. $P(D) = 2/6 = 0.333$. $P(NO) = 1/6 = 0.167$.

f. $P(N^C) = 1 - P(NC) = 1 - 0.333 = 0.667$.

3.169 a. The sample points of this experiment are the locations where toxic chemical incidents occurred. They are:

School laboratory, In Transit, Chemical plant, Non-chemical plant, and Other.

b. Reasonable probabilities would be the percents of the incidents changed to proportions.

$P(\text{School laboratory}) = 0.06$, $P(\text{In Transit}) = 0.26$, $P(\text{Chemical plant}) = 0.21$, $P(\text{Non-chemical plant}) = 0.35$, $P(\text{Other}) = 0.12$

c. $P(\text{School laboratory}) = 0.06$

d. $P(\text{Chemical plant or Non-chemical plant}) = 0.21 + 0.35 = 0.56$

e. $P(\text{Not occur In Transit}) = 1 - P(\text{In Transit}) = 1 - 0.26 = 0.74$

f. $P(\text{all 3 incidents occur in school laboratory})$
$= P(\text{School laboratory})P(\text{School laboratory})P(\text{School laboratory})$
$= 0.06(0.06)(0.06) = 0.000216$

g. No. The probability that the event in part f occurs is only 0.000216. This probability is extremely small. It is very unlikely that this event would occur.

3.171 a. The event $A \cap B$ is the event the outcome is black and odd. The event is $A \cap B$: {11, 13, 15, 17, 29, 31, 33, 35}

b. The event $A \cup B$ is the event the outcome is black or odd or both. The event $A \cup B$ is {2, 4, 6, 8, 10, 11, 13, 15, 17, 20, 22, 24, 26, 28, 29, 31, 33, 35, 1, 3, 5, 7, 9, 19, 21, 23, 25, 27}

c. Assuming all events are equally likely, each has a probability of $1/38$.

$$P(A) = 18\left(\frac{1}{38}\right) = \frac{18}{38} = \frac{9}{19}$$

$$P(B) = 18\left(\frac{1}{38}\right) = \frac{18}{38} = \frac{9}{19}$$

$$P(A \cap B) = 8\left(\frac{1}{38}\right) = \frac{8}{38} = \frac{4}{19}$$

$$P(A \cup B) = 28\left(\frac{1}{38}\right) = \frac{28}{38} = \frac{14}{19}$$

$$P(C) = 18\left(\frac{1}{38}\right) = \frac{18}{38} = \frac{9}{19}$$

d. The event $A \cap B \cap C$ is the event the outcome is odd and black and low. The event $A \cap B \cap C$ is {11, 13, 15, 17}.

e. $P(A \cup B) = P(A) + P(B) - P(A \cap B) = \frac{9}{19} + \frac{9}{19} - \frac{4}{19} = \frac{14}{19}$. Events A and B are not mutually exclusive because $P(A \cap B) \neq 0$.

f. $P(A \cap B \cap C) = 4\left(\frac{1}{38}\right) = \frac{4}{38} = \frac{2}{19}$

g. The event $A \cup B \cup C$ is the event the outcome is odd or black or low. The event $A \cup B \cup C$ is:

{1, 2, 3, ... , 29, 31, 33, 35}

or

{All simple events except 00, 0, 30, 32, 34, 36}

h. $P(A \cup B \cup C) = 32\left(\frac{1}{38}\right) = \frac{32}{38} = \frac{16}{19}$

3.173 a. $P(E_1 \mid error) = \frac{P(E_1 \cap error)}{P(error)}$

$$= \frac{P(error \mid E_1)P(E_1)}{P(error \mid E_1)P(E_1) + P(error \mid E_2)P(E_2) + P(error \mid E_3)P(E_3)}$$

$$= \frac{0.01(0.30)}{0.01(0.30) + 0.03(0.20) + 0.02(0.50)} = \frac{0.003}{0.003 + 0.006 + 0.01} = \frac{0.003}{0.019} = 0.158$$

b. $P(E_2 \mid error) = \frac{P(E_2 \cap error)}{P(error)}$

$$= \frac{P(error \mid E_2)P(E_2)}{P(error \mid E_1)P(E_1) + P(error \mid E_2)P(E_2) + P(error \mid E_3)P(E_3)}$$

$$= \frac{0.03(0.20)}{0.01(0.30) + 0.03(0.20) + 0.02(0.50)} = \frac{0.006}{0.003 + 0.006 + 0.01} = \frac{0.006}{0.019} = 0.316$$

c. $P(E_3 \mid error) = \frac{P(E_3 \cap error)}{P(error)}$

$$= \frac{P(error \mid E_3)P(E_3)}{P(error \mid E_1)P(E_1) + P(error \mid E_2)P(E_2) + P(error \mid E_3)P(E_3)}$$

$$= \frac{0.02(0.50)}{0.01(0.30) + 0.03(0.20) + 0.02(0.50)} = \frac{0.01}{0.003 + 0.006 + 0.01} = \frac{0.01}{0.019} = 0.526$$

d. If there was a serious error, the probability that the error was made by engineer 3 is 0.526. This probability is higher than for any of the other engineers. Thus engineer #3 is most likely responsible for the error.

3.175 a. From the problem, we have three students and three hospitals. Suppose we consider the three hospitals as three positions. Using the permutations rule, there are

$$P = P_3^3 = \frac{3!}{(3-3)!} = \frac{3 \cdot 2 \cdot 1}{1} = 6 \text{ different assignments possible.}$$

These assignments are:

$[S_1H_A \quad S_2H_B \quad S_3H_C]$ $[S_1H_A \quad S_3H_B \quad S_2H_C]$
$[S_2H_A \quad S_1H_B \quad S_3H_C]$ $[S_2H_A \quad S_3H_B \quad S_1H_C]$
$[S_3H_A \quad S_1H_B \quad S_2H_C]$ $[S_3H_A \quad S_2H_B \quad S_1H_C]$

b. Assuming that all of the above assignments are equally likely, the probability of any one of them is $1/6$. The probability that student #1 is assigned to hospital B is $2/6 = 1/3$.

3.177 a. Define the following event:

F_i: {Player makes a foul shot on i^{th} attempt}

$P(F_i) = 0.8$, $P(F_i^c) = 1 - P(F_i) = 1 - 0.8 = 0.2$

The event "the player scores on both shots" is $F_1 \cap F_2$. If the throws are independent, then:

$$P(F_1 \cap F_2) = P(F_1)P(F_2) = 0.8(0.8) = 0.64$$

The event "the player scores on exactly one shot" is $(F_1 \cap F_2^c) \cup (F_1^c \cap F_2)$

$$\begin{aligned} P\left[(F_1 \cap F_2^c) \cup (F_1^c \cap F_2)\right] &= P(F_1 \cap F_2^c) + P(F_1^c \cap F_2) \\ &= P(F_1)P(F_2^c) + P(F_1^c)P(F_2) \\ &= 0.8(0.2) + 0.2(0.8) = 0.16 + 0.16 = 0.32 \end{aligned}$$

The event "the player scores on neither shot" is $(F_1^c \cap F_2^c)$.

$$P(F_1^c \cap F_2^c) = P(F_1^c)P(F_2^c) = 0.2(0.2) = 0.04$$

b. We know $P(F_1) = 0.8$, $P(F_2 \mid F_1) = 0.9$, and $P(F_2 \mid F_1^c) = 0.7$

The probability the player scores on both shots is:

$$P(F_1 \cap F_2) = P(F_2 \mid F_1)P(F_1) = 0.9(0.8) = 0.72$$

The probability the player scores on exactly one shot is:

$$\begin{aligned} P(F_1 \cap F_2^c) + P(F_1^c \cap F_2) &= P(F_2^c \mid F_1)P(F_1) + P(F_2 \mid F_1^c)P(F_1^c) \\ &= \left[1 - P(F_2 \mid F_1)\right]P(F_1) + P(F_2 \mid F_1^c)P(F_1^c) \\ &= (1 - 0.9)(0.8) + 0.7(0.2) = 0.08 + 0.14 = 0.22 \end{aligned}$$

The probability the player scores on neither shot is:

$$P(F_1^c \cap F_2^c) = P(F_2^c \mid F_1^c)P(F_1^c) = \left[1 - P(F_2 \mid F_1^c)\right]P(F_1^c) = (1 - 0.7)(0.2) = 0.06$$

c. Two consecutive foul shots are probably dependent. The outcome of the second shot probably depends on the outcome of the first.

3.179 Define the following events:

A: {Seed carries single spikelets}
B: {Seed carries paired spikelets}
C: {Seed produces ears with single spikelets}
D: {Seed produces ears with paired spikelets}

From the exercise, $P(A) = 0.4$, $P(B) = 0.6$, $P(C \mid A) = 0.29$, $P(D \mid A) = 0.71$, $P(C \mid B) = 0.26$, and $P(D \mid B) = 0.74$.

a. $P(A \cap C) = P(C \mid A)P(A) = 0.29(0.4) = 0.116$

b. $P(D) = P(A \cap D) + P(B \cap D) = P(D \mid A)P(A) + P(D \mid B)P(B) = 0.71(0.4) + 0.74(0.6)$
$= 0.284 + 0.444 = 0.728$

3.181 Define the following events:

A_1: {Component 1 in System A works properly}
A_2: {Component 2 in System A works properly}
A_3: {Component 3 in System A works properly}
B_1: {Component 1 in System B works properly}
B_2: {Component 2 in System B works properly}
B_3: {Component 3 in System B works properly}
B_4: {Component 4 in System B works properly}
A: {System A works properly}
B: {System B works properly}
C: {Subsystem C works properly}
D: {Subsystem D works properly}

$P(A_1) = 1 - P(A_1^c) = 1 - 0.12 = 0.88$ $\quad$ $P(B_1) = 1 - P(B_1^c) = 1 - 0.1 = 0.9$
$P(A_2) = 1 - P(A_2^c) = 1 - 0.09 = 0.91$ $\quad$ $P(B_2) = 1 - P(B_2^c) = 1 - 0.1 = 0.9$
$P(A_3) = 1 - P(A_3^c) = 1 - 0.11 = 0.89$ $\quad$ $P(B_3) = 1 - P(B_3^c) = 1 - 0.1 = 0.9$
$P(B_4) = 1 - P(B_4^c) = 1 - 0.1 = 0.9$

a. $P(A) = P(A_1 \cap A_2 \cap A_3) = P(A_1)P(A_2)P(A_3) = 0.88(0.91)(0.89) = 0.7127$
(since the three components operate independently)

b. $P(A^c) = 1 - P(A) = 1 - 0.7127 = 0.2873$

c. Subsystem C works properly if both components B_1 and B_2 work properly.

$$P(C) = P(B_1 \cap B_2) = P(B_1)P(B_2) = 0.9(0.9) = 0.81$$
(since the 2 components operate independently)

Similarly, subsystem D works properly if both components B_3 and B_4 work properly.

$$P(D) = P(B_3 \cap B_4) = P(B_3)P(B_4) = 0.9(0.9) = 0.81$$

System B operates properly if either subsystem C or subsystem D operates properly.

$$P(B) = P(C \cup D) = P(C) + P(D) - P(C \cap D) = P(C) + P(D) - P(C)P(D)$$
$$= 0.81 + 0.81 - 0.81(0.81) = 0.9639$$

d. The probability that exactly one subsystem in System B fails is:

$$P(C \cap D^c) + P(C^c \cap D) = P(C)P(D^c) + P(C^c)P(D)$$
$$= 0.81(1 - 0.81) + (1 - 0.81)(0.81) = 0.1539 + 0.1539 = 0.3078$$

e. System B fails to operate only if both subsystems C and D fail.

$$P(B^c) = P(C^c \cap D^c) = P(C^c)P(D^c) = (1-0.81)(1-0.81) = 0.0361$$

f. If System B operates correctly 99% of the time, then it fails 1% of the time. The probability that one of the subsystems fails is $1-0.81 = 0.19$. The probability that n subsystems fail is 0.19^n. Thus, we must find n such that

$$0.19^n \le 0.01 \Rightarrow n \ge 3$$

3.183 Define the following events:

S: {System shuts down}
F_1: {Hardware failure}
F_2: {Software failure}
F_3: {Power failure}

From the exercise, we know $P(F_1) = 0.01$, $P(F_2) = 0.05$, and $P(F_3) = 0.02$. Also, $P(S \mid F_1) = 0.73$, $P(S \mid F_2) = 0.12$, and $P(S \mid F_3) = 0.88$.

The probability that the current shutdown is due to a hardware failure is:

$$P(F_1 \mid S) = \frac{P(F_1 \cap S)}{P(S)} = \frac{P(S \mid F_1)P(F_1)}{P(S \mid F_1)P(F_1) + P(S \mid F_2)P(F_2) + P(S \mid F_3)P(F_3)}$$
$$= \frac{0.73(0.01)}{0.73(0.01)+0.12(0.05)+0.88(0.02)} = \frac{0.0073}{0.0073+0.006+0.0176} = \frac{0.0073}{0.0309} = 0.2362$$

The probability that the current shutdown is due to a software failure is:

$$P(F_2 \mid S) = \frac{P(F_2 \cap S)}{P(S)} = \frac{P(S \mid F_2)P(F_2)}{P(S \mid F_1)P(F_1) + P(S \mid F_2)P(F_2) + P(S \mid F_3)P(F_3)}$$
$$= \frac{0.12(0.05)}{0.73(0.01)+0.12(0.05)+0.88(0.02)} = \frac{0.006}{0.0073+0.006+0.0176} = \frac{0.006}{0.0309} = 0.1942$$

The probability that the current shutdown is due to a power failure is:

$$P(F_3 \mid S) = \frac{P(F_3 \cap S)}{P(S)} = \frac{P(S \mid F_3)P(F_3)}{P(S \mid F_1)P(F_1) + P(S \mid F_2)P(F_2) + P(S \mid F_3)P(F_3)}$$
$$= \frac{0.88(0.02)}{0.73(0.01)+0.12(0.05)+0.88(0.02)} = \frac{0.0176}{0.0073+0.006+0.0176} = \frac{0.0176}{0.0309} = 0.5696$$

3.185 $P(\text{Interview or Grounded Theory}) = (5{,}079 + 537)/7{,}506 = 5{,}616/7{,}506 = 0.748$

3.187 a. The odds in favor of Smarty Jones are $\frac{1}{3}$ to $\left(1-\frac{1}{3}\right)$ or $\frac{1}{3}$ to $\frac{2}{3}$ or 1 to 2

b. If the odds are 1 to 1, $P(\text{Smarty Jones will win}) = \frac{1}{1+1} = \frac{1}{2} = 0.5$

c. If the odds against Smarty Jones winning are 3 to 2, the odds for Smarty Jones winning are 2 to 3. The probability Smarty Jones will win is $\frac{2}{2+3} = \frac{2}{5} = 0.4$

3.189 Define the following event:

A: {Antigens match}

From the problem, $P(A) = 0.25$ and $P(A^c) = 1 - P(A) = 1 - 0.25 = 0.75$

a. The probability that one sibling has a match is $P(A) = 0.25$.

b. The probability that all three will match is:

$$P(AAA) = P(A \cap A \cap A) = P(A)P(A)P(A) = 0.25^3 = 0.0156$$

c. The probability that none of the three match is:

$$P(A^c A^c A^c) = P(A^c \cap A^c \cap A^c) = P(A^c)P(A^c)P(A^c) = 0.75^3 = 0.4219$$

d. For this part, $P(A) = 0.001$ and $P(A^c) = 1 - P(A) = 1 - 0.001 = 0.999$

$P(A) = 0.001$

$P(AAA) = P(A \cap A \cap A) = P(A)P(A)P(A) = 0.001^3 = 0.000000001$

$P(A^c A^c A^c) = P(A^c \cap A^c \cap A^c) = P(A^c)P(A^c)P(A^c) = 0.999^3 = 0.9970$

3.191 We will work with the following events:

A: {Woman is pregnant}
B: {Pregnancy test is positive}

The given probabilities can be written as:

$P(A) = 0.75$, $P(B \mid A) = 0.99$, $P(B \mid A^c) = 0.02$

$$P(A \mid B) = \frac{P(A \cap B)}{P(B)} = \frac{P(B|A)P(A)}{P(A \cap B) + P(A^c \cap B)}$$

Then,

$$= \frac{P(B|A)P(A)}{P(B|A)P(A) + P(B|A^c)P(A^c)}$$

$$= \frac{(0.99)(0.75)}{(0.99)(0.75) + (0.02)(0.25)} = \frac{0.7425}{0.7475} \approx 0.993$$

3.193 We first determine the number of simple events for this experiment. According to the partitions rule, the number of ways of dealing 52 cards so that each of the four players receives 13 cards is:

$$\frac{52!}{13!13!13!13!} = 5.3644738\text{x}10^{28}$$

If the cards are well shuffled, then the probability of each simple event is:

$$\frac{1}{5.3644738x10^{-28}}$$

We now need to count the number of simple events that result in one player receiving all the diamonds, another all the hearts, another all the spades, and another all the clubs. To do this, first consider a particular ***ordering*** that yields the desired result:

Player #1 receives 13 diamonds
Player #2 receives 13 hearts
Player #3 receives 13 spades
Player #4 receives 13 clubs

This particular ***ordered*** event can happen in

$$\binom{13}{13}\binom{13}{13}\binom{13}{13}\binom{13}{13} = 1 \text{ way.}$$

However, the ***ordering*** of the players is not relevant to us.

Since there are $P_4^4 = \frac{4!}{(4-4)!} = 24$ possible orderings of the players, we conclude that the

probability of the rare event observed in Dubuque is $\frac{24}{5.3644738\times10^{28}} \approx 4.4739\text{x}10^{-28}$

3.195 First, we will list all possible sample points for placing a car (C) and 2 goats (G) behind doors #1, #2, and #3. If the first position corresponds to door #1, the second position corresponds to door #2, and the third position corresponds to door #3, the sample space is:

(C G G) (G C G) (G G C)

Now, suppose you pick door #1. Initially, the probability that you will win the car is $1/3$: only one of the sample points has a car behind door #1.
The host will now open a door behind which is a goat. If you pick door #1 in the first sample point (C G G), the host will open either door #2 or door #3. Suppose he opens door #3 (it really does not matter). If you pick door #1 in the second sample point (G C G), the host will open door #3. If you pick door #1 in the third sample point (G G C), the host will open door #2. Now, the new sample space will be:

(C G) (G C) (G C)

where the first position corresponds to door #1 (the one you chose) and the second position corresponds to the door that was not opened by the host.

Now, if you keep door #1, the probability that you win the car is $1/3$. However, if you switch to the remaining door, the probability that you win the car is now $2/3$. Based on these probabilities, it is to your advantage to switch doors.

The above could be repeated by selecting door #2 initially or door #3 initially. In either of these cases, again, the probability of winning the car is $1/3$ if you do not switch and $2/3$ if you switch. Thus, Marilyn was correct.

Discrete Random Variables

Chapter 4

4.1 A random variable is a rule that assigns one and only one value to each simple event of an experiment.

4.3
a. Since we can count the number of words spelled correctly, the random variable is discrete.

b. Since we can assume values in an interval, the amount of water flowing through the Hoover Dam in a day is continuous.

c. Since time is measured on an interval, the random variable is continuous.

d. Since we can count the number of bacteria per cubic centimeter in drinking water, this random variable is discrete.

e. Since we cannot count the amount of carbon monoxide produced per gallon of unleaded gas, this random variable is continuous.

f. Since weight is measured on an interval, weight is a continuous random variable.

4.5
a. The reaction time difference is continuous because it lies within an interval.

b. Since we can count the number of violent crimes, this random variable is discrete.

c. Since we can count the number of near misses in a month, this variable is discrete.

d. Since we can count the number of winners each week, this variable is discrete.

e. Since we can count the number of free throws made per game by a basketball team, this variable is discrete.

f. Since distance traveled by a school bus lies in some interval, this is a continuous random variable.

4.7 The values x can assume are 0, 1, 2, 3, 4, ... Thus, x is a discrete random variable.

4.9 Annual rainfall can take on any value in an interval. Thus, annual rainfall is a continuous random variable. The number of ant species can assume only a countable number of outcomes. Thus, number of ant species is a discrete random variable.

4.11 Answers will vary. An example of a discrete random variable of interest to a psychologist might be the number of times a person has violent dreams in a week.

4.13 Answers will vary. An example of a discrete random variable of interest to a hospital nurse might be the number of medications a patient receives in 24 hours.

4.15 Error rates are the proportions of errors made. Error rates can take on values between 0 and 1, inclusive. Therefore, error rates are continuous random variables and the difference between two error rates will also be continuous.

4.17 The probability distribution of a discrete random variable can be represented by a table, graph, or formula that specifies the probability associated with each possible value the random variable can assume.

4.19 a. $p(22) = 0.25$

b. $p(20) + p(24) = 0.15 + 0.20 = 0.35$

c. $P(x \le 23) = p(20) + p(21) + p(22) + p(23) = 0.15 + 0.10 + 0.25 + 0.30 = 0.80$

4.21 a. $P(x \le 12) = P(x = 10) + P(x = 11) + P(x = 12) = 0.2 + 0.3 + 0.2 = 0.7$

b. $P(x > 12) = 1 - P(x \le 12) = 1 - 0.7 = 0.3$

c. $P(x \le 14) = P(x = 10) + P(x = 11) + P(x = 12) + P(x = 13) + P(x = 14)$
$= 0.2 + 0.3 + 0.2 + 0.1 + 0.2 = 1$

d. $P(x = 14) = 0.2$

e. $P(x \le 11 \text{ or } x > 12) = P(x \le 11) + P(x > 12)$
$= P(x = 10) + P(x = 11) + P(x = 13) + P(x = 14)$
$= 0.2 + 0.3 + 0.1 + 0.2 = 0.8$

4.23 a. The simple events are (where H = head, T = tail):

	HHH	HHT	HTH	THH	HTT	THT	TTH	TTT
x = # heads	3	2	2	2	1	1	1	0

b. If each event is equally likely, then $P(\text{simple event}) = \frac{1}{n} = \frac{1}{8}$

$$p(3) = \frac{1}{8},\ p(2) = \frac{1}{8} + \frac{1}{8} + \frac{1}{8} = \frac{3}{8},\ p(1) = \frac{1}{8} + \frac{1}{8} + \frac{1}{8} = \frac{3}{8}, \text{ and } p(0) = \frac{1}{8}$$

c.

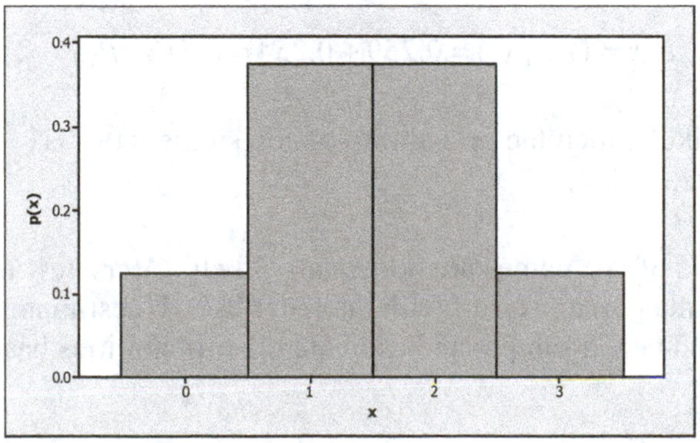

d. $P(x=2 \text{ or } x=3)=p(2)+p(3)=\frac{3}{8}+\frac{1}{8}=\frac{4}{8}=\frac{1}{2}$

4.25 a. The possible values of x are 0, 2, 3, and 4.

b. $P(x=0)=\frac{35}{106}=0.330$, $P(x=2)=\frac{58}{106}=0.547$, $P(x=3)=\frac{5}{106}=0.047$,

$P(x=4)=\frac{8}{106}=0.075$

The probability distribution of x is:

x	0	2	3	4
$p(x)$	0.330	0.547	0.047	0.075

c. $E(x)=\sum xp(x)=0(0.330)+2(0.547)+3(0.047)+4(0.075)=1.535$

The average number of legs per robot is 1.535.

4.27 a. The two properties of a probability distribution are
1) the probabilities for all values of x are greater than or equal to 0
2) the sum of the probabilities over all values of x is 1

For this problem, all probabilities are greater than 0 and
$\sum p(x)=0.17+0.10+0.11+\cdots+0.005=1.000$.

b. $P(x\geq 10)=P(x=10)+P(x=1)+\cdots+P(x=20)=0.02+0.02+\cdots+0.005=0.14$

4.29 a. There are 4 sample points: Boy-Boy, Boy-Girl, Girl-Boy, and Girl-Girl.

b. If all the sample points are equally likely, then each would have a probability of ¼.

c. $P(x=0)=p(0)=\frac{1}{4}$, $P(x=1)=p(1)=\frac{1}{4}+\frac{1}{4}=\frac{2}{4}=\frac{1}{2}$, $P(x=2)=p(2)=\frac{1}{4}$

d. Using the proportions in the table as approximate probabilities, the distribution of x is:

$P(x=0)=p(0)=0.222$, $P(x=1)=p(1)=0.259+0.254=0.513$, $P(x=2)=p(2)=0.265$

4.31 a. If each letter is equally likely, then the probability of any single letter is $1/26$. Thus, $P(x=i)=1/26$ for $i=1, 2, \ldots, 26$.

b. No. The beginning letters of last names are not equally likely. Very few names start with the letter 'Q' while many names start with the letter 'S'. To estimate the true probability of x, one could take a sample and estimate the probabilities based on the values in the sample.

4.33 a. $p(1)=(0.23)(0.77)^{1-1}=0.23(0.77)^{0}=0.23$. The probability that a contaminated cartridge is selected on the first sample is 0.23.

b. $p(5)=(0.23)(0.77)^{5-1}=0.23(0.77)^{4}=0.081$. The probability that the first contaminated cartridge will be selected on the 5th sample is 0.081.

c. $P(x\geq 2)=1-P(x\leq 1)=1-p(1)=1-0.23=0.77$. The probability that the first contaminated cartridge will be selected on the second trial or later is 0.77.

4.35 a. x is the sum of the grill numbers. To find the probability distribution for x, we divide the number of students choosing the particular grill display combination divided by the total number of students or 124. The table below shows the probability distribution for x.

Grill Display Combination	**Number of Students**	x	$p(x)$
1-2-3	35	6	35/124 = 0.282
1-2-4	8	7	8/124 = 0.065
1-2-5	42	8	42/124 = 0.339
2-3-4	4	9	4/124 = 0.032
2-3-5	1	10	1/124 = 0.008
2-4-5	34	11	34/124 = 0.274
Total	**124**		**1.000**

b. $P(x>10)=P(x=11)=0.274$.

4.37 a. The preference scores for committee {A, B, C} for each voter is shown in the table:

Voter:	*1*	*2*	*3*	*4*	*5*	*6*	*7*	*8*	*9*	*10*
	A	B	A	B	D	C	A	B	A	A
	C	C	B	C	E	E	C	C	B	C
	D	E	D	F	F	G	G	D	C	G
Preference Score	2	2	2	2	0	1	2	2	3	2

b. The values x can take on are 0, 1, 2, and 3. Because there are 10 voters, there are 10 possible outcomes, with each having probability $1/10$. The probability distribution of x is:

x	0	1	2	3
$p(x)$	0.10	0.10	0.70	0.10

c. $P(x>2)=P(x=3)=0.10$

d. No. The committee {A, B, C} is not preferred over all other committees. For this example, the committee {A, C, G} would be a "Condorcet" committee because it is preferred by 2 voters. No other committee is preferred by more than one voter.

4.39 a. Each point in the system can have one of 2 status levels, "free" or "obstacle". Define the following events:

A_F: {Point A is free} A_O: {Point A is obstacle}
B_F: {Point B is free} B_O: {Point B is obstacle}
C_F: {Point C is free} C_O: {Point C is obstacle}

Thus, the sample points for the space are:

$A_FB_FC_F$, $A_FB_FC_O$, $A_FB_OC_F$, $A_FB_OC_O$, $A_OB_FC_F$, $A_OB_FC_O$, $A_OB_OC_F$, $A_OB_OC_O$

The variable x is the total number of links that are "free". Thus, x can have values 0, 1, or 2.

b. Since it is stated that the probability of any point in the system having a "free" status is 0.5, the probability of any point having an "obstacle" status is also .5, Thus, the probability of each of the sample points above is $P(A_iB_iC_i) = 0.5(0.5)(0.5) = 0.125$.

The values of x, the number of free links in the system, for each sample point are listed below. A link is free if both the points are free. Thus, a link from A to B is free if A is free and B is free. A link from B to C is free if B is free and C is free.

Sample point	x	**Probability**
$A_FB_FC_F$	2	0.125
$A_FB_FC_O$	1	0.125
$A_FB_OC_F$	0	0.125
$A_FB_OC_O$	0	0.125
$A_OB_FC_F$	1	0.125
$A_OB_FC_O$	0	0.125
$A_OB_OC_F$	0	0.125
$A_OB_OC_O$	0	0.125

The probability distribution for x is:

x	**Probability**
0	0.625
1	0.250
2	0.125

4.41 The expected value of a random variable, $E(x)$, does not always equal a specific value of the random variable, x. The $E(x)$ is the mean of the probability distribution and does not have to equal a specific value of x.

4.43 a. $\mu = E(x) = \sum xp(x) = 1(0.2) + 2(0.4) + 4(0.2) + 10(0.2) = 0.2 + 0.8 + 0.8 + 2 = 3.8$

b. $\sigma^2 = E[(x-\mu)^2] = \sum (x-\mu)^2 p(x)$

$$= (1-3.8)^2(0.2) + (2-3.8)^2(0.4) + (4-3.8)^2(0.2) + (10-3.8)^2(0.2)$$
$$= 1.568 + 1.296 + 0.008 + 7.688 = 10.56$$

c. $\sigma = \sqrt{10.59} = 3.2496$

d. The average value of x over many trials is 3.8.

e. No. The random variable can only take on values 1, 2, 4, or 10.

f. Yes. It is possible that μ can be equal to an actual value of x.

4.45 a. It would seem that the mean of both would be 1 since they both are symmetric distributions centered at 1.

b. The distribution of x seems more variable since there appears to be greater probability for the two extreme values of 0 and 2 than there is in the distribution of y.

c. For x: $\mu = E(x) = \sum xp(x) = 0(0.3) + 1(0.4) + 2(0.3) = 0 + 0.4 + 0.6 = 1$

$$\sigma^2 = E[(x-\mu)^2] = \sum (x-\mu)^2 p(x)$$
$$= (0-1)^2(0.3) + (1-1)^2(0.4) + (2-1)^2(0.3) = 0.3 + 0 + 0.3 = 0.6$$

For y: $\mu = E(y) = \sum yp(y) = 0(0.1) + 1(0.8) + 2(0.1) = 0 + 0.8 + 0.2 = 1$

$$\sigma^2 = E[(y-\mu)^2] = \sum (y-\mu)^2 p(y)$$
$$= (0-1)^2(0.1) + (1-1)^2(0.8) + (2-1)^2(0.1) = 0.1 + 0 + 0.1 = 0.2$$

The variance for x is larger than that for y.

4.47 From Exercise 4.26, the probability distribution is:

x	1	2	3	4	5
$p(x)$	0.0000	0.0408	0.1735	0.6020	0.1837

$$\mu = E(x) = \sum xp(x) = 2(0.0408) + 3(0.1735) + 4(0.6020) + 5(0.1837) = 3.9286$$

Over a very large number of trials, the average driver-side crash rating is 3.9286 stars.

4.49 $\mu = E(x) = \sum xp(x) = 1(0.40) + 2(0.54) + 3(0.02) + 4(0.04) = 0.40 + 1.08 + 0.06 + 0.16 = 1.70$

The average number of insect eggs on a blade of water hyacinth is 1.70.

4.51 From Exercise 4.30, we found the probability distribution of x to be:

EVS score, x	$p(x)$	EVS score, x	$p(x)$
3	0.003	12	0.081
4	0.003	13	0.120
5	0.018	14	0.136
6	0.026	15	0.131
7	0.024	16	0.115
8	0.050	17	0.063
9	0.045	18	0.024
10	0.079	19	0.018
11	0.065	20	0.000

$$E(x) = \sum xp(x) = 3(0.003) + 4(0.003) + 5(0.018) + \cdots + 20(0.000) = 0.009 + 0.012 + 0.090 + \cdots + 0 = 12.831$$

The average EVS score for s randomly selected snake species in Mexico is 12.831.

4.53 a. Define the following events:
M: {Miami Beach, FL}
C: {Coney Island, NY}
S: {Surfside, CA}
B: {Monmouth Beach, NJ}
O: {Ocean City, NJ}
L: {Spring Lake, NJ}

The possible pairs of beach hotspots that can be selected are:

MC, MS, MB, MO, ML, CS, CB, CO, CL, SB, SO, SL, BO, BL, OL

b. If we are just selecting 2 hotspots from 6, each of the 15 combinations are equally likely. Thus, each would have a probability of 1/15.

c. The value of x for each of the pairs is:

Pair	x
MC	0
MS	0
MB	1
MO	0
ML	1
CS	0
CB	1
CO	0

Pair	x
CL	1
SB	1
SO	0
SL	1
BO	1
BL	2
OL	1

d. The probability distribution for x is:

x	$p(x)$
0	6/15
1	8/15
2	1/15

e. $\mu = E(x) = \sum xp(x) = 0(6/15) + 1(8/15) + 2(1/15) = 10/15 = 0.667$

The average total number of hotspots with a planar nearshore bar condition in each pair is 0.667.

4.55 For a \$5 bet, you will either win \$5 or lose \$5 (–\$5). The probability distribution for the net winnings is:

x	$p(x)$
–5	20/38
5	18/38

$$\mu = E(x) = \Sigma xp(x) = -5\left(\frac{20}{38}\right) + 5\left(\frac{18}{38}\right) = -\$0.263$$

Over a large number of trials, the average winning for a \$5 bet on red is -\$0.263.

4.57 The five characteristics of a binomial random variable are:

a. The experiment consists of n identical trials.
b. There are only two possible outcomes on each trial. We will denote one outcome by S (for Success) and the other by F (for Failure).
c. The probability of S remains the same from trial to trial. This probability is denoted by p, and the probability of F is denoted by q. Note that $q = 1 - p$.
d. The trials are independent.
e. The binomial random variable x is the number of S's in n trials.

4.59 a. There are $n = 5$ trials.

b. The value of p is $p = 0.7$.

4.61 a. $\frac{6!}{2!(6-2)!} = \frac{6!}{2!4!} = \frac{6 \cdot 5 \cdot 4 \cdot 3 \cdot 2 \cdot 1}{(2 \cdot 1)(4 \cdot 3 \cdot 2 \cdot 1)} = 15$

b. $\binom{5}{2} = \frac{5!}{2!(5-2)!} = \frac{5!}{2!3!} = \frac{5 \cdot 4 \cdot 3 \cdot 2 \cdot 1}{(2 \cdot 1)(3 \cdot 2 \cdot 1)} = 10$

c. $\binom{7}{0} = \frac{7!}{0!(7-0)!} = \frac{7!}{0!7!} = \frac{7 \cdot 6 \cdot 5 \cdot 4 \cdot 3 \cdot 2 \cdot 1}{(1)(7 \cdot 6 \cdot 5 \cdot 4 \cdot 3 \cdot 2 \cdot 1)} = 1$ (Note: $0! = 1$)

d. $\binom{6}{6} = \frac{6!}{6!(6-6)!} = \frac{6!}{6!0!} = \frac{6 \cdot 5 \cdot 4 \cdot 3 \cdot 2 \cdot 1}{(6 \cdot 5 \cdot 4 \cdot 3 \cdot 2 \cdot 1)(1)} = 1$

e. $\binom{4}{3} = \frac{4!}{3!(4-3)!} = \frac{4!}{3!1!} = \frac{4 \cdot 3 \cdot 2 \cdot 1}{(3 \cdot 2 \cdot 1)(1)} = 4$

4.63 a. $P(x=1)=\frac{5!}{1!4!}(0.2)^1(0.8)^4=\frac{5\cdot4\cdot3\cdot2\cdot1}{(1)(4\cdot3\cdot2\cdot1)}(0.2)^1(0.8)^4=5(0.2)^1(0.8)^4=0.4096$

b. $P(x=2)=\frac{4!}{2!2!}(0.6)^2(0.4)^2=\frac{4\cdot3\cdot2\cdot1}{(2\cdot1)(2\cdot1)}(0.6)^2(0.4)^2=6(0.6)^2(0.4)^2=0.3456$

c. $P(x=0)=\frac{3!}{0!3!}(0.7)^0(0.3)^3=\frac{3\cdot2\cdot1}{(1)(3\cdot2\cdot1)}(0.7)^0(0.3)^3=1(0.7)^0(0.3)^3=0.027$

d. $P(x=3)=\frac{5!}{3!2!}(0.1)^3(0.9)^2=\frac{5\cdot4\cdot3\cdot2\cdot1}{(3\cdot2\cdot1)(2\cdot1)}(0.1)^3(0.9)^2=10(0.1)^3(0.9)^2=0.0081$

e. $P(x=2)=\frac{4!}{2!2!}(0.4)^2(0.6)^2=\frac{4\cdot3\cdot2\cdot1}{(2\cdot1)(2\cdot1)}(0.4)^2(0.6)^2=6(0.4)^2(0.6)^2=0.3456$

f. $P(x=1)=\frac{3!}{1!2!}(0.9)^1(0.1)^2=\frac{3\cdot2\cdot1}{(1)(2\cdot1)}(0.9)^1(0.1)^2=3(0.9)^1(0.1)^2=0.027$

4.65 a. $P(x<10)=P(x\le9)=0$

b. $P(x\ge10)=1-P(x\le9)=1-0.002=0.998$

c. $P(x=2)=P(x\le2)-P(x\le1)=0.206-0.069=0.137$

4.67 x is a binomial random variable with $n=4$.

a. If the probability distribution of x is symmetric, $p(0)=p(4)$ and $p(1)=p(3)$.

Since $p(x)=\binom{n}{x}p^xq^{n-x}$ $\quad x=0, 1, \ldots, n$

When $n=4$,

$$\binom{4}{0}p^0q^4=\binom{4}{4}p^4q^0 \Rightarrow \frac{4!}{0!4!}p^0q^4=\frac{4!}{4!0!}p^4q^0 \Rightarrow q^4=p^4 \Rightarrow p=q$$

Since $p+q=1$, $p=0.5$

Therefore, the probability distribution of x is symmetric when $p=0.5$.

b. If the probability distribution of x is skewed to the right, then the mean is greater than the median. Therefore, there are more small values in the distribution (0, 1) than large values (3, 4). Therefore, p must be smaller than 0.5. Let $p = 0.2$ and the probability distribution of x will be skewed to the right.

c. If the probability distribution of x is skewed to the left, then the mean is smaller than the median. Therefore, there are more large values in the distribution (3, 4) than small values (0, 1). Therefore, p must be larger than 0.5. Let $p = 0.8$ and the probability distribution of x will be skewed to the left.

d. In part **a**, x is a binomial random variable with $n = 4$ and $p = 0.5$.

$$p(x) = \binom{4}{x}(0.5)^x(0.5)^{4-x} \qquad x = 0, 1, 2, 3, 4$$

$$p(0) = \binom{4}{0}(0.5)^0(0.5)^4 = \frac{4!}{0!4!}.5^4 = 1(0.5)^4 = 0.0625$$

$$p(1) = \binom{4}{1}(0.5)^1(0.5)^3 = \frac{4!}{1!3!}.5^4 = 4(0.5)^4 = 0.25$$

$$p(2) = \binom{4}{2}(0.5)^2(0.5)^2 = \frac{4!}{2!2!}(0.5)^4 = 6(0.5)^4 = 0.375$$

$p(3) = p(1) = 0.25$ (since the distribution is symmetric)

$p(4) = p(0) = 0.0625$

The probability distribution of x in tabular form is:

x	0	1	2	3	4
$p(x)$	0.0625	0.25	0.375	0.25	0.0625

$\mu = np = 4(0.5) = 2$

The graph of the probability distribution of x when $n = 4$ and $p = 0.5$ is as follows.

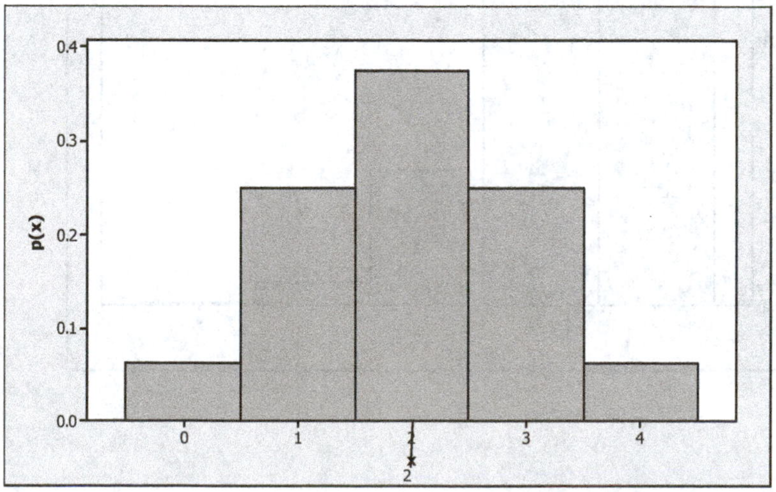

In part **b**, x is a binomial random variable with $n = 4$ and $p = 0.2$.

$$p(x) = \binom{4}{x}(0.2)^x(0.8)^{4-x} \quad x = 0, 1, 2, 3, 4$$

$$p(0) = \binom{4}{0}(0.2)^0(0.8)^4 = 1(1)(0.8)^4 = 0.4096$$

$$p(1) = \binom{4}{1}(0.2)^1(0.8)^3 = 4(0.2)(0.8)^3 = 0.4096$$

$$p(2) = \binom{4}{2}(0.2)^2(0.8)^2 = 6(0.2)^2(0.8)^2 = 0.1536$$

$$p(3) = \binom{4}{3}(0.2)^3(0.8)^1 = 4(0.2)^3(0.8) = 0.0256$$

$$p(4) = \binom{4}{4}(0.2)^4(0.8)^0 = 1(0.2)^4(1) = 0.0016$$

The probability distribution of x in tabular form is:

x	0	1	2	3	4
$p(x)$	0.4096	0.4096	0.1536	0.0256	0.0016

$$\mu = np = 4(0.2) = 0.8$$

The graph of the probability distribution of x when $n = 4$ and $p = 0.2$ is as follows:

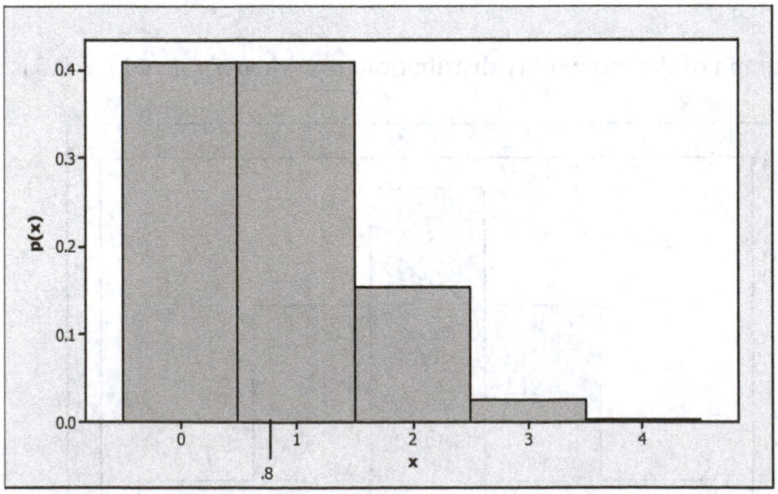

In part **c**, x is a binomial random variable with $n = 4$ and $p = 0.8$.

$$p(x) = \binom{4}{x}(0.8)^x(0.2)^{4-x} \quad x = 0, 1, 2, 3, 4$$

$$p(0) = \binom{4}{0}(0.8)^0(0.2)^4 = 1(1)(0.2)^4 = 0.0016$$

$$p(1) = \binom{4}{1}(0.8)^1(0.2)^3 = 4(0.8)(0.2)^3 = 0.0256$$

$$p(2) = \binom{4}{2}(0.8)^2(0.2)^2 = 6(0.8)^2(0.2)^2 = 0.1536$$

$$p(3) = \binom{4}{3}(0.8)^3(0.2)^1 = 4(0.8)^3(0.2) = 0.4096$$

$$p(4) = \binom{4}{4}(0.8)^4(0.2)^0 = 1(0.8)^4(1) = 0.4096$$

The probability distribution of x in tabular form is:

x	0	1	2	3	4
$p(x)$	0.0016	0.0256	0.1536	0.4096	0.4096

Note: The distribution of x when $n = 4$ and $p = 0.2$ is the reverse of the distribution of x when $n = 4$ and $p = 0.8$.

$$\mu = np = 4(0.8) = 3.2$$

The graph of the probability distribution of x when $n = 4$ and $p = 0.8$ is as follows:

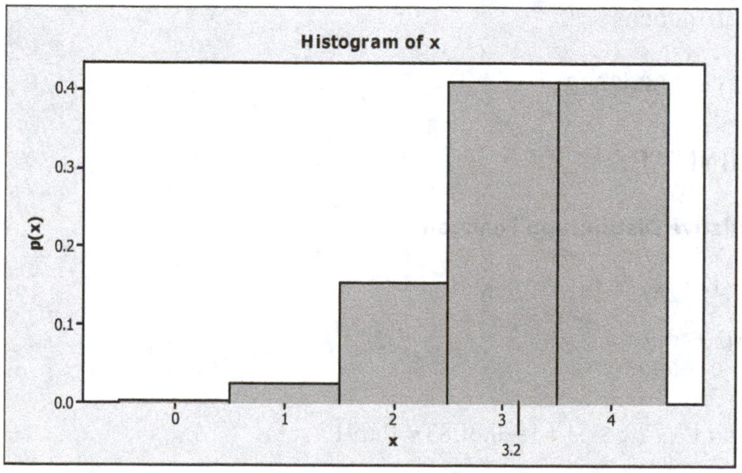

e. In general, when $p = 0.5$, a binomial distribution will be symmetric regardless of the value of n. When p is less than 0.5, the binomial distribution will be skewed to the right; and when p is greater than 0.5, it will be skewed to the left. (Refer to parts **a**, **b**, and **c**.)

4.69 a. A success would be an adult American who believes finding and picking up a penny is good luck.

b. For this experiment, there are $n = 20$ identical trials. Each trial can result in 2 possible outcomes (an adult American who believes finding and picking up a penny is good luck or an adult American who does not believe finding and picking up a penny is good luck). Let S = an adult American who believes finding and picking up a penny is good luck and F = an adult American who does not believe finding and picking up a penny is good luck. Then, $P(S) = p = 0.33$ and $P(F) = q = 1 - p = 1 - 0.33 = 0.67$. We can assume the trials are independent. Thus, the conditions of a binomial random variable hold.

c. $P(S) = p = 0.33$

d. Using MINITAB,

Cumulative Distribution Function

```
Binomial with n = 20 and p = 0.33

x  P( X ≤ x )
9    0.913422
```

$P(x < 10) = P(x \le 9) = 0.9134$

e. Using MINITAB,

Probability Density Function

```
Binomial with n = 20 and p = 0.33

 x  P( X = x )
17   0.0000022
```

$P(x = 17) = 0.0000022$

f. Using MINITAB,

Cumulative Distribution Function

```
Binomial with n = 20 and p = 0.33

x  P( X ≤ x )
5    0.308250
```

$P(x > 5) = 1 - P(x \le 5) = 1 - 0.3083 = 0.6917$

4.71 a. For this experiment, there are $n = 100$ identical trials. Each trial can result in 2 possible outcomes (mouse responds positively or mouse does not respond positively). Let S = mouse responds positively and F = mouse does not respond positively. Then, $P(S) = p = 0.4$ and $P(F) = q = 1 - 0.4 = 0.6$. We can assume the trials are independent. Thus, the conditions of a binomial random variable hold.

b. $E(x) = \mu = np = 100(0.4) = 40$

c. $V(x) = \sigma^2 = npq = 100(0.4)(0.6) = 24$

d. Since $p = 0.4$ is close to 0.5, the distribution should be fairly symmetric. Thus, we would expect most of the observations to fall within 2 standard deviations of the mean. The standard deviation is $\sigma = \sqrt{24} = 4.899$. The interval is

$$\mu \pm 2\sigma \Rightarrow 40 \pm 2(4.899) \Rightarrow 40 \pm 9.798 \Rightarrow (30.202,\ 49.798)$$

4.73 a. Let x = number of births in 1,000 that take place by Caesarian section. $E(x) = np = 1000(0.32) = 320$.

b. $\sigma = \sqrt{npq} = \sqrt{1000(0.32)(0.68)} = 14.7513$

c. Since p is not real small, the distribution of x will be fairly mound-shaped, so the Empirical Rule will apply. We know that approximately 95% of the observations will fall within 2 standard deviations of the mean. Thus,

$$\mu \pm 2\sigma \Rightarrow 320 \pm 2(14.7513) \Rightarrow 320 \pm 29.5026 \Rightarrow (290.4974,\ 349.5026)$$

In a sample of 1000 births, we would expect that somewhere between 291 and 349 will be Caesarian section births.

4.75 a. We will check the characteristics of a binomial random variable:

1. This experiment consists of $n = 15$ identical trials.

2. There are only 2 possible outcomes for each trial. Hotel guests are aware and participate in the hotel's conservation efforts (*S*) or not (*F*).

3. The probability of *S* remains the same from trial to trial. In this case, $p = P(S) = 0.66(0.72) = 0.4752$ for each trial.

4. The trials are independent. Whether one hotel guest is aware and participates in the hotel's conservation efforts should not affect whether another guest is aware and participates.

5. x = number of hotel guests who are aware and participate in the hotel's conservation efforts in 15 trials.

The characteristics of the binomial random variable are met.

b. Define the following events:
G: {hotel guest is aware of the hotel's "green" conservation program}
P: {hotel guest participates in the conservation program}

We are given $P(G) = 0.66$ and $P(P \mid G) = 0.72$.

$P(S)=P(P\cap G)=P(P|G)P(G)=0.66(0.72)=0.4752.$

c. $P(x\geq 10)=p(10)+p(11)+p(12)+p(13)+p(14)+p(15)$

$$=\binom{15}{10}(0.45)^{10}(0.55)^{15-10}+\binom{15}{11}(0.45)^{11}(0.55)^{15-11}+\binom{15}{12}(0.45)^{12}(0.55)^{15-12}$$

$$+\binom{15}{13}(0.45)^{13}(0.55)^{15-13}+\binom{15}{14}(0.45)^{14}(0.55)^{15-14}+\binom{15}{15}(0.45)^{15}(0.55)^{15-10}$$

$$=\frac{15!}{10!5!}(0.45)^{10}(0.55)^{5}+\frac{15!}{11!4!}(0.45)^{11}(0.55)^{4}+\frac{15!}{12!3!}(0.45)^{12}(0.55)^{3}$$

$$+\frac{15!}{13!2!}(0.45)^{13}(.55)^{2}+\frac{15!}{14!1!}(0.45)^{14}(.55)^{1}+\frac{15!}{15!0!}(0.45)^{15}(.55)^{0}$$

$$=0.0515+0.0191+0.0052+0.0010+0.0001+0.0000=0.0769$$

4.77 a. Let x = number of commissioners favoring an issue in 4 trials. Then x is a binomial random variable with $n=4$ and $p=0.5$.

$$P(x=2)=\binom{4}{2}(0.5)^2(0.5)^{4-2}=\frac{4!}{2!(4-2)!}(0.5)^2(0.5)^2=\frac{4(3)(2)(1)}{2(1)(2)(1)}(0.25)(0.25)=0.375$$

b. Let x = number of commissioners favoring an issue in 2 trials. Then x is a binomial random variable with $n=2$ and $p=0.5$.

$$P(x=1)=\binom{2}{1}(0.5)^1(0.5)^{2-1}=\frac{2!}{1!(2-1)!}(0.5)^1(0.5)^1=\frac{2(1)}{1(1)}(0.5)(0.5)=0.5$$

4.79 a. Let x = number of women out of 15 that have been abused. Then x is a binomial random variable with $n=15$, S = woman has been abused, F = woman has not been abused, $P(S)=p=\frac{1}{3}$ and $q=1-p=1-\frac{1}{3}=\frac{2}{3}$.

$$P(x\geq 4)=1-P(x<4)=1-P(x=0)-P(x=1)-P(x=2)-P(x=3)$$

$$=1-\frac{15!}{0!15!}\left(\frac{1}{3}\right)^0\left(\frac{2}{3}\right)^{15}-\frac{15!}{1!14!}\left(\frac{1}{3}\right)^1\left(\frac{2}{3}\right)^{14}-\frac{15!}{2!13!}\left(\frac{1}{3}\right)^2\left(\frac{2}{3}\right)^{13}-\frac{15!}{3!12!}\left(\frac{1}{3}\right)^3\left(\frac{2}{3}\right)^{12}$$

$$=1-0.0023-0.0171-0.0599-0.1299=0.7908$$

b. Using Table I, Appendix B, with $n = 15$ and $p = 0.1$,

$$P(x \geq 4) = 1 - P(x \leq 3) = 1 - 0.944 = 0.056$$

c. We sampled 15 women and actually found that 4 had been abused. If $p = 1/3$, the probability of observing 4 or more abused women is 0.7908. If $p = 0.1$, the probability of observing 4 or more abused women is only 0.056. If $p = 0.1$, we would have seen a very unusual event because the probability is so small (0.056). If $p = 1/3$, we would have seen an event that was very common because the probability was very large (0.7908). Since we normally do not see rare events, the true probability of abuse is probably close to 1/3.

4.81 a. If the psychic is just guessing, she has 1 chance out of 10 of guessing the correct decision. Thus, $p = 1/10 = 0.1$.

b. Let x = number of correct decisions in 7 trials. Then x is a binomial random variable with $n = 7$ and $p = 0.1$. The expected number of correct decisions in 7 trials is

$$E(x) = np = 7(0.1) = 0.7.$$

Thus, if the psychic is just guessing, we would expect her to make less than 1 correct decision in 7 trials.

c. $P(x=0) = \binom{7}{0}(0.1)^0(0.9)^7 = 0.4783$

d. Now, let $p = 0.5$. $P(x=0) = \binom{7}{0}(0.5)^0(0.5)^7 = 0.0078$

e. Yes. If the psychic is just guessing, the probability that she makes no correct decisions is .4783. Thus, if the psychic, in fact, made no correct decisions, this would not be an unusual event if she is guessing. However, on the other hand, if she does have ESP $(p = 0.5)$ and made no correct decisions, this would be a very unusual event $(P(x = 0) = 0.0078)$. The evidence indicates the psychic probably is just guessing and does not have ESP.

4.83 a. We must assume that the probability that a specific type of ball meets the requirements is always the same from trial to trial and the trials are independent. To use the binomial probability distribution, we need to know the probability that a specific type of golf ball meets the requirements.

b. For a binomial distribution,

$$\mu = np; \quad \sigma = \sqrt{npq}$$

In this example, $n = \text{two dozen} = 2(12) = 24$.

$p = 0.1$ (Success here means the golf ball *does not* meet standards.)
$q = 0.9$
$\mu = np = 24(0.10) = 2.4$; $\sigma = \sqrt{npq} = \sqrt{24(.10)(.90)} = 1.47$

c. In this situation,

p = Probability of success
= Probability golf ball *does* meet standards
$= 0.9$
$q = 1 - 0.9 = 0.10$
$n = 24$
$E(y) = \mu = np = 24(0.90) = 21.6$
$\sigma = \sqrt{npq} = \sqrt{24(0.10)(0.90)} = 1.47$ (Note that this is the same as in part **b.**)

4.85 The four characteristics of a Poisson random variable are:

a. The experiment consists of counting the number of times a certain event occurs during a given unit of time or in a given area or volume (or weight, distance, or any other unit of measurement).
b. The probability that an event occurs in a given unit of time, area, or volume is the same for all the units.
c. The number of events that occur in one unit of time, area, or volume is independent of the number that occur in other units.
d. The mean (or expected) number of events in each unit is denoted by the Greek letter lambda, λ.

4.87 For the Poisson probability distribution $p(x) = \dfrac{3^x e^{-3}}{x!}$ $(x = 0, 1, 2, \ldots)$

the value of λ is 3.

4.89 a. In order to graph the probability distribution, we need to know the probabilities for the possible values of x. Using MINITAB with $\lambda = 3$:

Probability Density Function

```
Poisson with mean = 3

 x   P( X = x )
 0     0.049787
 1     0.149361
 2     0.224042
 3     0.224042
 4     0.168031
 5     0.100819
 6     0.050409
 7     0.021604
 8     0.008102
 9     0.002701
10     0.000810
```

The probability distribution of x in graphical form is:

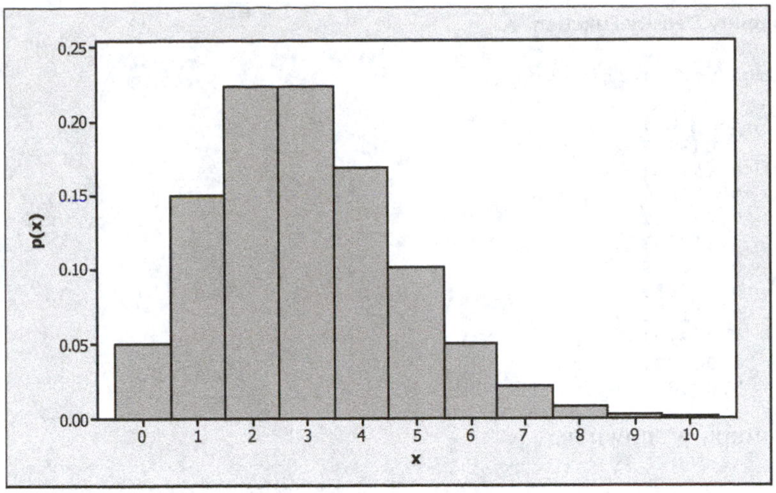

b. $\mu = \lambda = 3$

$\sigma^2 = \lambda = 3; \ \sigma = \sqrt{3} = 1.7321$

4.91 $\mu = \lambda = 1.5$

Using MINITAB:

Probability Density Function

Poisson with mean = 1.5

x	P(X = x)
0	0.223130
1	0.334695
2	0.251021
3	0.125511
4	0.047067
5	0.014120
6	0.003530
7	0.000756
8	0.000142
9	0.000024
10	0.000004

Cumulative Distribution Function

Poisson with mean = 1.5

x	P(X ≤ x)
0	0.22313
1	0.55783
2	0.80885
3	0.93436
4	0.98142
5	0.99554
6	0.99907
7	0.99983
8	0.99997
9	1.00000
10	1.00000

a. $P(x \le 3) = 0.9344$

b. $P(x \ge 3) = 1 - P(x \le 2) = 1 - 0.8089 = 0.1911$

c. $P(x = 3) = P(x \le 3) - P(x \le 2) = 0.9344 - 0.8089 = 0.1255$

d. $P(x = 0) = 0.2231$

e. $P(x > 0) = 1 - P(x = 0) = 1 - 0.2231 = 0.7769$

f. $P(x > 6) = 1 - P(x \le 6) = 1 - 0.9991 = 0.0009$

4.93 a. To graph the Poisson probability distribution with $\lambda = 3$, we need to calculate $p(x)$ for $x = 0$ to 10. Using MINITAB,

Probability Density Function

```
Poisson with mean = 3

 x   P( X = x )
 0     0.049787
 1     0.149361
 2     0.224042
 3     0.224042
 4     0.168031
 5     0.100819
 6     0.050409
 7     0.021604
 8     0.008102
 9     0.002701
10     0.000810
```

The graph is shown here.

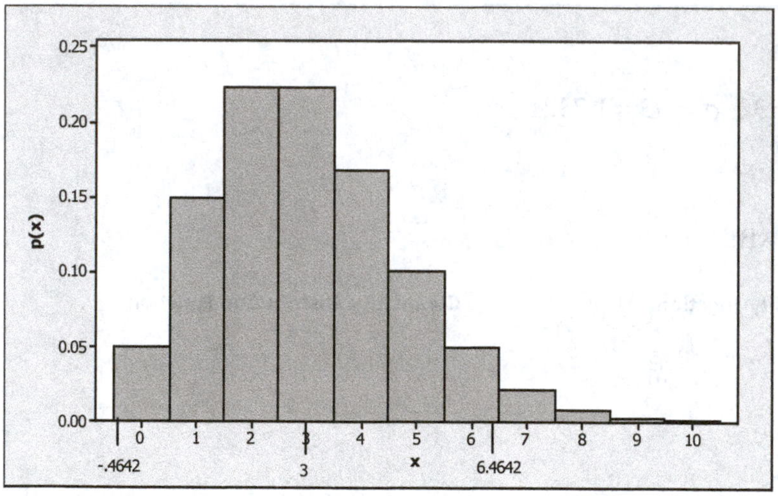

b. $\mu = \lambda = 3$

$\sigma = \sqrt{\lambda} = \sqrt{3} = 1.7321$

$\mu \pm 2\sigma \Rightarrow 3 \pm 2(1.7321) \Rightarrow 3 \pm 3.4642 \Rightarrow (-0.4642,\ 6.4642)$

c. $P(\mu - 2\sigma < x < \mu + 2\sigma) = P(-0.4642 < x < 6.4642) = P(0 \le x \le 6) = P(x \le 6)$

$= 0.0498 + 0.1494 + 0.2240 + 0.2240 + 0.1680 + 0.1008 + 0.0504 = 0.9664$

4.95 Using MINITAB with $\lambda = 1$:

Probability Density Function

```
Poisson with mean = 1

x   P( X = x )
0     0.367879
1     0.367879
2     0.183940
3     0.061313
```

a. $P(x = 0) = 0.3679$.

b. $P(x > 1) = 1 - P(x \le 1) = 1 - 0.3679 - 0.3679 = 0.2642$.

c. $P(x \le 2) = 0.3679 + 0.3679 + 0.1839 = 0.9197$.

4.97 Using MINITAB with $\lambda = 1.2$:

Probability Density Function

```
Poisson with mean = 1.2

x   P( X = x )
0     0.301194
1     0.361433
2     0.216860
3     0.086744
```

a. $P(x = 0) = 0.3012$

b. $P(x \ge 2) = 1 - P(x \le 1) = 1 - 0.3012 - 0.3614 = 0.3374$

4.99 a. $P(x = 0) = \dfrac{\lambda^0 e^{-\lambda}}{0!} = \dfrac{4.5^0 e^{-4.5}}{0!} = 0.0111$

b. $P(x = 1) = \dfrac{\lambda^1 e^{-\lambda}}{1!} = \dfrac{4.5^1 e^{-4.5}}{1!} = 0.0500$

c. $E(x) = \mu = \lambda = 4.5$; $\sigma = \sqrt{\lambda} = \sqrt{4.5} = 2.1213$

4.101 Using MINITAB with $\lambda = 5$:

Probability Density Function

```
Poisson with mean = 5

x   P( X = x )
0     0.006738
1     0.033690
2     0.084224
3     0.140374
```

a. $P(x < 3) = P(x \le 2) = 0.0067 + 0.0337 + 0.0842 = 0.1246$.

b. $E(x) = \mu = \lambda = 5$. The average number of blocked calls during the peak hour is 5.

c. $P(x = 0) = 0.0067$. The probability of no blocked calls in the peak hour is very small. Thus, it is not likely that there will be no blocked calls during the peak hour.

d. Since the probability that $x = 0$ is so small, we would infer that we have either seen a very rare event or that the value of λ is not 5 but a value smaller than 5.

4.103 a. $\sigma^2 = \lambda = 0.25$

b. $\mu = \lambda = 0.25$, $\sigma = \sqrt{0.25} = 0.5$

$$\mu \pm 2\sigma \Rightarrow 0.25 \pm 2(0.5) \Rightarrow 0.25 \pm 1 \Rightarrow (-0.75,\ 1.25)$$

$$P(-0.75 < x < 1.25) = P(x=0) + P(x=1) = \frac{\lambda^0 e^{-\lambda}}{0!} + \frac{\lambda^1 e^{-\lambda}}{1!}$$

$$= \frac{0.25^0 e^{-0.25}}{0!} + \frac{0.25^1 e^{-0.25}}{1!} = 0.7788 + 0.1947 = 0.9735$$

c. From part b, we found that the probability that the number of times the word "though" appears once or fewer in 1,000 words is 0.9735. From this, we can find the probability that the word "though" appears more than once in 1,000 words as:

$$P(x > 1) = 1 - P(x \le 1) = 1 - 0.9735 = 0.0265$$

If Davey Crockett actually wrote the article, the probability of observing the word "though" 2 times in the first 1,000 words is extremely unusual. We would conclude that Davey Crockett did not write the *Texas* narrative.

4.105 Let x = number of cars that will arrive within the next 30 minutes. If the average number of cars that arrive in 60 minutes is 10, the average number of cars that arrive in 30 minutes is $\lambda = 5$.

In order for no cars to be in line at closing time, no more than 1 car can arrive in the next 30 minutes. Using MINITAB with $\lambda = 5$:

Cumulative Distribution Function

```
Poisson with mean = 5

x  P( X ≤ x )
1   0.0404277
```

$P(x \le 1) = 0.0404$

Since this probability is so small, it is very likely that at least one car will be in line at closing time. The probability that at least one car will be in line at closing is $1 - 0.0404 = 0.9596$.

4.107 Sampling with replacement means that once an experimental unit has been observed, it is returned to the population before the next experimental unit is sampled. Thus, for each trial, each member of the population has a chance of being selected. The same experimental unit may be observed more than once.

Sampling without replacement means that once an experimental unit has been observed, it is not returned to the population before the next experimental unit is observed. Thus, an experimental unit can only be observed once in each sample.

4.109 A binomial random variable is characterized by n trials where the trials are identical. In order for the trials to be identical, we have to sample from an infinite population or sample from a finite population with replacement. The probability of success on any trial remains constant from trial to trial and the trials are independent. A hypergeometric random variable is characterized by n trials where the elements are selected from a finite population without replacement. The probability of success on any trial depends on what happened on previous trials, and thus the trials are not independent.

Both a binomial random variable and a hypergeometric random variable are characterized by n trials, 2 possible outcomes on each trial (S or F), and the random variable represents the number of successes in n trials.

4.111 a. $$P(x=1)=\frac{\binom{r}{x}\binom{N-r}{n-x}}{\binom{N}{n}}=\frac{\binom{3}{1}\binom{5-3}{3-1}}{\binom{5}{3}}=\frac{\frac{3!}{1!2!}\frac{2!}{2!0!}}{\frac{5!}{3!2!}}=\frac{3(1)}{10}=0.3$$

b. $$P(x=3)=\frac{\binom{r}{x}\binom{N-r}{n-x}}{\binom{N}{n}}=\frac{\binom{3}{3}\binom{9-3}{5-3}}{\binom{9}{5}}=\frac{\frac{3!}{3!0!}\frac{6!}{2!4!}}{\frac{9!}{5!4!}}=\frac{1(15)}{126}=0.119$$

c. $$P(x=2)=\frac{\binom{r}{x}\binom{N-r}{n-x}}{\binom{N}{n}}=\frac{\binom{2}{2}\binom{4-2}{2-2}}{\binom{4}{2}}=\frac{\frac{2!}{2!0!}\frac{2!}{0!2!}}{\frac{4!}{2!2!}}=\frac{1(1)}{6}=0.167$$

d. $$P(x=0)=\frac{\binom{r}{x}\binom{N-r}{n-x}}{\binom{N}{n}}=\frac{\binom{2}{0}\binom{4-2}{2-0}}{\binom{4}{2}}=\frac{\frac{2!}{0!2!}\frac{2!}{2!0!}}{\frac{4!}{2!2!}}=\frac{1(1)}{6}=0.167$$

4.113 From Exercise 4.112, the probability distribution of x in tabular form is:

x	$p(x)$
2	0.030
3	0.242
4	0.455
5	0.242
6	0.030

a. $P(x=1)=0$

b. $P(x=4)=0.455$

c. $P(x\le 4)=P(x=2)+P(x=3)+P(x=4)=0.030+0.242+0.455=0.727$

d. $P(x\ge 5)=P(x=5)+P(x=6)=0.242+0.030=0.272$

e. $P(x < 3) = P(x = 2) = 0.030$

f. $P(x \ge 8) = 0$

4.115 For this problem, $N = 100$, $n = 10$, and $x = 4$.

a. If the sample is drawn without replacement, the hypergeometric distribution should be used. The hypergeometric distribution requires that sampling be done without replacement.

b. If the sample is drawn with replacement, the binomial distribution should be used. The binomial distribution requires that sampling be done with replacement.

4.117 a. This is a hypergeometric distribution.

b. For this experiment, $r = 20$, $N = 100$, and $n = 3$.

c. $$p(0) = \frac{\binom{20}{0}\binom{80}{3-0}}{\binom{100}{3}} = \frac{\frac{20!}{0!20!}\frac{80!}{3!77!}}{\frac{100!}{3!97!}} = \frac{1\cdot\frac{80\cdot 79\cdot 78\cdots 1}{3\cdot 2\cdot 1\cdot 77\cdot 76\cdot 75\cdots 1}}{\frac{100\cdot 99\cdot 98\cdots 1}{3\cdot 2\cdot 1\cdot 97\cdot 96\cdot 95\cdots 1}} = \frac{82{,}160}{161{,}700} = 0.5081$$

d. $$p(1) = \frac{\binom{20}{1}\binom{80}{3-1}}{\binom{100}{3}} = \frac{\frac{20!}{1!19!}\frac{80!}{2!78!}}{\frac{100!}{3!97!}}$$

$$= \frac{\frac{20\cdot 19\cdot 18\cdots 1}{1\cdot 19\cdot 18\cdot 17\cdots 1}\cdot\frac{80\cdot 79\cdot 78\cdots 1}{2\cdot 1\cdot 78\cdot 77\cdot 76\cdots 1}}{\frac{100\cdot 99\cdot 98\cdots 1}{3\cdot 2\cdot 1\cdot 97\cdot 96\cdot 95\cdots 1}} = \frac{63{,}200}{161{,}700} = 0.3908$$

e. $$p(2) = \frac{\binom{20}{2}\binom{80}{3-2}}{\binom{100}{3}} = \frac{\frac{20!}{2!18!}\frac{80!}{1!79!}}{\frac{100!}{3!97!}}$$

$$= \frac{\frac{20\cdot 19\cdot 18\cdots 1}{2\cdot 1\cdot 18\cdot 17\cdot 16\cdots 1}\cdot\frac{80\cdot 79\cdot 78\cdots 1}{1\cdot 79\cdot 78\cdot 77\cdots 1}}{\frac{100\cdot 99\cdot 98\cdots 1}{3\cdot 2\cdot 1\cdot 97\cdot 96\cdot 95\cdots 1}} = \frac{15{,}200}{161{,}700} = 0.0940$$

f. $$p(3)=\frac{\binom{20}{3}\binom{80}{3-3}}{\binom{100}{3}}=\frac{\frac{20!}{3!17!}\frac{80!}{0!80!}}{\frac{100!}{3!97!}}$$

$$=\frac{\frac{20\cdot19\cdot18\cdots1}{3\cdot2\cdot1\cdot17\cdot16\cdot15\cdots1}\cdot\frac{80\cdot79\cdot78\cdots1}{1\cdot80\cdot79\cdot78\cdots1}}{\frac{100\cdot99\cdot98\cdots1}{3\cdot2\cdot1\cdot97\cdot96\cdot95\cdots1}}=\frac{1{,}140}{161{,}700}=0.0071$$

4.119 a. Let x = number of facilities chosen in a sample of 10 that treat hazardous waste on-site. For this problem, $N=209$, $r=8$, and $n=10$.

$$\mu=E(x)=\frac{rn}{N}=\frac{8(10)}{209}=0.383$$

If 10 facilities are repeatedly selected, the average number of facilities in the sample of 10 that treat hazardous waste on-site is 0.383.

b. $$P(x=4)=\frac{\binom{r}{x}\binom{N-r}{n-x}}{\binom{N}{n}}=\frac{\binom{8}{4}\binom{209-8}{10-4}}{\binom{209}{10}}=\frac{\frac{8!}{4!4!}\cdot\frac{201!}{6!195!}}{\frac{209!}{10!199!}}=0.0002$$

4.121 Let x = number of "clean" cartridges selected in 5 trials. For this problem, $N=158$, $r=122$, and $n=5$.

$$P(x=5)=\frac{\binom{r}{x}\binom{N-r}{n-x}}{\binom{N}{n}}=\frac{\binom{122}{5}\binom{36}{0}}{\binom{158}{5}}=\frac{\frac{122!}{5!117!}\frac{36!}{0!36!}}{\frac{158!}{5!153!}}=0.2693$$

4.123 Because the museums are selected without replacement, this is a hypergeometric distribution. For this exercise, $N=30$, $r=8$, and $n=5$.

$$P(x=0)=\frac{\binom{r}{x}\binom{N-r}{n-x}}{\binom{N}{n}}=\frac{\binom{8}{0}\binom{30-8}{5-0}}{\binom{30}{5}}=\frac{\frac{8!}{0!(8)!}\frac{22!}{5!(17)!}}{\frac{30!}{5!(25)!}}=0.1848$$

4.125 There are a total of $N=10$ professors, of which $r=4$ are from the Department of Engineering Physics. A total of $n=3$ professors are to be selected and x = the number of committee members from the Department of Engineering Physics selected from the 3. Because the selection is done without replacement, the distribution of x is the hypergeometric distribution. The appropriate formula is:

$$p(x)=\frac{\binom{r}{x}\binom{N-r}{n-x}}{\binom{N}{n}}=\frac{\binom{4}{x}\binom{6}{3-x}}{\binom{10}{3}}$$

4.127 Let x = number of grants awarded to the north side in 140 trials. The random variable x has a hypergeometric distribution with $N = 743$, $r = 601$, and $n = 140$.

a. $\mu = E(x) = \frac{nr}{N} = \frac{140(601)}{743} = 113.24$

$$\sigma^2 = \frac{r(N-r)n(N-n)}{N^2(N-1)} = \frac{601(743-601)140(743-140)}{743^2(743-1)} = 17.5884$$

$\sigma = \sqrt{17.5884} = 4.194$

b. If the grants were awarded at random, we would expect approximately 113 to be awarded to the north side. We observed 140. The z-score associated with 140 is:

$$z = \frac{x-\mu}{\sigma} = \frac{140-113.24}{4.194} = 6.38$$

Because this z-score is so large, it would be extremely unlikely to observe all 140 grants to the north side if they are randomly selected. Thus, we would conclude that the grants were not randomly selected.

4.129 a. Poisson

b. Binomial

c. Binomial

4.131 From Table I, Appendix B:

a. $P(x=14) = P(x \le 14) - P(x \le 13) = 0.584 - 0.392 = 0.192$

b. $P(x \le 12) = 0.228$

c. $P(x > 12) = 1 - P(x \le 12) = 1 - 0.228 = 0.772$

d. $P(9 \le x \le 18) = P(x \le 18) - P(x \le 8) = 0.992 - 0.005 = 0.987$

e. $P(8 < x < 18) = P(x \le 17) - P(x \le 8) = 0.965 - 0.005 = 0.960$

f. $\mu = np = 20(0.7) = 1.4$

$\sigma^2 = npq = 20(0.7)(0.3) = 4.2$, $\sigma = \sqrt{4.2} = 2.049$

g. $\mu \pm 2\sigma \Rightarrow 14 \pm 2(2.049) \Rightarrow 14 \pm 4.098 \Rightarrow (9.902,\ 18.098)$

$$P(9.902 < x < 18.098) = P(10 \le x \le 18) = P(x \le 18) - P(x \le 9) = 0.992 - 0.017 = 0.975$$

4.133 a. $P(x=1) = \binom{3}{1}(0.1)^1(0.9)^{3-1} = \frac{3!}{1!2!}(0.1)^1(0.9)^2 = 3(0.1)(0.81) = 0.243$

b. $P(x=4) = P(x \le 4) - P(x \le 3) = 0.238 - 0.107 = 0.131$ from Table I, Appendix B.

c. $P(x=0) = \binom{2}{0}(0.4)^0(0.6)^{2-0} = \frac{2!}{0!2!}(1)(0.6)^2 = 0.36$

d. $P(x=4) = P(x \le 4) - P(x \le 3) = 0.969 - 0.812 = 0.157$ from Table I, Appendix B.

e. $P(x=12) = P(x \le 12) - P(x \le 11) = 0.184 - 0.056 = 0.128$ from Table I, Appendix A.

f. $P(x=8) = P(x \le 8) - P(x \le 7) = 0.954 - 0.833 = 0.121$ from Table I, Appendix B.

4.135 a. Using MINITAB with $\lambda = 2$,

Probability Density Function

```
Poisson with mean = 2

x  P( X = x )
3    0.180447
```

$P(x=3) = 0.1804$

b. Using MINITAB with $\lambda = 1$,

Probability Density Function

```
Poisson with mean = 1

x  P( X = x )
4   0.0153283
```

$P(x=4) = 0.0153$

c. Using MINITAB with $\lambda = 0.5$,

Probability Density Function

```
Poisson with mean = 0.5

x  P( X = x )
2   0.0758163
```

$P(x=2) = 0.0758$

4.137 a. To find the probabilities associated with each value of x, we divide the frequency associated with each value by the total sample size, 743. The probabilities appear in the table:

First Digit	Frequency of Occurrence	Probability
1	109	109 / 743 = 0.1467
2	75	75 / 743 = 0.1009
3	77	77 / 743 = 0.1036
4	99	99 / 743 = 0.1332
5	72	72 / 743 = 0.0969
6	117	117 / 743 = 0.1575
7	89	89 / 743 = 0.1198
8	62	62 / 743 = 0.0834
9	43	43 / 743 = 0.0579
Total	**743**	1.0000

b. $\mu = E(x) = \sum xp(x) = 1(0.1467) + 2(0.1009) + 3(0.1036) + 4(0.1332) + 5(0.0969)$
$+6(0.1575) + 7(0.1198) + 8(0.0834) + 9(0.0579) = 4.6485$

c. Over a very large number of trials, the average first significant digit is 4.6485.

4.139 a. We will check the characteristics of a binomial random variable:

1. This experiment consists of $n = 20$ identical trials.
2. There are only 2 possible outcomes for each trial. SBIRS detects intruding object (*S*) or not (*F*).
3. The probability of S remains the same from trial to trial. In this case, $p = P(S) = 0.8$ for each trial.
4. The trials are independent. Whether SBIRS detects one intruding object should not affect whether it detects another.
5. x = number of intruding objects detected by SBIRS in 20 trials.

b. $n = 20$ and $p = 0.8$

c. Using the Table I, Appendix B with $n = 20$ and $p = 0.8$,

$P(x=15) = 1 - P(x \le 15) - P(x \le 14) = 0.370 - 0.196 = 0.174$.

d. $P(x \ge 15) = 1 - P(x \le 14) = 1 - 0.196 = 0.804$

e. $E(x) = \mu = E(x) = np = 20(0.8) = 16$. The average number of intruding objects detected in 20 trials is 16.

4.141 The random variable x would have a binomial distribution:

1. n identical trials. A sample of married women was taken and would be considered small compared to the entire population of married women, so the trials would be very close to identical.

2. There are only two possible outcomes for each trial. In this experiment, there are only two possible outcomes: the woman would marry the same man (*S*) or she would not (*F*).

3. The probability of Success stays the same from trial to trial. Here, $P(S) = 0.50$ and $P(F) = 0.50$. In reality, these probabilities would not be exactly the same from trial to trial, but rounded off to 4 decimal places, they would be the same.

4. The trials are independent. In this experiment, the trials would not be exactly independent because we would be sampling without replacement from a finite population. However, if the sample size is fairly small compared to the population size, the trials will be essentially independent.

5. The binomial random variable *x* would be the number of successes in *n* trials. For this experiment, *x* = the number of women out of those sampled who would marry the same man again.

Thus, *x* would possess (approximately) a binomial distribution.

4.143 a. $p(0)+p(1)+p(2)+p(3)+p(4)=0.09+0.30+0.37+0.20+0.04=1.00$

b. $P(x=3 \text{ or } 4)=p(3)+p(4)=0.20+0.04=0.24$

c. $P(x<2)=p(0)+p(1)=0.09+0.30=0.39$

d. $E(x)=\sum xp(x)=0(0.09)+1(0.30)+2(0.37)+3(0.20)+4(0.04)$
$=0+0.30+0.74+0.60+0.16=1.8$

In a random sample of 4 homes, the average number of homes with high dust mite levels is 1.8.

e. $\sigma^2 = E[(x-\mu)^2] = \sum (x-\mu)^2 p(x)$
$=(0-1.8)^2(0.09)+(1-1.8)^2(0.30)+(2-1.8)^2(0.37)$
$+(3-1.8)^2(0.20)+(4-1.8)^2(0.04)$
$=0.2916+0.1920+0.0148+0.2880+0.1936=0.98$
$\sigma=\sqrt{0.98}=0.9899$

f. $\mu \pm 2\sigma \Rightarrow 1.8 \pm 2(0.9899) \Rightarrow 1.8 \pm 1.9798 \Rightarrow (-0.1798,\ 3.7798)$

$P(-0.1798<x<3.7798)=p(0)+p(1)+p(2)+p(3)=0.09+0.03+0.37+0.20=0.96$

Chebyshev's rule says that the interval $\mu \pm 2\sigma$ will contain at least 0.75 of the data. The Empirical Rule says that approximately 0.95 of the data will be contained in the interval. Both Chebyshev's rule and the Empirical Rule fit the distribution.

4.145 a. From the problem, *x* is a binomial random variable with $n = 10$ and $p = 0.8$.

$P(x=3)=P(x\le 3)-P(x\le 2)=0.001-0.000=0.001$ (from Table I, Appendix B)

b. $P(x\le 7)=0.322$

c. $P(x>4)=1-P(x\le 4)=1-0.006=0.994$

4.147 Let x = number of extinct bird species in $n=10$ trials. Then x is a hypergeometric random variable with $N=132$, $r=38$, and $n=10$.

a. $$P(x=5)=\frac{\binom{38}{5}\binom{132-38}{10-5}}{\binom{132}{10}}=\frac{\frac{38!}{5!(33)!}\frac{94!}{5!(89)!}}{\frac{132!}{10!(122)!}}=\frac{501{,}942(54{,}891{,}018)}{3.12058705x10^{14}}=0.0883$$

b. $$P(x\le 1)=P(x=0)+P(x=1)=\frac{\binom{38}{0}\binom{132-38}{10-0}}{\binom{132}{10}}+\frac{\binom{38}{1}\binom{132-38}{10-0}}{\binom{132}{10}}$$

$$=\frac{\frac{38!}{0!(38)!}\cdot\frac{94!}{10!(84)!}}{\frac{132!}{10!(122)!}}+\frac{\frac{38!}{1!(37)!}\cdot\frac{94!}{9!(85)!}}{\frac{132!}{10!(122)!}}=\frac{9.041256842x10^{12}}{3.12058705x10^{14}}+\frac{4.041973645x10^{13}}{3.12058705x10^{14}}$$

$$=0.0290+0.1295=0.1585$$

4.149 Let x = number of parents who yell at their child before, during, or after a meet in 20 trials. Then x has a binomial distribution with $n=20$ and $p=0.05$.

$P(x\ge 1)=1-P(x=0)=1-0.358=0.642$ (Using Table I, Appendix B)

4.151 a. $\sigma=\sqrt{\lambda}=\sqrt{4}=2$

b. Using MINITAB with $\lambda=4$:

Cumulative Distribution Function

```
Poisson with mean = 4

 x   P( X ≤ x )
10     0.997160
```

$P(x>10)=1-P(x\le 10)=1-0.9972=0.0028$. Since the probability is so small (0.0028), it would be very unlikely that the plant would yield a value that would exceed the EPA limit.

4.153 Let x = number of slaughtered chickens in 5 that passes inspection with fecal contamination.

Then x is a binomial random variable with $n = 5$ and $p = 0.01$ (from Exercise 3.21.)

$P(x \geq 1) = 1 - P(x = 0) = 1 - 0.951 = 0.049$ (From Table I, Appendix B).

4.155 a. Because there are 20 possible outcomes that are all equally likely, the probability of any of the 20 numbers is 1/20. The probability distribution of x is:

$P(x = 5) = 1/20 = 0.05$; $P(x = 10) = 1/20 = 0.05$; etc.

x	5	10	15	20	25	30	35	40	45	50	55	60	65	70	75	80	85	90	95	100
$p(x)$	0.05	0.05	0.05	0.05	0.05	0.05	0.05	0.05	0.05	0.05	0.05	0.05	0.05	0.05	0.05	0.05	0.05	0.05	0.05	0.05

b. $$E(x) = \sum xp(x) = (5 - 52.5)^2(0.05) + (10 - 52.5)^2(0.05) + 15(0.05) + 20(0.05)$$
$$+ 25(0.05) + 30(0.05) + 35(0.05) + 40(0.05) + 45(0.05) + 50(0.05) + 55(0.05)$$
$$+ 60(0.05) + 65(0.05) + 70(0.05) + 75(0.05) + 80(0.05) + 85(0.05) + 90(0.05)$$
$$+ 95(0.05) + 100(0.05) = 52.5$$

c. $$\sigma^2 = E(x - \mu)^2 = \sum (x - \mu)^2 p(x) = (5 - 52.5)^2(0.05) + (10 - 52.5)^2(0.05)$$
$$+ (15 - 52.5)^2(0.05) + (20 - 52.5)^2(0.05) + (25 - 52.5)^2(0.05)$$
$$+ (30 - 52.5)^2(0.05) + (35 - 52.5)^2(0.05) + (40 - 52.5)^2(0.05)$$
$$+ (45 - 52.5)^2(0.05) + (50 - 52.5)^2(0.05) + (55 - 52.5)^2(0.05)$$
$$+ (60 - 52.5)^2(0.05) + (65 - 52.5)^2(0.05) + (70 - 52.5)^2(0.05)$$
$$+ (75 - 52.5)^2(0.05) + (80 - 52.5)^2(0.05) + (85 - 52.5)^2(0.05)$$
$$+ (90 - 52.5)^2(0.05) + (95 - 52.5)^2(0.05) + (100 - 52.5)^2(0.05)$$
$$= 831.25$$

$$\sigma = \sqrt{\sigma^2} = \sqrt{831.25} = 28.83$$

Since the uniform distribution is not mound-shaped, we will use Chebyshev's theorem to describe the data. We know that at least 8/9 of the observations will fall with 3 standard deviations of the mean and at least 3/4 of the observations will fall within 2 standard deviations of the mean. For this problem,

$\mu \pm 2\sigma \Rightarrow 52.5 \pm 2(28.83) \Rightarrow 52.5 \pm 57.66 \Rightarrow (-5.16,\ 110.16)$. Thus, at least 3/4 of the data will fall between -5.16 and 110.16. For our problem, all of the observations will fall within 2 standard deviations of the mean. Thus, x is just as likely to fall within any interval of equal length.

d. If a player spins the wheel twice, the total number of outcomes will be $20(20) = 400$. The sample space is:

5, 5	10, 5	15, 5	20, 5	25, 5...	100, 5
5,10	10,10	15,10	20,10	25,10...	100,10
5,15	10,15	15,15	20,15	25,15...	100,15
⋮	⋮	⋮	⋮	⋮	⋮
5,100	10,100	15,100	20,100	25,100...	100,100

Each of these outcomes are equally likely, so each has a probability of $1/400 = 0.0025$.

Now, let x equal the sum of the two numbers in each sample. There is one sample with a sum of 10, two samples with a sum of 15, three samples with a sum of 20, etc. If the sum of the two numbers exceeds 100, then x is zero. The probability distribution of x is:

x	$p(x)$	x	$p(x)$
0	0.5250	55	0.0250
10	0.0025	60	0.0275
15	0.0050	65	0.0300
20	0.0075	70	0.0325
25	0.0100	75	0.0350
30	0.0125	80	0.0375
35	0.0150	85	0.0400
40	0.0175	90	0.0425
45	0.0200	95	0.0450
50	0.0225	100	0.0475

e. We assumed that the wheel is fair, or that all outcomes are equally likely.

f. $$\mu = E(x) = \sum xp(x) = 0(0.5250) + 10(0.0025) + 15(0.0050) + \ldots + 100(0.0475) = 33.25$$

$$\sigma^2 = E(x-\mu)^2 = \sum (x-\mu)^2 p(x) = (0-33.25)^2(0.525) + (10-33.25)^2(0.0025) + (15-33.25)^2(0.0050) + \ldots + (100-33.25)^2(0.0475) = 1471.3125$$

$$\sigma = \sqrt{\sigma^2} = \sqrt{1471.3125} = 38.3577$$

g. $P(x=0) = 0.525$

h. Given that the player obtains a 20 on the first spin, the possible values for x (sum of the two spins) are 0 (player spins 85, 90, 95, or 100 on the second spin), 25, 30, ..., 100. In order to get an x of 25, the player would spin a 5 on the second spin. Similarly, the player would have to spin a 10 on the second spin in order to get an x of 30, etc. Since all of the outcomes are equally likely on the second spin, the distribution of x is:

x	$p(x)$	x	$p(x)$

0	0.20	65	0.05
25	0.05	70	0.05
30	0.05	75	0.05
35	0.05	80	0.05
40	0.05	85	0.05
45	0.05	90	0.05
50	0.05	95	0.05
55	0.05	100	0.05
60	0.05		

i. The probability that the players total score will exceed one dollar is the probability that x is zero. $P(x=0)=0.20$

j. Given that the player obtains a 65 on the first spin, the possible values for x (sum of the two spins) are 0 (player spins 40, 45, 50, up to 100 on second spin), 70, 75, 80,..., 100. In order to get an x of 70, the player would spin a 5 on the second spin. Similarly, the player would have to spin a 10 on the second spin in order to get an x of 75, etc. Since all of the outcomes are equally likely on the second spin, the distribution of x is:

x	$p(x)$
0	0.65
70	0.05
75	0.05
80	0.05
85	0.05
90	0.05
95	0.05
100	0.05

The probability that the players total score will exceed one dollar is the probability that x is zero. $P(x=0)=0.65$.

k. There are many possible answers. Notice that $P(x=0)$ on the second spin is equal to the value on the first spin divided by 100. If a player got a 20 on the first spin, then $P(x=0)$ on the second spin is $20/100=0.20$. If a player got a 65 on the first spin, then $P(x=0)$ on the second spin is $65/100=0.20$.

4.157 We know that $\mu=np=n(0.4)=0.4n$ and $\sigma=\sqrt{npq}=\sqrt{n(0.4)(0.6)}=\sqrt{0.24n}$. Since $p=0.4$ is close to 0.5, the distribution of x will be close to symmetrical. Therefore, we can use the Empirical Rule. We know that approximately 95% of all observations will fall within 2 standard deviations of the mean. Thus, approximately 97.5% of the observations will fall above two standard deviations below the mean. We will set 100 equal to the mean minus 2 standard deviations and solve for n:

$$0.4n - 2\sqrt{0.24n} = 100 \Rightarrow n - 5\sqrt{0.24n} = 250 \Rightarrow n - 250 = 5\sqrt{0.24n}$$

$$\Rightarrow n^2 - 500n + 62500 = 6n \Rightarrow n^2 - 506n + 62500 = 0$$

Now, using Pythagorean Theorem to solve for n, we get:

$$n = \frac{506 \pm \sqrt{506^2 - 4(62500)}}{2} = \frac{506 \pm 77.69}{2} \Rightarrow n = 214 \text{ or } n = 292$$

Since the mean is *np,* and for this exercise the mean must be greater than 100, we know that *n* cannot be 214. Thus, *n* has to be 292.

4.159 There will not be enough beds if x = the number of newly admitted patients exceeds 4. For $\lambda = 2.5$,

$$P(x>4) = 1 - P(x \le 4) = 1 - \{P(x=0) + P(x=1) + P(x=2) + P(x=3) + P(x=4)\}$$

$$= 1 - \left(\frac{2.5^0 \cdot e^{-2.5}}{0!} + \frac{2.5^1 \cdot e^{-2.5}}{1!} + \frac{2.5^2 \cdot e^{-2.5}}{2!} + \frac{2.5^3 \cdot e^{-2.5}}{3!} + \frac{2.5^4 \cdot e^{-2.5}}{4!} \right)$$

$$= 1 - (0.0821 + 0.2052 + 0.2565 + 0.2138 + 0.1336) = 1 - 0.8912 = 0.1088$$

Chapter 5

Continuous Random Variables

5.1 A uniform random variable is a continuous random variable that appears to have equally likely outcomes over its range of possible values.

5.3 a. For a uniform random variable,

$$f(x)=\begin{cases}\dfrac{1}{d-c}=\dfrac{1}{30-10}=\dfrac{1}{20} & 10\le x\le 30\\ 0 & \text{otherwise}\end{cases}$$

b. $$\mu=\frac{c+d}{2}=\frac{10+30}{2}=20$$

$$\sigma=\frac{d-c}{\sqrt{12}}=\frac{30-10}{\sqrt{12}}=5.774$$

c.

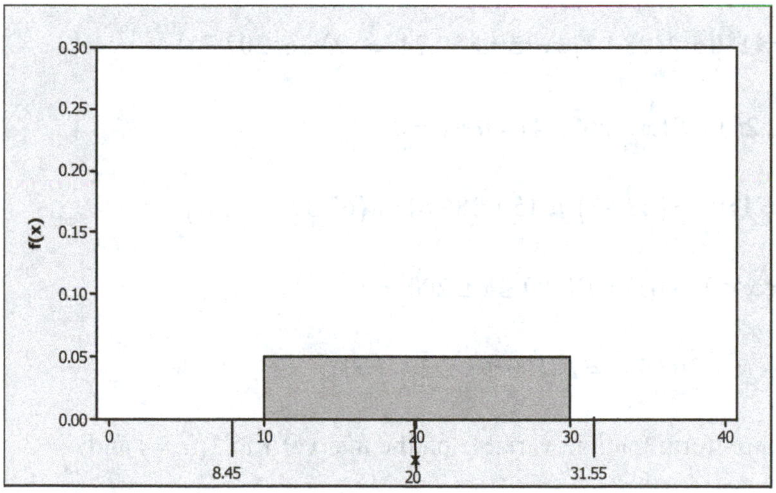

$$\mu\pm 2\sigma\Rightarrow 20\pm 2(5.774)\Rightarrow 20\pm 11.584\Rightarrow (8.452,\ 31.548)$$

5.5 a. For a uniform random variable,

$$f(x)=\begin{cases}\dfrac{1}{d-c}=\dfrac{1}{4-2}=\dfrac{1}{2} & 2\le x\le 4\\ 0 & \text{otherwise}\end{cases}$$

b. $$\mu=\frac{c+d}{2}=\frac{2+4}{2}=3 \qquad \sigma=\frac{d-c}{\sqrt{12}}=\frac{4-2}{\sqrt{12}}=0.577$$

c. $\mu \pm \sigma \Rightarrow 3 \pm 0.577 \Rightarrow (2.423, 3.577)$

$$P(2.423 \le x \le 3.577) = (3.577 - 2.423)\frac{1}{2} = 0.577$$

d. $P(x > 2.78) = (4 - 2.78)\frac{1}{2} = 0.61$

e. $P(2.4 \le x \le 3.7) = (3.7 - 2.4)\frac{1}{2} = 0.65$

f. $P(x < 2) = 0$

5.7 For the uniform random variable,

$$f(x) = \begin{cases} \frac{1}{d-c} = \frac{1}{200-100} = \frac{1}{100} & 100 \le x \le 200 \\ 0 & \text{otherwise} \end{cases}$$

a. $\mu = \frac{c+d}{2} = \frac{100+200}{2} = 150$ $\sigma = \frac{d-c}{\sqrt{12}} = \frac{200-100}{\sqrt{12}} = 28.87$

$$\mu \pm 2\sigma \Rightarrow 150 \pm 2(28.87) \Rightarrow 150 \pm 57.74 \Rightarrow (92.26, 207.74)$$

$$P(x < 92.26) + P(x > 207.74) = 0 + 0 = 0$$

b. $\mu \pm 3\sigma \Rightarrow 150 \pm 3(28.87) \Rightarrow 150 \pm 86.61 \Rightarrow (63.39, 236.61)$

$$P(63.9 < x < 236.61) = P(100 \le x \le 200) = 1$$

c. $P(92.26 < x < 207.74) = P(100 \le x \le 200) = 1$

5.9 a. Since x is a uniform random variable on the interval 1 to 3, $c = 1$ and $d = 3$.

$E(x) = \frac{c+d}{2} = \frac{1+3}{2} = \frac{4}{2} = 2$. The average amount of uranium in all reservoirs is 2 parts per million.

b. $P(2 < x < 2.5) = \frac{2.5-2}{3-1} = \frac{0.5}{2} = 0.25$

c. $P(x \le 1.75) = \frac{1.75-1}{3-1} = \frac{0.75}{2} = 0.375$

5.11 a. Let x be the number of anthrax spores detected. Then x has a uniform distribution on the interval from 0 to 10. Thus, $c = 0$ and $d = 10$.

$$f(x) = \begin{cases} \frac{1}{d-c} = \frac{1}{10-0} = \frac{1}{10} & \text{for } 0 \le x \le 10 \\ 0 & \text{otherwise} \end{cases}$$

$$P(x \le 8) = (8-0)\left(\frac{1}{10}\right) = \frac{8}{10} = 0.8$$

b. $$P(2 \le x \le 5) = (5-2)\left(\frac{1}{10}\right) = \frac{3}{10} = 0.3$$

5.13 a. Suppose no bolt-on trace elements are used. Let x be a uniform random variable on the interval 260 to 290. From this, $c = 260$ and $d = 290$.

$$f(x) = \begin{cases} \frac{1}{d-c} = \frac{1}{290-260} = \frac{1}{30} & 260 \le x \le 290 \\ 0 & \text{otherwise} \end{cases}$$

$$P(280 \le x \le 284) = (284-280)\left(\frac{1}{30}\right) = 4\left(\frac{1}{30}\right) = 0.1333$$

Suppose bolt-on trace elements are used. Let y be a uniform random variable on the interval 278 to 285. From this, $c = 278$ and $d = 285$.

$$f(y) = \begin{cases} \frac{1}{d-c} = \frac{1}{285-278} = \frac{1}{7} & 278 \le y \le 285 \\ 0 & \text{otherwise} \end{cases}$$

$$P(280 \le y \le 284) = (284-280)\left(\frac{1}{7}\right) = 4\left(\frac{1}{7}\right) = 0.5714$$

b. Suppose no bolt-on trace elements are used.

$$P(x \le 268) = (268-260)\left(\frac{1}{30}\right) = 8\left(\frac{1}{30}\right) = 0.2667$$

Suppose bolt-on trace elements are used.

$$P(x \le 268) = (0)\left(\frac{1}{7}\right) = 0$$

5.15 a. We are given that x is a uniform random variable on the interval from 100 to 115 days. Therefore, $c = 100$ and $d = 115$.

$$P(x > 110) = \frac{115 - 110}{115 - 100} = \frac{5}{15} = 0.333$$

b. $$P(x < 102) = \frac{102 - 100}{115 - 100} = \frac{2}{15} = 0.133$$

c. $$P(x > L) = 0.1 \Rightarrow \frac{115 - L}{115 - 100} = 0.1 \Rightarrow \frac{115 - L}{15} = 0.1 \Rightarrow 115 - L = 1.5 \Rightarrow L = 113.5$$

5.17 a. The amount dispensed by the beverage machine is a continuous random variable since it can take on any value between 6.5 and 7.5 ounces.

b. Since the amount dispensed is random between 6.5 and 7.5 ounces, x is a uniform random variable.

$$f(x) = \frac{1}{d - c} \qquad (c \le x \le d)$$

$$\frac{1}{d - c} = \frac{1}{7.5 - 6.5} = \frac{1}{1} = 1$$

Therefore, $f(x) = \begin{cases} 1 & (6.5 \le x \le 7.5) \\ 0 & \text{otherwise} \end{cases}$

The graph is as follows:

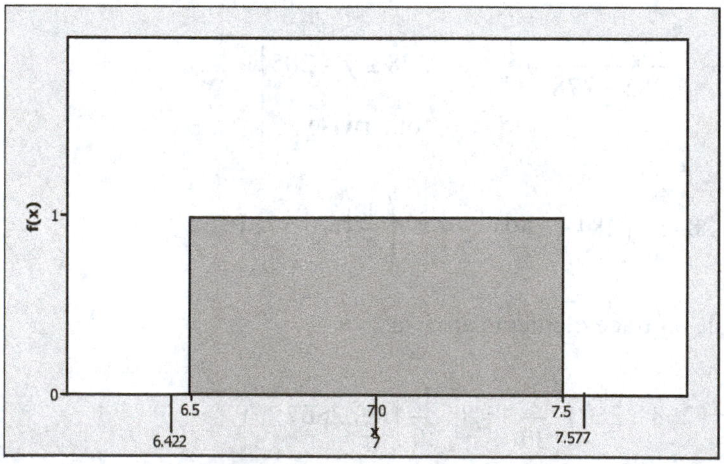

c. $$\mu = \frac{c + d}{2} = \frac{6.5 + 7.5}{2} = \frac{14}{2} = 7$$

$$\sigma = \frac{d - c}{\sqrt{12}} = \frac{7.5 - 6.5}{\sqrt{12}} = 0.2887$$

$$\mu \pm 2\sigma \Rightarrow 7 \pm 2(0.2887) \Rightarrow 7 \pm 0.577 \Rightarrow (6.422,\ 7.577)$$

d. $P(x \geq 7) = (7.5 - 7)(1) = 0.5$

e. $P(x < 6) = 0$

f. $P(6.5 \leq x \leq 7.25) = (7.25 - 6.5)(1) = 0.75$

g. The probability that the next bottle filled will contain more than 7.25 ounces is:

$$P(x > 7.25) = (7.5 - 7.25)(1) = 0.25$$

The probability that the next 6 bottles filled will contain more than 7.25 ounces is:

$$P\left[(x>7.25)\cap(x>7.25)\cap(x>7.25)\cap(x>7.25)\cap(x>7.25)\cap(x>7.25)\right]$$
$$= \left[P(x > 7.25)\right]^6 = 0.25^6 = 0.0002$$

5.19 Let x = number of inches a gouge is from one end of the spindle. Then x has a uniform distribution with $f(x)$ as follows:

$$f(x) = \begin{cases} \dfrac{1}{d-c} = \dfrac{1}{18-0} = \dfrac{1}{18} & 0 \leq x \leq 18 \\ 0 & \text{otherwise} \end{cases}$$

In order to get at least 14 consecutive inches without a gouge, the gouge must be within 4 inches of either end. Thus, we must find:

$$P(x<4) + P(x>14) = (4-0)(1/18) + (18-14)(1/18) = 4/18 + 4/18 = 8/14 = 0.4444$$

5.21 A normal distribution is a bell-shaped curve with the center of the distribution at μ.

5.23 A normal distribution with $\mu = 0$ and $\sigma = 1$ is a standard normal distribution.

5.25 a. $P(z = 1) = 0$, since a single point does not have an area.

b. $P(z \leq 1) = P(z \leq 0) + P(0 < z \leq 1)$
$= A_1 + A_2$
$= 0.5 + 0.3413 = 0.8413$
(Table II, Appendix B)

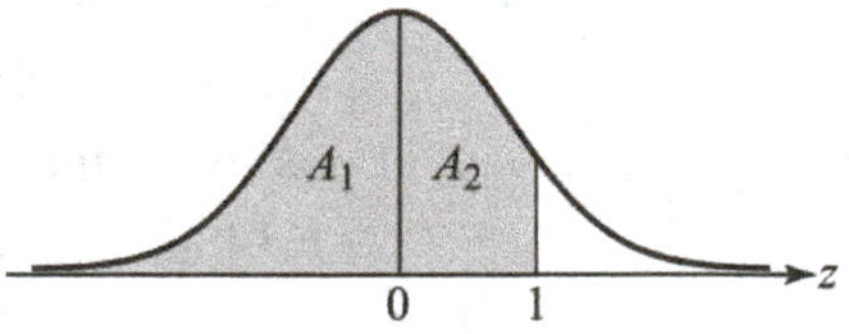

c. $P(z < 1) = P(z \leq 1) = 0.8413$ (Refer to part **b.**)

d. $P(z > 1) = 1 - P(z \leq 1) = 1 - 0.8413 = 0.1587$ (Refer to part **b.**)

e. $P(-1 \le z \le 1) = P(-1 \le z \le 0) + P(0 \le z \le 1)$
$= A_1 + A_2$
$= 0.3413 + 0.3413 = 0.6826$

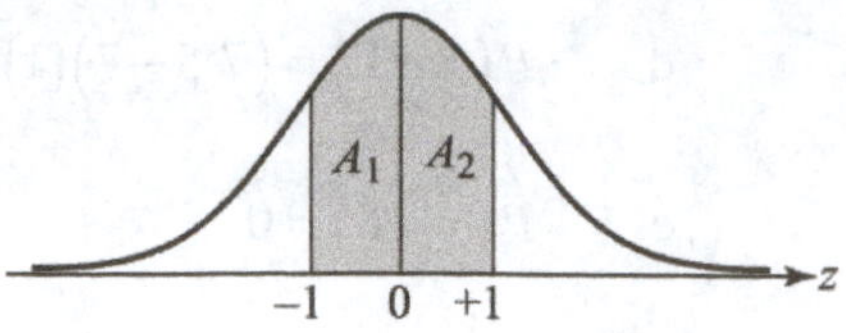

f. $P(-2 \le z \le 2) = P(-2 \le z \le 0) + P(0 \le z \le 2)$
$= A_1 + A_2$
$= 0.4772 + 0.4772 = 0.9544$

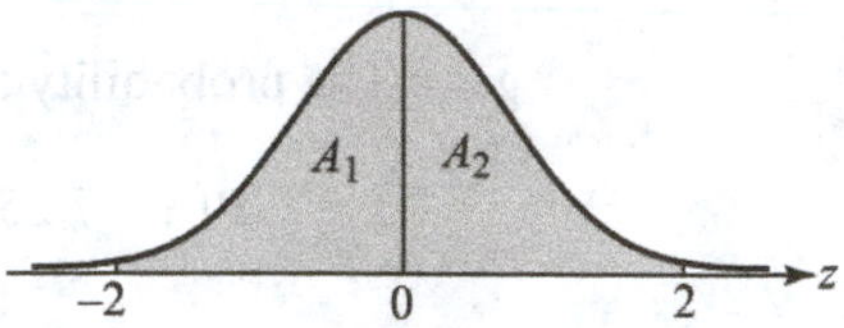

g. $P(-2.16 \le z \le 0.55) = P(-2.16 \le z \le 0) + P(0 \le z \le 0.55)$
$= A_1 + A_2$
$= 0.4846 + 0.2088 = 0.6934$

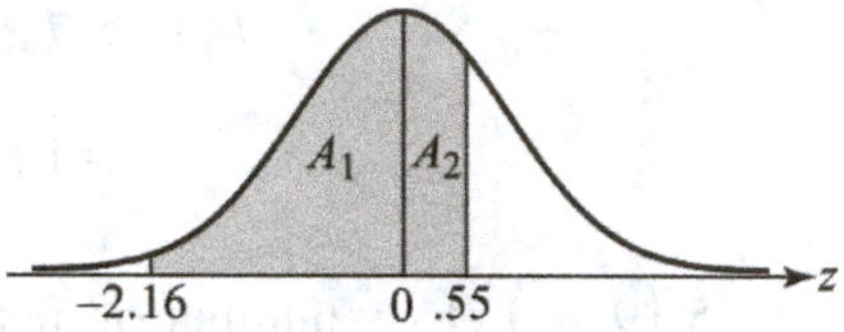

h. $P(-0.42 < z < 1.96) = P(-0.42 \le z \le 0) + P(0 \le z \le 1.96)$
$= A_1 + A_2$
$= 0.1628 + 0.4750 = 0.6378$

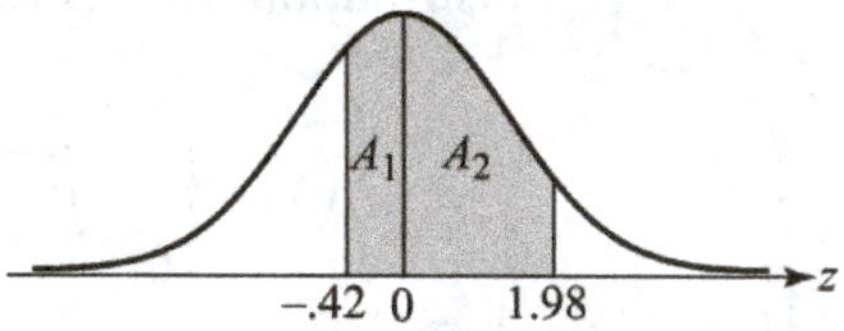

5.27 a. $z = 1$

b. $z = -1$

c. $z = 0$

d. $z = -2.5$

e. $z = 3$

5.29 Using Table II, Appendix B:

a. $P(z \ge z_0) = 0.05$
$A_1 = .5 - .05 = .4500$
Looking up the area 0.4500 in Table II gives $z_0 = 1.645$.

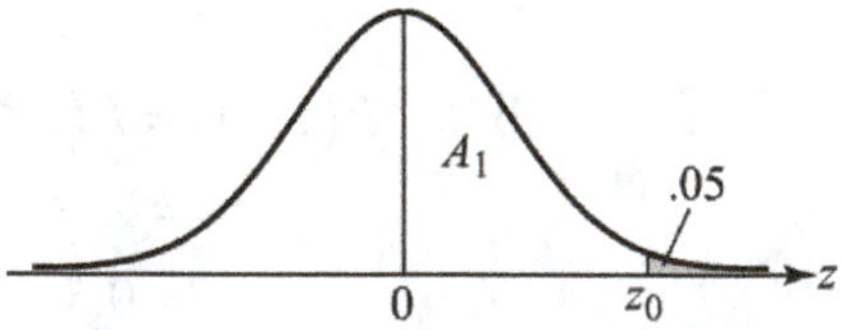

b. $P(z \ge z_0) = 0.025$
$A_1 = 0.5 - 0.025 = 0.4750$
Looking up the area 0.4750 in Table II gives $z_0 = 1.96$.

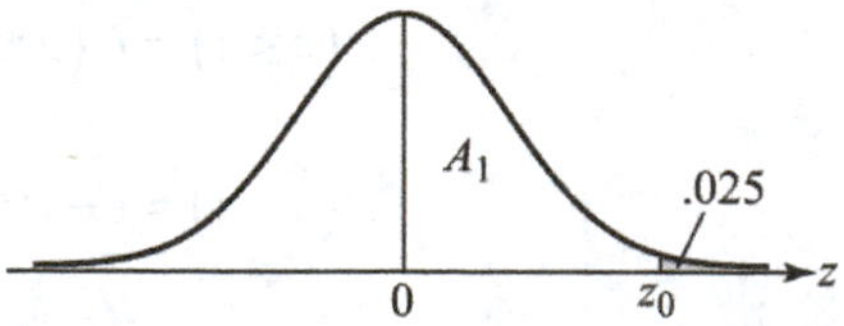

c. $P(z \le z_0) = 0.025$

$A_1 = .5 - .025 = .4750$

Looking up the area 0.4750 in Table II gives $z = 1.96$. Since z_0 is to the left of 0, $z_0 = -1.96$.

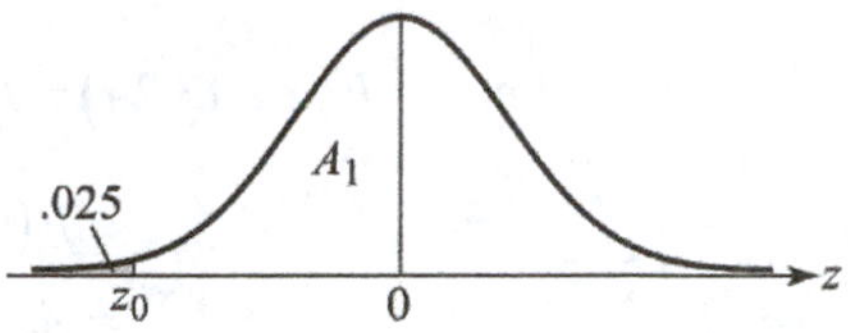

d. $P(z \ge z_0) = 0.10$

$A_1 = 0.5 - 0.10 = 0.4000$

Looking up the area 0.4000 in Table II gives $z_0 = 1.28$.

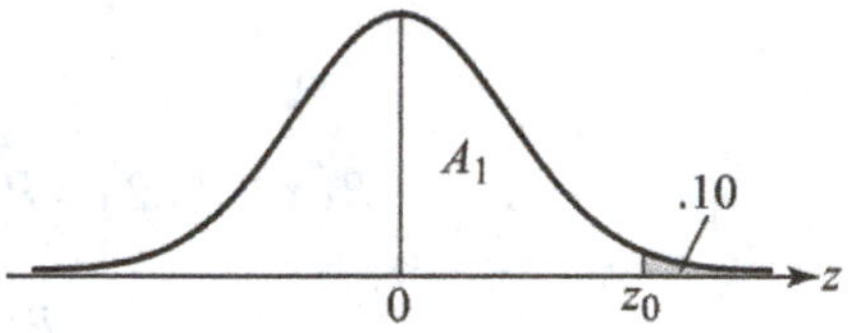

e. $P(z > z_0) = 0.10$

$A_1 = 0.5 - 0.10 = 0.4000$

$z_0 = 1.28$ (same as in **d**)

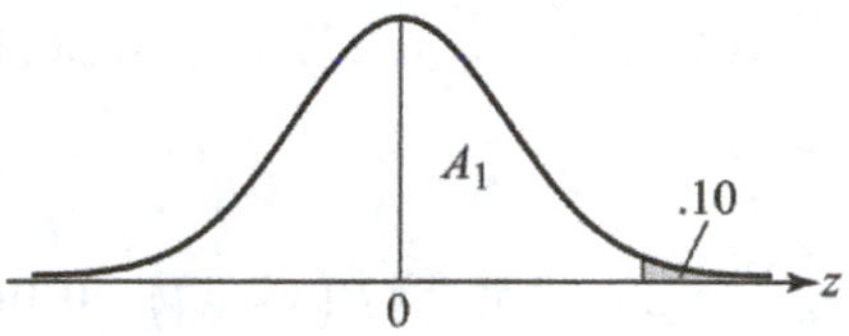

5.31 a. $P(10 \le x \le 12) = P\left(\frac{10-11}{2} \le z \le \frac{12-11}{2}\right)$

$= P(-0.50 \le z \le 0.50)$

$= A_1 + A_2$

$= 0.1915 + 0.1915 = 0.3830$

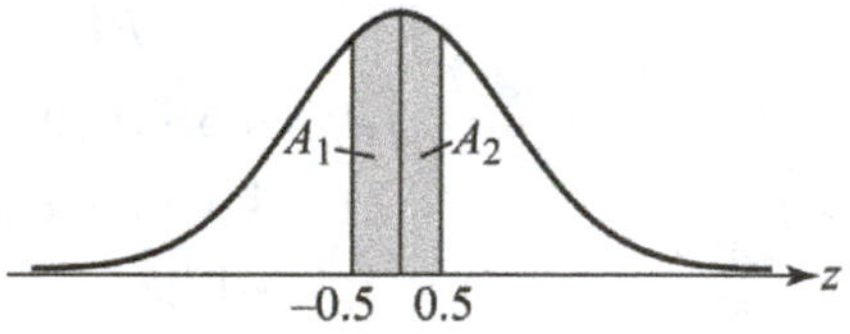

b. $P(6 \le x \le 10) = P\left(\frac{6-11}{2} \le z \le \frac{10-11}{2}\right)$

$= P(-2.50 \le z \le -0.50)$

$= P(-2.50 \le z \le 0) - P(-0.50 \le z \le 0)$

$= 0.4938 - 0.1915 = 0.3023$

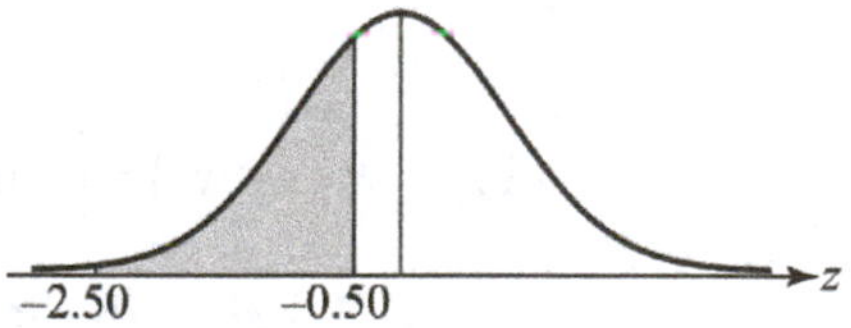

c. $P(13 \le x \le 16) = P\left(\frac{13-11}{2} \le z \le \frac{16-11}{2}\right)$

$= P(1.00 \le z \le 2.50)$

$= P(0 \le z \le 2.50) - P(0 \le z \le 1.00)$

$= 0.4938 - 0.3413 = 0.1525$

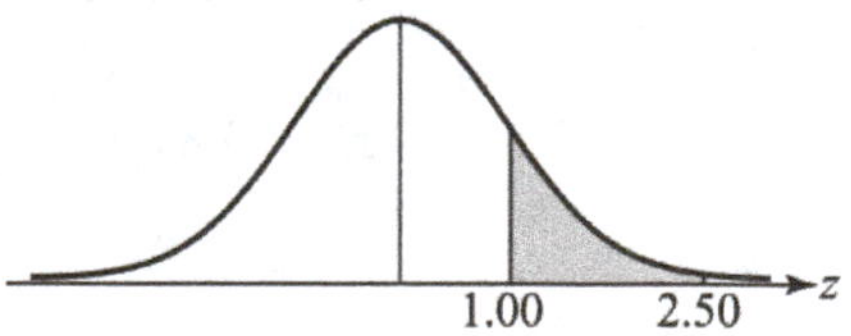

d. $P(7.8 \le x \le 12.6) = P\left(\frac{7.8-11}{2} \le z \le \frac{12.6-11}{2}\right)$

$= P(-1.60 \le z \le 0.80)$

$= A_1 + A_2$

$= 0.4452 + 0.2881 = 0.7333$

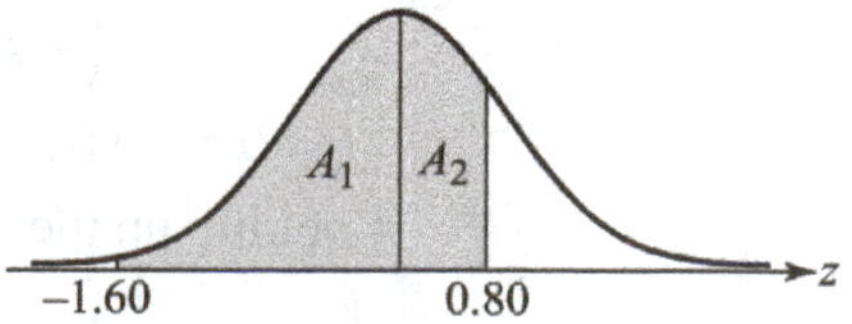

e. $P(x \geq 13.24) = P\left(z \geq \frac{13.24-11}{2}\right)$

$= P(z \geq 1.12)$

$= A_2 = 0.5 - A_1$

$= 0.5000 - 0.3868 = 0.1314$

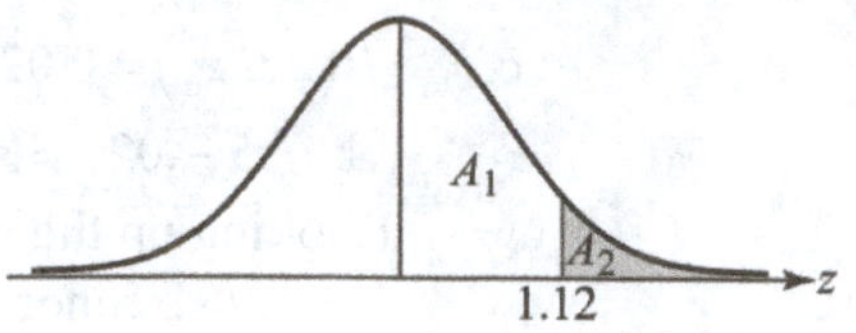

f. $P(x \geq 7.62) = P\left(z \geq \frac{7.62-11}{2}\right)$

$= P(z \geq -1.69)$

$= A_1 + A_2$

$= 0.4545 + 0.5000 = 0.9545$

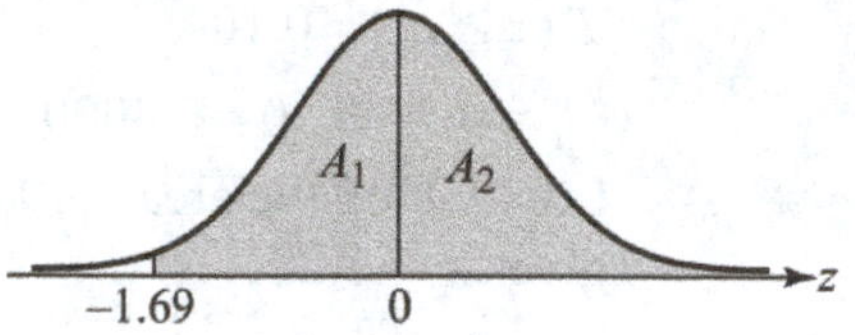

5.33 a. $P(x < x_0) = 0.10 \Rightarrow P\left(z < \frac{x_0 - 30}{8}\right)$

$P(z < z_0) = 0.10$

$A_1 = 0.5 - 0.10 = 0.4000$

Looking up the area 0.4000 in Table II gives $z_0 = 1.28$. Since z_0 is to the left of 0, $z_0 = -1.28$.

$z_0 = -1.28 = \frac{x_0 - 30}{8} \Rightarrow x_0 = 8(-1.28) + 30 = 19.76$

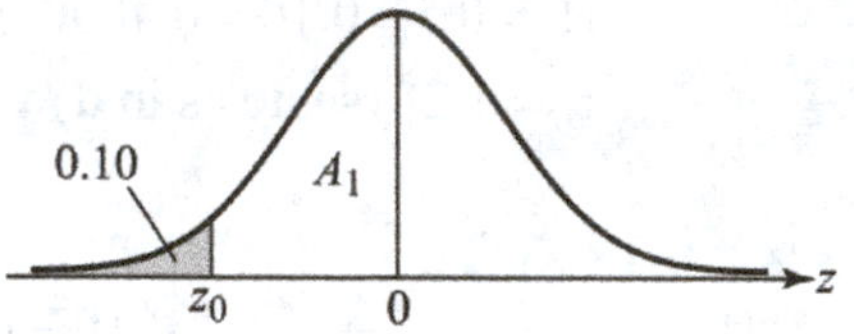

b. $P(x < x_0) = 0.80 \Rightarrow P\left(z < \frac{x_0 - 30}{8}\right)$

$P(z < z_0) = 0.80$

$A_1 = 0.80 - 0.5 = 0.3000$

Looking up the area 0.3000 in Table II gives $z_0 = 0.84$.

$z_0 = 0.84 = \frac{x_0 - 30}{8} \Rightarrow x_0 = 8(0.84) + 30 = 36.72$

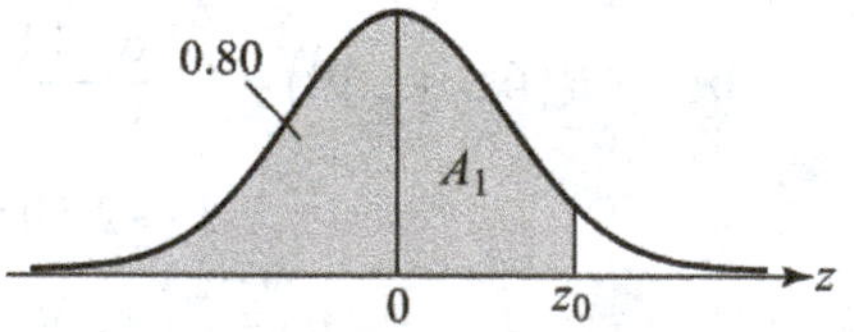

c. $P(x > x_0) = 0.01 \Rightarrow P\left(z > \frac{x_0 - 30}{8}\right)$

$P(z > z_0) = 0.01$

$A_1 = 0.5 - 0.01 = 0.4900$

Looking up the area 0.4900 in Table II gives $z_0 = 2.33$.

$z_0 = 2.33 = \frac{x_0 - 30}{8} \Rightarrow x_0 = 8(2.33) + 30 = 48.64$

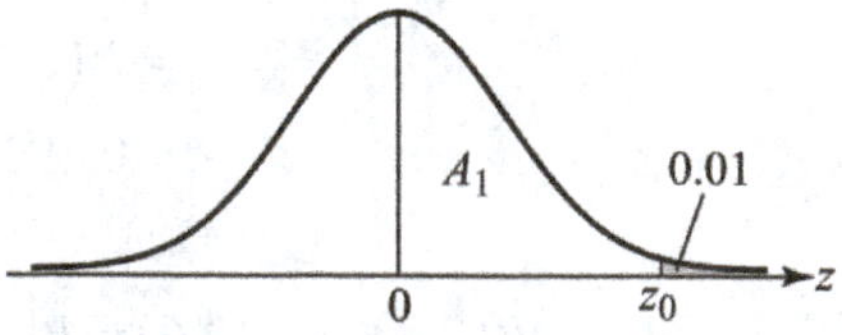

5.35 The random variable x has a normal distribution with $\sigma = 25$.

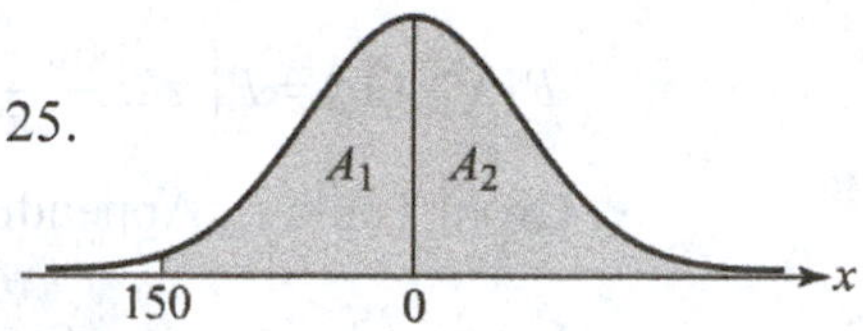

We know $P(x > 150) = 0.90$. So, $A_1 + A_2 = 0.90$.
Since $A_2 = 0.50$, $A_1 = 0.90 - 0.50 = 0.40$. Look up the area 0.40 in the body of Table II; (take the closest value) $z_0 = -1.28$.

To find μ, substitute all the values into the z-score formula:

$$z = \frac{x-\mu}{\sigma} \Rightarrow -1.28 = \frac{150-\mu}{25} \Rightarrow \mu = 150 + 25(1.28) = 182$$

5.37 a. $z = \frac{x-\mu}{\sigma} = \frac{10-11}{4} = \frac{-1}{4} = -0.25$

b. $P(10 < x < 15) = P\left(\frac{10-11}{4} < z < \frac{15-11}{4}\right) = P(-0.25 < z < 1)$
$= P(-0.25 < z < 0) + P(0 < z < 1)$
$= 0.0987 + 0.3413 = 0.4400$
(Using Table II, Appendix B)

c. $P(x > 20) = P\left(z > \frac{20-11}{4}\right) = P(z > 2.25)$
$= 0.5 - P(0 < z < 2.25) = 0.5 - 0.4878 = 0.0122$
(Using Table II, Appendix B)

5.39 a. $P(1.3 < x < 4.0) = P\left(\frac{1.3-2.2}{0.9} < z < \frac{4-2.2}{0.9}\right) = P(-1 < z < 2) = 0.3413 + 0.4772 = 0.8185$
(from Table II, Appendix B)

b. $P(x > 1) = P\left(z > \frac{1-2.2}{0.9}\right) = P(z > -1.33) = 0.5 + 0.4082 = 0.9082$
(from Table II, Appendix B)

c. $P(x > T) = 0.35 \Rightarrow P\left(z > \frac{T-2.2}{0.9}\right) = P(z > z_0)$
$A_1 = 0.5 - 0.35 = 0.15$
Looking up the area 0.15 in Table II gives $z_0 = 0.39$.
$z_0 = 0.39 = \frac{T-2.2}{0.9} \Rightarrow T = 0.9(0.39) + 2.2 = 2.551$

5.41 a. Let x = choice score for shoppers in the flexed arm condition. Then x has a normal distribution with $\mu = 59$ and $\sigma = 5$.

$$P(x \geq 60) = P\left(z \geq \frac{60-59}{5}\right) = P(z \geq 0.2) = 0.5 - 0.0793 = 0.4207$$

(from Table II, Appendix B)

b. Let y = choice score for shoppers in the extended arm condition. Then y has a normal distribution with $\mu = 43$ and $\sigma = 5$. Using MINITAB,

Cumulative Distribution Function

```
Normal with mean = 43 and standard deviation = 5

 x  P( X ≤ x )
60    0.999663
```

$$P(y \geq 60) = 1 - P(y \leq 60) = 1 - 0.999663 = 0.000337$$

5.43 Let y = perception of light flicker. Then x has a normal distribution with $\mu = 2.2$ and $\sigma = 0.5$.

$$P(x \geq 3) = P\left(z \geq \frac{3-2.2}{0.5}\right) = P(z \geq 1.6) = 0.5 - 0.4452 = 0.0548 \text{ (from Table II, Appendix B)}$$

5.45 a. Let x = number of sags per week. Then x has a normal distribution with $\mu = 353$ and $\sigma = 30$. Using MINITAB,

Cumulative Distribution Function

```
Normal with mean = 353 and standard deviation = 30

  x  P( X ≤ x )
400    0.941404
```

$$P(x < 400) = 0.941404$$

b. Let y = number of swells per week. Then y has a normal distribution with $\mu = 184$ and $\sigma = 25$. Using MINITAB,

Cumulative Distribution Function

```
Normal with mean = 184 and standard deviation = 25

  x  P( X ≤ x )
100   0.0003897
```

$$P(y > 100) = 1 - P(y < 100) = 1 - 0.0003897 = 0.9996103$$

5.47 a. Let x = ambulance response time at Station A. Then x has a normal distribution with $\mu = 7.5$ and $\sigma = 2.5$. Using Table II, Appendix B,

$P(x \leq 9) = P\left(z \leq \frac{9-7.5}{2.5}\right) = P(z \leq 0.60) = 0.5 + 0.2257 = 0.7257$. Since this probability is less than the requirement of 0.9, the regulations are not met at EMS station A.

b. $P(x \le 2) = P\left(z \le \frac{2-7.5}{2.5}\right) = P(z \le -2.20) = 0.5 - 0.4861 = 0.0139$. If the call was serviced by Station A, the probability of a response time of 2 minutes or less would be extremely unusual. Thus, we would conclude that the call was probably not serviced by Station A.

5.49 Let x = number of additional Electoral College votes. Then x has a normal distribution with $\mu = 241.5$ and $\sigma = 49.8$. If the candidate wins the popular vote in California, then he/she needs only $270 - 55 = 215$ additional votes.

$$P(x \ge 215) = P\left(z \ge \frac{215-241.5}{49.8}\right) = P(z \ge -0.53) = 0.5 + 0.2019 = 0.7019$$

5.51 a. If z is a standard normal random variable,

$Q_L = z_L$ is the value of the standard normal distribution which has 25% of the data to the left and 75% to the right.

Find z_L such that $P(z < z_L) = 0.25$

$A_1 = 0.50 - 0.25 = 0.25$.

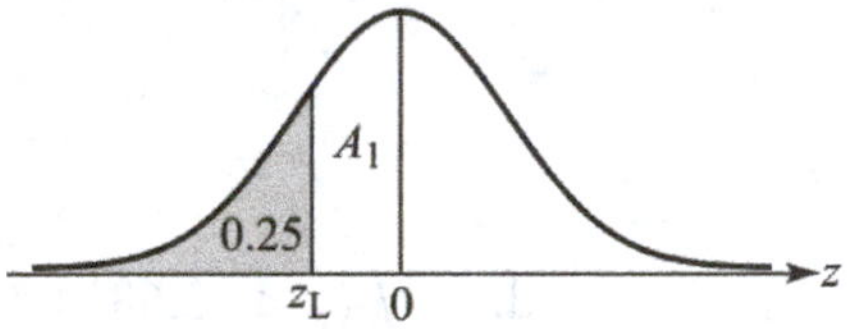

Look up the area $A_1 = 0.25$ in the body of Table II of Appendix B; $z_L = -0.67$ (taking the closest value). If interpolation is used, −0.675 would be obtained.

$Q_U = z_U$ is the value of the standard normal distribution which has 75% of the data to the left and 25% to the right.

Find z_U such that $P(z < z_U) = 0.75$

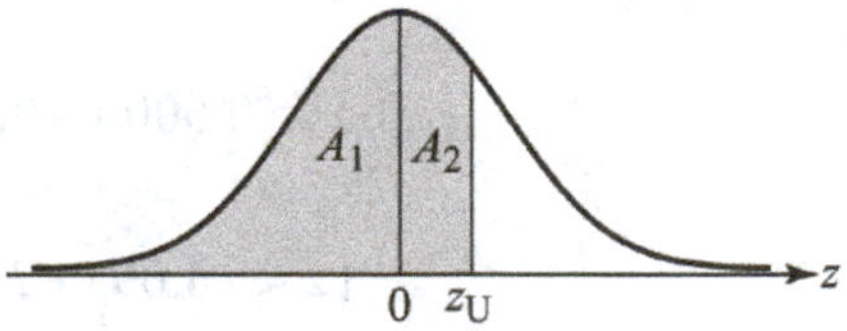

$$A_1 + A_2 = P(z \le 0) + P(0 \le z \le z_U)$$
$$= 0.5 + P(0 \le z \le z_U) = 0.75$$

Therefore, $P(0 \le z \le z_U) = 0.25$.

Look up the area 0.25 in the body of Table II of Appendix B; $z_U = 0.67$ (taking the closest value). If interpolation is used, 0.675 would be obtained.

b. Recall that the inner fences of a box plot are located $1.5(Q_U - Q_L)$ outside the hinges (Q_L and Q_U).

To find the lower inner fence,

$$Q_L - 1.5(Q_U - Q_L) = -0.67 - 1.5[0.67 - (-0.67)]$$
$$= -0.67 - 1.5(1.34)$$
$$= -2.68 \quad (-2.70 \text{ if } z_L = -0.675 \text{ and } z_U = 0.675)$$

The upper inner fence is:

$$Q_U + 1.5(Q_U - Q_L) = 0.67 + 1.5[0.67 - (-0.67)]$$
$$= 0.67 + 1.5(1.34)$$
$$= 2.68 \quad (2.70 \text{ if } z_L = -0.675 \text{ and } z_U = 0.675)$$

c. Recall that the outer fences of a box plot are located $3(Q_U - Q_L)$ outside the hinges (Q_L and Q_U).

To find the lower outer fence,

$$Q_L - 3(Q_U - Q_L) = -0.67 - 3[0.67 - (-0.67)]$$
$$= -0.67 - 3(1.34)$$
$$= -4.69 \quad (-4.725 \text{ if } z_L = -0.675 \text{ and } z_U = 0.675)$$

The upper outer fence is:

$$Q_U + 3(Q_U - Q_L) = 0.67 + 3[0.67 - (-0.67)]$$
$$= 0.67 + 3(1.34)$$
$$= 4.69 \quad (4.725 \text{ if } z_L = -0.675 \text{ and } z_U = 0.675)$$

d. $P(z < -2.68) + P(z > 2.68)$

$= 2P(z > 2.68)$

$= 2(0.5000 - 0.4963)$

$= 2(0.0037) = 0.0074$

(Table II, Appendix B)

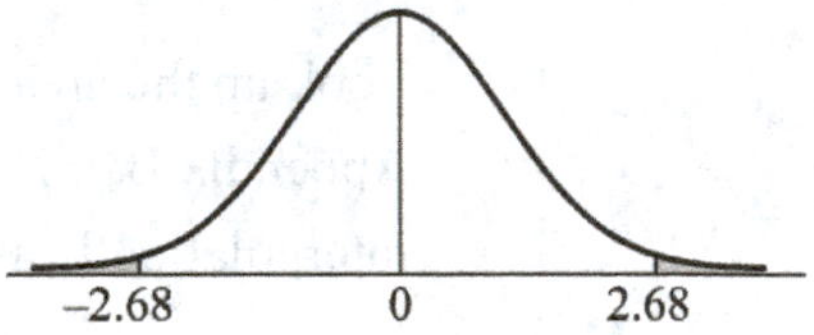

(or $2(0.5000 - 0.4965) = 0.0070$ if -2.70 and 2.70 are used)

$P(z < -4.69) + P(z > 4.69)$

$= 2P(z > 4.69)$

$\approx 2(0.5000 - 0.5000) \approx 0$

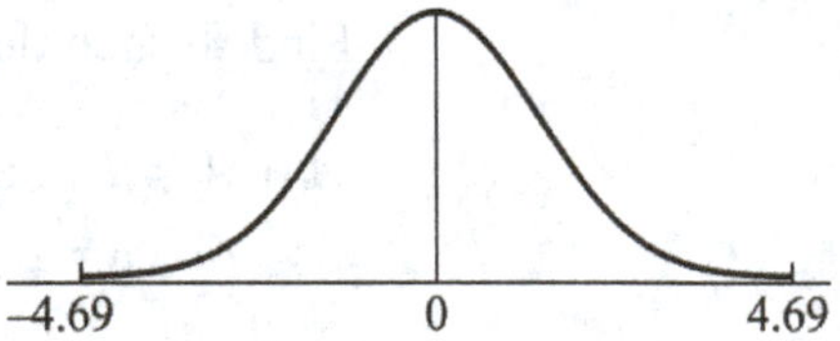

e. In a normal probability distribution, the probability of an observation being beyond the inner fences is only 0.0074 and the probability of an observation being beyond the outer fences is approximately zero. Since the probabilities are so small, there should not be any observations beyond the inner and outer fences. Therefore, they are probably outliers.

5.53 a-e. Let x = height of women over 20 years old. We are given that $\mu = 64$ and that the distribution of x is normal. However, we are given no information about the standard deviation. The answers to parts *a* through *e* will depend on what each person uses as the standard deviation.

f-a. Find x_L and x_U such that $P(x_L < x < x_U) = 0.50$.

First, we need to find z_0 such that $P(-z_0 < z < z_0) = 0.50$. Since we know that the center of the z distribution is 0, half of the area or $0.50/2 = 0.25$ will be between $-z_0$ and 0 and half will be between 0 and z_0.

Look up 0.25 in the body of Table II, Appendix B to find $z_0 = 0.67$.

We know $P(x_L < x < x_U) = P\left(\frac{x_L - 64}{2.6} < z < \frac{x_U - 64}{2.6}\right) = P(-0.67 < z < 0.67) = 0.50$

Thus,

$$\frac{x_L - 64}{2.6} = -0.67 \text{ and } \frac{x_U - 64}{2.6} = 0.67$$
$$\Rightarrow x_L = 64 - 0.67(2.6) = 64 - 1.74 = 62.26$$

$$\Rightarrow x_U = 64 + 0.67(2.6) = 64 + 1.74 = 65.74$$

f-b. Find x_L and x_U such that $P(x_L < x < x_U) = 0.75$.

First, we need to find z_0 such that $P(-z_0 < z < z_0) = 0.75$. Since we know that the center of the z distribution is 0, half of the area or $0.75/2 = 0.375$ will be between $-z_0$ and 0 and half will be between 0 and z_0.

Look up 0.375 in the body of Table II, Appendix B to find $z_0 = 1.15$.

We know $P(x_L < x < x_U) = P\left(\frac{x_L - 64}{2.6} < z < \frac{x_U - 64}{2.6}\right) = P(-1.15 < z < 1.15) = 0.75$

Thus,

$$\frac{x_L - 64}{2.6} = -1.15 \text{ and } \frac{x_U - 64}{2.6} = 1.15$$
$$\Rightarrow x_L = 64 - 1.15(2.6) = 64 - 2.99 = 61.01$$

$$\Rightarrow x_U = 64 + 1.15(2.6) = 64 + 2.99 = 66.99$$

f-c. Find x_L and x_U such that $P(x_L < x < x_U) = 0.90$.

First, we need to find z_0 such that $P(-z_0 < z < z_0) = 0.90$. Since we know that the center of the z distribution is 0, half of the area or $0.90/2 = 0.45$ will be between $-z_0$ and 0 and half will be between 0 and z_0.

Look up 0.45 in the body of Table II, Appendix B to find $z_0 = 1.645$.

We know $P(x_L < x < x_U) = P\left(\frac{x_L - 64}{2.6} < z < \frac{x_U - 64}{2.6}\right) = P(-1.645 < z < 1.645) = 0.90$

Thus,

$$\frac{x_L - 64}{2.6} = -1.645 \text{ and } \frac{x_U - 64}{2.6} = 1.645$$

$$\Rightarrow x_L = 64 - 1.645(2.6) = 64 - 4.28 = 59.72$$

$$\Rightarrow x_U = 64 + 1.645(2.6) = 64 + 4.28 = 68.28$$

f-d. Find x_L and x_U such that $P(x_L < x < x_U) = 0.95$.

First, we need to find z_0 such that $P(-z_0 < z < z_0) = 0.95$. Since we know that the center of the z distribution is 0, half of the area or $0.95/2 = 0.475$ will be between $-z_0$ and 0 and half will be between 0 and z_0.

Look up 0.475 in the body of Table II, Appendix B to find $z_0 = 1.96$.

We know $P(x_L < x < x_U) = P\left(\frac{x_L - 64}{2.6} < z < \frac{x_U - 64}{2.6}\right) = P(-1.96 < z < 1.96) = 0.95$

Thus,

$$\frac{x_L - 64}{2.6} = -1.96 \text{ and } \frac{x_U - 64}{2.6} = 1.96$$

$$\Rightarrow x_L = 64 - 1.96(2.6) = 64 - 5.10 = 58.90$$

$$\Rightarrow x_U = 64 + 1.96(2.6) = 64 + 5.10 = 69.10$$

f-e. Find x_L and x_U such that $P(x_L < x < x_U) = 0.99$.

First, we need to find z_0 such that $P(-z_0 < z < z_0) = 0.99$. Since we know that the center of the z distribution is 0, half of the area or $0.99/2 = 0.495$ will be between $-z_0$ and 0 and half will be between 0 and z_0.

We look up 0.495 in the body of Table II, Appendix B to find $z_0 = 2.575$.

We know $P(x_L < x < x_U) = P\left(\frac{x_L - 64}{2.6} < z < \frac{x_U - 64}{2.6}\right) = P(-2.575 < z < 2.575) = 0.99$

Thus,

$$\frac{x_L - 64}{2.6} = -2.575 \text{ and } \frac{x_U - 64}{2.6} = 2.575$$

$$\Rightarrow x_L = 64 - 2.575(2.6) = 64 - 6.70 = 57.30$$

$$\Rightarrow x_U = 64 + 2.575(2.6) = 64 + 6.70 = 70.70$$

5.55 Several statistical techniques are based on the assumption that the population is approximately normally distributed. Thus, it is important to determine whether the sample data come from a normal population so that the techniques can be properly applied.

5.57 a. The proportion of measurements that one would expect to fall in the interval $\mu \pm \sigma$ is about 0.68.

b. The proportion of measurements that one would expect to fall in the interval $\mu \pm 2\sigma$ is about 0.95.

c. The proportion of measurements that one would expect to fall in the interval $\mu \pm 3\sigma$ is about 0.997.

5.59 If the data are normally distributed, then the normal probability plot should be an approximate straight line. Of the three plots, only plot c implies that the data are normally distributed. The data points in plot c form an approximately straight line. In both plots a and b, the plots of the data points do not form a straight line.

5.61 a. Using MINITAB, the stem-and-leaf display is:

Stem-and-Leaf Display: Data

```
Stem-and-leaf of Data  N  = 40
Leaf Unit = 10

 14    0  00012223333344
(10)   0  5556666889
 16    1  001334
 10    1  5689
 6     2  01
 4     2
 4     3  000
 1     3
 1     4  0
```

The distribution from the stem-and-leaf display looks like it is skewed to the right. Thus, it does not appear that the data are normally distributed.

b. Using MINITAB, the descriptive statistics are:

Descriptive Statistics: Data

```
Variable    N  Mean   StDev  Minimum    Q1   Median     Q3  Maximum
Data       40 105.1    95.8     1.00  37.5     65.0  152.8    401.0
```

The lower quartile is $Q_L = 37.5$, the upper quartile is $Q_S = 152.8$, and $s = 95.8$.

c. The interquartile range is $\text{IQR} = Q_U - Q_L = 152.8 - 37.5 = 115.3$.
$\text{IQR} / s = 115.3 / 95.8 = 1.204$. If the data are approximately normal, then this ratio should be 1.3. Since the actual ratio is 1.204 which is close to 1.3, it appears that the data could be approximately normal.

d. Using MINITAB, the normal probability plot is:

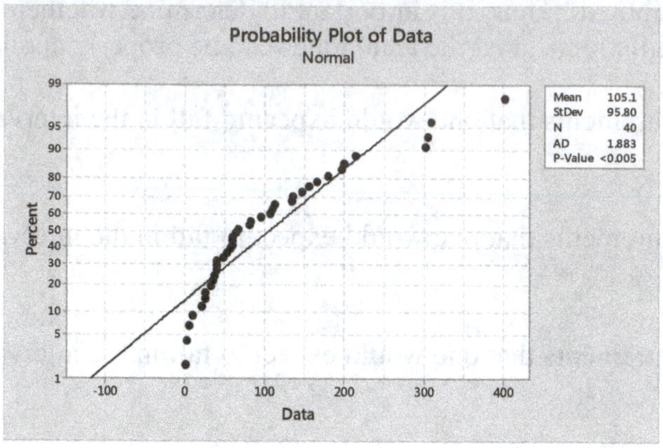

Since the data points do not fall along the straight line, the data do not appear to be normally distributed.

5.63 From the histogram, the data are somewhat skewed to the right. This indicates that the data may not be normally distributed.

The interval $\bar{x} \pm s \Rightarrow 78.19 \pm 63.24 \Rightarrow (14.95, 141.43)$ contains 55 of the 71 observations. The proportion is $55/71 = 0.775$. This is greater than the 0.68 from the Empirical Rule.

The interval $\bar{x} \pm 2s \Rightarrow 78.19 \pm 2(63.24) \Rightarrow 78.19 \pm 126.48 \Rightarrow (-48.29, 204.67)$ contains 65 of the 71 observations. The proportion is $65/71 = 0.915$. This is less than the 0.95 from the Empirical Rule.

The interval $\bar{x} \pm 2s \Rightarrow 78.19 \pm 3(63.24) \Rightarrow 78.19 \pm 189.72 \Rightarrow (-111.53, 267.91)$ contains 71 of the 71 observations. The proportion is $71/71 = 1.0$. This is very close to the 0.997 from the Empirical Rule. Based on the Empirical Rule, it appears that the data may not be normally distributed.

The interquartile range is $\text{IQR} = 64.99$ and $s = 63.24$. $\text{IQR}/s = 64.99/63.24 = 1.028$. This is a somewhat smaller than the 1.3 that we would expect if the data were normal. Thus, there is evidence that the data may not be normally distributed.

Using MINITAB, the normal probability plot is:

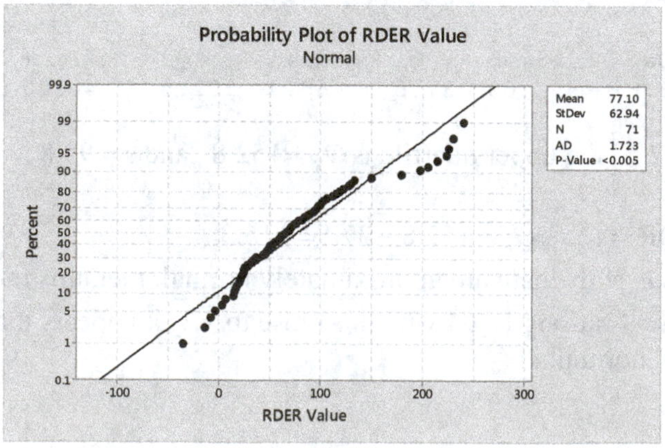

Because the points do not follow the line very well, there is evidence that the data may not be normally distributed.

From all of the tests, there is evidence that the data are not normally distributed. We would not recommend using the normal distribution.

5.65 The interquartile range is $IQR = Q_U - Q_L = 6 - 1 = 5$. $IQR / s = 5 / 6.09 = 0.82$. This is much less than the 1.3 that we would expect if the data were normal. It appears that these data are not approximately normal. In addition, $\bar{x} = 4.71$ and $s = 6.09$. Since the number of updates cannot be negative, the data must be skewed to the right since one standard deviation below the mean would be a negative number which is impossible.

5.67 a. From the graph, the histogram of the difference between the elevation estimates is too peaked. This means that there are too many observations in the very center of the distribution and not enough observations in the tails for the distribution to be normal.

b. The interval $\mu \pm 2\sigma$ would be $0.28 \pm 2(1.6) \Rightarrow 0.28 \pm 3.2 \Rightarrow (-2.92, 3.48)$. From the graph, almost all the observations are between are between -2 and 2. Thus, the interval $(-2.92, 3.48)$ will contain more that 95% of the 400 elevation differences.

5.69 **Without calcium/gypsum**:
Using MINITAB, the histogram is:

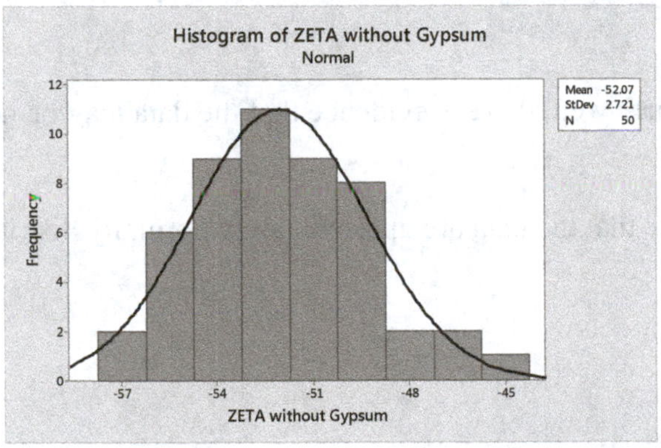

The data look fairly mound-shaped. The data may be normally distributed.

Using MINITAB, the descriptive statistics are:

Descriptive Statistics: ZETA without Gypsum

Variable	N	Mean	StDev	Minimum	Q1	Median	Q3	Maximum	IQR
ZETA no Gypsum	50	-52.070	2.721	-57.600	-53.900	-52.250	-50.200	-45.700	3.700

The interval $\bar{x} \pm s \Rightarrow -52.070 \pm 2.721 \Rightarrow (-54.791, -49.349)$ contains 36 of the 50 observations. The proportion is $36 / 50 = 0.72$. This is somewhat greater than the 0.68 from the Empirical Rule.

The interval $\bar{x} \pm s \Rightarrow -52.070 \pm 2(2.721) \Rightarrow -52.070 \pm 5.442 \Rightarrow (-57.512, -46.628)$ contains 47 of the 50 observations. The proportion is $47 / 50 = 0.94$. This is fairly close to the 0.95 from the Empirical Rule.

The interval $\bar{x} \pm s \Rightarrow -52.070 \pm 3(2.721) \Rightarrow -52.070 \pm 8.163 \Rightarrow (-60.233, -43.907)$ contains 50 of the 50 observations. The proportion is $50/50 = 1.0$. This is very close to the 0.997 from the Empirical Rule. Based on the Empirical Rule, it appears that the data may be normally distributed.

The interquartile range is $IQR = 3.7$ and $s = 2.721$. $IQR/s = 3.7/2.721 = 1.360$. This is fairly close to the 1.3 that we would expect if the data were normal. Thus, there is evidence that the data may be normally distributed.

Using MINITAB, the normal probability plot is:

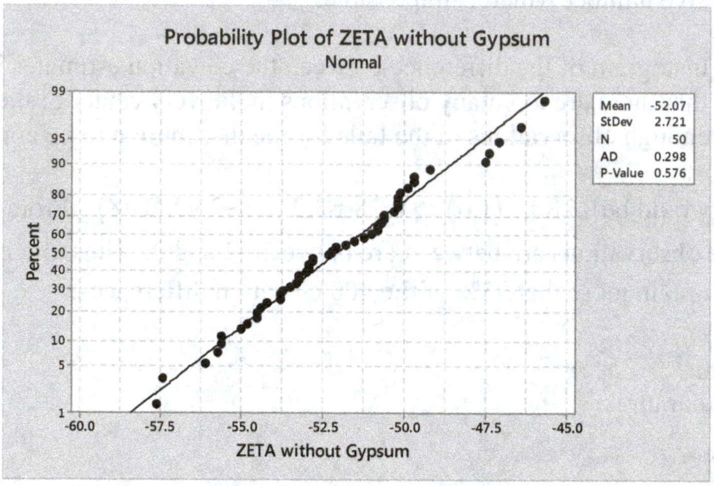

Because the points follow the line fairly well, there is evidence that the data may be normally distributed.

From all of the tests, there is evidence that the data are approximately normally distributed.

With calcium/gypsum:
Using MINITAB, the histogram is:

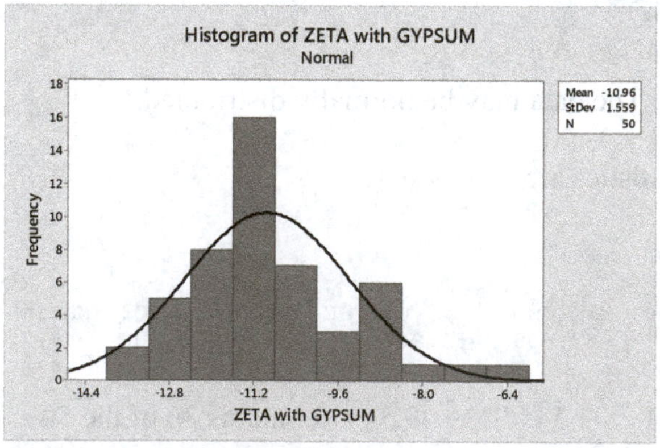

The data are approximately mound-shaped. The data may be normally distributed.

Using MINITAB, the descriptive statistics are:

Descriptive Statistics: ZETA with GYPSUM

Variable	N	Mean	StDev	Minimum	Q1	Median	Q3	Maximum	IQR
ZETA with GYPSUM	50	-10.958	1.559	-13.600	-12.100	-11.300	-10.075	-6.800	2.025

The interval $\bar{x} \pm s \Rightarrow -10.958 \pm 1.559 \Rightarrow (-12.517, -9.399)$ contains 34 of the 50 observations. The proportion is $34/50 = 0.68$. This is the same as the 0.68 from the Empirical Rule.

The interval $\bar{x} \pm 2s \Rightarrow -10.958 \pm 2(1.559) \Rightarrow -10.958 \pm 3.118 \Rightarrow (-14.076, -7.840)$ contains 48 of the 50 observations. The proportion is $48/50 = 0.96$. This is fairly close to the 0.95 from the Empirical Rule.

The interval $\bar{x} \pm 3s \Rightarrow -10.958 \pm 3(1.559) \Rightarrow -10.958 \pm 3.118 \Rightarrow (-15.635, -6.281)$ contains 50 of the 50 observations. The proportion is $50/50 = 1.0$. This is very close to the 0.997 from the Empirical Rule.
Based on the Empirical Rule, it appears that the data may be normally distributed.

The interquartile range is $\text{IQR} = 2.025$ and $s = 1.559$. $\text{IQR}/s = 2.025/1.559 = 1.299$. This is very close to the 1.3 that we would expect if the data were normal. Thus, there is evidence that the data may be normally distributed.

Using MINITAB, the normal probability plot is:

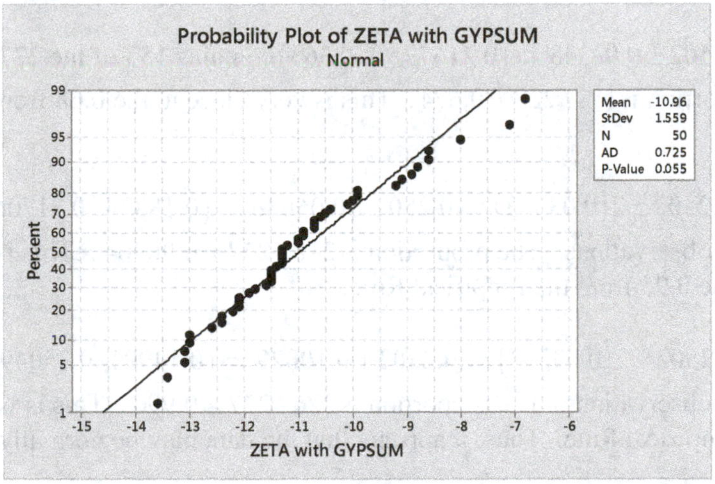

Because the points fall fairly close to the line, there is evidence that the data may be normally distributed.

From all of the tests, there is evidence that the data are approximately normally distributed.

Both data sets are approximated by the normal distribution.

5.71 **American League**

To determine if the distribution of batting averages is approximately normal, we will run through the tests. Using MINITAB, a histogram of the data with a normal curve included is:

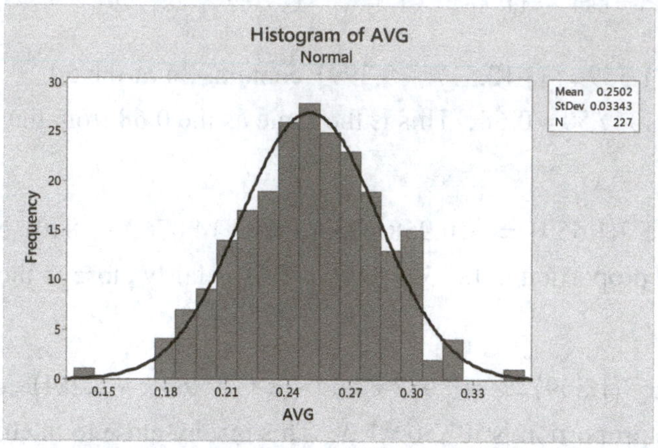

From the graph, the data appear to be approximately mound-shaped. The data may be normally distributed.

Using MINITAB, the descriptive statistics are:

Descriptive Statistics: AVG

```
Variable    N     Mean     StDev  Minimum       Q1   Median       Q3  Maximum
AVG       227  0.25020  0.03343  0.14200  0.22600  0.25100  0.27300  0.34800
```

The interval $\bar{x} \pm s \Rightarrow 0.2502 \pm 0.03343 \Rightarrow (0.21677,\ 0.28363)$ contains 153 of the 227 observations. The proportion is $153/227 = 0.674$. This is very close to the 0.68 from the Empirical Rule.

The interval $\bar{x} \pm 2s \Rightarrow 0.2502 \pm 2(0.03343) \Rightarrow 0.2502 \pm 0.06686 \Rightarrow (0.18334,\ 0.31706)$ contains 218 of the 227 observations. The proportion is $218/227 = 0.96904$. This is somewhat larger than the 0.95 from the Empirical Rule.

The interval $\bar{x} \pm 3s \Rightarrow 0.2502 \pm 3(0.03343) \Rightarrow 0.2502 \pm 0.10029 \Rightarrow (0.14991,\ 0.35049)$ contains 226 of the 227 observations. The proportion is $226/227 = 0.996$. This is very close to the 0.997 from the Empirical Rule. Thus, it appears that the data may be normally distributed.

The lower quartile is $Q_L = 0.226$ and the upper quartile is $Q_U = 0.273$. The interquartile range is $\text{IQR} = Q_U - Q_L = 0.273 - 0.226 = 0.047$. From the printout, $s = 0.03343$.
$\text{IQR}/s = 0.047/0.03343 = 1.406$. This is a fairly close to the 1.3 that we would expect if the data were normal. Thus, there is evidence that the data may be normally distributed.

Using MINITAB, the normal probability plot is:

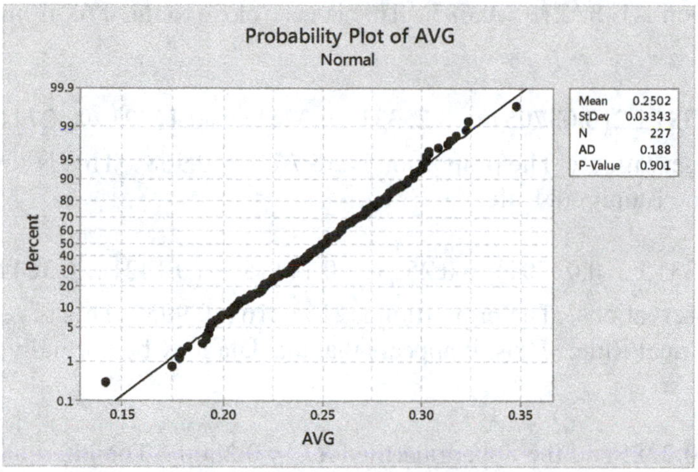

The data are very close to a straight line. Thus, it appears that the data may be normally distributed.

From all of the indicators, it appears that the batting averages from the American League come from an approximate normal distribution.

National League

To determine if the distribution of batting averages is approximately normal, we will run through the tests. Using MINITAB, a histogram of the data with a normal curve included is:

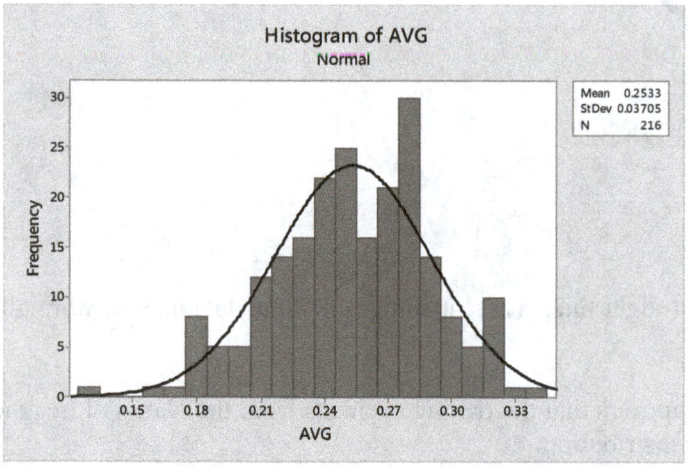

From the graph, the data appear to be approximately mound-shaped. The data may be normally distributed.

Using MINITAB, the descriptive statistics are:

Descriptive Statistics: AVG

```
Variable    N     Mean    StDev  Minimum       Q1   Median       Q3  Maximum
AVG       216  0.25331  0.03705  0.12700  0.22850  0.25400  0.28000  0.34500
```

The interval $\bar{x} \pm s \Rightarrow 0.25331 \pm 0.03705 \Rightarrow (0.21626,\ 0.29036)$ contains 148 of the 216 observations. The proportion is $148/216 = 0.685$. This is very close to the 0.68 from the Empirical Rule.

The interval $\bar{x} \pm 2s \Rightarrow 0.25331 \pm 2(0.03705) \Rightarrow 0.25331 \pm 0.0741 \Rightarrow (0.17921,\ 0.32741)$ contains 209 of the 216 observations. The proportion is $209/216 = 0.968$. This is somewhat larger than the 0.95 from the Empirical Rule.

The interval $\bar{x} \pm 3s \Rightarrow 0.25331 \pm 3(0.03705) \Rightarrow 0.25331 \pm 0.11115 \Rightarrow (0.14216,\ 0.36446)$ contains 215 of the 216 observations. The proportion is $215/216 = 0.995$. This is very close to the 0.997 from the Empirical Rule. Thus, it appears that the data may be normally distributed.

The lower quartile is $Q_L = 0.2285$ and the upper quartile is $Q_U = 0.2800$. The interquartile range is $\text{IQR} = Q_U - Q_L = 0.2800 - 0.2285 = 0.0515$. From the printout, $s = 0.03705$. $\text{IQR}/s = 0.0515/0.03705 = 1.39$. This is a fairly close to the 1.3 that we would expect if the data were normal. Thus, there is evidence that the data may be normally distributed.

Using MINITAB, the normal probability plot is:

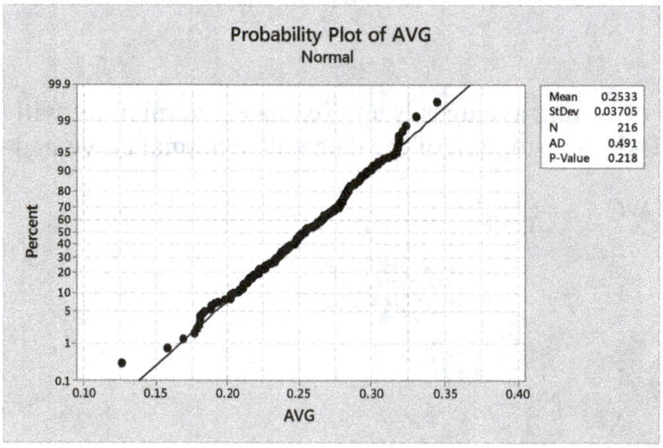

The data are very close to a straight line. Thus, it appears that the data may be normally distributed.

From all of the indicators, it appears that the batting averages from the National League come from an approximate normal distribution.

5.73 **Group A**:
Using MINITAB, the histogram is:

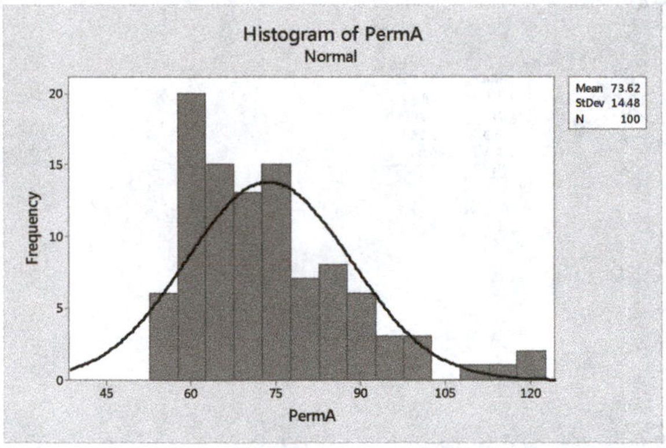

The data are not approximately mound-shaped. The data may not be normally distributed.

Using MINITAB, the descriptive statistics are:

Descriptive Statistics: PermA

Variable	N	Mean	StDev	Minimum	Q1	Median	Q3	Maximum	IQR
PermA	100	73.62	14.48	55.20	62.00	70.45	81.42	122.40	19.42

The interval $\bar{x} \pm s \Rightarrow 73.62 \pm 14.48 \Rightarrow (59.14, 88.10)$ contains 76 of the 100 observations. The proportion is $76/100 = 0.76$. This is the much greater than the 0.68 from the Empirical Rule.

The interval $\bar{x} \pm 2s \Rightarrow 73.62 \pm 2(14.48) \Rightarrow 73.62 \pm 28.96 \Rightarrow (44.66, 102.58)$ contains 96 of the 100 observations. The proportion is $96/100 = 0.96$. This is fairly close to the 0.95 from the Empirical Rule.

The interval $\bar{x} \pm 3s \Rightarrow 73.62 \pm 3(14.48) \Rightarrow 73.62 \pm 43.44 \Rightarrow (30.18, 117.06)$ contains 97 of the 100 observations. The proportion is $97/100 = 0.97$. This is much smaller than the 0.997 from the Empirical Rule.
Based on the Empirical Rule, it appears that the data may not be normally distributed.

The interquartile range is $\text{IQR} = 19.42$ and $s = 14.48$. $\text{IQR}/s = 19.42/14.48 = 1.341$. This is very close to the 1.3 that we would expect if the data were normal. Thus, there is evidence that the data may be normally distributed.

Using MINITAB, the normal probability plot is:

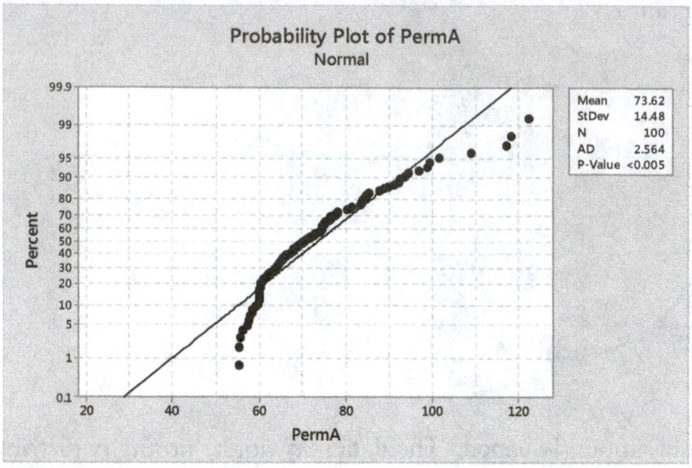

Because several points fall off the line, there is evidence that the data may not be normally distributed.

From all of the tests, there is evidence that the data are not approximately normally distributed.

Group B:
Using MINITAB, the histogram is:

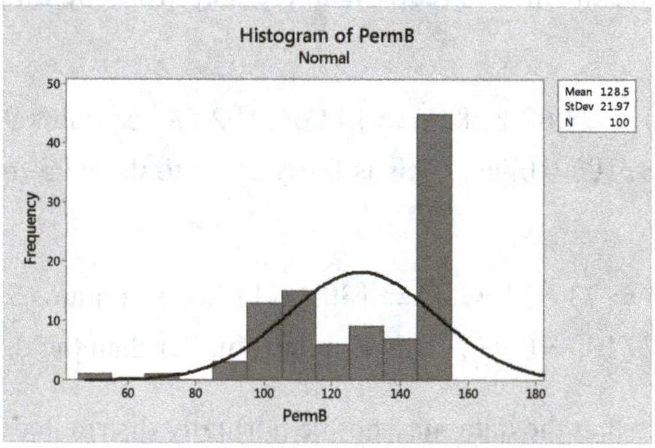

The data are not approximately mound-shaped. The data may not be normally distributed.

Using MINITAB, the descriptive statistics are:

Descriptive Statistics: PermB

Variable	N	Mean	StDev	Minimum	Q1	Median	Q3	Maximum	IQR
PermB	100	128.54	21.97	50.40	108.65	139.30	147.02	150.00	38.37

The interval $\bar{x} \pm s \Rightarrow 128.54 \pm 21.97 \Rightarrow (106.57, 150.51)$ contains 81 of the 100 observations. The proportion is $81/100 = 0.81$. This is the much greater than the 0.68 from the Empirical Rule.

The interval $\bar{x} \pm 2s \Rightarrow 128.54 \pm 2(21.97) \Rightarrow 128.54 \pm 43.94 \Rightarrow (84.60, 172.48)$ contains 98 of the 100 observations. The proportion is $98/100 = 0.98$. This is greater than the 0.95 from the Empirical Rule.

The interval $\bar{x} \pm 3s \Rightarrow 128.54 \pm 3(21.97) \Rightarrow 128.54 \pm 65.91 \Rightarrow (62.63, 194.45)$ contains 99 of the 100 observations. The proportion is $99/100 = 0.99$. This is close to the 0.997 from the Empirical Rule.
Based on the Empirical Rule, it appears that the data may not be normally distributed.

The interquartile range is $\text{IQR} = 38.37$ and $s = 21.97$. $\text{IQR}/s = 38.37/21.97 = 1.746$. This is much greater than the 1.3 that we would expect if the data were normal. Thus, there is evidence that the data may not be normally distributed.

Using MINITAB, the normal probability plot is:

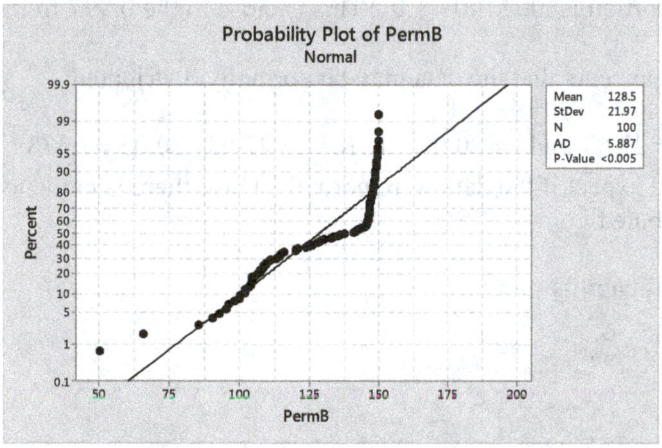

Because several points fall off the line, there is evidence that the data may not be normally distributed.

From all of the tests, there is evidence that the data are not approximately normally distributed.

Group C:
Using MINITAB, the histogram is:

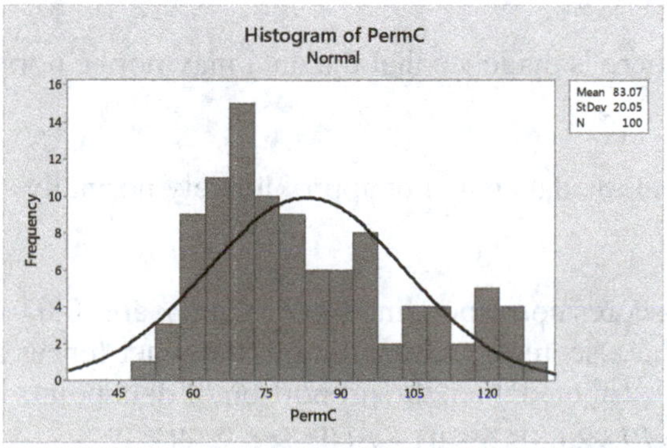

The data are not approximately mound-shaped. The data may not be normally distributed.

Using MINITAB, the descriptive statistics are:

Descriptive Statistics: PermC

Variable	N	Mean	StDev	Minimum	Q1	Median	Q3	Maximum	IQR
PermC	100	83.07	20.05	52.20	67.72	78.65	95.35	129.00	27.63

The interval $\bar{x} \pm s \Rightarrow 83.07 \pm 20.05 \Rightarrow (63.02, 103.12)$ contains 66 of the 100 observations. The proportion is $66/100 = 0.66$. This is close to the 0.68 from the Empirical Rule.

The interval $\bar{x} \pm 2s \Rightarrow 83.07 \pm 2(20.05) \Rightarrow 83.07 \pm 40.10 \Rightarrow (42.97, 123.17)$ contains 96 of the 100 observations. The proportion is $96/100 = 0.96$. This is close to the 0.95 from the Empirical Rule.

The interval $\bar{x} \pm 3s \Rightarrow 83.07 \pm 3(20.05) \Rightarrow 83.07 \pm 60.15 \Rightarrow (22.92, 143.22)$ contains 100 of the 100 observations. The proportion is $100/100 = 1.0$. This is close to the 0.997 from the Empirical Rule.
Based on the Empirical Rule, it appears that the data may be normally distributed.

The interquartile range is $\text{IQR} = 27.63$ and $s = 20.05$. $\text{IQR}/s = 27.63/20.05 = 1.378$. This is close to the 1.3 that we would expect if the data were normal. Thus, there is evidence that the data may be normally distributed.

Using MINITAB, the normal probability plot is:

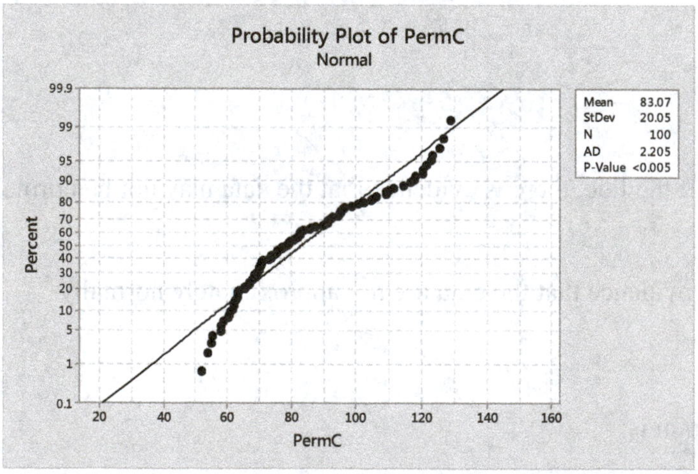

Because several points fall off the line, there is evidence that the data may not be normally distributed.

From all of the tests, there is evidence that the data may not approximately normally distributed.

5.75 From exercise 5.51, we found that the z-scores corresponding to Q_L and Q_U are -0.67 and 0.67, respectively. We also found that the z-scores corresponding to the inner fences are -2.68 and 2.68. If the data are normally distributed, then the proportion of data points between the inner fences is $P(-2.68 \le z \le 2.68) = P(-2.68 \le z \le 0) + P(0 \le z \le 2.68)$

$= 0.4963 + 0.4963 = 0.9926$ (using Table II, Appendix B). If the data are normally distributed, then it is almost impossible to get observations outside the inner fences. For genotype CC, there are 3 observations outside the inner fences. Therefore, it is very unlikely that the distribution of M for the genotype CC is normal. For genotype CT, there are 2 observations outside the inner fences. Therefore, it is very unlikely that the distribution of M for the genotype CT is normal. For genotype TT, there are no observations outside the inner fences, and the distribution looks fairly symmetric. Therefore, it is likely that the distribution of M for the genotype TT is normal.

5.77 When n is large, there often are not Binomial Tables for finding probabilities. If n is large, then finding the probabilities of events by using the formulas can be quite tedious and time consuming. The normal approximation gives very good estimates of the true probabilities.

5.79 a. In order to approximate the binomial distribution with the normal distribution, the interval $\mu \pm 3\sigma \Rightarrow np \pm 3\sqrt{npq}$ should lie in the range 0 to n.

When $n = 25$ and $p = 0.4$,

$$np \pm 3\sqrt{npq} \Rightarrow 25(0.4) \pm 3\sqrt{25(0.4)(1-0.4)}$$
$$\Rightarrow 10 \pm 3\sqrt{6} \Rightarrow 10 \pm 7.3485 \Rightarrow (2.6515, 17.3485)$$

Since the interval calculated does lie in the range 0 to 25, we can use the normal approximation.

b. $\mu = np = 25(0.4) = 10$ $\qquad \sigma^2 = npq = 25(0.4)(0.6) = 6$

c. $P(x \geq 9) = 1 - P(x \leq 8) = 1 - 0.274 = 0.726$ (using Table II, Appendix B)

d. $P(x \geq 9) \approx P\left(z \geq \frac{(9-0.5)-10}{\sqrt{6}}\right)$

$= P(z \geq -0.61) = 0.5000 + 0.2291 = 0.7291$

(Using Table II in Appendix B.)

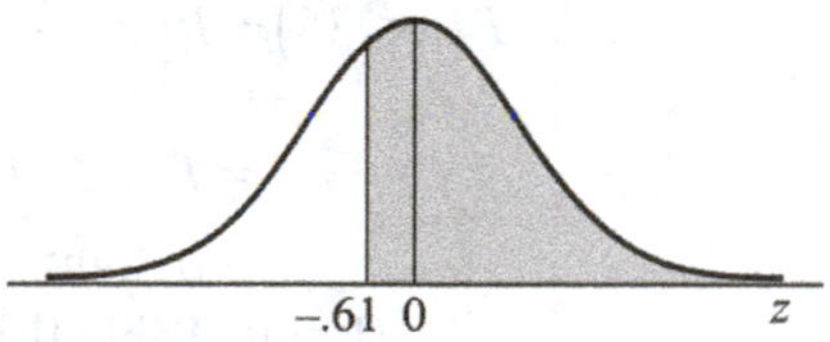

5.81 a. Using Table I, $P(x \leq 11) = 0.345$

$$\mu = np = 25(0.5) = 12.5, \quad \sigma = \sqrt{npq} = \sqrt{25(0.5)(0.5)} = 2.5$$

Using the normal approximation,

$$P(x \leq 11) \approx P\left(z \leq \frac{(11+0.5)-12.5}{2.5}\right) = P(z \leq -.40) = 0.5 - 0.1554 = 0.3446$$

(from Table II, Appendix B)

b. Using Table I, $P(x \geq 16) = 1 - P(x \leq 15) = 1 - 0.885 = 0.115$
Using the normal approximation,

$$P(x \geq 16) \approx P\left(z \geq \frac{(16-0.5)-12.5}{2.5}\right) = P(z \geq 1.2) = 0.5 - 0.3849 = 0.1151$$

(from Table II, Appendix B)

c. Using Table I, $P(8 \leq x \leq 16) = P(x \leq 16) - P(x \leq 7) = 0.946 - 0.022 = 0.924$
Using the normal approximation,

$$P(8 \leq x \leq 16) \approx P\left(\frac{(8-0.5)-12.5}{2.5} \leq z \leq \frac{(16+0.5)-12.5}{2.5}\right)$$
$$= P(-2.0 \leq z \leq 1.6) = 0.4772 + 0.4452 = 0.9224$$

(from Table II, Appendix B)

5.83 x is a binomial random variable with $n = 100$ and $p = 0.4$.

$$\mu \pm 3\sigma \Rightarrow np \pm 3\sqrt{npq} \Rightarrow 100(0.4) \pm 3\sqrt{100(0.4)(1-0.4)}$$
$$\Rightarrow 40 \pm 3(4.8990) \Rightarrow (25.303, 54.697)$$

Since the interval lies in the range 0 to 100, we can use the normal approximation to approximate the probabilities.

a. $$P(x \leq 35) \approx P\left(z \leq \frac{(35+0.5)-40}{4.899}\right)$$
$$= P(z \leq -0.92)$$
$$= 0.5000 - 0.3212 = 0.1788$$
(Using Table II in Appendix B.)

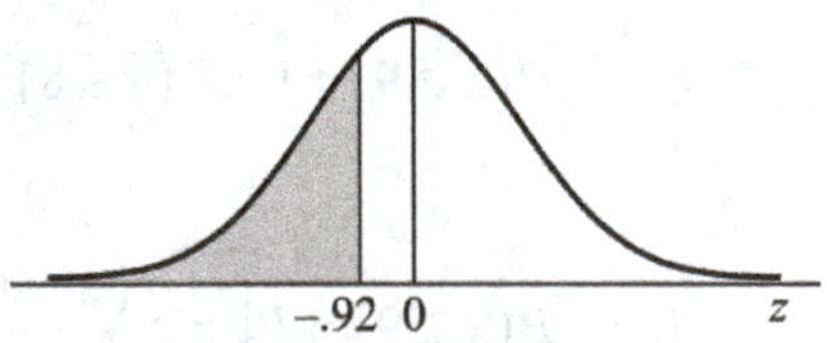

b. $P(40 \leq x \leq 50)$

$$\approx P\left(\frac{(40-0.5)-40}{4.899} \leq z \leq \frac{(50+0.5)-40}{4.899}\right)$$
$$= P(-0.10 \leq z \leq 2.14)$$
$$= P0(-.10 \leq z \leq 0) + P(0 \leq z \leq 2.14)$$
$$= 0.0398 + 0.4838 = 0.5236$$
(Using Table II in Appendix B.)

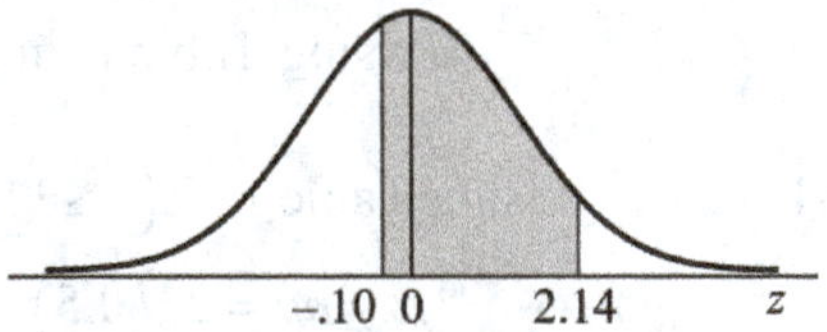

c. $P(x \geq 38) \approx P\left(z \geq \frac{(38-0.5)-40}{4.899}\right)$

$= P(z \geq -0.51)$

$= 0.5000 + 0.1950 = 0.6950$

(Using Table II in Appendix B.)

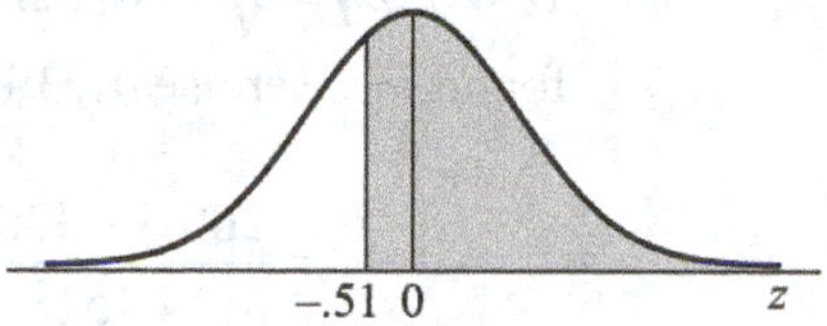

5.85 a. We know that x is a binomial random variable with $n = 1,000$ and $p = 0.35$. The mean or expected value of x is $\mu = np = 1,000(0.35) = 350$.

b. The standard deviation of x is $\sigma^2 = npq = 1,000(0.35)(0.65) = 227.5$.

c. $P(x < 325) = P\left(z < \frac{(325-0.5)-350}{\sqrt{227.5}}\right) = P(z < -1.69) = 0.5 - 0.4545 = 0.0455$

(Using Table II, Appendix B.)

d. $P(x \geq 390) = P\left(z \geq \frac{(390-0.5)-350}{\sqrt{227.5}}\right) = P(z \geq 2.62) = 0.5 - 0.4956 = 0.0044$

(Using Table II, Appendix B.)
Because the probability of observing 390 or more U.S. adults in a sample of 1,000 who do not work during summer vacation is so small, we would not expect to see this.

5.87 a. Let x = number of bottles that contain tap water in 65 trials. Then x is a binomial random variable with $n = 65$ and $p = 0.25$.

$E(x) = \mu = np = 65(0.25) = 16.25$

b. $\sigma = \sqrt{npq} = \sqrt{65(0.25)(0.75)} = \sqrt{12.1875} = 3.4911$

c. $z = \frac{x-\mu}{\sigma} = \frac{20-16.25}{3.4911} = 1.07$

d. $\mu \pm 3\sigma \Rightarrow 16.25 \pm 3(3.4911) \Rightarrow 16.25 \pm 10.4733 \Rightarrow (5.7767,\ \ 26.7233)$ Since the interval lies in the range 0 to 65, we can use the normal approximation to approximate the binomial probability.

$P(x \geq 20) = P\left(z \geq \frac{(20-0.5)-16.25}{3.4911}\right) = P(z \geq 0.93) = 0.5 - 0.3238 = 0.1762$

(Using Table II, Appendix B)

5.89 a. $\mu = E(x) = np = 1000(0.32) = 320$. This is the same value that was found in Exercise 4.73 a.

b. $\sigma=\sqrt{npq}=\sqrt{1000(0.32)(1-0.32)}=\sqrt{217.6}=14.751$ This is the same value that was found in Exercise 4.73 b.

c. $z=\dfrac{x-\mu}{\sigma}=\dfrac{200.5-320}{14.751}=-8.10$

d. $\mu\pm3\sigma\Rightarrow320\pm3(14.751)\Rightarrow320\pm44.253\Rightarrow(275.474,\ 364.253)$

Since the above is completely contained in the interval 0 to 1000, the normal approximation is valid.

$$P(x\le200)\approx P\left(z\le\frac{(200+0.5)-320}{14.751}\right)=P(z\le-8.10)\approx0.5-0.5=0$$

(Using Table II, Appendix B)

5.91 Let x = number of guppies that survive in 300 trials. Then x is a binomial random variable with $n=300$ and $p=0.6$.

$\mu\pm3\sigma\Rightarrow np\pm3\sqrt{npq}\Rightarrow300(0.6)\pm3\sqrt{300(0.6)(1-0.6)}\Rightarrow180\pm3\sqrt{72}\Rightarrow180\pm25.456$
$\Rightarrow(154.544,\ \ 205.456)$

Since the interval lies in the range 0 to 300, we can use the normal approximation to approximate the binomial probability.

$$P(x<100)=P\left(z<\frac{(100-0.5)-180}{\sqrt{72}}\right)=P(z<-9.49)\approx0.5-0.5=0$$

(Using Table II, Appendix B)

5.93 Let x = number of guests who participate in the conservation program. Then x is a binomial random variable with $n=200$ and $p=0.45$.

$\mu=np=200(0.45)=90$ and $\sigma=\sqrt{npq}=\sqrt{200(0.45)(0.55)}=\sqrt{49.5}=7.0356$

$\mu\pm3\sigma\Rightarrow90\pm3(7.0356)\Rightarrow90\pm21.1068\Rightarrow(68.8932,\ 111.1068)$

Since the above is completely contained in the interval 0 to 200, the normal approximation is valid.

$$P(x>110)=P\left(z>\frac{(110+0.5)-90}{7.0356}\right)=P(z>2.91)=0.5-0.4982=0.0018$$

Since this probability is so small, it is very unlikely that the director's claim is correct.

5.95 Let x = number of women who change their last names upon marriage in 500 trials. Then x is a binomial random variable with $n=500$ and $p=0.9$.

$\mu = np = 500(0.9) = 450$ and $\sigma = \sqrt{npq} = \sqrt{500(0.9)(0.1)} = \sqrt{45} = 6.7082$

$\mu \pm 3\sigma \Rightarrow 450 \pm 3(6.7082) \Rightarrow 450 \pm 20.1246 \Rightarrow (429.8754,\ 470.1246)$

Since the above is completely contained in the interval 0 to 500, the normal approximation is valid.

$$P(x < 400) = P\left(z < \frac{(400 - 0.5) - 450}{6.7082}\right) = P(z < -7.53) \approx 0.5 - 0.5 = 0.0$$

(Using Table II, Appendix B)

Because this probability is so small, we would infer that the probability of a European female changing her name upon marriage is not 0.9, but something smaller than 0.9.

5.97 The amount of time or distance between occurrences of random events can be described by the exponential probability distribution. This distribution has the property that the mean is equal to the standard deviation.

5.99 a. If $\theta = 1$, $a = 1$, then $e^{-a/\theta} = e^{-1/1} = e^{-1} = 0.367879$

b. If $\theta = 1$, $a = 2.5$, then $e^{-a/\theta} = e^{-2.5/1} = e^{-2.5} = 0.082085$

c. If $\theta = 0.4$, $a = 3$, then $e^{-a/\theta} = e^{-3/0.4} = e^{-7.5} = 0.000553$

d. If $\theta = 0.2$, $a = 0.3$, then $e^{-a/\theta} = e^{-0.3/0.2} = e^{-1.5} = 0.223130$

5.101 a. $P(x \le 4) = 1 - P(x > 4) = 1 - e^{-4/2.5} = 1 - e^{-1.6} = 1 - 0.201897 = 0.798103$

b. $P(x > 5) = e^{-5/2.5} = e^{-2} = 0.135335$

c. $P(x \le 2) = 1 - P(x > 2) = 1 - e^{-2/2.5} = 1 - e^{-0.8} = 1 - 0.449329 = 0.550671$

d. $P(x > 3) = e^{-3/2.5} = e^{-1.2} = 0.301194$

5.103 $\mu = \theta = 1,\ \sigma = \theta = 1$

$\mu \pm 2\sigma \Rightarrow 1 \pm 2(1) \Rightarrow 1 \pm 2 \Rightarrow (-1,\ 3)$

$P(-1 < x < 3) = P(0 < x < 3) = 1 - P(x \ge 3) = 1 - e^{-3/1} = 1 - e^{-1} = 1 - 0.049787 = 0.950213$

5.105 a. $P(x > 2) = e^{-2/2.5} = e^{-0.8} = 0.449329$

b. $P(x < 5) = 1 - P(x \ge 5) = 1 - e^{-5/2.5} = 1 - e^{-2} = 1 - 0.135335 = 0.864665$

5.107 a. Let x = time between successive email notifications. Then x has an exponential distribution with $\mu = \theta = 95$. Two minutes is equal to 120 seconds.

$P(x \geq 120) = e^{-120/95} = e^{-1.263} = 0.282760$

b. Using MINITAB, the descriptive statistics are:

Descriptive Statistics: INTTIME

Variable	N	Mean	StDev	Minimum	Q1	Median	Q3	Maximum
INTTIME	267	95.52	91.54	1.86	30.59	70.88	133.34	513.52

The sample mean is 95.52 which is very close to 95. In addition, the standard deviation is 91.54, which is also close to 95, which is the standard deviation of the exponential distribution.

Using MINITAB, a histogram of the data with an exponential distribution with mean 95superimposed on the graph is:

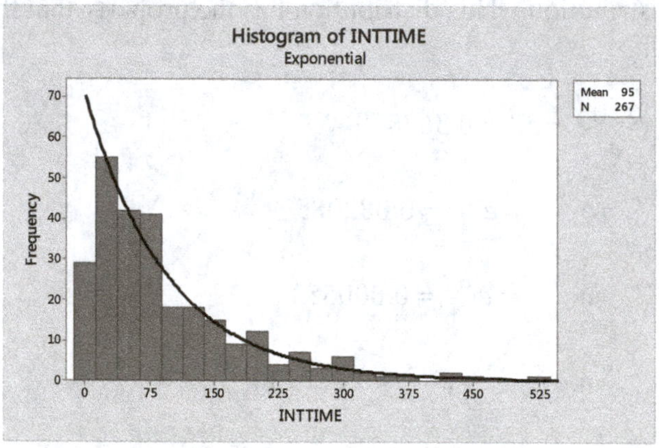

Based on the sample mean and standard deviation along with the graph, it appears that the data follow an exponential distribution with $\theta = 95$.

5.109 a. Given $f(x) = 0.04e^{-0.04x}$. Therefore, $\frac{1}{\theta} = 0.04 \Rightarrow \theta = 25$.

$E(x) = \mu = \theta = 25$.

b. $\sigma^2 = \theta^2 = 25^2 = 625$

c. By the Empirical Rule, $P(\mu - 2\sigma < x < \mu + 2\sigma) \approx 0.95$.

d. $\sigma = \theta = 25$. Therefore, $\mu - 2\sigma = 25 - 2(25) = -25$ and $\mu + 2\sigma = 25 + 2(25) = 75$.
The actual probability is
$P(\mu - 2\sigma < x < \mu + 2\sigma) = P(-25 < x < 75) = P(x > -25) - P(x > 75) = P(x > 0) - P(x > 75)$
$= e^{-0.04(0)} - e^{-0.04(75)} = 1 - 0.0498 = 0.9502$

This probability is much greater than what we would expect from the Empirical Rule.

e. By definition, $F(x) = P(X \leq x)$. Thus, $R(t) = 1 - F(t) = 1 - P(x \leq t)$.

$R(5)=1-F(5)=1-P(x\leq 5)=P(x>5)=e^{-0.04(5)}=0.8187$.

In words, $R(t)$ is the probability that the time between major earthquakes will be more than t years. Thus, the probability that the time between major earthquakes will be greater than 5 years is 0.8187.

5.111 a. For this exercise, $\mu=\theta=17$. To graph the distribution, we will pick several values of x and find the value of $f(x)$, where x = time between arrivals of the smaller craft at the pier.

$$f(x)=\frac{1}{\theta}e^{-x/\theta}=\frac{1}{17}e^{-x/17}$$

$$f(1)=\frac{1}{17}e^{-1/17}=0.0555 \qquad f(3)=\frac{1}{17}e^{-3/17}=0.0493$$

$$f(5)=\frac{1}{17}e^{-5/17}=0.0438 \qquad f(7)=\frac{1}{17}e^{-7/17}=0.0390$$

$$f(10)=\frac{1}{17}e^{-10/17}=0.0327 \qquad f(15)=\frac{1}{17}e^{-15/17}=0.0243$$

$$f(20)=\frac{1}{17}e^{-20/17}=0.0181 \qquad f(25)=\frac{1}{17}e^{-25/17}=0.0135$$

The graph is:

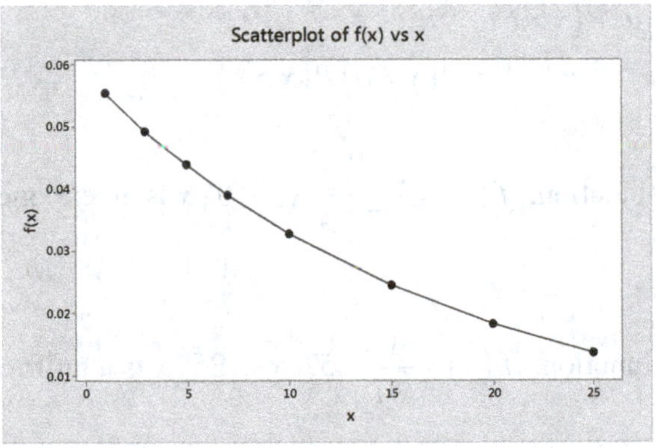

b. We want to find the probability that the time between arrivals is less than 15 minutes.

$$P(x<15)=1-P(x\geq 15)=1-e^{-15/17}=1-0.4138=0.5862$$

5.113 a. For $\theta=250$, $P(x>a)=e^{-a/250}$

For $a=300$ and $b=200$, show $P(x>a+b)\geq P(x>a)P(x>b)$

$$P(x>300+200)=P(x>500)=e^{-500/250}=e^{-2}=0.135335$$

$$P(x>300)P(x>200)=e^{-300/250}e^{-200/250}=e^{-1.2}e^{-0.8}=0.301194(0.449329)=0.135335$$

Since $P(x>300+200)=P(x>300)P(x>200)$, then
$P(x>300+200)\geq P(x>300)P(x>200)$

Also, show $P(x>300+200)\leq P(x>300)P(x>200)$. Since we already showed that
$P(x>300+200)=P(x>300)P(x>200)$,
then $P(x>300+200)\leq P(x>300)P(x>200)$.

b. Let $a=50$ and $b=100$. Show $P(x>a+b)\geq P(x>a)P(x>b)$

$$P(x>50+100)=P(x>150)=e^{-150/250}=e^{-0.6}=0.548812$$

$$P(x>50)P(x>100)=e^{-50/250}e^{-100/250}=e^{-0.2}e^{-0.4}=0.818731(0.670320)=0.548812$$

Since $P(x>50+150)=P(x>50)P(x>100)$, then
$P(x>50+150)\geq P(x>50)P(x>100)$

Also, show $P(x>a+b)\leq P(x>a)P(x>b)$. Since we already showed that
$P(x>50+150)=P(x>50)P(x>100)$, then $P(x>50+150)\leq P(x>50)P(x>100)$.

c. Show $P(x>a+b)\geq P(x>a)P(x>b)$
$P(x>a+b)=e^{-(a+b)/250}=e^{-a/250}e^{-b/250}=P(x>a)P(x>b)$

5.115 a. For the probability density function, $f(x)=\frac{e^{-x/7}}{7}$, $x>0$, x is an exponential random variable.

b. For the probability density function, $f(x)=\frac{1}{20}$, $5<x<25$, x is a uniform random variable.

c. For the probability function, $f(x)=\frac{e^{-0.5[(x-10)/5]^2}}{5\sqrt{2\pi}}$, x is a normal random variable.

5.117 a. $P(z\leq 2.1)=A_1+A_2=P(z<0)+P(0\leq z\leq 2.1)$
$=0.5+0.4821=0.9821$

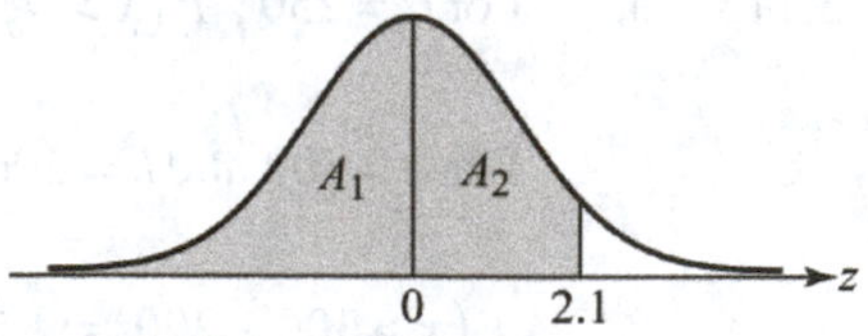

b. $P(z\geq 2.1)=A_2=0.5-A_1=0.5-P(0\leq z\leq 2.1)$
$=0.5-0.4821=0.0179$

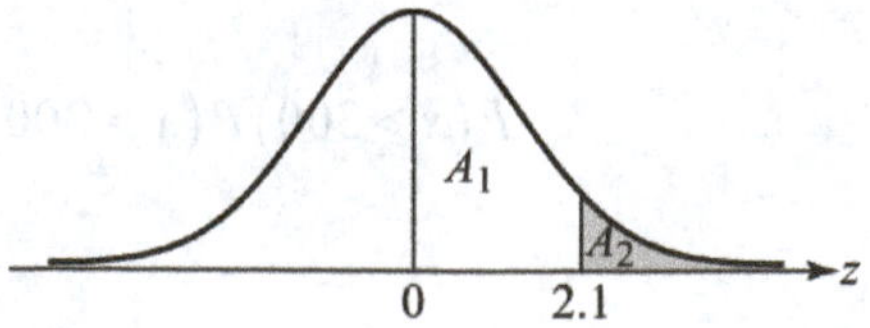

c. $P(z \geq -1.65) = A_1 + A_2 = P(-1.65 \leq z \leq 0) + 0.5$
$= 0.4505 + 0.5 = 0.9505$

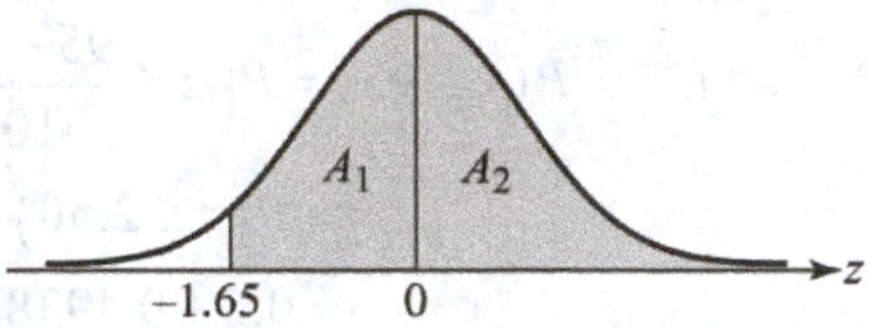

d. $P(-2.13 \leq z \leq -0.41)$
$= P(-2.13 \leq z \leq 0) - P(-0.41 \leq z \leq 0)$
$= 0.4834 - 0.1591 = 0.3243$

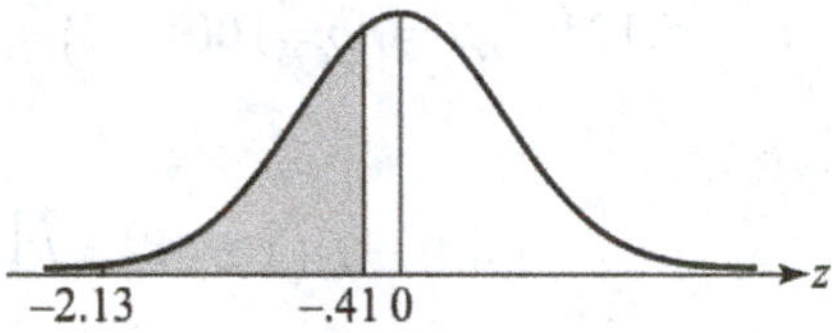

e. $P(-1.45 \leq z \leq 2.15) = A_1 + A_2$
$= P(-1.45 \leq z \leq 0) - P(0 \leq z \leq 2.15)$
$= 0.4265 + 0.4842 = 0.9107$

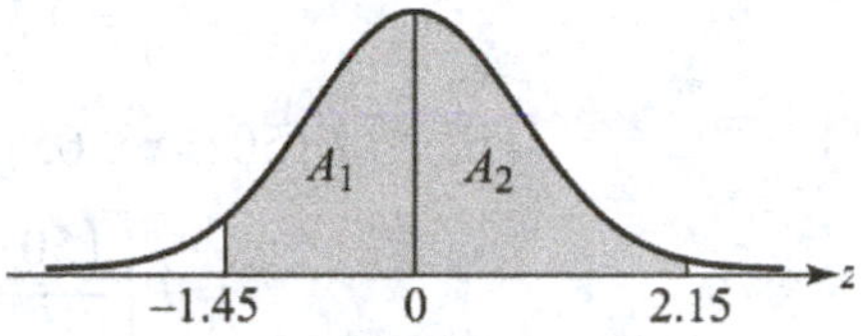

f. $P(z \leq -1.43) = A_1 = 0.5 - A_2$
$= 0.5 - P(-1.43 \leq z \leq 0)$
$= 0.5 - 0.4236 = 0.0764$

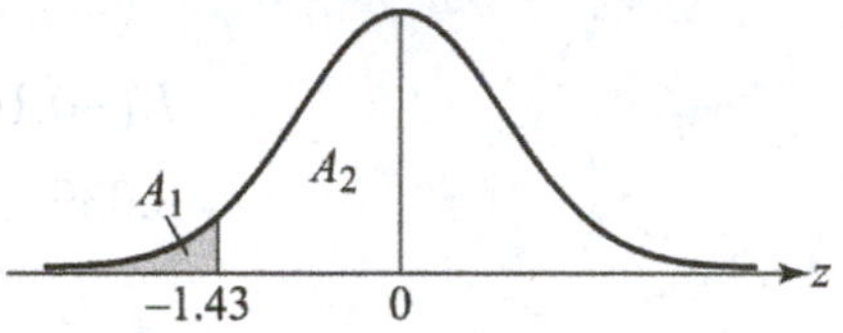

5.119 a. $P(x \leq 75) = P\left(z \leq \frac{75-70}{10}\right)$
$= P(z \leq 0.50)$
$= 0.5 + 0.1915 = 0.6915$

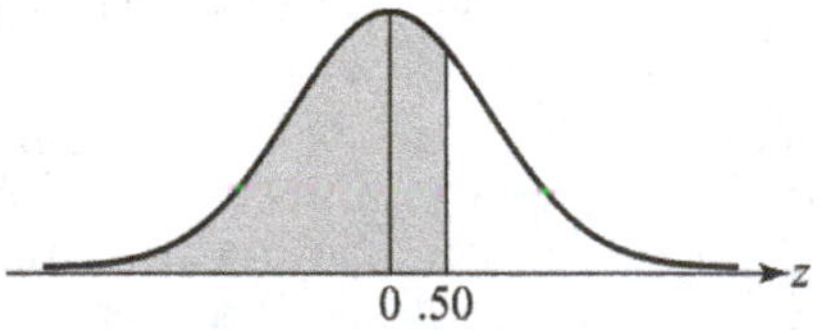

b. $P(x \geq 90) = P\left(z \geq \frac{90-70}{10}\right)$
$= P(z \geq 2.00)$
$= 0.5 - 0.4772 = 0.0228$

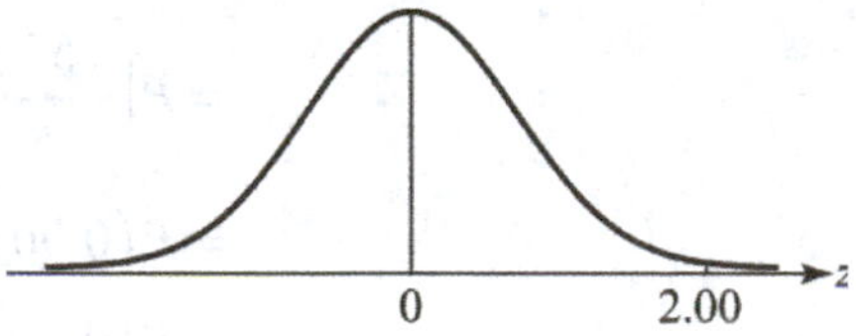

c. $P(60 \leq x \leq 75) = P\left(\frac{60-70}{10} \leq z \leq \frac{75-70}{10}\right)$
$= P(-1.00 \leq z \leq 0.50)$
$= 0.3413 + 0.1915 = 0.5328$

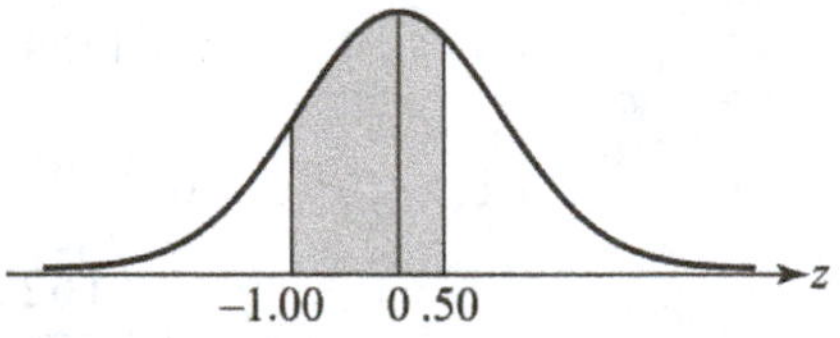

d. This is the probability of the complement of the event in part **a**. Therefore:
$P(x > 75) = 1 - P(x \leq 75) = 1 - 0.6915 = 0.3085$

e. $P(x = 75) = 0$ (True for any continuous random variable.)

f. $P(x \le 95) = P\left(z \le \frac{95-70}{10}\right)$

$= P(z \le 2.50)$

$= 0.5 + 0.4938 = 0.9938$

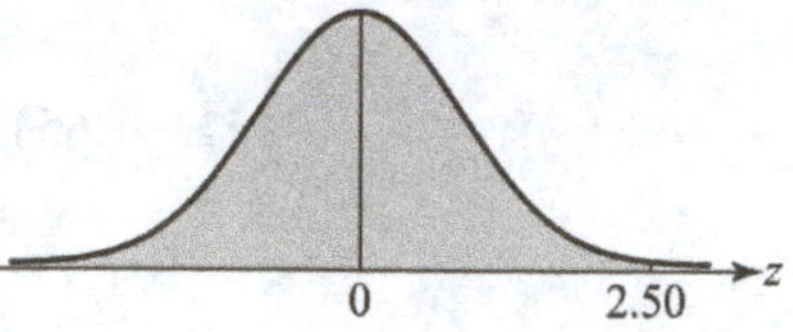

5.121 $\mu = np = 100(0.5) = 50$, $\sigma = \sqrt{npq} = \sqrt{100(0.5)(0.5)} = 5$

a. $P(x \le 48) = P\left(z \le \frac{(48+0.5)-50}{5}\right)$

$= P(z \le -0.30)$

$= 0.5 - 0.1179 = 0.3821$

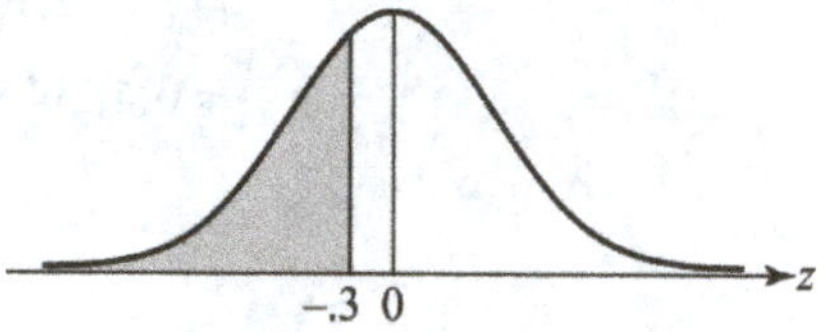

b. $P(50 \le x \le 65)$

$= P\left(\frac{(50-0.5)-50}{5} \le z \le \frac{(65+0.5)-50}{5}\right)$

$= P(-0.10 \le z \le 3.10)$

$= 0.0398 + 0.5 = 0.5398$

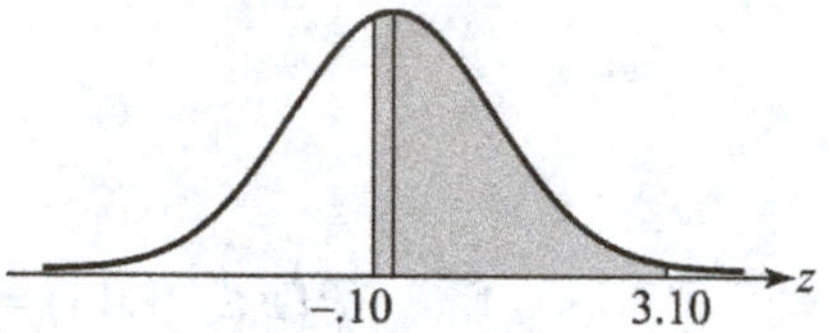

c. $P(x \ge 70) = P\left(z \ge \frac{(70-0.5)-50}{5}\right)$

$= P(z \ge 3.90)$

$= 0.5 - 0.5 = 0$

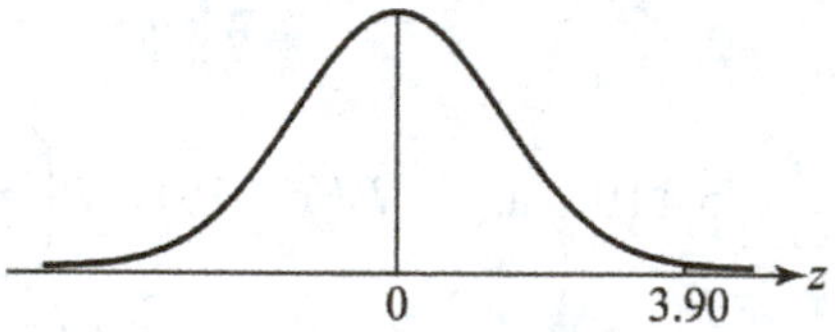

d. $P(55 \le x \le 58)$

$= P\left(\frac{(55-0.5)-50}{5} \le z \le \frac{(58+0.5)-50}{5}\right)$

$= P(0.90 \le z \le 1.70)$

$= P(0 \le z \le 1.70) - P(0 \le z \le 0.90)$

$= 0.4554 - 0.3159 = 0.1395$

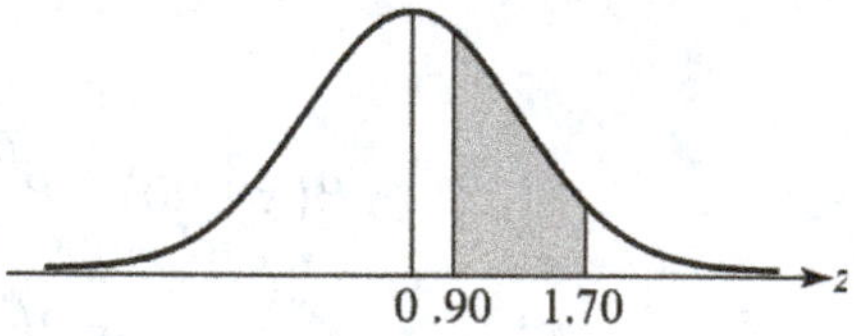

e. $P(x = 62)$

$= P\left(\frac{(62-0.5)-50}{5} \le z \le \frac{(62+0.5)-50}{5}\right)$

$= P(2.30 \le z \le 2.50)$

$= P(0 \le z \le 2.50) - P(0 \le z \le 2.30)$

$= 0.4938 - 0.4893 = 0.0045$

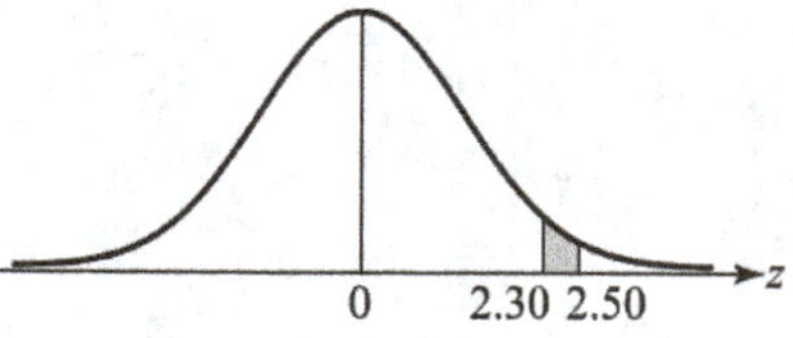

f. $P(x \le 49 \text{ or } x \ge 72)$

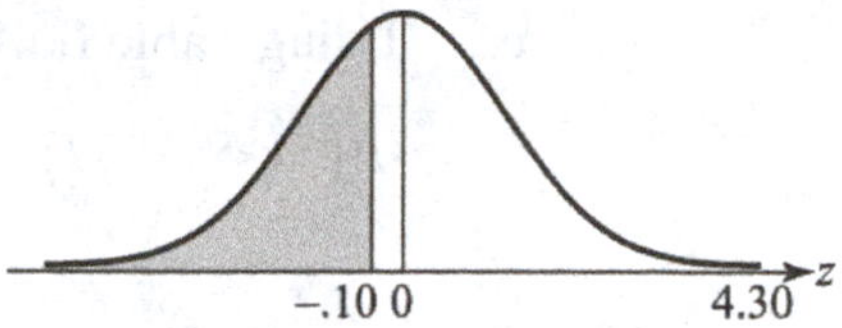

$= P\left(z \le \frac{(49+0.5)-50}{5}\right) + P\left(z \ge \frac{(72-0.5)-50}{5}\right)$

$= P(z \le -0.10) + P(z \ge 4.30)$

$= (0.5 - 0.0398) + (0.5 - 0.5) = 0.4602$

5.123 a. $P(x \le 1) = 1 - P(x > 1) = 1 - e^{-1/3} = 1 - 0.716531 = 0.283469$ (using calculator)

b. $P(x > 1) = e^{-1/3} = 0.716531$

c. $P(x = 1) = 0$ (*x* is a continuous random variable. There is no probability associated with a single point.)

d. $P(x \le 6) = 1 - P(x > 6) = 1 - e^{-6/3} = 1 - e^{-2} = 1 - 0.135335 = 0.864665$

e. $P(2 \le x \le 10) = P(x \ge 2) - P(x > 10) = e^{-2/3} - e^{-10/3}$
$= 0.513417 - 0.035674 = 0.477743$

5.125 Based only on the histogram provided, I would say that the data are probably not normally distributed. There are several observations that are very small compared to the rest of the data and some that are very large compared to the rest of the data.

5.127 Let x = score for first time students on mathematics achievement test. Then x is a normal random variable with $\mu = 77$ and $\sigma = 7.3$.

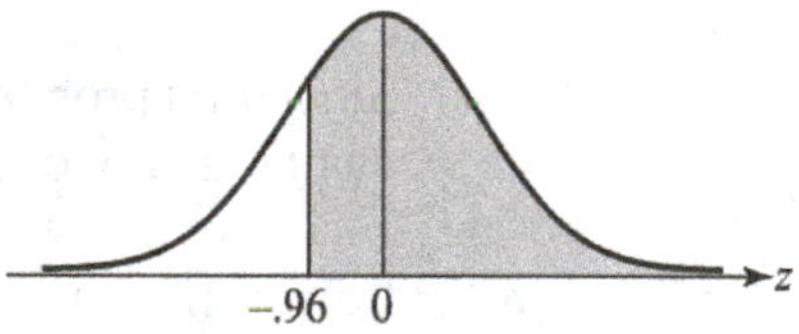

$P(x \ge 70) = P\left(z \ge \frac{70-77}{7.3}\right)$

$= P(z \ge -0.96)$

$= 0.3315 + 0.5 = 0.8315$

Thus, 83.15% of students will pass the test the first time.

5.129 a. Let x = alkalinity level of water specimens collected from the Han River.

Using Table II, Appendix B,

$P(x > 45) = P\left(z > \frac{45-50}{3.2}\right) = P(z > -1.56) = 0.5 + 0.4406 = 0.9406$.

b. Using Table II, Appendix B,

$P(x < 55) = P\left(z < \frac{55-50}{3.2}\right) = P(z < 1.56) = 0.5 + 0.4406 = 0.9406$.

c. Using Table II, Appendix B,

$$P(51<x<52)=P\left(\frac{51-50}{3.2}<z<\frac{52-50}{3.2}\right)=P(0.31<z<0.63)$$
$$=0.2357-0.1217=0.1140$$

5.131 a. $n=10{,}000$ and $p=0.15$

$$E(x)=\mu=np=10{,}000(0.15)=1{,}500$$

$$\sigma^2=npq=10{,}000(0.15)(0.85)=1{,}275$$

b. $\sigma=\sqrt{1{,}275}=35.707$

Using the normal approximation to the binomial,

$$P(x>1{,}600)\approx P\left(z>\frac{(1{,}600+0.5)-1{,}500}{35.707}\right)=P(z>2.81)=0.5-0.4975=0.0025$$
(from Table II, Appendix B)

c. Using the normal approximation to the binomial,

$$P(x>6{,}500)\approx P\left(z>\frac{(6{,}500+0.5)-1{,}500}{35.707}\right)=P(z>140.04)\approx 0.5-0.5=0$$
(from Table II, Appendix B)

Since this probability is essentially 0, we would not expect to see more than 6,500 deaths in any one year.

5.133 a. Let x = transmission delay of an RSVP linked wireless device. Using Table II, Appendix B,

$$P(x<57)=P\left(z\le\frac{57-48.5}{8.5}\right)=P(z\le 1.00)=0.5000+0.3413=0.8413$$

b. $$P(40<x<60)=P\left(\frac{40-48.5}{8.5}<z<\frac{60-48.5}{8.5}\right)=P(-1.00<z<1.35)$$
$$=0.3413+0.4115=0.7528$$

5.135 The IQR $=Q_U-Q_L=34-14=20$. IQR / $s=20/12.8=1.5625$. If the data are normally distributed, this ratio should be close to 1.3. Since 1.5625 is fairly close to 1.3, this indicates that the data are approximately normal.

5.137 a. Using Table II, Appendix B, with $\mu=9.06$ and $\sigma=2.11$,

$$P(x<6)=\left(z<\frac{6-9.06}{2.11}\right)=P(z<-1.45)=0.5-P(-1.45\le z\le 0)$$
$$=0.5-0.4265=0.0735$$

b. $P(8<x<10)=P\left(\frac{8-9.06}{2.11}<z<\frac{10-9.06}{2.11}\right)=P(-0.50<z<0.45)$

$$=P(-0.50<z<0)+P(0<z<0.45)=0.1915+0.1736=0.3651$$

c. $P(x<x_0)=0.2000$.

If $P(x<x_0)=0.2000$, then $P\left(z<\frac{x_0-9.06}{2.11}\right)=P(z<z_0)=0.2000$

If $P(z<z_0)=0.2000$, then $P(z_0<z<0)=0.3000$. Looking up 0.3000 in Table II, the *z*-score is 0.84. Since $z_0<0$, $z_0=-0.84$. Now, we must convert z_0 back to an *x* score.

$$z_0=\frac{x_0-9.06}{2.11}\Rightarrow -0.84=\frac{x_0-9.06}{2.11}\Rightarrow x_0=2.11(-0.84)+9.06=7.29$$

5.139 $c=0$, $d=4$

$$f(x)=\begin{cases}\frac{1}{d-c}=\frac{1}{4-0}=\frac{1}{4} & 0\le x\le 4\\ 0 & \text{otherwise}\end{cases}$$

$$\mu=\frac{c+d}{2}=\frac{0+4}{2}=2,\ \sigma=\frac{d-c}{\sqrt{12}}=\frac{4-0}{\sqrt{12}}=1.155$$

$$15 \text{ seconds}=\frac{15}{60}=0.25 \text{ minutes}$$

It takes 0.25 minutes to go from floor 2 to floor 1. Thus, we must find the probability that the waiting time is less than $1.5-0.25=1.25$.

$$P(x<1.25)=(1.25-0)\frac{1}{4}=1.25\left(\frac{1}{4}\right)=0.3125$$

5.141 Let x = velocity of a galaxy located within the galaxy cluster A2142. Then x is a normal random variable with mean $\mu=27{,}117$ km/s and standard deviation $\sigma=1{,}280$ km/s.

$$P(x\le 24{,}350)=P\left(z\le\frac{24{,}350-27{,}117}{1{,}280}\right)=P(z\le -2.16)=0.5-0.4846=0.0154$$

Since the probability of observing a galaxy from the galaxy cluster A2142 with a velocity of 24,350 km/s or slower is so small $(p=.0154)$, it would be very unlikely that the galaxy observed came from the galaxy cluster A2142.

5.143 a. Let x = number of abused women in a sample of 150. The random variable x is a binomial random variable with $n=150$ and $p=1/3$. Thus, for the normal approximation,

$$\mu=np=150(1/3)=50 \text{ and } \sigma=\sqrt{npq}=\sqrt{150(1/3)(2/3)}=5.7735$$

$$\mu \pm 3\sigma \Rightarrow 50 \pm 3(5.7735) \Rightarrow 50 \pm 17.3205 \Rightarrow (32.6795, 67.3205)$$

Since this interval lies in the range from 0 to $n = 150$, the normal approximation is appropriate.

$$P(x > 75) \approx P\left(z > \frac{(75+0.5)-50}{5.7735}\right) = P(z > 4.42) \approx 0.5 - 0.5 = 0$$

(Using Table II, Appendix B.)

b. $$P(x < 50) \approx P\left(z < \frac{(50-0.5)-50}{5.7735}\right) = P(z < -0.09) \approx 0.5 - 0.0359 = 0.4641$$

c. $$P(x < 30) \approx P\left(z < \frac{(30-0.5)-50}{5.7735}\right) = P(z < -3.55) \approx 0.5 - 0.5 = 0$$

Since the probability of seeing fewer than 30 abused women in a sample of 150 is so small $(p \approx 0)$, it would be very unlikely to see this event.

5.145 First, we will graph the data. Using MINITAB, a histogram of the data with a normal distribution superimposed on the graph is:

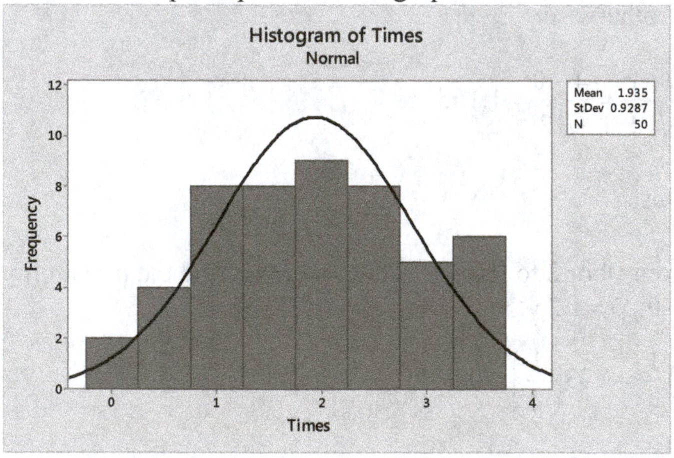

The data are somewhat mound shaped. The data could be normally distributed.

The descriptive statistics of the data are:

Descriptive Statistics: Times

Variable	N	Mean	StDev	Minimum	Q1	Median	Q3	Maximum
Times	50	1.935	0.929	0.0100	1.218	1.835	2.645	3.500

We know if the data are normally distributed, then about 68% of the observations will fall within 1 standard deviations of the mean, about 95% of the observations will fall within 2 standard deviations of the mean, and about all of the observations will fall within 3 standard deviations of the mean.

$\bar{x} \pm s \Rightarrow 1.935 \pm 0.929 \Rightarrow (1.006, 2.864)$. 33 of the 50 or $33/50 = 0.66$ of the observations are within 1 standard deviation of the mean. This is close to the 0.68 if the data are normal.

$\bar{x} \pm 2s \Rightarrow 1.935 \pm 2(0.929) \Rightarrow 1.935 \pm 1.858 \Rightarrow (0.077, 3.793)$. 49 of the 50 or $49/50 = 0.98$ of the observations are within 2 standard deviations of the mean. This is higher than the 0.95 if the data are normal.

$\bar{x} \pm 3s \Rightarrow 1.935 \pm 3(0.929) \Rightarrow 1.935 \pm 2.787 \Rightarrow (-0.852, 4.722)$. 50 of the 50 or $50/50 = 1.00$ of the observations are within 3 standard deviations of the mean. This is equal to the 1.00 if the data are normal.

The data do not appear to be exactly normal.

$\text{IQR} = 3.5 - 1.218 = 2.282$ and $s = 0.929$. $\frac{IQR}{s} = \frac{2.282}{0.929} = 2.456$. This is much larger than the 1.3 expected is the data are normal. This indicates that the data are probably not normal. Finally, the probability plot of the data using MINITAB is:

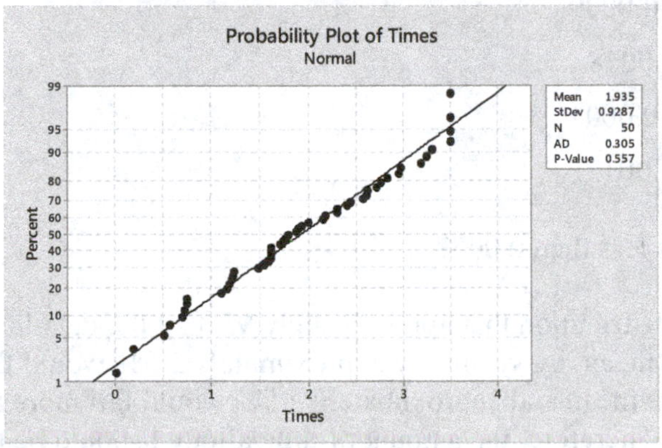

If the data are normal, then the normal probability plot should be a straight line. On the right side of the graph, the plot of the data is almost a vertical line. There is some indication that the data are not normally distributed.

Thus, it appears that the data may not be normal.

5.147 a. Using Table II, Appendix B, with $\mu = 50$ and $\sigma = 12$,

$$P(x > d) = 0.30 \Rightarrow P\left(z > \frac{d-50}{12}\right) = P(z > z_0) = 0.30$$

$A_1 = 0.5 - 0.30 = 0.2000$
Looking up the area 0.2000 in Table II gives $z_0 = 0.52$.

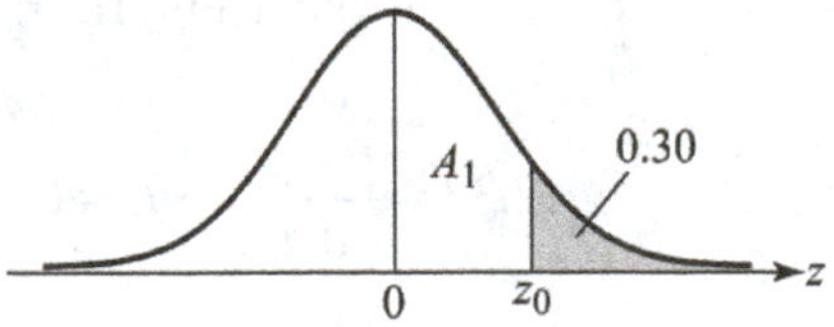

$$z_0 = 0.52 = \frac{d-50}{12} \Rightarrow d = 12(0.52) + 50 = 56.24 \text{ cm}$$

b. Let x = breast height diameter of the western hemlock. Then x has an exponential distribution with a mean of 30 cm.

$$P(x > 25) = e^{-25/30} = e^{-0.833333} = 0.434598$$

5.149 a. (i) $P(0<z<1.2)\approx z(4.4-z)/10=1.2(4.4-1.2)/10=0.384$

(ii) $P(0<z<2.5)\approx 0.49$

(iii) $P(z>0.8)=0.5-P(0<z<0.8)\approx 0.5-0.8(4.4-0.8)/10=0.5-0.288=0.212$

(iv) $P(z<1.0)=0.5+P(0<z<1.0)=0.5+1(4.4-1)/10=0.5+0.34=0.84$

b. Using Table II, Appendix B:

(i) $P(0<z<1.2)=0.3849$

(ii) $P(0<z<2.5)=0.4938$

(iii) $P(z>0.8)=0.5-P(0<z<0.8)=0.5-0.2881=0.2119$

(iv) $P(z<1.0)=0.5+P(0<z<1.0)=0.5+0.3413=0.8413$

c. For each part, the absolute error is:

(i) Error $=|0.384-0.3849|=0.0009$

(ii) Error $=|0.49-0.4938|=0.0038$

(iii) Error $=|0.212-0.2119|=0.0001$

(iv) Error $=|0.84-0.8413|=0.0013$

In all cases, the absolute error is less than 0.0052.

5.151 Since we know for a mound-shaped distribution that approximately 95% of the data lies within two standard deviations of the mean, by symmetry approximately 2.5% would fall below 2 standard deviations below the mean and approximately 2.5% would fall more than 2 standard deviations above the mean. Therefore, the amount of time allotted should be two standard deviations above the mean. This is:

$\mu+2\sigma=40+2(6)=52$ minutes

5.153 From Table II, Appendix B, with $\sigma=.4$:

$P(x>6)=0.01$

$$P(x>6)=P\left(z>\frac{6-\mu}{0.4}\right)=P(z>z_0)=0.01$$

$A_1=0.5-0.01=0.4900$

From Table II, $z_0=2.33$

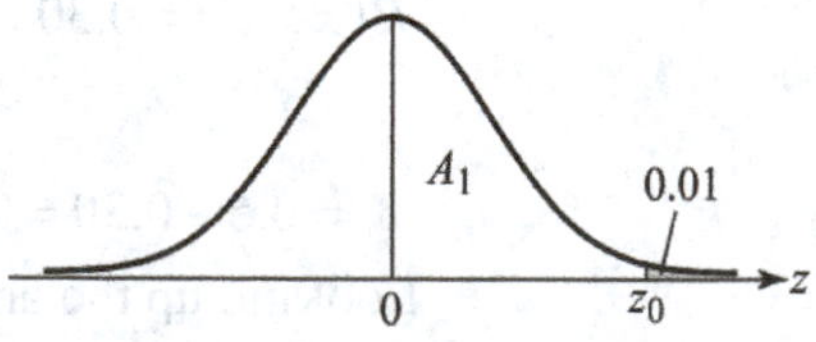

$$2.33=\frac{6-\mu}{0.4}\Rightarrow\mu=6-2.33(0.4)=5.068$$

5.155 Using Table II, Appendix B, with $\mu=.27$ and $\sigma=.04$,

$$P(x\le 0.14)=P\left(z\le\frac{0.14-0.27}{0.04}\right)=P(z<-3.25)\approx 0.5-0.5=0.0000$$

Since the probability of observing a reading of 0.14 or smaller is so small ($p \approx .0000$) if the pressure is 0.1 MPa, it would be extremely unlikely that this reading was obtained at a pressure of 0.1 MPa.

5.157 a. Class I (very bright) consists of those with IQs above the 95th percentile. Thus, there would be 0.05 or 5% of the people in Class I.

Class II (bright) consists of those with IQs between the 75th and 95th percentile. Thus, there would be $0.95 - 0.75 = 0.20$ or 20% of the people in Class II.

Class III (normal) consists of those with IQs between the 25th and 75th percentile. Thus, there would be $0.75 - 0.25 = 0.50$ or 50% of the people in Class III.
Class IV (dull) consists of those with IQs between the 5th and 25th percentile. Thus, there would be $0.25 - 0.05 = 0.20$ or 20% of the people in Class IV.

Class V (very dull) consists of those with IQs below the 5th percentile. Thus, there would be 0.05 or 5% of the people in Class V.

b. Suppose we pick a pair of percentiles and compute the corresponding *z*-scores. Suppose we compute the *z*-scores for the 50th and 55th percentiles.

The *z*-score corresponding to the 50th percentile is $z = 0$. The *z*-score corresponding to the 55th percentile is $P(z < z_0) = 0.55 \Rightarrow z_0 = 0.13$.

Now, suppose we pick another pair of percentiles that are also 5 points apart. Suppose we compute the *z*-scores for the 94th and 99th percentiles.

The *z*-score corresponding to the 94th percentile is $P(z < z_0) = 0.94 \Rightarrow z_0 = 1.555$

The *z*-score corresponding to the 99th percentile is $P(z < z_0) = 0.99 \Rightarrow z_0 = 2.33$

The *z*-scores corresponding to the 50th and 55th percentiles are 0 and 0.13. The difference is $0.13 - 0 = 0.13$. The *z*-scores corresponding to the 94th and 99th percentiles are 1.555 and 2.33. The difference is $2.33 - 1.555 = 0.775$. Even though the differences in the percentiles are the same, the differences in the *z*-scores are much different. The information in the *z*-scores is more informative.

c. If the distribution of the IQs is skewed to the right, then the tail to the right is much longer than the tail to the left. If the distribution of the IQs is skewed to the left, then the tail to the left is much longer than the tail to the right. However, the proportions in the 5 cognitive classes would not differ, regardless of the shape of the distribution.

Class I (very bright) consists of those with IQs above the 95th percentile. Thus, there would be 0.05 or 5% of the people in Class I.

Class II (bright) consists of those with IQs between the 75th and 95th percentile. Thus, there would be $0.95 - 0.75 = 0.20$ or 20% of the people in Class II.

Class III (normal) consists of those with IQs between the 25th and 75th percentile. Thus, there would be $0.75 - 0.25 = 0.50$ or 50% of the people in Class III.

Class IV (dull) consists of those with IQs between the 5th and 25th percentile. Thus, there would be $0.25 - 0.05 = 0.20$ or 20% of the people in Class IV.

Class V (very dull) consists of those with IQs below the 5th percentile. Thus, there would be 0.05 or 5% of the people in Class V.

Chapter 6

Sampling Distributions

6.1 A population parameter is a numerical descriptive measure of a population. Because it is based on the observations in a population, its value is almost always unknown.

A sample statistic is a numerical descriptive measure of a sample. It is calculated from the observations in the sample.

6.3 a–b. The different samples of $n = 2$ with replacement and their means are:

Possible Samples	$\bar{x}$	Possible Samples	$\bar{x}$
0, 0	0	4, 0	2
0, 2	1	4, 2	3
0, 4	2	4, 4	4
0, 6	3	4, 6	5
2, 0	1	6, 0	3
2, 2	2	6, 2	4
2, 4	3	6, 4	5
2, 6	4	6, 6	6

c. Since each sample is equally likely, the probability of any 1 being selected is $\frac{1}{4}\left(\frac{1}{4}\right) = \frac{1}{16}$

d.
$$P(\bar{x}=0)=\frac{1}{16}$$
$$P(\bar{x}=1)=\frac{1}{16}+\frac{1}{16}=\frac{2}{16}$$
$$P(\bar{x}=2)=\frac{1}{16}+\frac{1}{16}+\frac{1}{16}=\frac{3}{16}$$
$$P(\bar{x}=3)=\frac{1}{16}+\frac{1}{16}+\frac{1}{16}+\frac{1}{16}=\frac{4}{16}$$
$$P(\bar{x}=4)=\frac{1}{16}+\frac{1}{16}+\frac{1}{16}=\frac{3}{16}$$
$$P(\bar{x}=5)=\frac{1}{16}+\frac{1}{16}=\frac{2}{16}$$
$$P(\bar{x}=6)=\frac{1}{16}$$

$\bar{x}$	$p(\bar{x})$
0	1/16
1	2/16
2	3/16
3	4/16
4	3/16
5	2/16
6	1/16

e.

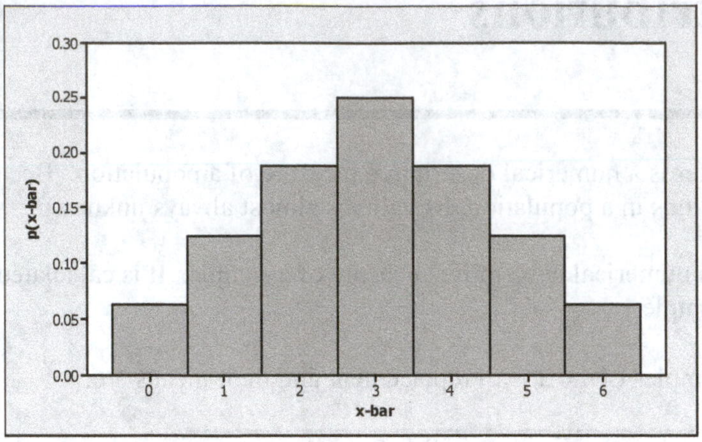

6.5 If the observations are independent of each other, then

$$P(1,1) = p(1)p(1) = 0.2(0.2) = 0.04$$
$$P(1,2) = p(1)p(2) = 0.2(0.3) = 0.06$$
$$P(1,3) = p(1)p(3) = 0.2(0.2) = 0.04$$

etc.

a.

Possible Samples	$\bar{x}$	$p(\bar{x})$	Possible Samples	$\bar{x}$	$p(\bar{x})$
1, 1	1	0.04	3, 4	3.5	0.04
1, 2	1.5	0.06	3, 5	4	0.02
1, 3	2	0.04	4, 1	2.5	0.04
1, 4	2.5	0.04	4, 2	3	0.06
1, 5	3	0.02	4, 3	3.5	0.04
2, 1	1.5	0.06	4, 4	4	0.04
2, 2	2	0.09	4, 5	4.5	0.02
2, 3	2.5	0.06	5, 1	3	0.02
2, 4	3	0.06	5, 2	3.5	0.03
2, 5	3.5	0.03	5, 3	4	0.02
3, 1	2	0.04	5, 4	4.5	0.02
3, 2	2.5	0.06	5, 5	5	0.01
3, 3	3	0.04			

Summing the probabilities, the probability distribution of $\bar{x}$ is:

$\bar{x}$	$p(\bar{x})$
1	0.04
1.5	0.12
2	0.17
2.5	0.20
3	0.20
3.5	0.14
4	0.08
4.5	0.04
5	0.01

b.

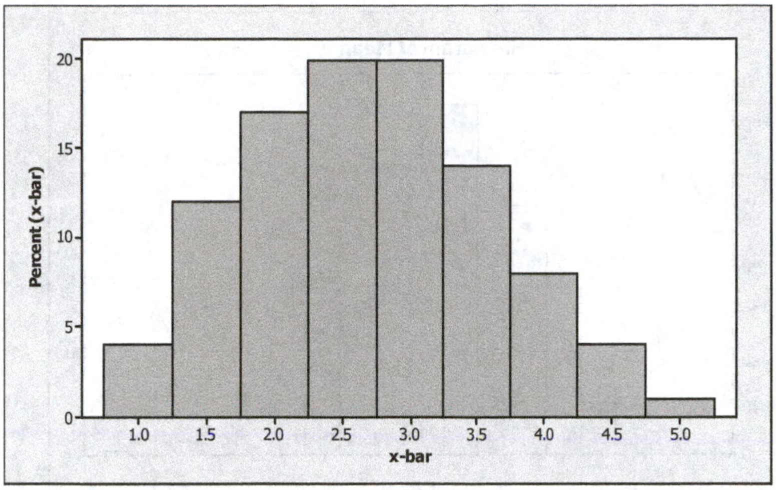

c. $P(\bar{x} \geq 4.5) = 0.04 + 0.01 = 0.05$

d. No. The probability of observing $\bar{x} = 4.5$ or larger is small $(p = 0.05)$.

6.7 a. For a sample of size $n = 2$, the sample mean and sample median are exactly the same. Thus, the sampling distribution of the sample median is the same as that for the sample mean (see Exercise 6.5**a**).

b. The probability histogram for the sample median is identical to that for the sample mean (see Exercise 6.5**b**).

6.9 a. Answers will vary. MINITAB was used to generate 500 samples of size $n = 25$ observations from a uniform population from 00 to 99. The first 10 samples along with the sample means are shown in the table below:

Sample	Observations																									Mean	Var
1	29	57	34	57	82	63	57	94	59	28	29	48	29	12	88	7	93	25	67	69	84	59	28	90	48	53.44	679.76
2	83	27	46	42	18	40	52	5	81	38	58	28	30	16	57	60	45	92	82	19	30	12	42	16	13	41.28	604.79
3	68	15	75	26	46	69	46	81	97	23	38	17	7	47	32	85	98	20	69	8	28	33	7	34	61	45.2	809.75
4	60	38	42	88	85	1	58	5	74	36	4	73	68	99	17	65	13	41	49	97	15	6	95	65	9	48.12	1066.11
5	62	76	37	6	70	14	95	94	12	18	63	65	33	37	44	86	71	67	35	10	17	53	21	27	65	47.12	761.44
6	53	56	92	5	57	13	57	13	7	91	2	76	98	21	93	44	91	74	30	17	99	77	49	86	5	52.24	1178.44
7	38	54	18	62	31	70	72	85	50	57	8	44	48	8	21	50	90	75	2	86	50	1	87	43	49	47.96	741.54
8	54	29	55	23	50	90	66	46	9	30	39	29	33	67	67	93	60	46	28	2	11	63	96	61	46	47.72	631.63
9	32	30	33	92	57	53	40	88	18	60	9	3	7	34	98	62	25	13	67	3	98	96	40	66	25	45.96	976.29
10	7	24	53	4	5	48	10	21	52	23	82	51	55	76	56	44	81	32	57	27	36	84	37	22	49	41.44	582.84

Using MINTAB, the histogram of the sample means is:

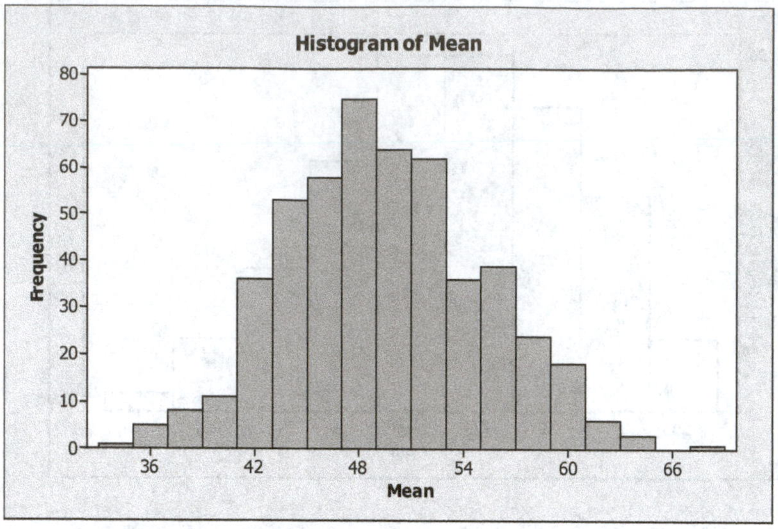

b. The sample variances for the first 10 samples are also shown in the table in part a. Using MINITAB, the histogram of the 500 sample variances is:

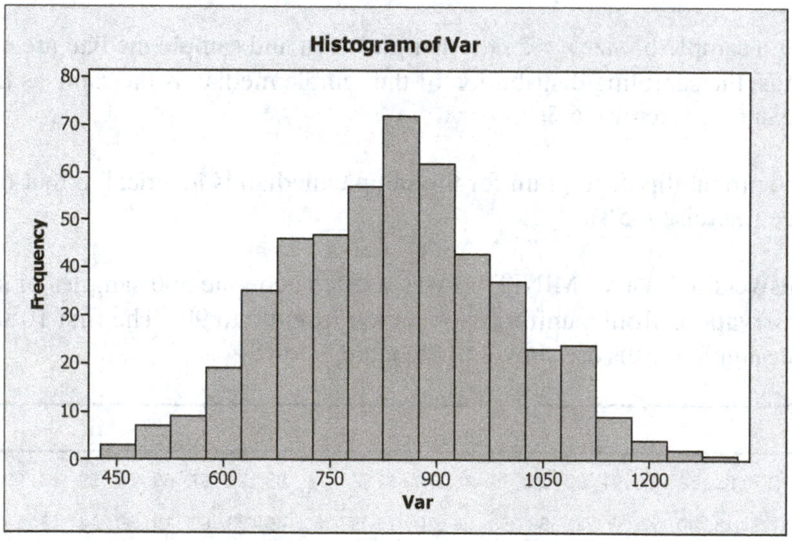

6.11 If the sampling distribution of a sample statistic has a mean equal to the population parameter the statistic is intended to estimate, the statistic is said to be an unbiased estimator of the parameter. If the mean of the sampling distribution is not equal to the parameter, the statistic is said to be a biased estimator of the parameter.

6.13 The properties of an ideal estimator are unbiased and minimum variance. Ideally, when we are estimating a parameter, we would like to use an unbiased estimator. In addition, of all unbiased estimators, we would like to choose the estimator with the minimum variance.

6.15 a. $\mu \sum xp(x) = 2\left(\frac{1}{3}\right) + 4\left(\frac{1}{3}\right) + 9\left(\frac{1}{3}\right) = \frac{15}{3} = 5$

b. The possible samples of size $n = 3$, the sample means, and the probabilities are:

Possible Samples	$\bar{x}$	$p(\bar{x})$	m	Possible Samples	$\bar{x}$	$p(\bar{x})$	m
2, 2, 2	2	1/27	2	4, 4, 4	4	1/27	4
2, 2, 4	8/3	1/27	2	4, 4, 9	17/3	1/27	4
2, 2, 9	13/3	1/27	2	4, 9, 2	5	1/27	4
2, 4, 2	8/3	1/27	2	4, 9, 4	17/3	1/27	4
2, 4, 4	10/3	1/27	4	4, 9, 9	22/3	1/27	9
2, 4, 9	5	1/27	4	9, 2, 2	13/3	1/27	2
2, 9, 2	13/3	1/27	2	9, 2, 4	5	1/27	4
2, 9, 4	5	1/27	4	9, 2, 9	20/3	1/27	9
2, 9, 9	20/3	1/27	9	9, 4, 2	5	1/27	4
4, 2, 2	8/3	1/27	2	9, 4, 4	17/3	1/27	4
4, 2, 4	10/3	1/27	4	9, 4, 9	22/3	1/27	9
4, 2, 9	5	1/27	4	9, 9, 2	20/3	1/27	9
4, 4, 2	10/3	1/27	4	9, 9, 4	22/3	1/27	9
				9, 9, 9	9	1/27	9

The sampling distribution of $\bar{x}$ is:

$\bar{x}$	$p(\bar{x})$
2	1/27
8/3	3/27
10/3	3/27
4	1/27
13/3	3/27
5	6/27
17/3	3/27
20/3	3/27
22/3	3/27
9	1/27
	27/27

$$E(\bar{x}) = \sum \bar{x}p(\bar{x}) = 2\left(\frac{1}{27}\right) + \frac{8}{3}\left(\frac{3}{27}\right) + \frac{10}{3}\left(\frac{3}{27}\right) + 4\left(\frac{1}{27}\right) + \frac{13}{3}\left(\frac{3}{27}\right)$$
$$+5\left(\frac{6}{27}\right) + \frac{17}{3}\left(\frac{3}{27}\right) + \frac{20}{3}\left(\frac{3}{27}\right) + \frac{22}{3}\left(\frac{3}{27}\right) + 9\left(\frac{1}{27}\right)$$
$$= \frac{2}{27} + \frac{8}{27} + \frac{10}{27} + \frac{4}{27} + \frac{13}{27} + \frac{30}{27} + \frac{17}{27} + \frac{20}{27} + \frac{22}{27} + \frac{9}{27} = \frac{135}{27} = 5$$

Since $\mu = 5$ in part **a**, and $E(\bar{x}) = \mu = 5$, $\bar{x}$ is an unbiased estimator of μ.

c. The median was calculated for each sample and is shown in the table in part **b**. The sampling distribution of m is:

m	$p(m)$
2	7/27
4	13/27
9	7/27
	27/27

$$E(m)=\sum mp(m)=2\left(\frac{7}{27}\right)+4\left(\frac{13}{27}\right)+9\left(\frac{7}{27}\right)=\frac{14}{27}+\frac{52}{27}+\frac{63}{27}=\frac{129}{27}=4.778$$

Because $E(m)=4.778\neq\mu=5$, m is a biased estimator of μ.

d. Use the sample mean, $\bar{x}$. It is an unbiased estimator.

6.17 Answers will vary. MINITAB was used to generate 500 samples of size $n=25$ observations from a uniform population from 1 to 50. The first 10 samples along with the sample means and medians are shown in the table below:

Sample	Observations																								Mean	Median
1	28	27	11	19	50	30	47	26	9	33	50	15	21	41	31	41	35	32	32	17	6	32	39	34	21	29.08	31
2	8	4	32	32	3	45	18	9	40	3	42	21	44	50	42	14	24	10	36	6	15	47	26	48	28	25.88	26
3	6	20	27	1	50	14	21	37	46	23	1	34	42	47	24	46	8	29	18	28	40	39	49	33	23	28.24	28
4	45	12	26	13	40	17	11	43	8	35	20	8	44	48	13	46	49	17	47	27	5	45	9	21	36	27.4	26
5	40	38	25	37	47	2	17	40	32	6	22	30	23	2	18	22	14	6	22	3	43	47	16	35	35	24.88	23
6	17	8	43	27	21	5	18	45	31	15	2	38	22	18	7	9	3	35	23	45	24	39	38	35	37	24.20	23
7	40	1	22	29	6	8	22	20	36	18	45	16	29	9	6	3	49	34	24	40	27	5	49	11	30	23.16	22
8	25	3	44	34	29	6	33	32	43	6	43	24	49	14	37	8	46	44	1	12	36	18	30	25	4	25.84	29
9	7	33	36	41	30	13	17	19	14	36	20	39	41	20	15	38	12	37	14	9	19	2	37	15	8	22.88	19
10	4	46	49	49	45	49	24	3	25	22	27	28	23	17	14	6	35	5	20	34	4	41	9	15	3	23.88	23

Using MINITAB, side-by side histograms of the means and medians of the 500 samples are:

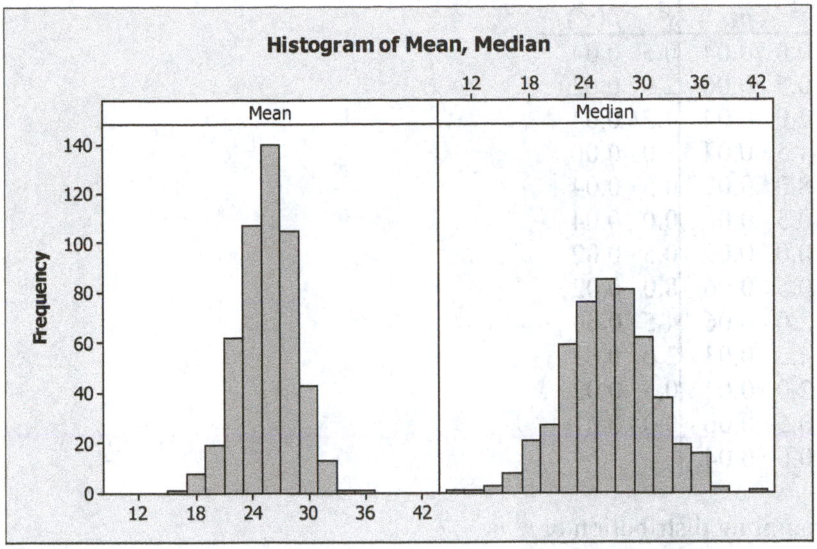

a. Yes, it appears that $\bar{x}$ and the median are unbiased estimators of the population mean. The centers of both distributions above appear to be around 25 to 26. In fact, the mean of the sampling distribution of $\bar{x}$ is 25.65 and the mean of the sampling distribution of the median is 25.73.

b. The sampling distribution of the median has greater variation because it is more spread out than the sampling distribution of $\bar{x}$.

6.19 a. Refer to the solution to Exercise 6.5. The values of s^2 and the corresponding probabilities are listed below:

$$s^2 = \frac{\sum x^2 - \frac{\left(\sum x\right)^2}{n}}{n-1}$$

For sample 1, 1; $s^2 = \dfrac{2 - \frac{2^2}{2}}{1} = 0$

For sample 1, 2: $s^2 = \dfrac{5 - \frac{3^2}{2}}{1} = .5$

The rest of the values are calculated and shown:

s^2	$p(s^2)$	s^2	$p(s^2)$
0.0	0.04	0.5	0.04
0.5	0.06	2.0	0.02
2.0	0.04	4.5	0.04
4.5	0.04	2.0	0.06
8.0	0.02	0.5	0.04
0.5	0.06	0.0	0.04
0.0	0.09	0.5	0.02
0.5	0.06	8.0	0.02
2.0	0.06	4.5	0.03
4.5	0.03	2.0	0.02
2.0	0.04	0.5	0.02
0.5	0.06	0.0	0.01
0.0	0.04		

The sampling distribution of s^2 is:

s^2	$p(s^2)$
0.0	.22
0.5	.36
2.0	.24
4.5	.14
8.0	.04

b. $$\sigma^2 = \sum (x-\mu)^2 p(x) = (1-2.7)^2(0.2) + (2-2.7)^2(0.3) + (3-2.7)^2(0.2) + (4-2.7)^2(0.2) + (5-2.7)^2(0.1) = 1.61$$

c. $$E(s^2) = \sum s^2 p(s^2) = 0(0.22) + 0.5(0.36) + 2(0.24) + 4.5(0.14) + 8(0.04) = 1.61$$

d. The sampling distribution of s is listed below, where $s = \sqrt{s^2}$:

s	$p(s)$
0.000	0.22
0.707	0.36
1.414	0.24
2.121	0.14
2.828	0.04

e. $$E(s) = \sum sp(s) = 0(0.22) + 0.707(0.36) + 1.41(0.24) + 2.1212(0.14) + 2.828(0.04) = 1.00394$$

Since $E(s) = 1.00394$ is not equal to $\sigma = \sqrt{\sigma^2} = \sqrt{1.61} = 1.269$, s is a biased estimator of σ.

6.21 The symbol $\mu_{\bar{x}}$ represents the mean of the sampling distribution of $\bar{x}$. The symbol $\sigma_{\bar{x}}$ represents the standard deviation of the sampling distribution of $\bar{x}$.

6.23 The standard deviation of the sampling distribution of $\bar{x}$, $\sigma_{\bar{x}}$, is the standard deviation of the population from which the sample is selected divided by $\sqrt{n}$.

6.25 The Central Limit Theorem states: Consider a random sample of n observations selected from a population (*any* population) with mean μ and standard deviation σ. Then, when n is sufficiently large, the sampling distribution of $\bar{x}$ will be approximately a normal distribution with mean $\mu_{\bar{x}} = \mu$ and standard deviation $\sigma_{\bar{x}} = \sigma/\sqrt{n}$. The larger the sample size, the better will be the normal approximation to the sampling distribution of $\bar{x}$.

6.27 a. $\mu_{\bar{x}} = \mu = 100, \sigma_{\bar{x}} = \dfrac{\sigma}{\sqrt{n}} = \dfrac{\sqrt{100}}{\sqrt{4}} = 5$

b. $\mu_{\bar{x}} = \mu = 100, \sigma_{\bar{x}} = \dfrac{\sigma}{\sqrt{n}} = \dfrac{\sqrt{100}}{\sqrt{25}} = 2$

c. $\mu_{\bar{x}} = \mu = 100, \sigma_{\bar{x}} = \dfrac{\sigma}{\sqrt{n}} = \dfrac{\sqrt{100}}{\sqrt{100}} = 1$

d. $\mu_{\bar{x}} = \mu = 100, \sigma_{\bar{x}} = \dfrac{\sigma}{\sqrt{n}} = \dfrac{\sqrt{100}}{\sqrt{50}} = 1.414$

e. $\mu_{\bar{x}} = \mu = 100, \sigma_{\bar{x}} = \dfrac{\sigma}{\sqrt{n}} = \dfrac{\sqrt{100}}{\sqrt{500}} = 0.447$

f. $\mu_{\bar{x}} = \mu = 100, \sigma_{\bar{x}} = \dfrac{\sigma}{\sqrt{n}} = \dfrac{\sqrt{100}}{\sqrt{1000}} = 0.316$

6.29 a. $\mu = \sum xp(x) = 1(0.1) + 2(0.4) + 3(0.4) + 8(0.1) = 2.9$

$$\sigma^2 = \sum (x-\mu)^2 p(x) = (1-2.9)^2(0.1) + (2-2.9)^2(0.4) + (3-2.9)^2(0.4)$$
$$+ (8-2.9)^2(0.1)$$
$$= 0.361 + 0.324 + 0.004 + 2.601 = 3.29$$

$\sigma = \sqrt{3.29} = 1.814$

b. The possible samples, values of $\bar{x}$, and associated probabilities are listed:

Possible Samples	$\bar{x}$	$p(\bar{x})$	Possible Samples	$\bar{x}$	$p(\bar{x})$
1, 1	1	0.01	3, 1	2	0.04
1, 2	1.5	0.04	3, 2	2.5	0.16
1, 3	2	0.04	3, 3	3	0.16
1, 8	4.5	0.01	3, 8	5.5	0.04
2, 1	1.5	0.04	8, 1	4.5	0.01
2, 2	2	0.16	8, 2	5	0.04
2, 3	2.5	0.16	8, 3	5.5	0.04
2, 8	5	0.04	8, 8	8	0.01

The sampling distribution of $\bar{x}$ is:

$\bar{x}$	$p(\bar{x})$
1	.01
1.5	.08
2	.24
2.5	.32
3	.16
4.5	.02
5	.08
5.5	.08
8	.01
	1.00

c. $$\mu_{\bar{x}} = E(\bar{x}) = \sum \bar{x}p(\bar{x}) = 1(0.01)+1.5(0.08)+2(0.24)+2.5(0.32)+3(0.16)+4.5(0.02)$$
$$5(0.08)+5.5(0.08)+8(0.01) = 2.9 = \mu$$

$$\sigma_{\bar{x}}^2 = \sum(\bar{x}-\mu_{\bar{x}})^2 p(\bar{x}) = (1-2.9)^2(0.01)+(1.5-2.9)^2(0.08)+(2-2.9)^2(0.24)$$
$$+(2.5-2.9)^2(0.32)+(3-2.9)^2(0.16)+(4.5-2.9)^2(0.02)$$
$$+(5-2.9)^2(0.08)+(5.5-2.9)^2(0.08)+(8-2.9)^2(0.01)$$
$$=0.0361+0.1568+0.1944+0.0512+0.0016+0.0512+0.3528$$
$$+0.5408+0.2601=1.645$$

$$\sigma_{\bar{x}} = \sqrt{1.645} = 1.283$$
$$\sigma_{\bar{x}} = \frac{\sigma}{\sqrt{n}} = \frac{1.814}{\sqrt{2}} = 1.283$$

6.31 a. $\mu_{\bar{x}} = \mu = 30$, $\sigma_{\bar{x}} = \dfrac{\sigma}{\sqrt{n}} = \dfrac{16}{\sqrt{100}} = 1.6$

b. By the Central Limit Theorem, the sampling distribution of $\bar{x}$ is approximately normal.

c. $P(\bar{x} \geq 28) = P\left(z \geq \frac{28-30}{1.6}\right) = P(z \geq -1.25) = 0.5 + 0.3944 = 0.8944$

d. $P(22.1 \leq \bar{x} \leq 26.8) = P\left(\frac{22.1-30}{1.6} \leq z \leq \frac{26.8-30}{1.6}\right) = P(-4.94 \leq z \leq -2)$
$= 0.5 - 0.4772 = 0.0228$

e. $P(\bar{x} \leq 28.2) = P\left(z \leq \frac{28.2-30}{1.6}\right) = P(z \leq -1.125) = 0.5 - \left(\frac{0.3868 + 0.3708}{2}\right)$
$= 0.5 - 0.3697 = 0.1303$

f. $P(\bar{x} \geq 27.0) = P\left(z \geq \frac{27.0-30}{1.6}\right) = P(z \geq -1.88) = 0.5 + 0.4699 = 0.9699$

6.33 Answers will vary. MINITAB was used to generate the samples. The side-by-side histograms of the distributions of $\bar{x}$ were generated by MINITAB.

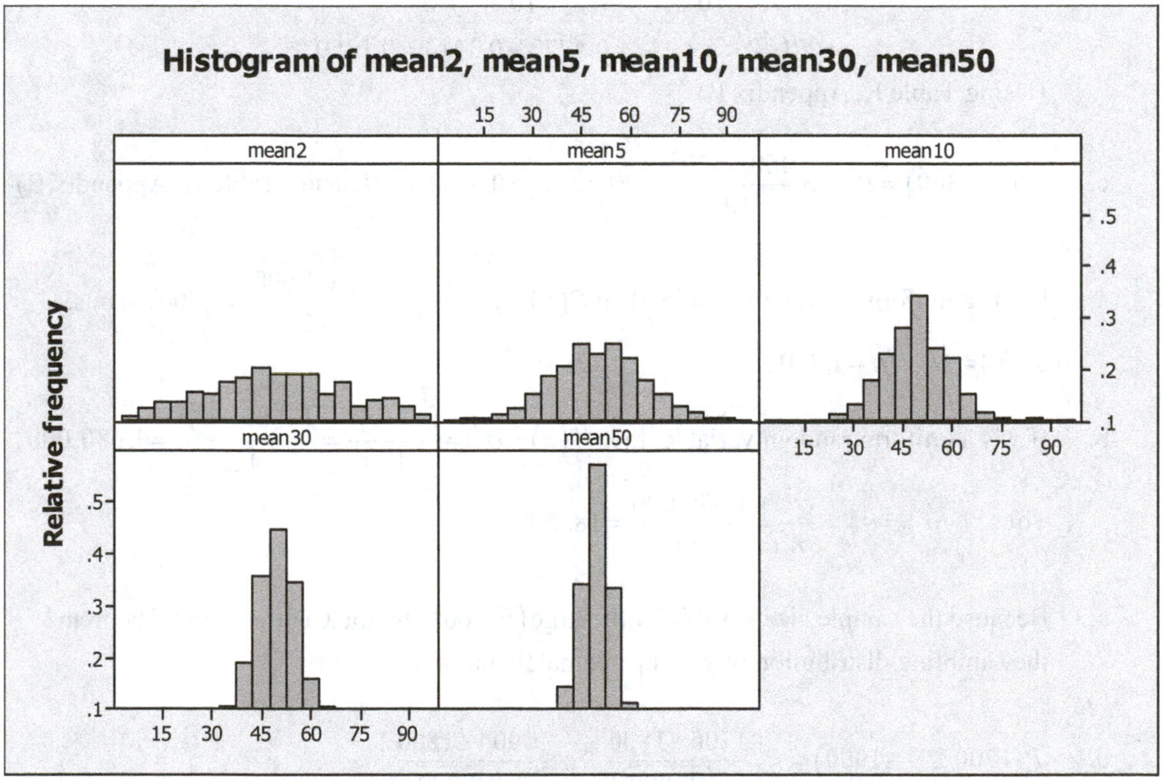

It appears that each of the distributions is centered around the same value, approximately 49.5. All of the distributions appear to be symmetric. As the sample size increases, the spread of the sampling distribution of $\bar{x}$ decreases.

6.35 a. $E(\bar{x}) = \mu_{\bar{x}} = \mu = 353$. The mean of the sampling distribution of $\bar{x}$ with $n = 45$ is equal to the mean of the distribution being sampled from which is 353.

b. $V(\bar{x}) = \sigma_{\bar{x}}^2 = \frac{\sigma^2}{n} = \frac{30^2}{45} = 20$

c. Because the sample size is large $(n = 45)$, by the Central Limit Theorem, the sampling distribution of $\bar{x}$ is approximately normal.

d. $P(\bar{x} > 400) = P\left(z > \frac{400 - 353}{\sqrt{20}}\right) = P(z > 10.51) \approx 0.5 - 0.5 = 0$. It would be extremely unlikely to observe a sample mean number of sags that exceeds 400.

6.37 a. By the Central Limit Theorem, the sampling distribution of $\bar{x}$ is approximately normal. The mean of the $\bar{x}$ distribution is $\mu_{\bar{x}} = \mu = 320$ and the standard deviation of the $\bar{x}$ distribution is $\sigma_{\bar{x}} = \frac{\sigma}{\sqrt{n}} = \frac{100}{\sqrt{100}} = \frac{100}{10} = 10$.

b. $P(300 < \bar{x} < 310) = P\left(\frac{300 - 320}{10} < z < \frac{310 - 320}{10}\right)$

$= P(-2 < z < -1) = 0.4772 - 0.3413 = 0.1359$

(Using Table II, Appendix B)

c. $P(\bar{x} > 360) = P\left(z > \frac{360 - 320}{10}\right) = P(z > 4) \approx 0.5 - 0.5 = 0$ (using Table II, Appendix B)

6.39 a. If x is a uniform random variable, then $E(x) = \mu = \frac{c + d}{2} = \frac{0 + 3,600}{2} = 1,800$. Thus, $E(\bar{x}) = \mu_{\bar{x}} = \mu = 1,800$.

b. If x is a uniform random variable, then $V(x) = \sigma^2 = \frac{(d - c)^2}{12} = \frac{(3600 - 0)^2}{12} = 1,080,000$.

Thus, $V(\bar{x}) = \sigma_{\bar{x}}^2 = \frac{\sigma^2}{n} = \frac{1,080,000}{60} = 18,000$.

c. Because the sample size is sufficiently large $(n = 60)$, by the Central Limit Theorem, the sampling distribution of $\bar{x}$ is approximately normal.

d. $P(1700 \le \bar{x} \le 1900) = P\left(\frac{1700 - 1800}{\sqrt{18,000}} \le z \le \frac{1900 - 1800}{\sqrt{18,000}}\right)$

$= P(-0.75 \le z \le 0.75) = 0.2734 + 0.2734 = 0.5468$

e. $P(\bar{x} > 2000) = P\left(z > \frac{2000 - 1800}{\sqrt{18,000}}\right) = P(z > 1.49) = 0.5 - 0.4319 = 0.0681$

6.41 Let $\bar{x}$ = sample mean amount of miraculin produced. By the Central Limit Theorem, the sampling distribution of $\bar{x}$ is approximately normal with $\mu_{\bar{x}} = \mu = 105.3$ and

$$\sigma_{\bar{x}} = \frac{\sigma}{\sqrt{n}} = \frac{8.0}{\sqrt{64}} = 1.$$

$$P(\bar{x} < 103) = P\left(z < \frac{103 - 105.3}{1}\right) = P(z < -2.30) = 0.5 - 0.4893 = 0.0107$$

(using Table II, Appendix B)

Since this probability is so small, we would not expect to observe a value of $\bar{x}$ less than 103 micro-grams per gram of fresh weight.

6.43 a. $E(\bar{x}) = \mu_{\bar{x}} = \mu = 0.10$ and $V(\bar{x}) = \sigma_{\bar{x}}^2 = \frac{\sigma^2}{n} = \frac{(0.10)^2}{50} = 0.0002$

b. By the Central Limit Theorem, the sampling distribution of $\bar{x}$ is approximately normal if the sample size is sufficiently large. In this case, $n = 50$ is considered large.

c. $P(\bar{x} > 0.13) = P\left(z > \frac{0.13 - 0.10}{\sqrt{0.0002}}\right) = P(z > 2.12) = 0.5 - 0.4830 = 0.0170$

6.45 a. By the Central Limit Theorem, the sampling distribution of $\bar{x}$ is approximately normal if the sample size is sufficiently large. In this case, $n = 326$ is considered large.

$$\mu_{\bar{x}} = \mu = 6, \quad \sigma_{\bar{x}} = \frac{\sigma}{\sqrt{n}} = \frac{10}{\sqrt{326}} = 0.5538$$

$$P(\bar{x} > 7.5) = P\left(z > \frac{7.5 - 6}{0.5538}\right) = P(z > 2.71) = 0.5 - 0.4966 = 0.0034$$

b. $P(\bar{x} > 300) = P\left(z > \frac{300 - 6}{0.5538}\right) = P(z > 530.88) \approx 0.5 - 0.5 = 0$

Since it is essentially impossible to get a sample mean of 300 if the population mean was 6, we would conclude that the mean PFOA in people who live near DuPont's Teflon-making facility is not 6 ppb but something much larger.

6.47 **Hand Rubbing**

$$\mu_{\bar{x}} = \mu = 35, \quad \sigma_{\bar{x}} = \frac{\sigma}{\sqrt{n}} = \frac{59}{\sqrt{50}} = 8.344$$

$$P(\bar{x} < 30) = P\left(z < \frac{30 - 35}{8.344}\right) = P(z < -0.60) = 0.5 - 0.2257 = 0.2743$$

(using Table II, Appendix B.)

Hand Washing

$$\mu_{\bar{x}} = \mu = 69, \quad \sigma_{\bar{x}} = \frac{\sigma}{\sqrt{n}} = \frac{106}{\sqrt{50}} = 14.991$$

$$P(\bar{x} < 30) = P\left(z < \frac{30-69}{14.991}\right) = P(z < -2.60) = 0.5 - 0.4953 = 0.0047$$

(using Table II, Appendix B.)

This sample of workers probably came from the hand rubbing group. The probability of seeing a mean of 30 or smaller for this group is 0.2743, while the probability of observing a mean of 30 or less for the hand washers is 0.0047. This would be an extremely rare event if the workers were hand washers because the probability is so small. It would not be a very unusual event for the hand rubbers because the probability is not that small.

6.49 a. $\mu_{\hat{p}} = p = 0.2$ and $\sigma_{\hat{p}} = \sqrt{\frac{p(1-p)}{n}} = \sqrt{\frac{0.2(1-0.2)}{50}} = 0.0566$

b. $\mu_{\hat{p}} = p = 0.2$ and $\sigma_{\hat{p}} = \sqrt{\frac{p(1-p)}{n}} = \sqrt{\frac{0.2(1-0.2)}{1{,}000}} = 0.0126$

c. $\mu_{\hat{p}} = p = 0.2$ and $\sigma_{\hat{p}} = \sqrt{\frac{p(1-p)}{n}} = \sqrt{\frac{0.2(1-0.2)}{400}} = 0.02$

6.51 a. $\mu_{\hat{p}} = p = 0.3$ and $\sigma_{\hat{p}} = \sqrt{\frac{p(1-p)}{n}} = \sqrt{\frac{0.3(1-0.3)}{80}} = 0.0512$

b. The sampling distribution of $\hat{p}$ will be approximately normal since the sample size is sufficiently large.

c. $$z = \frac{\hat{p}-p}{\sqrt{\frac{p(1-p)}{n}}} = \frac{0.35-0.3}{\sqrt{\frac{0.3(1-0.3)}{80}}} = 0.98$$

d. $$P(\hat{p} > 0.35) = P\left(z > \frac{0.35-0.3}{\sqrt{\frac{0.3(1-0.3)}{80}}}\right) = P(z > 0.98) = 0.5 - 0.3365 = 0.1635$$

6.53. We would not expect to see any values of $\hat{p}$ more than 3 standard deviations below or above the mean value of $\hat{p}$.

$$z = \frac{\hat{p}-p}{\sqrt{\frac{p(1-p)}{n}}} \Rightarrow -3 = \frac{\hat{p}-0.4}{\sqrt{\frac{0.4(1-0.4)}{1500}}} \Rightarrow -3\sqrt{\frac{0.4(1-0.4)}{1500}} = \hat{p}-0.4$$

$$\Rightarrow -0.0379 = \hat{p}-0.4 \Rightarrow \hat{p} = 0.3621$$

The smallest value we would expect for $\hat{p}$ would be 0.3621.

$$z = \frac{\hat{p}-p}{\sqrt{\frac{p(1-p)}{n}}} \Rightarrow 3 = \frac{\hat{p}-0.4}{\sqrt{\frac{0.4(1-0.4)}{1500}}} \Rightarrow 3\sqrt{\frac{0.4(1-0.4)}{1500}} = \hat{p}-0.4 \Rightarrow 0.0379 = \hat{p}-0.4 \Rightarrow \hat{p} = 0.4379$$

The largest value we would expect for $\hat{p}$ would be .4379.

6.55 a. $\mu_{\hat{p}} = p = 0.4$ and $\sigma_{\hat{p}} = \sqrt{\frac{p(1-p)}{n}} = \sqrt{\frac{0.4(1-0.4)}{106}} = 0.0476$

b. The sampling distribution of $\hat{p}$ will be approximately normal because the sample size is sufficiently large.

c. $$P(\hat{p} > 0.59) = P\left(z > \frac{0.59-0.4}{\sqrt{\frac{0.4(1-0.4)}{106}}}\right) = P(z > 3.99) = 0.5 - 0.49997 = 0.00003$$

(Using Table II, Appendix B)

d. $\hat{p} = \frac{x}{n} = \frac{63}{106} = 0.59$. We found in part c that $P(\hat{p} > 0.59) = 0.00003$. Since this probability is so small, it casts doubt on the assumption that 40% of all social robots are designed with legs, but no wheels.

6.57 By the Central Limit theorem, the sampling distribution of $\hat{p}$ will be approximately normal because the sample size is sufficiently large, with $\mu_{\hat{p}} = p = 0.35$ and $\sigma_{\hat{p}} = \sqrt{\frac{p(1-p)}{n}} = \sqrt{\frac{0.35(1-0.35)}{500}} = 0.0213$.

a. $$P(0.5 < \hat{p} < 0.6) = P\left(\frac{0.5-0.35}{\sqrt{\frac{0.35(1-0.35)}{500}}} < z < \frac{0.6-0.35}{\sqrt{\frac{0.35(1-0.35)}{500}}}\right)$$

$$= P(7.04 < z < 11.74) \approx 0.5 - 0.5 = 0 \text{ (using Table II, Appendix B)}$$

b. $P(\hat{p} > 0.4) = P\left(z > \dfrac{0.4 - 0.35}{\sqrt{\dfrac{0.35(1-0.35)}{500}}}\right) = P(z > 2.35) = 0.5 - 0.4906 = 0.0094$

(using Table II, Appendix B)

6.59 a. For cereal crops, by the Central Limit Theorem, the sampling distribution of $\hat{p}$ will be approximately normal because the sample size is sufficiently large, with $\mu_{\hat{p}} = p = 0.45$ and $\sigma_{\hat{p}} = \sqrt{\dfrac{p(1-p)}{n}} = \sqrt{\dfrac{0.45(1-0.45)}{100}} = 0.0497$. Finding the probability that at least 50 incidents involve cereal crops is the same things as the probability that the sample proportion $\hat{p}$ is at least $50/100 = 0.50$.

$$P(\hat{p} > 0.50) = P\left(z > \dfrac{0.5 - 0.45}{\sqrt{\dfrac{0.45(1-0.45)}{100}}}\right) = P(z > 1.01) = 0.5 - 0.3438 = 0.1562$$

(using Table II, Appendix B)

b. For legumes, by the Central Limit Theorem, the sampling distribution of $\hat{p}$ will be approximately normal because the sample size is sufficiently large, with $\mu_{\hat{p}} = p = 0.20$ and $\sigma_{\hat{p}} = \sqrt{\dfrac{p(1-p)}{n}} = \sqrt{\dfrac{0.20(1-0.20)}{100}} = 0.04$. Finding the probability that at least 50 incidents involve cereal crops is the same things as the probability that the sample proportion $\hat{p}$ is at least $50/100 = 0.50$.

$$P(\hat{p} > 0.50) = P\left(z > \dfrac{0.5 - 0.20}{\sqrt{\dfrac{0.20(1-0.20)}{100}}}\right) = P(z > 7.5) = 0.5 - 0.5 = 0$$

(using Table II, Appendix B)

c. For orchards, by the Central Limit Theorem, the sampling distribution of $\hat{p}$ will be approximately normal because the sample size is sufficiently large, with $\mu_{\hat{p}} = p = 0.05$ and $\sigma_{\hat{p}} = \sqrt{\dfrac{p(1-p)}{n}} = \sqrt{\dfrac{0.05(1-0.05)}{100}} = 0.0218$. Finding the probability that at least 50 incidents involve cereal crops is the same things as the probability that the sample proportion $\hat{p}$ is at least $50/100 = 0.50$.

$$P(\hat{p} > 0.50) = P\left(z > \dfrac{0.5 - 0.05}{\sqrt{\dfrac{0.05(1-0.05)}{100}}}\right) = P(z > 20.64) = 0.5 - 0.5 = 0$$

(Using Table II, Appendix B)

6.61 From Exercise 4.75, we defined the following events:

P: {hotel guest is aware of conservation program}
A: {hotel guest participates in conservation efforts}

Then, the probability of a hotel guest being aware and participating in the hotel's conservation efforts is $p = P(P \mid A)P(A) = 0.72(0.66) = 0.4752$.

By the Central Limit theorem, the sampling distribution of $\hat{p}$ will be approximately normal since the sample size is sufficiently large, with $\mu_{\hat{p}} = p = 0.4752$ and

$$\sigma_{\hat{p}} = \sqrt{\frac{p(1-p)}{n}} = \sqrt{\frac{0.4752(1-0.4752)}{100}} = 0.0499 .$$

$$P\left(\hat{p} < \frac{42}{100}\right) = P(\hat{p} < 0.42) = P\left(z < \frac{0.42 - 0.4752}{\sqrt{\frac{0.4752(1-0.4752)}{100}}}\right) = P(z < -1.11) = 0.5 - 0.3665 = 0.1335$$

(Using Table II, Appendix B)

6.63 The statement "The sample mean, $\bar{x}$, will always be equal to μ_x" is false. The sample mean is used to estimate the population mean, but the probability that it is equal to the population mean is 0.

6.65 The statement "The standard error of $\bar{x}$ will always be small than σ" is true. The standard error of $\bar{x}$ is $\sigma_{\bar{x}} = \frac{\sigma}{\sqrt{n}}$. Thus, the standard error of $\bar{x}$ will always be smaller than σ.

6.67 This statement is true.

6.69 a. "The sampling distribution of the sample statistic A" is the probability distribution of the variable A.

b. "A" is an unbiased estimator of α because the mean of the sampling distribution of A is α.

c. If both A and B are unbiased estimators of α, then the statistic whose standard deviation is smaller is a better estimator of α.

d. No. The Central Limit Theorem applies only to the sample mean. If A is the sample mean, $\bar{x}$, and n is sufficiently large, then the Central Limit Theorem will apply. However, both A and B cannot be sample means. Thus, we cannot apply the Central Limit Theorem to both A and B.

6.71 By the Central Limit Theorem, the sampling distribution of $\bar{x}$ is approximately normal.

$$\mu_{\bar{x}} = \mu = 19.6, \quad \sigma_{\bar{x}} = \frac{3.2}{\sqrt{68}} = 0.388$$

a. $P(\bar{x} \le 19.6) = P\left(z \le \frac{19.6 - 19.6}{0.388}\right) = P(z \le 0) = 0.5$ (using Table II, Appendix B)

b. $P(\bar{x} \le 19) = P\left(z \le \frac{19 - 19.6}{0.388}\right) = P(z \le -1.55) = 0.5 - 0.4394 = 0.0606$

(using Table II, Appendix B)

c. $P(\bar{x} \ge 20.1) = P\left(z \ge \frac{20.1 - 19.6}{0.388}\right) = P(z \ge 1.29) = 0.5 - 0.4015 = 0.0985$

(using Table II, Appendix B)

d. $P(19.2 \le \bar{x} \le 20.6) = P\left(\frac{19.2 - 19.6}{0.388} < z < \frac{20.6 - 19.6}{0.388}\right)$

$= P(-1.03 \le z \le 2.58) = 0.3485 + 0.4951 = 0.8436$

(using Table II, Appendix B)

6.73 a. By the Central Limit theorem, the sampling distribution of $\hat{p}$ will be approximately normal since the sample size is sufficiently large, with $\mu_{\hat{p}} = p = 0.8$ and

$$\sigma_{\hat{p}} = \sqrt{\frac{p(1-p)}{n}} = \sqrt{\frac{0.8(1-0.8)}{300}} = 0.0231.$$

$$P(\hat{p} < 0.83) = P\left(z < \frac{0.83 - 0.8}{\sqrt{\frac{0.8(1-0.8)}{300}}}\right) = P(z < 1.30) = 0.5 + 0.4032 = 0.9032$$

b. $P(\hat{p} > 0.75) = P\left(z > \frac{0.75 - 0.8}{\sqrt{\frac{0.8(1-0.8)}{300}}}\right) = P(z > -2.16) = 0.5 + 0.4846 = 0.9846$

c. $P(0.79 < \hat{p} < 0.81) = P\left(\frac{0.79 - 0.8}{\sqrt{\frac{0.8(1-0.8)}{300}}} < z < \frac{0.81 - 0.8}{\sqrt{\frac{0.8(1-0.8)}{300}}}\right) = P(-0.43 < z < 0.43)$

$= 0.1664 + 0.1664 = 0.3328$

6.75 Answers will vary. One hundred samples of size $n = 2$ were selected from a uniform distribution on the interval from 0 to 10. The process was repeated for samples of size $n = 5$, $n = 10$, $n = 30$, and $n = 50$. For each sample, the value of $\bar{x}$ was computed. Using MINITAB, the histograms for each set of 100 $\bar{x}$'s were constructed:

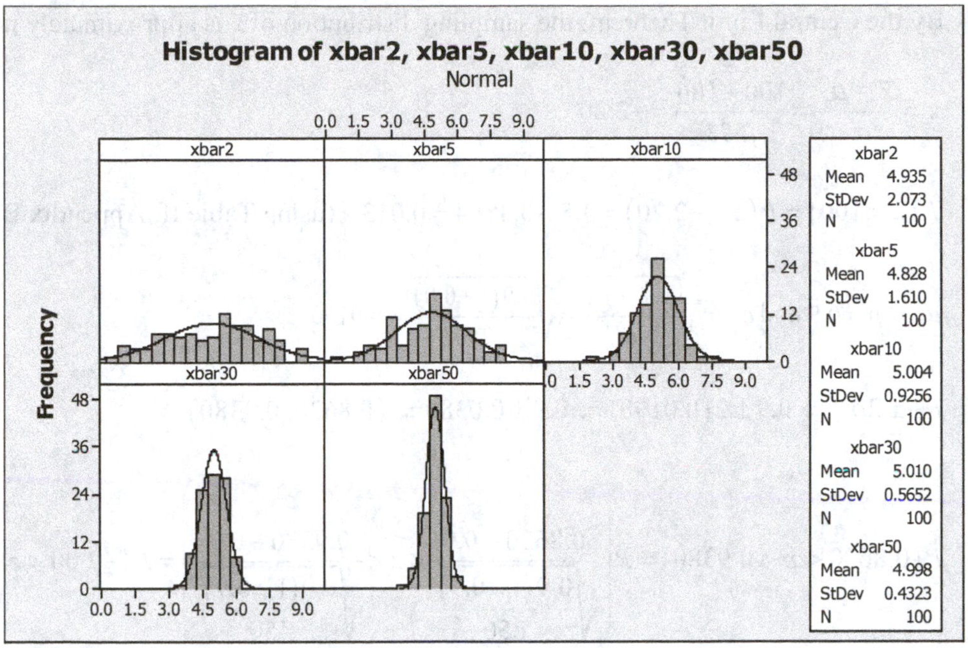

For small sizes of *n*, the sampling distributions of $\overline{x}$ are somewhat normal. As *n* increases, the sampling distributions of $\overline{x}$ become more normal.

6.77 Given: $\mu = 100$ and $\sigma = 10$

n	1	5	10	20	30	40	50
$\frac{\sigma}{\sqrt{n}}$	10	4.472	3.162	2.236	1.826	1.581	1.414

The graph of $\frac{\sigma}{\sqrt{n}}$ against n is given here:

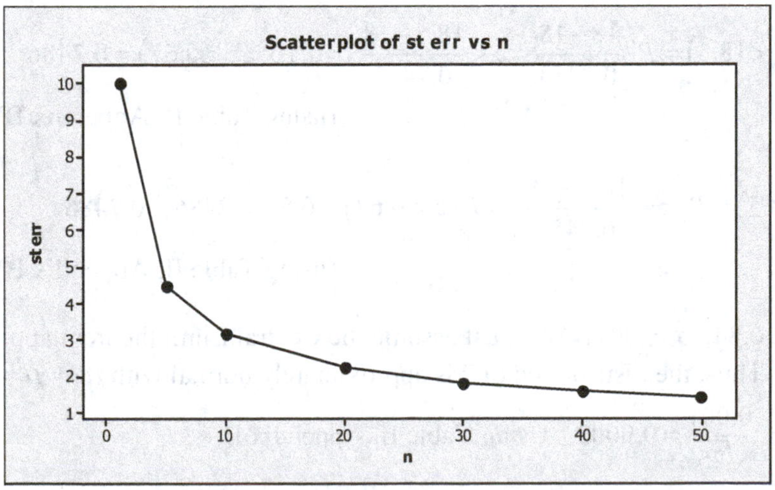

6.79 a. $\mu_{\overline{x}}$ is the mean of the sampling distribution of $\overline{x}$. $\mu_{\overline{x}} = \mu = 106$.

b. $\sigma_{\overline{x}}$ is the standard deviation of the sampling distribution of $\overline{x}$. $\sigma_{\overline{x}} = \frac{\sigma}{\sqrt{n}} = \frac{16.4}{\sqrt{36}} = 2.73$

c. By the Central Limit Theorem, the sampling distribution of $\bar{x}$ is approximately normal.

d. $$z = \frac{\bar{x} - \mu_{\bar{x}}}{\sigma_{\bar{x}}} = \frac{100 - 106}{2.73} = -2.20$$

e. $P(\bar{x} < 100) = P(z < -2.20) = 0.5 - 0.4864 = 0.0136$ (using Table II, Appendix B.)

6.81 a. $\mu_{\hat{p}} = p = 0.9$ and $\sigma_{\hat{p}} = \sqrt{\frac{p(1-p)}{n}} = \sqrt{\frac{0.9(1-0.9)}{250}} = 0.0190$

b. $\mu_{\hat{p}} \pm 2\sigma_{\hat{p}} \Rightarrow 0.9 \pm 2(0.0190) \Rightarrow 0.9 \pm 0.0380 \Rightarrow (0.8620,\ 0.9380)$

c. $$P(0.8620 < \hat{p} < 0.9380) = P\left(\frac{0.8620 - 0.9}{\sqrt{\frac{0.9(1-0.9)}{250}}} < z < \frac{0.9380 - 0.9}{\sqrt{\frac{0.9(1-0.9)}{250}}}\right) = P(-2.00 < z < 2.00)$$
$$= 0.4772 + 0.4772 = 0.9544 \text{ (using Table II, Appendix B.)}$$

6.83 a. Let $\bar{x}$ = sample mean FNE score. By the Central Limit Theorem, the sampling distribution of $\bar{x}$ is approximately normal with

$$\mu_{\bar{x}} = \mu = 18 \text{ and } \sigma_{\bar{x}} = \frac{\sigma}{\sqrt{n}} = \frac{5}{\sqrt{45}} = .7453.$$

$$P(\bar{x} > 17.5) = P\left(z > \frac{17.5 - 18}{0.7453}\right) = P(z > -0.67) = 0.5 + 0.2486 = 0.7486$$

(using Table II, Appendix B)

b. $$P(18 < \bar{x} < 18.5) = P\left(\frac{18 - 18}{0.7453} < z < \frac{18.5 - 18}{0.7453}\right) = P(0 < z < 0.67) = 0.2486$$

(using Table II, Appendix B)

c. $$P(\bar{x} < 18.5) = P\left(z < \frac{18.5 - 18}{0.7453}\right) = P(z < 0.67) = 0.5 + 0.2486 = 0.7486$$

(using Table II, Appendix B)

6.85 From Exercise 6.84, $\sigma = .001$. We must assume the Central Limit theorem applies (n is only 25). Thus, the distribution of $\bar{x}$ is approximately normal with $\mu_{\bar{x}} = \mu = 0.501$ and $\sigma_{\bar{x}} = \frac{\sigma}{\sqrt{n}} = \frac{0.001}{\sqrt{25}} = 0.0002$. Using Table II, Appendix B,

$$P(\bar{x} < 0.4994) + P(\bar{x} > 0.5006) = P\left(z < \frac{0.4994 - 0.501}{0.0002}\right) + P\left(z > \frac{0.5006 - 0.501}{0.0002}\right)$$
$$= P(z < -8) + P(z > -2) = (0.5 - 0.5) + (0.5 + 0.4772) = 0.9772$$

6.87 For this problem, $\mu_{\bar{x}} = \mu = 4.59$ and $\sigma_{\bar{x}} = \frac{\sigma}{\sqrt{n}} = \frac{2.95}{\sqrt{50}} = 0.4172$

a. $P(\bar{x} \geq 6) = P\left(z \geq \frac{6-4.59}{0.4172}\right) = P(z \geq 3.38) \approx 0.5 - 0.5 = 0$

(using Table II, Appendix B)

Since the probability of observing a sample mean CAHS score of 6 or higher is so small (p is essentially 0), we would not expect to see a sample mean of 6 or higher.

b. If a psychologist actually observed $x = 6.2$, we would conclude that he has seen an extremely unusual event, or the population from which the sample was drawn probably has a mean greater than 4.59.

6.89 a. By the Central Limit Theorem, the sampling distribution of $\bar{x}$ is approximately normal if the sample size is sufficiently large. In this case, $n = 50$ is considered large.

$$\mu_{\bar{x}} = \mu = 0.53, \quad \sigma_{\bar{x}} = \frac{\sigma}{\sqrt{n}} = \frac{0.193}{\sqrt{50}} = 0.0273$$

b. $P(\bar{x} > 0.58) = P\left(z > \frac{0.58-0.53}{0.0273}\right) = P(z > 1.83) = 0.5 - 0.4664 = 0.0336$

(using Table II, Appendix B)

c. $P(\bar{x} > 0.59 \mid \mu = 0.58) = P\left(z > \frac{0.59-0.58}{0.0273}\right) = P(z > 0.37) = 0.5 - 0.1443 = 0.3557$

(using Table II, Appendix B) The above is the probability *after tensioning and loading.*

$P(\bar{x} > 0.59 \mid \mu = 0.53) = P\left(z > \frac{0.59-0.53}{0.0273}\right) = P(z > 2.20) = 0.5 - 0.4772 = 0.0228$

(using Table II, Appendix B) The above is the probability *before tensioning.*

Since the probability *after tensioning* is not small, this would not be an unusual event. Since the probability before tensioning is very small, this would be an unusual event. The sample measurements were probably obtained after tensioning and loading.

6.91 Even though the number of flaws per piece of siding has a Poisson distribution, the Central Limit Theorem implies that the distribution of the sample mean will be approximately normal with $\mu_{\bar{x}} = \mu = 2.5$ and $\sigma_{\bar{x}} = \frac{\sigma}{\sqrt{n}} = \frac{\sqrt{2.5}}{\sqrt{35}} = 0.2673$. Therefore,

$$P(\bar{x} > 2.1) = P\left(z > \frac{2.1-2.5}{\sqrt{2.5}/\sqrt{35}}\right) = P(z > -1.50) = 0.5 + 0.4332 = 0.9332$$

6.93 Answers will vary. We are to assume that the fecal bacteria concentrations of water specimens follow an approximate normal distribution. Now, suppose that the distribution of the fecal bacteria concentration at a beach is normal with a true mean of 360 with a standard deviation of 40. If only a single sample was selected, then the probability of getting an observation at the 400 level or higher would be:

$$P(x \geq 400) = P\left(z \geq \frac{400-360}{40}\right) = P(z \geq 1) = 0.5 - 0.3413 = 0.1587$$

(Using Table II, Appendix B)

Thus, even if the water is safe, the beach would be closed approximately 15.87% of the time.

On the other hand, if the mean was 440 and the standard deviation was still 40, then the probability of getting a single observation less than the 400 level would also be .1587. Thus, the beach would remain open approximately 15.78% of the time when it should be closed.

Now, suppose we took a random sample of 64 water specimens. The sampling distribution of $\bar{x}$ is approximately normal by the Central Limit Theorem with $\mu_{\bar{x}} = \mu$ and

$$\sigma_{\bar{x}} = \frac{\sigma}{\sqrt{n}} = \frac{40}{\sqrt{64}} = 5.$$

If $\mu = 360$, $P(\bar{x} \geq 400) = P\left(z \geq \frac{400-360}{5}\right) = P(z \geq 8) \approx 0.5 - 0.5 = 0$. Thus, the beach would never be shut down if the water was actually safe if we took samples of size 64.

If $\mu = 440$, $P(\bar{x} \leq 400) = P\left(z \leq \frac{400-440}{5}\right) = P(z \leq -8) \approx 0.5 - 0.5 = 0$. Thus, the beach would never be left open if the water was actually unsafe if we took samples of size 64.

The single sample standard can lead to unsafe decisions or inconvenient decisions, but is much easier to collect than samples of size 64.

Chapter 7

Inferences Based on a Single Sample
Estimation with Confidence Intervals

7.1 The unknown population parameter (e.g., mean or proportion) that we are interested in estimating is called the target parameter.

7.3 An interval estimator estimates μ with a range of values, while a point estimator estimates μ with a single point.

7.5 Yes. As long as the sample size is sufficiently large, the Central Limit Theorem says the distribution of $\bar{x}$ is approximately normal regardless of the original distribution.

7.7 a. For $\alpha = 0.10$, $\alpha/2 = 0.10/2 = 0.05$. $z_{\alpha/2} = z_{0.05}$ is the z-score with 0.05 of the area to the right of it. The area between 0 and $z_{0.05}$ is $0.5 - 0.05 = 0.4500$. Using Table II, Appendix B, $z_{0.05} = 1.645$.

b. For $\alpha = 0.01$, $\alpha/2 = 0.01/2 = 0.005$. $z_{\alpha/2} = z_{0.005}$ is the z-score with 0.005 of the area to the right of it. The area between 0 and $z_{0.005}$ is $0.5 - 0.005 = 0.4950$. Using Table II, Appendix B, $z_{0.005} = 2.58$.

c. For $\alpha = 0.05$, $\alpha/2 = 0.05/2 = 0.025$. $z_{\alpha/2} = z_{0.025}$ is the z-score with 0.025 of the area to the right of it. The area between 0 and $z_{0.025}$ is $0.5 - 0.025 = 0.4750$. Using Table II, Appendix B, $z_{0.025} = 1.96$.

d. For $\alpha = 0.20$, $\alpha/2 = 0.20/2 = 0.10$. $z_{\alpha/2} = z_{0.10}$ is the z-score with 0.10 of the area to the right of it. The area between 0 and $z_{0.10}$ is $0.5 - 0.10 = 0.4000$. Using Table II, Appendix B, $z_{0.10} = 1.28$.

7.9 a. For confidence coefficient 0.95, $\alpha = 0.05$ and $\alpha/2 = 0.05/2 = 0.025$. From Table II, Appendix B, $z_{0.025} = 1.96$. The confidence interval is:

$$\bar{x} \pm z_{0.025}\frac{\sigma}{\sqrt{n}} \Rightarrow 28 \pm 1.96\frac{20}{\sqrt{75}} \Rightarrow 28 \pm 4.53 \Rightarrow (23.47, 32.53)$$

b. $$\bar{x} \pm z_{0.025}\frac{\sigma}{\sqrt{n}} \Rightarrow 102 \pm 1.96\frac{20}{\sqrt{200}} \Rightarrow 102 \pm 2.77 \Rightarrow (99.23, 104.77)$$

c. $$\bar{x} \pm z_{0.025}\frac{\sigma}{\sqrt{n}} \Rightarrow 15 \pm 1.96\frac{20}{\sqrt{100}} \Rightarrow 15 \pm 3.92 \Rightarrow (11.08, 18.92)$$

d. $$\bar{x} \pm z_{0.025}\frac{\sigma}{\sqrt{n}} \Rightarrow 4.05 \pm 1.96\frac{20}{\sqrt{100}} \Rightarrow 4.05 \pm 3.92 \Rightarrow (0.13, 7.97)$$

e. No. Since the sample size in each part was large (n ranged from 75 to 200), the Central Limit Theorem indicates that the sampling distribution of $\bar{x}$ is approximately normal.

7.11 a. For confidence coefficient 0.95, $\alpha = 0.05$ and $\alpha/2 = 0.05/2 = 0.025$. From Table II, Appendix B, $z_{0.025} = 1.96$. The confidence interval is:

$$\bar{x} \pm z_{0.025}\frac{s}{\sqrt{n}} \Rightarrow 83.2 \pm 1.96\frac{6.4}{\sqrt{100}} \Rightarrow 83.2 \pm 1.25 \Rightarrow (81.95,\ 84.45)$$

b. The confidence coefficient of 0.95 means that in repeated sampling, 95% of all confidence intervals constructed will include μ.

c. For confidence coefficient 0.99, $\alpha = 0.01$ and $\alpha/2 = 0.01/2 = 0.005$. From Table II, Appendix B, $z_{0.005} = 2.58$. The confidence interval is:

$$\bar{x} \pm z_{0.005}\frac{s}{\sqrt{n}} \Rightarrow 83.2 \pm 2.58\frac{6.4}{\sqrt{100}} \Rightarrow 83.2 \pm 1.65 \Rightarrow (81.55,\ 84.85)$$

d. As the confidence coefficient increases, the width of the confidence interval also increases.

e. Yes. Since the sample size is 100, the Central Limit Theorem applies. This ensures the distribution of $\bar{x}$ is normal, regardless of the original distribution.

7.13 a. The confidence coefficient is 0.95.

b. For the officers diagnosed with hypertension, the interval is (4.1, 124.5). We are 95% confident that the true mean HRV for officers diagnosed with hypertension is between 4.1 and 124.5. For the officers who are not hypertensive, the interval is (148.0, 192.6). We are 95% confident that the true mean HRV for officers not hypertensive is between 148.0 and 192.6.

c. In repeated sampling, 95% of all confidence intervals constructed in this manner will contain the true mean.

d. To reduce the width of the confidence interval, you should use a smaller confidence coefficient. The interval will be smaller, thus, there will be less confidence.

7.15 a. The point estimate for the average number of latex gloves used per week by all healthcare workers with a latex allergy is $\bar{x} = 19.3$.

b. For confidence coefficient 0.95, $\alpha = 0.05$ and $\alpha/2 = 0.05/2 = 0.025$. From Table II, Appendix B, $z_{0.025} = 1.96$. The 95% confidence interval is:

$$\bar{x} \pm z_{0.025}\sigma_{\bar{x}} \Rightarrow \bar{x} \pm 1.96\frac{\sigma}{\sqrt{n}} \Rightarrow 19.3 \pm 1.96\frac{11.9}{\sqrt{46}} \Rightarrow 19.3 \pm 3.44 \Rightarrow (15.86,\ 22.74)$$

c. We are 95% confident that the average number of latex gloves used per week by all healthcare workers with a latex allergy is between 15.86 and 22.74.

d. We must assume that we have a random sample from the target population and that the sample size is sufficiently large.

7.17 For confidence coefficient 0.99, $\alpha = 0.01$ and $\alpha/2 = 0.01/2 = 0.005$. From Table II, Appendix B, $z_{0.005} = 2.58$. The confidence interval is:

$$\bar{x} \pm z_{0.005}\frac{s}{\sqrt{n}} \Rightarrow 39 \pm 2.58\frac{6}{\sqrt{100}} \Rightarrow 39 \pm 1.55 \Rightarrow (37.45,\ 40.55)$$

We are 99% confident that the true mean WR score for all convicted drug dealers is between 37.45 and 40.55.

7.19 a. Some preliminary calculations are:

$$\bar{x} = \frac{\sum x}{n} = \frac{181.56}{504} = 0.3602$$

$$s^2 = \frac{\sum x^2 - \frac{\left(\sum x\right)^2}{n}}{n-1} = \frac{74.6546 - \frac{181.56^2}{504}}{504-1} = \frac{9.24977143}{503} = 0.018389$$

$$s = \sqrt{0.018389} = 0.1356$$

For confidence coefficient 0.90, $\alpha = 0.10$ and $\alpha/2 = 0.10/2 = 0.05$. From Table II, Appendix A, $z_{0.05} = 1.645$. The 90% confidence interval is:

$$\bar{x} \pm z_{0.05}\sigma_{\bar{x}} \Rightarrow \bar{x} \pm 1.645\frac{\sigma}{\sqrt{n}} \Rightarrow 0.3602 \pm 1.645\frac{0.1356}{\sqrt{504}} \Rightarrow 0.3602 \pm 0.0099$$

$$\Rightarrow (0.3503,\ 0.3701)$$

b. We are 90% confident that the true mean visible albedo value of all Canadian Arctic ice ponds is between 0.3503 and 0.3701.

In repeated sampling 90% of all intervals constructed in the same manner will contain the true mean and 10% will not.

c. **First-year Ice**:

Some preliminary calculations are:

$$\bar{x} = \frac{\sum x}{n} = \frac{26.63}{88} = 0.3026$$

$$s^2 = \frac{\sum x^2 - \frac{\left(\sum x\right)^2}{n}}{n-1} = \frac{10.6923 - \frac{26.63^2}{88}}{88-1} = \frac{2.633698864}{87} = 0.03027$$

$$s = \sqrt{0.03027} = 0.1740$$

The 90% confidence interval is:

$$\overline{x} \pm z_{0.05}\sigma_{\overline{x}} \Rightarrow \overline{x} \pm 1.645\frac{\sigma}{\sqrt{n}} \Rightarrow 0.3026 \pm 1.645\frac{0.1740}{\sqrt{88}} \Rightarrow 0.3026 \pm 0.0305$$
$$\Rightarrow (0.2721,\ \ 0.3331)$$

We are 90% confident that the true mean visible albedo value of all First-year Canadian Arctic ice ponds is between 0.2721 and 0.3331.

Landfast Ice:

Some preliminary calculations are:

$$\overline{x} = \frac{\sum x}{n} = \frac{71.04}{196} = 0.3624$$

$$s^2 = \frac{\sum x^2 - \frac{\left(\sum x\right)^2}{n}}{n-1} = \frac{30.161 - \frac{71.04^2}{196}}{196-1} = \frac{4.41262449}{195} = 0.02263$$

$$s = \sqrt{0.02263} = 0.1504$$

The 90% confidence interval is:

$$\overline{x} \pm z_{0.05}\sigma_{\overline{x}} \Rightarrow \overline{x} \pm 1.645\frac{\sigma}{\sqrt{n}} \Rightarrow 0.3624 \pm 1.645\frac{0.1504}{\sqrt{196}} \Rightarrow 0.3624 \pm 0.0177$$
$$\Rightarrow (0.3447,\ \ 0.3801)$$

We are 90% confident that the true mean visible albedo value of all Landfast Canadian Arctic ice ponds is between 0.3447 and 0.3801.

Multi-year Ice:

Some preliminary calculations are:

$$\overline{x} = \frac{\sum x}{n} = \frac{83.89}{220} = 0.3813$$

$$s^2 = \frac{\sum x^2 - \frac{\left(\sum x\right)^2}{n}}{n-1} = \frac{33.8013 - \frac{83.89^2}{220}}{220-1} = \frac{1.81251773}{219} = 0.00827$$

$$s = \sqrt{0.00827} = 0.0910$$

The 90% confidence interval is:

$$\bar{x} \pm z_{0.05}\sigma_{\bar{x}} \Rightarrow \bar{x} \pm 1.645\frac{\sigma}{\sqrt{n}} \Rightarrow 0.3813 \pm 1.645\frac{0.0910}{\sqrt{220}} \Rightarrow 0.3813 \pm 0.0101$$
$$\Rightarrow (0.3712,\ 0.3914)$$

We are 90% confident that the true mean visible albedo value of all Multi-year Canadian Arctic ice ponds is between 0.3712 and 0.3914.

7.21 a. Using MINITAB, the 95% confidence interval for μ is:

One-Sample Z

```
The assumed standard deviation = 0.15

N     Mean    SE Mean        95% CI
55   1.9600    0.0202   (1.9204, 1.9996)
```

We are 95% confident that the true mean facial WHR for all CEOs at publically traded Fortune 500 firms is between 1.9204 and 1.9996.

b. No. Because 2.2 is not in the 95% confidence interval, it is not a likely value for the true value of the mean facial WHR. Therefore, we would not recommend using the value $\mu = 2.2$.

7.23 a. The parameter of interest is the mean effect size for all psychological studies of personality and aggressive behavior.

b. No, the distribution does not look normal. The data appear to be skewed to the right. The shape of this distribution is not of interest because the sample size is large, $n = 109$. By the Central Limit Theorem, the sampling distribution of $\bar{x}$ is normal regardless of the original distribution.

c. From the printout, the 95% confidence interval for μ is $(0.4786, 0.8167)$. We are 95% confident that the true mean effect size is between 0.4786 and 0.8167.

d. Yes. Since 0 is not contained in the 95% confidence interval, it is not a likely value for the true mean effect size. Since all the values in the 95% confident interval are above 0, the researchers are justified in concluding the true effect size mean is greater than 0 or that those who score high on a personality test are more aggressive than those who score low.

7.25 a. For confidence coefficient 0.99, $\alpha = 0.01$ and $\alpha / 2 = 0.01/2 = 0.005$. From Table II, Appendix B, $z_{0.005} = 2.58$. The confidence interval is:

$$\bar{x} \pm z_{0.005} \frac{s}{\sqrt{n}} \Rightarrow 1.13 \pm 2.58 \frac{2.21}{\sqrt{72}} \Rightarrow 1.13 \pm 0.672 \Rightarrow (0.458, 1.802)$$

We are 99% confident that the true mean number of pecks made by chickens pecking at blue string is between 0.458 and 1.802.

b. Yes, there is evidence that chickens are more apt to peck at white string. The mean number of pecks for white string is 7.5. Since 7.5 is not in the 99% confidence interval for the mean number of pecks at blue string, it is not a likely value for the true mean for blue string.

7.27 a. For confidence coefficient 0.95, $\alpha = 0.05$ and $\alpha / 2 = 0.05/2 = 0.025$. From Table II, Appendix B, $z_{0.025} = 1.96$.

For Males, the 95% confidence interval is:

$$\bar{x} \pm z_{0.025}\sigma_{\bar{x}} \Rightarrow \bar{x} \pm 1.96 \frac{\sigma}{\sqrt{n}} \Rightarrow 16.79 \pm 1.96 \frac{13.57}{\sqrt{128}} \Rightarrow 16.79 \pm 2.35 \Rightarrow (14.44, \ 19.14)$$

For Females, the 95% confidence interval is:

$$\bar{x} \pm z_{0.025}\sigma_{\bar{x}} \Rightarrow \bar{x} \pm 1.96 \frac{\sigma}{\sqrt{n}} \Rightarrow 10.79 \pm 1.96 \frac{11.53}{\sqrt{184}} \Rightarrow 10.79 \pm 1.67 \Rightarrow (9.12, \ 12.46)$$

b. Since the two intervals are independent, the probability that at least one of the 2 confidence intervals will not contain the population mean is equal to 1 minus the probability that neither of the 2 confidence intervals will not contain the population mean. Thus the probability is:

P(at least one interval will not contain mean)
= 1 – *P*(neither interval will not contain the mean)
= 1 – *P*(both will contain the mean) $= 1 - 0.95(0.95) = 1 - 0.9025 = 0.0975$.

c. The 95% confidence interval for the males is $(14.44, \ 19.14)$. The 95% confidence interval for the females is $(9.12, \ 12.46)$. Since all of the values in the male interval are larger than all the values in the female interval, we can infer that the males consume the most alcohol, on average, per week.

7.29 The two problems (and corresponding solutions) with using a small sample to estimate μ are:

1. The shape of the sampling distribution of the sample mean $\bar{x}$ now depends on the shape of the population that is sampled. The Central Limit Theorem no longer applies since the sample size is small. However, if the sampled population is normal or approximately normal, then the sampling distribution of $\bar{x}$ is always normal.

2. The population standard deviation σ is almost always unknown. Although it is still true that $\sigma_{\bar{x}} = \dfrac{\sigma}{\sqrt{n}}$, the sample standard deviation s may provide a poor approximation for σ when the sample size is small. Instead of using $z = \dfrac{\bar{x}-\mu}{\frac{\sigma}{\sqrt{n}}}$, we use $t = \dfrac{\bar{x}-\mu}{\frac{s}{\sqrt{n}}}$ as the statistic.

7.31 a. If x is normally distributed, the sampling distribution of $\bar{x}$ is normal, regardless of the sample size. If n is large, then the test statistic is z. If n is small, then the test statistic would be t.

b. If nothing is known about the distribution of x, the sampling distribution of $\bar{x}$ is approximately normal if n is sufficiently large. If n is not large, the distribution of $\bar{x}$ is unknown if the distribution of x is not known.

7.33 a. $P(t \geq t_0) = 0.025$ where $df = 10$

$t_0 = 2.228$

b. $P(t \geq t_0) = 0.01$ where $df = 17$

$t_0 = 2.567$

c. $P(t \leq t_0) = 0.005$ where $df = 6$

Because of symmetry, the statement can be rewritten

$P(t \geq -t_0) = 0.005$ $P(t \geq -t_0) = .005$ where $df = 6$

$t_0 = -3.707$

d. $P(t \leq -t_0) = 0.05$ where $df = 13$

$t_0 = -1.771$

7.35 First, we must compute $\bar{x}$ and s.

$$\bar{x} = \frac{\sum x}{n} = \frac{30}{6} = 5 \qquad s^2 = \frac{\sum x^2 - \frac{\left(\sum x\right)^2}{n}}{n-1} = \frac{176 - \frac{(30)^2}{6}}{6-1} = \frac{26}{5} = 5.2$$

$$s = \sqrt{5.2} = 2.2804$$

a. For confidence coefficient 0.90, $\alpha = 0.10$ and $\alpha/2 = 0.10/2 = 0.05$. From Table III, Appendix B, with $df = n-1 = 6-1 = 5$, $t_{0.05} = 2.015$. The 90% confidence interval is:

$$\bar{x} \pm t_{0.05}\frac{s}{\sqrt{n}} \Rightarrow 5 \pm 2.015\frac{2.2804}{\sqrt{6}} \Rightarrow 5 \pm 1.876 \Rightarrow (3.124,\ 6.876)$$

b. For confidence coefficient 0.95, $\alpha = 0.05$ and $\alpha/2 = 0.05/2 = 0.025$. From Table III, Appendix B, with $df = n - 1 = 6 - 1 = 5$, $t_{0.025} = 2.571$. The 95% confidence interval is:

$$\bar{x} \pm t_{0.025}\frac{s}{\sqrt{n}} \Rightarrow 5 \pm 2.571\frac{2.2804}{\sqrt{6}} \Rightarrow 5 \pm 2.394 \Rightarrow (2.606,\ 7.394)$$

c. For confidence coefficient 0.99, $\alpha = 0.01$ and $\alpha/2 = 0.01/2 = 0.005$. From Table III, Appendix B, with $df = n - 1 = 6 - 1 = 5$, $t_{0.005} = 4.032$. The 99% confidence interval is:

$$\bar{x} \pm t_{0.005}\frac{s}{\sqrt{n}} \Rightarrow 5 \pm 4.032\frac{2.2804}{\sqrt{6}} \Rightarrow 5 \pm 3.754 \Rightarrow (1.246,\ 8.754)$$

d. a) For confidence coefficient 0.90, $\alpha = 0.10$ and $\alpha/2 = 0.10/2 = 0.05$. From Table III, Appendix B, with $df = n - 1 = 25 - 1 = 24$, $t_{0.05} = 1.711$. The 90% confidence interval is:

$$\bar{x} \pm t_{0.05}\frac{s}{\sqrt{n}} \Rightarrow 5 \pm 1.711\frac{2.2804}{\sqrt{25}} \Rightarrow 5 \pm .780 \Rightarrow (4.220,\ 5.780)$$

b) For confidence coefficient 0.95, $\alpha = 0.05$ and $\alpha/2 = 0.05/2 = 0.025$. From Table III, Appendix B, with $df = n - 1 = 25 - 1 = 24$, $t_{0.025} = 2.064$. The 95% confidence interval is:

$$\bar{x} \pm t_{0.025}\frac{s}{\sqrt{n}} \Rightarrow 5 \pm 2.064\frac{2.2804}{\sqrt{25}} \Rightarrow 5 \pm .941 \Rightarrow (4.059,\ 5.941)$$

c) For confidence coefficient 0.99, $\alpha = 0.01$ and $\alpha/2 = 0.01/2 = 0.005$. From Table III, Appendix B, with $df = n - 1 = 25 - 1 = 24$, $t_{0.005} = 2.797$. The 99% confidence interval is:

$$\bar{x} \pm t_{0.005}\frac{s}{\sqrt{n}} \Rightarrow 5 \pm 2.797\frac{2.2804}{\sqrt{25}} \Rightarrow 5 \pm 1.276 \Rightarrow (3.724,\ 6.276)$$

Increasing the sample size decreases the width of the confidence interval.

7.37 a. Using MINITAB, the descriptive statistics are:

Descriptive Statistics: PAI

```
Variable  N   Mean  StDev
PAI       8  45.63   6.59
```

The sample mean is $\bar{x} = 45.63$.

b. From the output in part a, the sample standard deviation is $s = 6.59$.

c. For confidence coefficient 0.95, $\alpha = 0.05$ and $\alpha/2 = 0.05/2 = 0.025$. From Table III, Appendix B, with $df = n - 1 = 8 - 1 = 7$, $t_{0.025} = 2.365$. The 95% confidence interval is:

$$\bar{x} \pm t_{0.025}\frac{s}{\sqrt{n}} \Rightarrow 45.63 \pm 2.365\frac{6.59}{\sqrt{8}} \Rightarrow 45.63 \pm 5.510 \Rightarrow (40.120,\ 51.140)$$

d. In order for the interval in part c to be valid, the population of PAI values for all music performance anxiety studies should be normally distributed.

e. In repeated sampling, 95% of the confidence intervals will contain the true value of μ.

7.39 Using MINITAB, the 95% confidence interval is:

One-Sample T

N	Mean	StDev	SE Mean	95% CI
12	3643	1187	343	(2889, 4397)

The 95% confidence interval is $(2,889, 4,397)$. We are 95% confident that the true mean level of radon exposure in tombs in the Valley of Kings is between 2,889 and 4,397 Bq/m^3.

7.41 a. The target parameter is μ = the mean trap spacing for the population of red spiny lobster fishermen fishing in Baja California Sur, Mexico.

b. Using MINITAB, the descriptive statistics are:

Descriptive Statistics: Spacing

Variable	N	Mean	StDev	Minimum	Q1	Median	Q3	Maximum
Spacing	7	89.86	11.63	70.00	82.00	93.00	99.00	105.00

The point estimate for μ is $\bar{x} = 89.86$.

c. We do not know what the distribution of the population of trap spacings is. The Central Limit Theorem also does not apply because the sample size $(n = 7)$ is too small. The sample standard deviation, *s*, is no longer necessarily a good estimate for the population standard deviation σ.

d. For confidence coefficient 0.95, $\alpha = 0.05$ and $\alpha / 2 = 0.05 / 2 = 0.025$. From Table III, Appendix B, with $df = n - 1 = 7 - 1 = 6$, $t_{0.025} = 2.447$. The confidence interval is:

$$\bar{x} \pm t_{0.025,6} \frac{s}{\sqrt{n}} \Rightarrow 89.86 \pm 2.447 \frac{11.63}{\sqrt{7}} \Rightarrow 89.86 \pm 10.756 \Rightarrow (79.104,\ 100.616)$$

e. We are 95% confident that the mean trap spacing for the population of red spiny lobster fishermen fishing in Baja California Sur, Mexico is between 79.104 and 100.616 meters.

f. We must assume that the population of trap spacings is normal and that a random sample of trapping spacings was selected.

7.43 a. From the printout, the 95% confidence interval is $(652.76,\ 817.40)$.

b. For confidence coefficient 0.95, $\alpha = 0.05$ and $\alpha / 2 = 0.05 / 2 = 0.025$. From Table III, Appendix B, with $df = n - 1 = 12 - 1 = 11$, $t_{0.025} = 2.201$. The confidence interval is:

$$\bar{x} \pm t_{0.025,11}\frac{s}{\sqrt{n}} \Rightarrow 735.08 \pm 2.201\frac{129.565}{\sqrt{12}} \Rightarrow 735.08 \pm 82.32 \Rightarrow (652.76,\ 817.40)$$

c. We are 95% confident that the true mean number of minutes of daylight per day in Sharon, PA is between 652.76 and 817.40 minutes.

d. If a random sample of size 12 was taken from the population, then we could see most observations in one season, like summer or winter. Since the amount of daylight is greatly affected by the time of the year, this might not be the best plan. By sampling one observation from each month, we are guaranteed that the sample will be fairly representative. However, it will not be random.

7.45 For confidence coefficient 0.90, $\alpha = 0.10$ and $\alpha/2 = 0.10/2 = 0.05$. From Table III, Appendix B, with $df = n-1 = 17-1 = 16$, $t_{0.05} = 1.746$. The confidence interval is:

$$\bar{x} \pm t_{0.05,16}\frac{s}{\sqrt{n}} \Rightarrow 0.11 \pm 1.746\frac{0.19}{\sqrt{17}} \Rightarrow 0.11 \pm 0.08 \Rightarrow (0.03,\ 0.19)$$

We are 90% confident that the true mean score difference for all amusiacs is between 0.03 and 0.19.

We must assume that the population of score differences is approximately normal and that the data were randomly selected.

7.47 a. Using MINITAB, the results are:

One-Sample T: Velocity

```
Variable   N     Mean    StDev  SE Mean         95% CI
Velocity  25  0.26208  0.04669  0.00934  (0.24281, 0.28135)
```

The 95% confidence interval is $(0.24281,\ 0.28135)$. We are 95% confident that the true mean bubble rising velocity is between 0.24281 and 0.28135.

b. No. The mean value of 0.338 does not fall in the 95% confidence interval. Thus, it is not a likely value for the true mean. Thus, it would be very unlikely that these data were generated at this sparging rate.

7.49 a. **Both Untreated**: For confidence coefficient 0.90, $\alpha = 0.10$ and $\alpha/2 = 0.10/2 = 0.05$. From Table III, Appendix B, with $df = n-1 = 29-1 = 28$, $t_{0.05} = 1.701$. The 90% confidence interval is:

$$\bar{x} \pm t_{0.05,28}\frac{s}{\sqrt{n}} \Rightarrow 20.9 \pm 1.701\frac{3.34}{\sqrt{29}} \Rightarrow 20.9 \pm 1.055 \Rightarrow (19.845,\ 21.955)$$

Male Treated: For confidence coefficient 0.90, $\alpha = 0.10$ and $\alpha/2 = 0.10/2 = 0.05$. From Table III, Appendix B, with $df = n-1 = 23-1 = 22$, $t_{0.05} = 1.717$. The 90% confidence interval is:

$$\bar{x} \pm t_{0.05,22}\frac{s}{\sqrt{n}} \Rightarrow 20.3 \pm 1.717\frac{3.50}{\sqrt{23}} \Rightarrow 20.3 \pm 1.253 \Rightarrow (19.047,\ 21.553)$$

Female Treated: For confidence coefficient 0.90, $\alpha = 0.10$ and $\alpha/2 = 0.10/2 = 0.05$. From Table III, Appendix B, with $df = n-1 = 18-1 = 17$, $t_{0.05} = 1.740$. The 90% confidence interval is:

$$\bar{x} \pm t_{0.05,17}\frac{s}{\sqrt{n}} \Rightarrow 22.9 \pm 1.740\frac{4.37}{\sqrt{18}} \Rightarrow 22.9 \pm 1.792 \Rightarrow (21.108,\ 24.692)$$

Both Treated: For confidence coefficient 0.90, $\alpha = 0.10$ and $\alpha/2 = 0.10/2 = 0.05$. From Table III, Appendix B, with $df = n-1 = 21-1 = 20$, $t_{0.05} = 1.725$. The 90% confidence interval is:

$$\bar{x} \pm t_{0.05,20}\frac{s}{\sqrt{n}} \Rightarrow 18.6 \pm 1.725\frac{2.11}{\sqrt{21}} \Rightarrow 18.6 \pm 0.794 \Rightarrow (17.806,\ 19.394)$$

b. The female/male pair that appears to produce the highest mean number of eggs is the **Female treated** pair. The 90% confidence interval for this pair is the highest. This interval does overlap with the **Both treated** pair and the **Males treated** pair, but just barely.

7.51 a. For confidence coefficient 0.95, $\alpha = 0.05$ and $\alpha/2 = 0.05/2 = 0.025$. From Table III, Appendix B, with $df = n-1 = 15-1 = 14$, $t_{0.025} = 2.145$. The confidence interval is:

$$\bar{x} \pm t_{0.025}\frac{s}{\sqrt{n}} \Rightarrow 37.3 \pm 2.145\frac{13.9}{\sqrt{15}} \Rightarrow 37.3 \pm 7.70 \Rightarrow (29.60,\ 45.00)$$

b. We are 95% confident that the true mean adrenocorticotropin level in sleepers one-hour prior to anticipated waking is between 29.60 and 45.00

c. Suppose we assume that the true mean adrenocorticotropin level of sleepers three hours before anticipated wake-up time is 25.5. Since the 95% confidence interval for the true mean adrenocorticotropin level in sleepers one-hour prior to anticipated waking $(29.60,\ 45.00)$ does not contain 25.5, there is evidence to indicate that the true mean adrenocorticotropin level of sleepers one hour before anticipated wake-up time is greater than that of sleepers three hours prior to anticipated waking.

7.53 An unbiased estimator is one in which the mean of the sampling distribution is the parameter of interest, i.e., $E(\hat{p}) = p$.

7.55 a. The sample size is large enough if both $n\hat{p} \geq 15$ and $n\hat{q} \geq 15$.

$n\hat{p}=196(0.64)=125.44$ and $n\hat{q}=196(0.36)=70.56$. Since both of these numbers are greater than or equal to 15, the sample size is sufficiently large to conclude the normal approximation is reasonable.

b. For confidence coefficient 0.95, $\alpha=0.05$ and $\alpha/2=0.05/2=0.025$. From Table II, Appendix B, $z_{0.025}=1.96$. The confidence interval is:

$$\hat{p}\pm z_{0.025}\sqrt{\frac{\hat{p}\hat{q}}{n}}\Rightarrow 0.64\pm 1.96\sqrt{\frac{0.64(0.36)}{196}}\Rightarrow 0.64\pm 0.067\Rightarrow (0.537,\ 0.707)$$

c. We are 95% confident the true value of p is between 0.372 and 0.468.

d. "95% confidence" means that if repeated samples of size 196 were selected from the population and 95% confidence intervals formed, 95% of all confidence intervals will contain the true value of p.

7.57 The sample size is large enough if both $n\hat{p}\geq 15$ and $n\hat{q}\geq 15$.

a. $n\hat{p}=500(0.05)=25$ and $n\hat{q}=500(0.95)=475$. Since both of these numbers are greater than or equal to 15, the sample size is sufficiently large to conclude the normal approximation is reasonable.

b. $n\hat{p}=100(0.05)=5$ and $n\hat{q}=100(0.95)=95$. Since the first number is not greater than 15, the sample size is not sufficiently large to conclude the normal approximation is reasonable.

c. $n\hat{p}=10(0.5)=5$ and $n\hat{q}=10(0.5)=5$. Since neither of these numbers is greater than or equal to 15, the sample size is not sufficiently large to conclude the normal approximation is reasonable.

d. $n\hat{p}=10(0.3)=3$ and $n\hat{q}=10(0.7)=7$. Since neither of these numbers is greater than or equal to 15, the sample size is not sufficiently large to conclude the normal approximation is reasonable.

7.59 a. Of the 1,003 observations, 506 admitted having paid to download music $\Rightarrow \hat{p}=\frac{506}{1,003}=0.504$.

b. The sample size is large enough if both $n\hat{p}\geq 15$ and $n\hat{q}\geq 15$.

$n\hat{p}=1,003(0.504)=506$ and $n\hat{q}=1,003(0.496)=497$. Since both of these numbers are greater than or equal to 15, the sample size is sufficiently large to conclude the normal approximation is reasonable.

For the confidence coefficient 0.90, $\alpha = 0.10$ and $\alpha / 2 = 0.10 / 2 = 0.05$. From Table II, Appendix B, $z_{0.05} = 1.645$. The confidence interval is:

$$\hat{p} \pm z_{0.05}\sqrt{\frac{\hat{p}\hat{q}}{n}} \Rightarrow 0.504 \pm 1.645\sqrt{\frac{0.504(0.496)}{1,003}} \Rightarrow 0.504 \pm 0.026 \Rightarrow (0.478, 0.530)$$

c. We are 90% confident the true proportion of all adults living in the U.S. who have paid to download music is between 0.478 and 0.530.

d. In repeated sampling, 90% of all intervals constructed in a similar manner will contain the true proportion.

7.61 a. The population of interest is all America adults.

b. The sample in the study is the 1,000 American adults surveyed.

c. The parameter of interest is $p =$ proportion of all adults who say Starbucks coffee is overpriced.

d. The sample size is large enough if both $n\hat{p} \geq 15$ and $n\hat{q} \geq 15$. For this problem, $\hat{p} = .73$. $n\hat{p} = 1000(0.73) = 730$ and $n\hat{q} = 1000(0.27) = 270$. Since both of these numbers are greater than or equal to 15, the sample size is sufficiently large to conclude the normal approximation is reasonable.
For confidence coefficient 0.95, $\alpha = 0.05$ and $\alpha / 2 = 0.05 / 2 = 0.025$. From Table II, Appendix B, $z_{0.025} = 1.96$. The 95% confidence interval is:

$$\hat{p} \pm z_{0.025}\sqrt{\frac{pq}{n}} \approx \hat{p} \pm 1.96\sqrt{\frac{\hat{p}\hat{q}}{n}} \Rightarrow 0.73 \pm 1.96\sqrt{\frac{0.73(0.27)}{1000}} \Rightarrow 0.73 \pm 0.028 \Rightarrow (0.702, 0.758)$$

We are 95% confident that the true proportion of American adults who think Starbucks coffee is overpriced is between 0.702 and 0.758.

7.63 a. The population of interest is the set of gun ownerships (Yes or No) of all adults in the U.S.

b. The parameter of interest is the true percentage or proportion, p, of all adults in the U.S. who own at least one gun.

c. The estimate of the population proportion is $\hat{p} = .26$. The estimate of the population percentage is $0.26(100\%) = 26\%$.

d. The sample size is large enough if both $n\hat{p} \geq 15$ and $n\hat{q} \geq 15$.

$n\hat{p} = 2,770(0.26) = 720.2$ and $n\hat{q} = 2,770(0.74) = 2,049.8$. Since both of these numbers are greater than or equal to 15, the sample size is sufficiently large to conclude the normal approximation is reasonable.

For confidence coefficient 0.99, $\alpha = 0.01$ and $\alpha/2 = 0.01/2 = 0.005$. From Table II, Appendix B, $z_{0.005} = 2.58$. The 99% confidence interval is:

$$\hat{p} \pm z_{0.005}\sqrt{\frac{pq}{n}} \Rightarrow \hat{p} \pm 2.58\sqrt{\frac{\hat{p}\hat{q}}{n}} \Rightarrow 0.26 \pm 2.58\sqrt{\frac{0.26(0.74)}{2,770}} \Rightarrow 0.26 \pm 0.02 \Rightarrow (0.24,\ 0.28)$$

The 99% confidence interval for the true percentage is $(24\%,\ 28\%)$.

e. We are 99% confident that the true percentage of adults in the U.S who own at least one gun is between 24% and 28%.

f. "99% confidence" means that in repeated sampling 99% of all confidence intervals constructed in this manner will contain the true percentage.

7.65 a. The sample size is large enough if both $n\hat{p} \geq 15$ and $n\hat{q} \geq 15$.

$n\hat{p} = 1,000(0.63) = 630$ and $n\hat{q} = 1,000(0.37) = 370$. Since both of the numbers are greater than or equal to 15, the sample size is sufficiently large to conclude the normal approximation is reasonable.

For confidence coefficient 0.95, $\alpha = 0.05$ and $\alpha/2 = 0.05/2 = 0.025$. From Table II, Appendix B, $z_{0.025} = 1.96$. The 95% confidence interval is:

$$\hat{p} \pm z_{0.025}\sqrt{\frac{pq}{n}} \Rightarrow \hat{p} \pm 1.96\sqrt{\frac{\hat{p}\hat{q}}{n}} \Rightarrow 0.63 \pm 1.96\sqrt{\frac{0.63(0.37)}{1,000}} \Rightarrow 0.63 \pm 0.030 \Rightarrow (0.600,\ 0.660)$$

b. Yes, we would be surprised. Since 0.70 is not in the 95% confidence interval, it is not a likely value for the true value of the population proportion of adults who would choose to sleep when home sick.

7.67 a. Of the 106 observations, 63 robots were designed with legs but no wheels $\Rightarrow \hat{p} = \frac{63}{106} = 0.594$.

The sample size is large enough if both $n\hat{p} \geq 15$ and $n\hat{q} \geq 15$.

$n\hat{p} = 106(0.594) = 63$ and $n\hat{q} = 106(0.406) = 43$. Since both of these numbers are greater than or equal to 15, the sample size is sufficiently large to conclude the normal approximation is reasonable.

For the confidence coefficient 0.99, $\alpha = 0.01$ and $\alpha/2 = 0.01/2 = 0.005$. From Table II, Appendix B, $z_{0.005} = 2.58$. The confidence interval is:

$$\hat{p} \pm z_{0.005}\sqrt{\frac{\hat{p}\hat{q}}{n}} \Rightarrow 0.594 \pm 2.58\sqrt{\frac{0.594(0.406)}{106}} \Rightarrow 0.594 \pm 0.123 \Rightarrow (0.471,\ 0.717)$$

We are 99% confident the true proportion of all social robots designed with legs and no wheels is between 0.471 and 0.717.

b. Because 40% does not fall in the 99% confidence interval, the assumption is probably not valid.

7.69 The estimate of the true proportion of all U. S. teenagers who have used at least one informal element in a school writing assignment is $\hat{p}=\frac{448}{700}=0.64$.

The sample size is large enough if both $n\hat{p}\geq 15$ and $n\hat{q}\geq 15$. For this problem, $\hat{p}=0.64$. $n\hat{p}=700(0.64)=448$ and $n\hat{q}=700(0.36)=252$. Since both of these numbers are greater than or equal to 15, the sample size is sufficiently large to conclude the normal approximation is reasonable.

For confidence coefficient 0.99, $\alpha=0.01$ and $\alpha/2=0.01/2=0.005$. From Table II, Appendix B, $z_{0.005}=2.58$. The 99% confidence interval is:

$$\hat{p}\pm z_{0.005}\sqrt{\frac{pq}{n}}\approx\hat{p}\pm 2.58\sqrt{\frac{\hat{p}\hat{q}}{n}}\Rightarrow 0.64\pm 2.58\sqrt{\frac{0.64(0.36)}{700}}\Rightarrow 0.64\pm 0.047\Rightarrow(0.593,\ 0.687)$$

We are 99% confident that the true proportion of all U. S. teenagers who have used at least one informal element in a school writing assignment is between 0.593 and 0.687.

7.71 a. The estimate of the true proportion of all mountain casualties that require a femoral shaft splint is $\hat{p}=\frac{1}{333}=0.003$.

The sample size is large enough if both $n\hat{p}\geq 15$ and $n\hat{q}\geq 15$. For this problem, $\hat{p}=0.003$. $n\hat{p}=333(0.003)=0.999$ and $n\hat{q}=333(0.997)=332.001$. Since the first of these numbers is less than to 15, the sample size is not sufficiently large to conclude the normal approximation is reasonable.

b. The Wilson adjusted sample proportion is $\tilde{p}=\frac{x+2}{n+4}=\frac{1+2}{333+4}=\frac{3}{337}=0.009$.

For confidence coefficient 0.95, $\alpha=0.05$ and $\alpha/2=0.05/2=0.025$. From Table II, Appendix B, $z_{0.025}=1.96$. The Wilson adjusted 95% confidence interval is:

$$\tilde{p}\pm z_{0.025}\sqrt{\frac{\tilde{p}(1-\tilde{p})}{n+4}}\Rightarrow 0.009\pm 1.96\sqrt{\frac{0.009(0.991)}{333+4}}\Rightarrow 0.009\pm 0.010\Rightarrow(-0.001,\ 0.019)$$

We are 95% confident that the true proportion of all mountain casualties that require a femoral shaft splint is between 0 and 0.019.

7.73 We will find 95% confidence interval for both proportions.

<u>Eye Shadow</u>:

Of the 131 observations, 12 were diagnosed with a nickel allergy $\Rightarrow\hat{p}=\frac{12}{131}=0.092$.

The sample size is large enough if both $n\hat{p} \geq 15$ and $n\hat{q} \geq 15$.

$n\hat{p} = 131(0.092) = 12$ and $n\hat{q} = 131(0.908) = 119$. Because $n\hat{p} < 15$, the sample size is not sufficiently large to conclude the normal approximation is reasonable. We will use Wilson's adjustment.

Using Wilson's adjustment, $\tilde{p} = \dfrac{x+2}{n+4} = \dfrac{12+2}{131+4} = \dfrac{14}{135} = 0.104$.

For the confidence coefficient 0.95, $\alpha = 0.05$ and $\alpha / 2 = 0.05 / 2 = 0.025$. From Table II, Appendix B, $z_{0.025} = 1.96$. The confidence interval is:

$$\tilde{p} \pm z_{0.025}\sqrt{\frac{\tilde{p}\tilde{q}}{n+4}} \Rightarrow 0.104 \pm 1.96\sqrt{\frac{0.104(0.896)}{131+4}} \Rightarrow 0.104 \pm 0.051 \Rightarrow (0.053,\ 0.155)$$

We are 95% confident that the true proportion of women with dermatitis from using eye shadow who have a nickel allergy is between 0.053 and 0.155.

<u>**Mascara**</u>:

Of the 250 observations, 25 were diagnosed with a nickel allergy $\Rightarrow \hat{p} = \dfrac{25}{250} = 0.10$.

The sample size is large enough if both $n\hat{p} \geq 15$ and $n\hat{q} \geq 15$.

$n\hat{p} = 2500(0.10) = 25$ and $n\hat{q} = 250(0.90) = 225$. Since both of these numbers are greater than or equal to 15, the sample size is sufficiently large to conclude the normal approximation is reasonable.

For the confidence coefficient 0.95, $\alpha = 0.05$ and $\alpha / 2 = 0.05 / 2 = 0.025$. From Table II, Appendix B, $z_{0.025} = 1.96$. The confidence interval is:

$$\hat{p} \pm z_{0.025}\sqrt{\frac{\hat{p}\hat{q}}{n}} \Rightarrow 0.10 \pm 1.96\sqrt{\frac{0.10(0.90)}{250}} \Rightarrow 0.10 \pm 0.037 \Rightarrow (0.063,\ 0.137)$$

We are 95% confident that the true proportion of women with dermatitis from using mascara who have a nickel allergy is between 0.063 and 0.137.

Because both confidence intervals contain 0.12, we cannot determine which group is referenced.

7.75 The statement "For a specified sampling error SE, increasing the confidence level $(1-\alpha)$ will lead to a larger n when determining sample size" is True.

7.77 To compute the necessary sample size, use

$$n = \frac{(z_{\alpha/2})^2 \sigma^2}{(SE)^2} \text{ where } \alpha = 1 - 0.95 = 0.05 \text{ and } \alpha/2 = 0.05/2 = 0025$$

From Table II, Appendix B, $z_{0.025} = 196$. Thus, $n = \frac{(1.96)^2(5.4)}{(0.2)^2} = 518.616 \approx 519$

You would need to take 519 samples.

7.79 a. $\text{Range} = 39 - 31 = 8$. $\sigma \approx \frac{\text{Range}}{4} = \frac{8}{4} = 2$

For confidence coefficient 0.90, $\alpha = 0.10$ and $\alpha/2 = 0.10/2 = 0.05$. From Table II, Appendix A, $z_{0.05} = 1.645$.

The sample size is $n = \frac{z^2_{\alpha/2}\sigma^2}{(SE)^2} = \frac{(1.645)^2(2)^2}{(0.15)^2} = 481.07 \approx 482$

b. $\sigma \approx \frac{\text{Range}}{6} = \frac{8}{6} = 1.333$

The sample size is $n = \frac{z^2_{\alpha/2}\sigma^2}{(SE)^2} = \frac{(1.645)^2(1.333)^2}{(0.15)^2} = 213.7 \approx 214$

7.81 For confidence coefficient 0.90, $\alpha = 0.10$ and $\alpha/2 = 0.10/2 = 0.05$. From Table II, Appendix A, $z_{0.05} = 1.645$.

We know $\hat{p}$ is in the middle of the interval, so $\hat{p} = \frac{0.54 + 0.26}{2} = 0.4$

The confidence interval is $\hat{p} \pm z_{0.05}\sqrt{\frac{\hat{p}\hat{q}}{n}} \Rightarrow 0.4 \pm 1.645\sqrt{\frac{0.4(0.6)}{n}}$

We know $0.4 - 1.645\sqrt{\frac{0.4(0.6)}{n}} = 0.26$

$$\Rightarrow 0.4 - \frac{0.8059}{\sqrt{n}} = 0.26$$

$$\Rightarrow 0.4 - 0.26 = \frac{0.8059}{\sqrt{n}} \Rightarrow \sqrt{n} = \frac{0.8059}{0.14} = 5.756$$

$$\Rightarrow n = 5.756^2 = 33.1 \approx 34$$

7.83 a. The width of a confidence interval is $2(SE) = 2z_{\alpha/2}\frac{\sigma}{\sqrt{n}}$

For confidence coefficient 0.95, $\alpha = 0.05$ and $\alpha/2 = 0.05/2 = 0.025$. From Table II, Appendix B, $z_{0.025} = 1.96$.

For $n = 16$, $W = 2z_{\alpha/2}\frac{\sigma}{\sqrt{n}} = 2(1.96)\frac{1}{\sqrt{16}} = 0.98$

For $n = 25$, $W = 2z_{\alpha/2}\frac{\sigma}{\sqrt{n}} = 2(1.96)\frac{1}{\sqrt{25}} = 0.784$

For $n = 49$, $W = 2z_{\alpha/2}\frac{\sigma}{\sqrt{n}} = 2(1.96)\frac{1}{\sqrt{49}} = 0.56$

For $n = 100$, $W = 2z_{\alpha/2}\frac{\sigma}{\sqrt{n}} = 2(1.96)\frac{1}{\sqrt{100}} = 0.392$

For $n = 400$, $W = 2z_{\alpha/2}\frac{\sigma}{\sqrt{n}} = 2(1.96)\frac{1}{\sqrt{400}} = 0.196$

b.

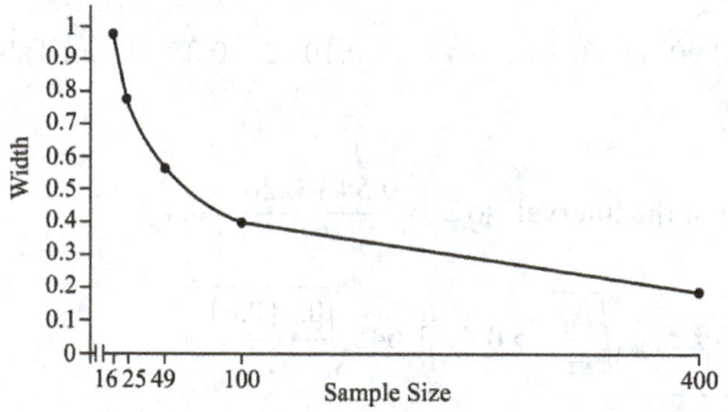

7.85 a. No. If we increase the confidence level, the width of the interval will increase.

b. Yes. Increasing the sample size will reduce the width of the confidence interval.

c. For confidence coefficient 0.90, $\alpha = 0.10$ and $\alpha/2 = 0.10/2 = 0.05$. From Table II, Appendix B, $z_{0.05} = 1.645$. From Exercise 7.46, $s = 2.351$.

$$n = \frac{z_{\alpha/2}^2\sigma^2}{(SE)^2} = \frac{(1.645)^2(2.351)^2}{(0.5)^2} = 59.8 \approx 60$$

Thus, we will need to sample approximately 60 shafts.

7.87 From Exercise 7.41, the sample standard deviation is $s = 11.63$. We will use this to estimate the population standard deviation. For the confidence coefficient 0.95, $\alpha = 0.05$ and $\alpha/2 = 0.05/2 = 0.025$. From Table II, Appendix B, $z_{0.025} = 1.96$. The necessary sample size is:

$$n = \frac{z_{\alpha/2}^2 \sigma^2}{(SE)^2} = \frac{(1.96)^2 (11.63)^2}{5^2} = 20.78 \approx 21$$

Thus, we will need to sample approximately 21 teams of fishermen.

7.89 From Exercise 7.67, the sample proportion was $\hat{p} = 0.594$. We will use this as our estimate of the population proportion, *p*. For confidence coefficient 0.99, $\alpha = 0.01$ and $\alpha/2 = 0.01/2 = 0.005$. From Table II, Appendix B, $z_{0.005} = 2.58$.

$$n = \frac{z_{\alpha/2}^2 pq}{(SE)^2} = \frac{(2.58)^2 (0.594)(0.406)}{(0.075)^2} = 285.4 \approx 286$$

Thus, we will need to sample approximately 286 social robots.

7.91 From Exercise 7.45, the standard deviation was $s = 0.19$. We will use this to estimate the population standard deviation. For confidence coefficient 0.90, $\alpha = 0.10$ and $\alpha/2 = 0.10/2 = 0.05$. From Table II, Appendix B, $z_{0.05} = 1.645$. The necessary sample size is:

$$n = \frac{z_{\alpha/2}^2 \sigma^2}{(SE)^2} = \frac{(1.645)^2 (0.19)^2}{(0.05)^2} = 39.075 \approx 40$$

Thus, we will need to sample approximately 40 amusiacs.

7.93 From Exercise 7.68, the sample proportion was $\hat{p} = 0.818$. We will use this as our estimate of the population proportion, *p*. For confidence coefficient 0.95, $\alpha = 0.05$ and $\alpha/2 = 0.05/2 = 0.025$. From Table II, Appendix B, $z_{0.025} = 1.96$.

$$n = \frac{z_{\alpha/2}^2 pq}{(SE)^2} = \frac{(1.96)^2 (0.818)(0.182)}{(0.05)^2} = 228.8 \approx 229$$

Thus, we will need to sample approximately 229 aircraft bird strikes.

7.95 From Exercise 7.72, our estimate of p is $\hat{p} = 0.021$.. We will use this to estimate the sample size. For confidence coefficient 0.99 $\alpha = 0.01$ and $\alpha/2 = 0.01/2 = 0.005$. From Table II, Appendix B, $z_{0.005} = 2.58$.

$$n = \frac{z_{\alpha/2}^2 pq}{(SE)^2} = \frac{(2.58)^2 (0.021)(0.979)}{(0.04)^2} = 85.5 \approx 86$$

We would need to sample 86 U.S. adults.

Now, the sample from Japan may not be representative because college students were used. To be extremely conservative, we could estimate the needed sample size by estimating p with 0.5. Now the needed sample size would be:

$$n = \frac{z_{\alpha/2}^2 pq}{(SE)^2} = \frac{(2.58)^2 (0.5)(0.5)}{(0.04)^2} = 1040.06 \approx 1041$$

This is much larger than the sample size found using the estimate from Exercise 7.72.

7.97 In Exercise 7.73, the sample proportion for those using eye shadow was $\hat{p} = 0.092$ and the sample proportion for those using mascara was $\hat{p} = 0.10$. We will use $\hat{p} = 0.10$ to estimate the population proportion. For confidence coefficient 0.95, $\alpha = 0.05$ and $\alpha / 2 = 0.05 / 2 = 0.025$. From Table II, Appendix B, $z_{0.025} = 1.96$.

$$n = \frac{z_{\alpha/2}^2 pq}{(SE)^2} = \frac{(1.96)^2 (0.10)(0.90)}{(0.03)^2} = 384.2 \approx 385$$

Thus, we will need to sample approximately 385 women.

7.99 The sampling distribution used to find interval estimates for σ^2 is the chi-square distribution.

7.101 The degrees of freedom associated with the chi-square sampling distribution for a sample of size n is $n-1$.

7.103 a. For confidence level 0.90, $\alpha = 0.10$ and $\alpha / 2 = 0.10 / 2 = 0.05$. From Table IV, Appendix B, with $df = n-1 = 50-1 = 49$, $\chi^2_{0.05,49} \approx 67.5048$ and $\chi^2_{0.95,49} \approx 34.7642$. The 90% confidence interval is:

$$\frac{(n-1)s^2}{\chi^2_{0.05}} \le \sigma^2 \le \frac{(n-1)s^2}{\chi^2_{0.95}} \Rightarrow \frac{(50-1)2.5^2}{67.5048} \le \sigma^2 \le \frac{(50-1)2.5^2}{34.7642} \Rightarrow 4.537 \le \sigma^2 \le 8.809$$

b. For confidence level 0.90, $\alpha = 0.10$ and $\alpha / 2 = 0.10 / 2 = 0.05$. From Table IV, Appendix B, with $df = n-1 = 15-1 = 14$, $\chi^2_{0.05,14} = 23.6848$ and $\chi^2_{0.95,14} = 6.57063$. The 90% confidence interval is:

$$\frac{(n-1)s^2}{\chi^2_{0.05}} \le \sigma^2 \le \frac{(n-1)s^2}{\chi^2_{0.95}} \Rightarrow \frac{(15-1)(0.02)^2}{23.6848} \le \sigma^2 \le \frac{(15-1)(0.02)^2}{6.57063} \Rightarrow 0.00024 \le \sigma^2 \le 0.00085$$

c. For confidence level 0.90, $\alpha = 0.10$ and $\alpha/2 = 0.10/2 = 0.05$. From Table IV, Appendix B, with $df = n-1 = 22-1 = 21$, $\chi^2_{0.05,21} = 32.6705$ and $\chi^2_{0.95,21} = 11.5913$. The 90% confidence interval is:

$$\frac{(n-1)s^2}{\chi^2_{0.05}} \le \sigma^2 \le \frac{(n-1)s^2}{\chi^2_{0.95}} \Rightarrow \frac{(22-1)(31.6)^2}{32.6705} \le \sigma^2 \le \frac{(22-1)(31.6)^2}{11.5913} \Rightarrow 641.86 \le \sigma^2 \le 1{,}809.09$$

d. For confidence level 0.90, $\alpha = 0.10$ and $\alpha/2 = 0.10/2 = 0.05$. From Table IV, Appendix B, with $df = n-1 = 5-1 = 4$, $\chi^2_{0.05,4} = 9.48773$ and $\chi^2_{0.95,4} = 0.710721$. The 90% confidence interval is:

$$\frac{(n-1)s^2}{\chi^2_{0.05}} \le \sigma^2 \le \frac{(n-1)s^2}{\chi^2_{0.95}} \Rightarrow \frac{(5-1)(1.5)^2}{9.48773} \le \sigma^2 \le \frac{(5-1)(1.5)^2}{0.710721} \Rightarrow 0.94859 \le \sigma^2 \le 12.6632$$

7.105 Using MINITAB, the descriptive statistics are:

Descriptive Statistics: x

```
Variable  N  Mean  StDev  Minimum    Q1  Median    Q3  Maximum
x         6  6.17   3.31     2.00  2.75    6.50  8.75    11.00
```

For confidence level 0.95, $\alpha = 0.05$ and $\alpha/2 = 0.05/2 = 0.025$. From Table IV, Appendix B, with $df = n-1 = 6-1 = 5$, $\chi^2_{0.025,5} = 12.8325$ and $\chi^2_{0.975,5} = 0.831211$. The 95% confidence interval is:

$$\frac{(n-1)s^2}{\chi^2_{0.025}} \le \sigma^2 \le \frac{(n-1)s^2}{\chi^2_{0.975}} \Rightarrow \frac{(6-1)(3.31)^2}{12.8325} \le \sigma^2 \le \frac{(6-1)(3.31)^2}{0.831211} \Rightarrow 4.2689 \le \sigma^2 \le 65.9044$$

7.107 a. From the printout, the 95% confidence interval for the variance is: $(8.6,\ 45.7)$. We are 95% confident that the true variance of rebound lengths is between 8.6 and 45.7.

b. From the printout, the 95% confidence interval for the standard deviation is: $(2.94,\ 6.76)$. We are 95% confident that the true standard deviation of rebound lengths is between 2.94 and 6.76.

c. We must assume that the sample was a random sample from the target population and that the target population is approximately normal.

7.109 a. For confidence level 0.95, $\alpha = 0.05$ and $\alpha/2 = 0.05/2 = 0.025$. From Table IV, Appendix B, with $df = n-1 = 55-1 = 54$, $\chi^2_{0.025,54} \approx 71.4202$ and $\chi^2_{0.975,54} \approx 32.3574$. The 95% confidence interval for σ^2 is:

$$\frac{(n-1)s^2}{\chi^2_{0.025}} \le \sigma^2 \le \frac{(n-1)s^2}{\chi^2_{0.975}} \Rightarrow \frac{(55-1)(0.15)^2}{71.4202} \le \sigma^2 \le \frac{(55-1)(0.15)^2}{32.3574} \Rightarrow 0.0170 \le \sigma^2 \le 0.0375$$

The 95% confidence interval for σ is: $\left(\sqrt{0.0170}, \sqrt{0.0375}\right) \Rightarrow (0.1304, 0.1936)$. We are 95% confident that the true population standard deviation of WHR values is between 0.1304 and 0.1936.

b. In order for the interval to be valid, the population of WHR values should be approximately normally distributed. A sketch of the required distribution is:

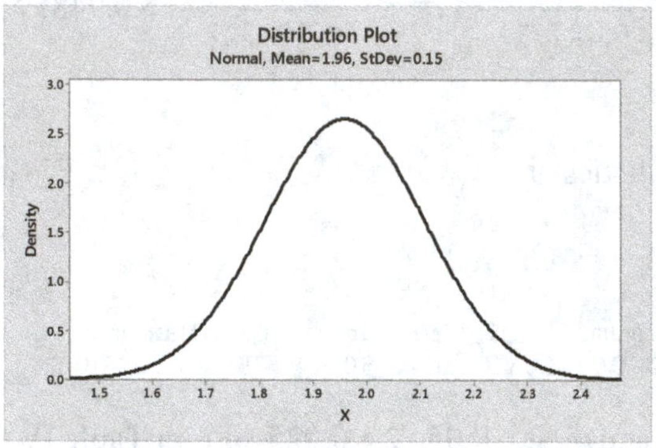

7.111 For confidence level 0.95, $\alpha = 0.05$ and $\alpha/2 = 0.05/2 = 0.025$. From Table IV, Appendix B, with $df = n-1 = 6-1 = 5$, $\chi^2_{0.025,5} = 12.8325$ and $\chi^2_{0.975,5} = 0.831211$. The 95% confidence interval for σ^2 is:

$$\frac{(n-1)s^2}{\chi^2_{0.025}} \le \sigma^2 \le \frac{(n-1)s^2}{\chi^2_{0.975}} \Rightarrow \frac{(6-1)(0.011)^2}{12.8325} \le \sigma^2 \le \frac{(6-1)(0.011)^2}{0.831211} \Rightarrow 0.000047 \le \sigma^2 \le 0.000728$$

The 95% confidence interval for σ is: $\left(\sqrt{0.000047}, \sqrt{0.000728}\right) \Rightarrow (0.0069, 0.0270)$. We are 95% confident that the true population standard deviation of the internal oil content for sweet potato chips is between 0.0069 and 0.0270.

7.113 Using MINITAB, the descriptive statistics are:

Descriptive Statistics: Depth

```
Variable   N    Mean  StDev  Minimum      Q1  Median      Q3  Maximum
Depth     18  16.499  1.970   13.250  15.285  16.160  18.123   19.700
```

For confidence level 0.95, $\alpha = 0.05$ and $\alpha/2 = 0.05/2 = 0.025$. From Table IV, Appendix B, with $df = n-1 = 18-1 = 17$, $\chi^2_{0.025,17} = 30.1910$ and $\chi^2_{0.975,17} = 7.56418$. The 95% confidence interval for the variance is:

$$\frac{(n-1)s^2}{\chi^2_{0.025}} \le \sigma^2 \le \frac{(n-1)s^2}{\chi^2_{0.975}} \Rightarrow \frac{(18-1)(1.97)^2}{30.1910} \le \sigma^2 \le \frac{(18-1)(1.97)^2}{7.56418} \Rightarrow 2.185 \le \sigma^2 \le 8.722$$

The 95% confidence interval for the standard deviation is:

$$\sqrt{2.185} \le \sigma \le \sqrt{8.722} \Rightarrow 1.478 \le \sigma \le 2.953$$

We are 95% confident that the true standard deviation of the molar depths for the population of cheek teeth in extinct primates is between 1.478 and 2.953.

We must assume that a random sample of observations was selected from a normal distribution.

Using MINITAB, a histogram of the data is:

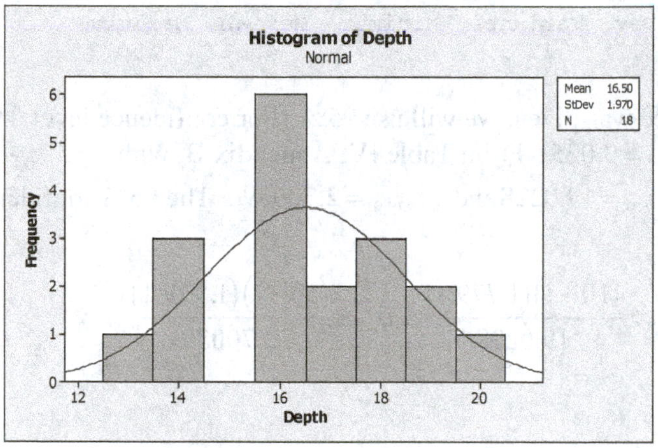

The data do not look particularly normal. The confidence interval for the standard deviation may not be valid.

7.115 From Exercise 7.41, $s = 11.63$.

For confidence level 0.99, $\alpha = 0.01$ and $\alpha / 2 = 0.01/2 = 0.005$. From Table IV, Appendix B, with $df = n-1 = 7-1 = 6$, $\chi^2_{0.005,6} = 18.5476$ and $\chi^2_{0.995,6} = 0.675727$. The 99% confidence interval for the variance is:

$$\frac{(n-1)s^2}{\chi^2_{0.005}} \le \sigma^2 \le \frac{(n-1)s^2}{\chi^2_{0.995}} \Rightarrow \frac{(7-1)(11.63)^2}{18.5476} \le \sigma^2 \le \frac{(7-1)(11.63)^2}{0.675727} \Rightarrow 43.755 \le \sigma^2 \le 1{,}200.990$$

We are 99% confident that the true variance of the trap spacing measurements of red spiny lobster fishermen fishing in Baja California Sur, Mexico is between 43.755 and 1,200.990.

7.117 a. Answers will vary. Using MINITAB, a random sample of 10 observations from this population is:

OBS Num	INTTIME
40	57.73
168	30.72

122	64.591
91	4.626
164	3.239
15	150.81
178	32.377
23	58.684
244	27.165
199	58.128

b. The descriptive statistics are:

Descriptive Statistics: Sample

```
Variable    N  Mean   StDev  Variance  Minimum  Median  Maximum
Sample     10  48.8    42.2    1779.1      3.2    45.1    150.8
```

Since no level of confidence was given, we will use 95%. For confidence level 0.95, $\alpha = 0.05$ and $\alpha/2 = 0.05/2 = 0.025$. From Table IV, Appendix B, with $df = n-1 = 10-1 = 9$, $\chi^2_{0.025,9} = 19.0228$ and $\chi^2_{0.975,9} = 2.70039$. The 95% confidence interval for σ^2 is:

$$\frac{(n-1)s^2}{\chi^2_{0.025}} \le \sigma^2 \le \frac{(n-1)s^2}{\chi^2_{0.975}} \Rightarrow \frac{(10-1)(1,779.1)}{19.0228} \le \sigma^2 \le \frac{(10-1)(1,779.1)}{2.70039}$$
$$\Rightarrow 841.72 \le \sigma^2 \le 5,929.48$$

We are 95% confident that the true population variance of interarrival times is between 841.72 and 5,929.48.

c. The true population variance is $\sigma^2 = 8,348.026$. This value does not fall in the 95% confidence interval. One reason why it may not fall in the confidence interval is because the interval may not be valid. One assumption necessary for the interval to be valid is that the population being sampling from is normal. In this case, the population of interarrival times is not normal, but skewed to the right.

7.119 a. The target parameter for the average score on the SAT is μ.

b. The target parameter for the mean time waiting at a supermarket checkout lane is μ.

c. The target parameter for the proportion of voters in favor of legalizing marijuana is p.

b. The target parameter for the percentage of NFL players who have ever made the Pro Bowl is p.

e. The target parameter for the dropout rate of American college students is p.

f. The target parameter for the variation in IQ scores of sociopaths is σ^2.

7.121 a. From Table IV, Appendix B, with $df = n-1 = 10-1 = 9$ and $\alpha/2 = 0.05/2 = 0.025$,

$\chi^2_{0.025,9} = 19.0228$ and $\chi^2_{0.975,9} = 2.70039$.

b. From Table IV, Appendix B, with $df = n-1 = 20-1 = 19$ and $\alpha/2 = 0.05/2 = 0.025$, $\chi^2_{0.025,19} = 32.8523$ and $\chi^2_{0.975,19} = 8.90655$.

c. Using MINITAB with $df = n-1 = 50-1 = 49$ and $\alpha/2 = 0.01/2 = 0.005$, $\chi^2_{0.005,49} = 78.2307$ and $\chi^2_{0.995,49} = 27.2493$.

7.123 a. The point estimate for the proportion of measurements in the population with characteristic A is $\hat{p} = \frac{x}{n} = \frac{227}{400} = 0.5675$.

The sample size is large enough if both $n\hat{p} \geq 15$ and $n\hat{q} \geq 15$.

$n\hat{p} = 400(0.5675) = 227$ and $n\hat{q} = 400(0.4325) = 173$. Since both of the numbers are greater than or equal to 15, the sample size is sufficiently large to conclude the normal approximation is reasonable.

For confidence coefficient 0.95, $\alpha = 0.05$ and $\alpha/2 = 0.05/2 = 0.025$. From Table II, Appendix B, $z_{0.025} = 1.96$. The 95% confidence interval is:

$$\hat{p} \pm z_{0.025}\sigma_{\hat{p}} \Rightarrow \hat{p} \pm 1.96\sqrt{\frac{\hat{p}\hat{q}}{n}} \Rightarrow 0.5675 \pm 1.96\sqrt{\frac{0.5675(0.4325)}{400}}$$

$$\Rightarrow 0.5675 \pm 0.0486 \Rightarrow (0.5189,\ 0.6161)$$

We are 95% confident that the proportion of measurements in the population with characteristic A is between 0.5189 and 0.6161.

b. We will use $\hat{p} = 0.5675$ to estimate p.

For confidence coefficient 0.95, $\alpha = 0.05$ and $\alpha/2 = 0.05/2 = 0.025$. From Table II, Appendix B, $z_{0.025} = 1.96$.

The sample size is $n = \frac{(z_{0.025})^2 pq}{(SE)^2} = \frac{(1.96)^2(0.5675)(0.4325)}{(0.02)^2} = 2{,}357.2 \approx 2{,}358$.

7.125 The parameters of interest for the problems are:

(1) The question requires a categorical response. One parameter of interest might be the proportion, p, of all Americans over 18 years of age who think their health is generally very good or excellent.

(2) A parameter of interest might be the mean number of days, μ, in the previous 30 days that all Americans over 18 years of age felt that their physical health was not good because of injury or illness.

(3) A parameter of interest might be the mean number of days, μ, in the previous 30 days that all Americans over 18 years of age felt that their mental health was not good because of stress, depression, or problems with emotions.

(4) A parameter of interest might be the mean number of days, μ, in the previous 30 days that all Americans over 18 years of age felt that their physical or mental health prevented them from performing their usual activities.

7.127 First, we compute $\hat{p}$: $\hat{p} = \frac{x}{n} = \frac{183}{837} = 0.219$

The sample size is large enough if both $n\hat{p} \geq 15$ and $n\hat{q} \geq 15$.

$n\hat{p} = 837(0.219) = 183.303$ and $n\hat{q} = 837(0.781) = 653.697$. Since both of the numbers are greater than or equal to 15, the sample size is sufficiently large to conclude the normal approximation is reasonable.

For confidence coefficient 0.90, $\alpha = 0.10$ and $\alpha/2 = 0.10/2 = 0.05$. From Table II, Appendix B, $z_{0.05} = 1.645$. The confidence interval is:

$$\hat{p} \pm z_{0.05}\sqrt{\frac{pq}{n}} \approx \hat{p} \pm 1.645\sqrt{\frac{\hat{p}\hat{q}}{n}} \Rightarrow 0.219 \pm 1.645\sqrt{\frac{0.219(0.781)}{837}} \Rightarrow 0.219 \pm 0.024 \Rightarrow (0.195,\ \ 0.243)$$

We are 95% confident that the population proportion of all pottery artifacts at Phylakopi that are painted is between 0.195 and 0.243.

7.129 a. The sample size may not be large enough. The sample size is large enough if both $n\hat{p} \geq 15$ and $n\hat{q} \geq 15$.

$$\hat{p} = \frac{x}{n} = \frac{52}{60} = 0.867$$

$n\hat{p} = 60(0.867) = 52$ and $n\hat{q} = 60(0.133) = 8$. Since the second number is not greater than or equal to 15, the sample size is not sufficiently large to conclude the normal approximation is reasonable.

b. For confidence coefficient 0.90, $\alpha = 0.10$ and $\alpha/2 = 0.10/2 = 0.05$. From Table II, Appendix A, $z_{0.05} = 1.645$. We will use the sample proportion to estimate the population proportion:

$$\hat{p} = \frac{x}{n} = \frac{52}{60} = 0.867$$

$$n = \frac{z_{\alpha/2}^2 pq}{(SE)^2} = \frac{(1.645)^2(0.867)(0.133)}{(0.05)^2} = 124.8 \approx 125$$

7.131 a. For confidence coefficient 0.90, $\alpha = 0.10$ and $\alpha/2 = 0.10/2 = 0.05$. From Table III, Appendix A, with $df = n-1 = 25-1 = 24$, $t_{0.05} = 1.711$. The 90% confidence interval is:

$$\bar{x} \pm t_{0.05,24}\frac{s}{\sqrt{n}} \Rightarrow 75.4 \pm 1.711\frac{10.9}{\sqrt{25}} \Rightarrow 75.4 \pm 3.73 \Rightarrow (71.67,\ \ 79.13)$$

We are 90% confident that the true mean breaking strength of white wood is between 71.67 and 79.13 MPa's.

b. We will use $s = 10.9$ to estimate the population standard deviation. For confidence coefficient 0.90, $\alpha = 0.10$ and $\alpha/2 = 0.10/2 = 0.05$. From Table II, Appendix B, $z_{0.05} = 1.645$.

$$n = \frac{z_{\alpha/2}^2\sigma^2}{(SE)^2} = \frac{(1.645)^2(10.9)^2}{4^2} = 20.09 \approx 21$$

7.133 a. The point estimate for the mean amount of cesium in lichen specimens collected in Alaska is $\bar{x} = 0.009027$.

b. For confidence coefficient 0.95, $\alpha = 0.05$ and $\alpha/2 = 0.05/2 = 0.025$. From Table III, Appendix B, with $df = n-1 = 9-1 = 8$, $t_{0.025} = 2.306$.

c. The 95% confidence interval is:

$$\bar{x} \pm t_{0.025}\frac{s}{\sqrt{n}} \Rightarrow 0.009027 \pm 2.306\frac{0.004854}{\sqrt{9}} \Rightarrow 0.009027 \pm 0.003731 \Rightarrow (0.005296,\ \ 0.012758)$$

d. This interval is the same as that found on the printout.

e. We are 95% confident that the mean amount of cesium in lichen specimens collected in Alaska is between 0.005296 and 0.012758 microcuries per milliliter.

f. We will use $s = 0.005$ to estimate the population standard deviation. For confidence coefficient 0.95, $\alpha = 0.05$ and $\alpha/2 = 0.05/2 = 0.025$. From Table II, Appendix B, $z_{0.025} = 1.96$.

The sample size is $n = \dfrac{(z_{.025})^2\sigma^2}{(SE)^2} = \dfrac{(1.96)^2(0.005)^2}{(0.001)^2} = 96.04 \approx 97$.

7.135 Some preliminary calculations:

$$\bar{x} = \frac{\sum x}{n} = \frac{6.44}{6} = 1.073 \qquad s^2 = \frac{\sum x^2 - \frac{\left(\sum x\right)^2}{n}}{n-1} = \frac{7.1804 - \frac{6.44^2}{6}}{6-1} = 0.0536$$

$s = \sqrt{0.0536} = 0.2316$

a. For confidence coefficient 0.95, $\alpha = 0.05$ and $\alpha/2 = 0.05/2 = 0.025$. From Table III, Appendix B, with $df = n-1 = 6-1 = 5$, $t_{0.025} = 2.571$. The confidence interval is:

$$\bar{x} \pm t_{0.025}\frac{s}{\sqrt{n}} \Rightarrow 1.073 \pm 2.571\frac{0.2316}{\sqrt{6}} \Rightarrow 1.073 \pm 0.243 \Rightarrow (0.830, 1.316)$$

We are 95% confident that the true average decay rate of fine particles produced from oven cooking or toasting is between 0.830 and 1.316.

b. The phrase "95% confident" means that in repeated sampling, 95% of all confidence intervals constructed in this manner will contain the true mean.

c. For confidence level 0.95, $\alpha = 0.05$ and $\alpha/2 = 0.05/2 = 0.025$. From Table IV, Appendix B, with $df = n-1 = 6-1 = 5$, $\chi^2_{0.025,5} = 12.8325$ and $\chi^2_{0.975,5} = 0.831211$. The 95% confidence interval for the variance is:

$$\frac{(n-1)s^2}{\chi^2_{0.025}} \le \sigma^2 \le \frac{(n-1)s^2}{\chi^2_{0.975}} \Rightarrow \frac{(6-1)(0.2316)^2}{12.8325} \le \sigma^2 \le \frac{(6-1)(0.2316)^2}{0.831211} \Rightarrow 0.0209 \le \sigma^2 \le 0.3227$$

The 95% confidence interval for the standard deviation is:

$\sqrt{0.0209} \le \sigma \le \sqrt{0.3227} \Rightarrow 0.1446 \le \sigma \le 0.5681$

We are 95% confident that the true standard deviation of the decay rate is between 0.1446 and 0.5681.

d. In order for the inferences above to be valid, the distribution of decay rates must be normally distributed.

e. For confidence coefficient 0.95, $\alpha = 0.05$ and $\alpha/2 = 0.05/2 = 0.025$. From Table II, Appendix B, $z_{0.025} = 1.96$.

$$n = \frac{z^2_{\alpha/2}\sigma^2}{(SE)^2} = \frac{(1.96)^2(0.2316)^2}{(0.04)^2} = 128.786 \approx 129$$

We would need to sample 129 decay rates.

7.137 a. First, we compute $\hat{p}$: $\hat{p} = \frac{x}{n} = \frac{665}{1007} = .660$

The sample size is large enough if both $n\hat{p} \ge 15$ and $n\hat{q} \ge 15$.

$n\hat{p}=1{,}007(0.660)=665$ and $n\hat{q}=1{,}007(0.340)=342$. Since both of the numbers are greater than or equal to 15, the sample size is sufficiently large to conclude the normal approximation is reasonable.

For confidence coefficient 0.95, $\alpha=0.05$ and $\alpha/2=0.05/2=0.025$. From Table II, Appendix B, $z_{0.025}=1.96$. The confidence interval is:

$$\hat{p}\pm z_{0.025}\sqrt{\frac{pq}{n}}\approx\hat{p}\pm1.96\sqrt{\frac{\hat{p}\hat{q}}{n}}\Rightarrow0.660\pm1.96\sqrt{\frac{0.660(0.340)}{1007}}\Rightarrow0.660\pm0.029\Rightarrow(0.631,\ 0.689)$$

We are 95% confident that the proportion of U.S. workers who take their lunch to work is between 0.631 and 0.689.

b. First, we compute $\hat{p}$: $\hat{p}=\frac{x}{n}=\frac{200}{665}=0.301$

The sample size is large enough if both $n\hat{p}\ge15$ and $n\hat{q}\ge15$.

$n\hat{p}=665(0.301)=200$ and $n\hat{q}=665(0.699)=465$. Since both of the numbers are greater than or equal to 15, the sample size is sufficiently large to conclude the normal approximation is reasonable.

For confidence coefficient 0.95, $\alpha=0.05$ and $\alpha/2=0.05/2=0.025$. From Table II, Appendix B, $z_{0.025}=1.96$. The confidence interval is:

$$\hat{p}\pm z_{0.025}\sqrt{\frac{pq}{n}}\approx\hat{p}\pm1.96\sqrt{\frac{\hat{p}\hat{q}}{n}}\Rightarrow0.301\pm1.96\sqrt{\frac{0.301(0.699)}{665}}\Rightarrow0.301\pm0.035\Rightarrow(0.266,\ 0.336)$$

We are 95% confident that the proportion of U.S. workers who take their lunch to work who brown-bag their lunch is between 0.266 and 0.336.

7.139 a. For confidence coefficient 0.99, $\alpha=0.01$ and $\alpha/2=0.01/2=0.005$. From Table II, Appendix B, $z_{0.005}=2.58$. The confidence interval is:

$$\bar{x}\pm z_{0.005}\frac{s}{\sqrt{n}}\Rightarrow0.044\pm2.58\frac{0.884}{\sqrt{197}}\Rightarrow0.044\pm0.162\Rightarrow(-0.118,\ 0.206)$$

We are 99% confident that the mean inbreeding coefficient for this species of wasps is between −0.118 and 0.206.

b. A coefficient of 0 indicates that the wasp has no tendency to inbreed. Since 0 is in the 99% confidence interval, it is a likely value for the true mean inbreeding coefficient. Thus, there is no evidence to indicate that this species of wasps has a tendency to inbreed.

7.141 For confidence coefficient 0.90, $\alpha=0.10$ and $\alpha/2=0.10/2=0.05$. From Table II, Appendix A, $z_{0.05}=1.645$.

The sample size is $n=\frac{(z_{\alpha/2})^2\sigma^2}{(SE)^2}=\frac{(1.645)^2(10)^2}{1^2}=270.6\approx271$

7.143 a. The confidence interval might lead to an erroneous inference because the sample size used is probably too small. Recall that the sample size is large enough if both $n\hat{p} \ge 15$ and $n\hat{q} \ge 15$. For this problem, $\hat{p} = \frac{10}{18} = 0.556$. $n\hat{p} = 18(0.556) = 10.008$ and $n\hat{q} = 18(0.444) = 7.992$. Neither of these two values is greater than 15. Thus, the sample size is not sufficiently large to conclude the normal approximation is reasonable.

b. For confidence coefficient 0.95, $\alpha = 0.05$ and $\alpha / 2 = 0.05 / 2 = 0.025$. From Table II, Appendix B, $z_{0.025} = 1.96$. From the previous study, we will use $\hat{p} = \frac{10}{18} = .556$ to estimate p.

$$n = \frac{z_{\alpha/2}^2 pq}{(SE)^2} = \frac{(1.96)^2 (0.556)(0.444)}{(0.04)^2} = 592.72 \approx 593$$

7.145 a. The point estimate for the fraction of the entire market that refuses to purchase bars is:

$$\hat{p} = \frac{x}{n} = \frac{23}{244} = 0.094$$

b. The sample size is large enough if both $n\hat{p} \ge 15$ and $n\hat{q} \ge 15$.

$n\hat{p} = 244(0.094) = 23$ and $n\hat{q} = 244(0.906) = 221$. Since both of the numbers are greater than or equal to 15, the sample size is sufficiently large to conclude the normal approximation is reasonable.

c. For confidence coefficient 0.95, $\alpha = 0.05$ and $\alpha / 2 = 0.05 / 2 = 0.025$. From Table II, Appendix B, $z_{0.025} = 1.96$. The confidence interval is:

$$\hat{p} \pm z_{0.025}\sqrt{\frac{\hat{p}\hat{q}}{n}} \Rightarrow 0.094 \pm 1.96\sqrt{\frac{0.094(0.906)}{224}} \Rightarrow 0.094 \pm 0.037 \Rightarrow (0.057, 0.131)$$

d. The best estimate of the true fraction of the entire market that refuses to purchase bars six months after the poisoning is 0.094. We are 95% confident the true fraction of the entire market that refuses to purchase bars six months after the poisoning is between 0.057 and 0.131.

7.147 The point estimate for the proportion of health care workers with latex allergy who suspect that he/she has the allergy is $\hat{p} = \frac{x}{n} = \frac{36}{83} = 0.434$.

The sample size is large enough if both $n\hat{p} \ge 15$ and $n\hat{q} \ge 15$.

$n\hat{p} = 83(0.434) = 36$ and $n\hat{q} = 83(0.566) = 47$. Since both of the numbers are greater than or equal to 15, the sample size is sufficiently large to conclude the normal approximation is reasonable.

Since no confidence limit is given, we will use 95%. For confidence coefficient 0.95, $\alpha = 0.05$ and $\alpha/2 = 0.05/2 = 0.025$. From Table II, Appendix B, $z_{0.025} = 1.96$. The 95% confidence interval is:

$$\hat{p} \pm z_{0.025}\sigma_{\hat{p}} \Rightarrow \hat{p} \pm 1.96\sqrt{\frac{\hat{p}\hat{q}}{n}} \Rightarrow 0.434 \pm 1.96\sqrt{\frac{0.434(0.566)}{83}} \Rightarrow 0.434 \pm 0.107 \Rightarrow (0.327,\ 0.541)$$

We are 95% confident that the proportion of health care workers with latex allergy who suspect that he/she has the allergy is between 0.327 and 0.541.

7.149 a. From Chebyshev's Rule, we know that at least $1 - \frac{1}{k^2}$ of the observations will fall within k standard deviations of the mean. We need to find k so that $1 - \frac{1}{k^2} = 0.60$.

$$1 - \frac{1}{k^2} = 0.60 \Rightarrow 0.40 = \frac{1}{k^2} \Rightarrow k^2 = \frac{1}{0.40} = 2.5 \Rightarrow k = \sqrt{2.5} = 1.58$$

$$s = \frac{80^{th} \text{ percentile} - 60^{th} \text{ percentile}}{2k} = \frac{73{,}000 - 35{,}100}{2(1.58)} = 11{,}993.67$$

For confidence coefficient 0.98, $\alpha = 0.02$ and $\alpha/2 = 0.02/2 = 0.01$. From Table II, Appendix B, $z_{0.01} = 2.33$. The sample size required is:

$$n = \frac{(z_{0.01})^2 \sigma^2}{(SE)^2} = \frac{(2.33)^2 (11{,}993.67)^2}{2{,}000^2} = 195.23 \approx 196$$

b. See part **a**.

c. We must assume that any sample selected will be random.

7.151 a. As long as the sample is random (and thus representative), a reliable estimate of the mean weight of all the scallops can be obtained.

b. The government is using only the sample mean to make a decision. Rather than using a point estimate, they should probably use a confidence interval to estimate the true mean weight of the scallops.

c. We will form a 95% confidence interval for the mean weight of the scallops. Using MINITAB, the descriptive statistics are:

Descriptive Statistics: Weight

Variable	N	Mean	StDev	Minimum	Q1	Median	Q3	Maximum
Weight	18	0.9317	0.0753	0.8400	0.8800	0.9100	9800	1.1400

For confidence coefficient 0.95, $\alpha = 0.05$ and $\alpha/2 = 0.05/2 = 0.025$. From Table III, Appendix B, with $df = n - 1 = 18 - 1 = 17$, $t_{0.025} = 2.110$.

The 95% confidence interval is:

$$\overline{x} \pm t_{0.025}\frac{s}{\sqrt{n}} \Rightarrow 0.9317 \pm 2.110\frac{0.0753}{\sqrt{18}} \Rightarrow 0.9317 \pm 0.0374 \Rightarrow (0.8943,\ \ 0.9691)$$

We are 95% confident that the true mean weight of the scallops is between 0.8943 and 0.9691. Recall that the weights have been scaled so that a mean weight of 1 corresponds to 1/36 of a pound. Since the above confidence interval does not include 1, we have sufficient evidence to indicate that the minimum weight restriction was violated.

Chapter 8

Inferences Based on a Single Sample
Tests of Hypothesis

8.1 The null hypothesis is the "status quo" hypothesis, while the alternative hypothesis is the research hypothesis.

8.3 The "level of significance" of a test is α. This is the probability that the test statistic will fall in the rejection region when the null hypothesis is true.

8.5 The four possible results are:

1. Rejecting the null hypothesis when it is true. This would be a Type I error.
2. Accepting the null hypothesis when it is true. This would be a correct decision.
3. Rejecting the null hypothesis when it is false. This would be a correct decision.
4. Accepting the null hypothesis when it is false. This would be a Type II error.

8.7 When you reject the null hypothesis in favor of the alternative hypothesis, this does not prove the alternative hypothesis is correct. We are $100(1-\alpha)\%$ confident that there is sufficient evidence to conclude that the alternative hypothesis is correct.

We know the hypothesis-testing process will lead to this conclusion (reject H_0) incorrectly only $100\alpha\%$ of the time when H_0 is true.

8.9 a. Probability of Type I error
$= P(z > 1.96) = 0.5 - 0.4750 = 0.0250$
(From Table II, Appendix B)

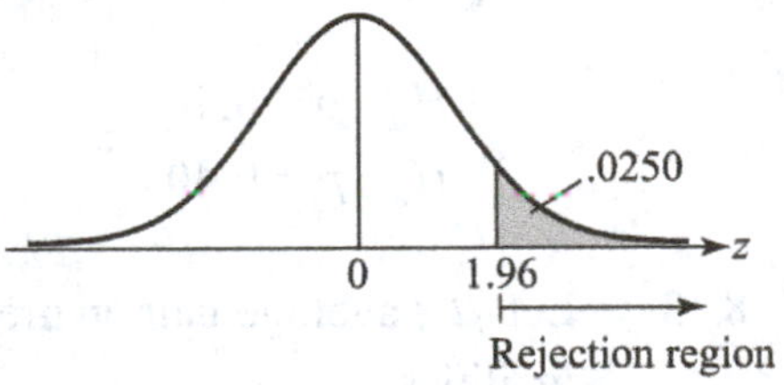

b. Probability of Type I error
$= P(z > 1.645) = 0.5 - 0.4500 = 0.0500$
(From Table II, Appendix B)

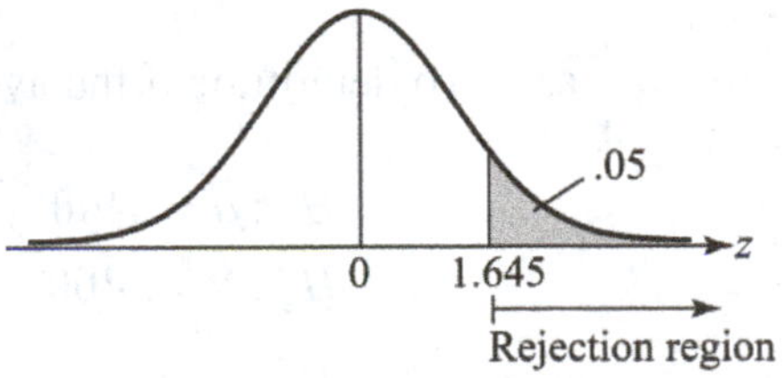

c. Probability of Type I error
$= P(z > 2.575) = 0.5 - 0.4950 = 0.0050$
(From Table II, Appendix B)

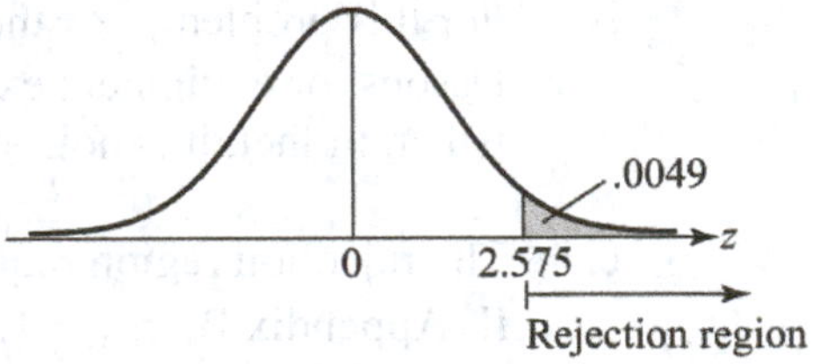

d. Probability of Type I error
$= P(z < -1.28) = 0.5 - 0.3997 = 0.1003$
(From Table II, Appendix B)

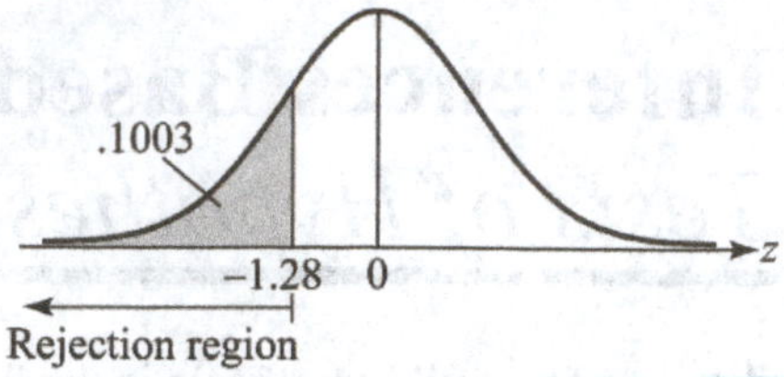

e. Probability of Type I error
$= P(z < -1.645) + P(z > 1.645)$
$= 0.5 - 0.4500 + 0.5 - 0.4500 = 0.05 + 0.05 = 0.10$
(From Table II, Appendix B)

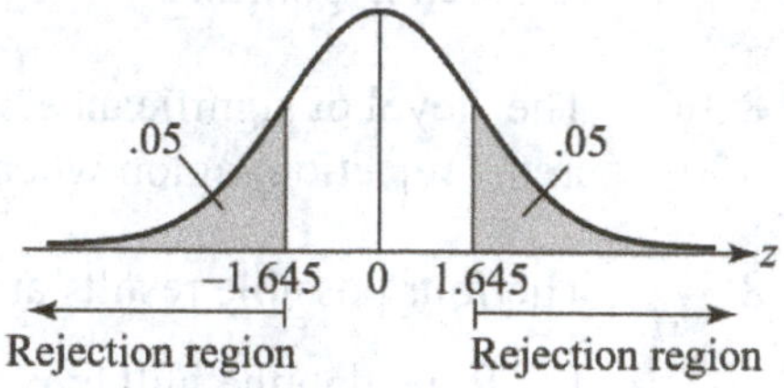

f. Probability of Type I error
$= P(z < -2.575) + P(z > 2.575)$
$= 0.5 - 0.4950 + 0.5 - 0.4950 = 0.0050 + 0.0050$
$= 0.0100$ (From Table II, Appendix B)

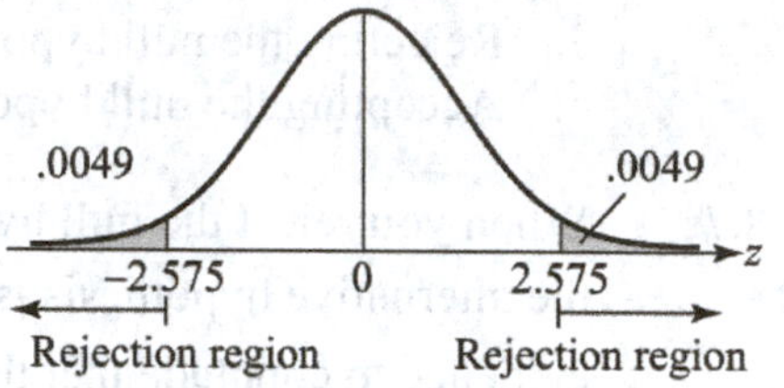

8.11 Let p = proportion of American adults who choose football as their favorite sport. To see if this proportion is different from 0.40, we test:

$H_0: p = 0.40$
$H_a: p \neq 0.40$

8.13 Let μ = average gain in green fees, lessons, or equipment expenditures for participating facilities.

a. To determine if the average gain exceeds \$2,400, we test:

$H_0: \mu = 2400$
$H_a: \mu > 2400$

b. For this problem, α = the probability of concluding the average gain in green fees, lessons, or equipment expenditures for participating facilities is greater than \$2,400 when, in fact, it is not.

c. The rejection region requires $\alpha = 0.05$ in the upper tail of the z distribution. From Table II, Appendix B, $z_{0.05} = 1.645$. The rejection region is $z > 1.645$.

8.15 Let p = error rate of the DNA-reading device. To determine if the error rate of the DNA-reading device is less than 5%, we test:

$H_0: p = 0.05$
$H_a: p < 0.05$

8.17 a. The parameter of interest is p, the true proportion of times the avatar can detect deceitful speech correctly.

b. To determine if the true proportion of times the avatar can detect deceitful speech correctly, we test:

$$H_0 : p = 0.75$$
$$H_a : p \neq 0.75$$

c. A Type I error would be saying the true proportion of times the avatar can detect deceitful speech correctly is not 0.75, when, in fact, the true proportion is 0.75.

d. A Type II error would be saying the true proportion of times the avatar can detect deceitful speech correctly is 0.75, when, in fact, the true proportion is not 0.75.

8.19 a. To determine if the average level of mercury uptake in wading birds in the Everglades today is less than 15 parts per million, we test:

$$H_0 : \mu = 15$$
$$H_a : \mu < 15$$

b. A Type I error is rejecting H_0 when H_0 is true. In terms of this problem, we would be concluding that the average level of mercury uptake in wading birds in the Everglades today is less than 15 parts per million, when in fact, the average level of mercury uptake in wading birds in the Everglades today is equal to 15 parts per million.

c. A Type II error is accepting H_0 when H_0 is false. In terms of this problem, we would be concluding that the average level of mercury uptake in wading birds in the Everglades today is equal to 15 parts per million, when in fact, the average level of mercury uptake in wading birds in the Everglades today is less than 15 parts per million.

8.21 a. The null hypothesis is: H_0: No intrusion occurs

b. The alternative hypothesis is: H_a: Intrusion occurs

c. $\alpha =$ Probability of a Type I error = Probability of rejecting H_o when it is true = P(system provides warning when no intrusion occurs) $= 1/1000 = 0.001$.

$\beta =$ Probability of a Type II error = Probability of accepting H_o when it is false = P(system does not provide warning when an intrusion occurs) $= 500/1000 = 0.5$.

8.23 We will reject H_0 if the p-value $< \alpha$.

a. $0.06 \not< 0.05$, do not reject H_0.

b. $0.10 \not< 0.05$, do not reject H_0.

c. $0.01 < 0.05$, reject H_0.

d. $0.001 < 0.05$, reject H_0.

e. $0.251 \not< 0.05$, do not reject H_0.

f. $0.042 < 0.05$, reject H_0.

8.25 $z = \dfrac{\bar{x} - \mu_0}{\sigma_{\bar{x}}} \approx \dfrac{49.4 - 50}{4.1/\sqrt{100}} = -1.46$

p-value $= p = P(z \geq -1.46) = 0.5 + 0.4279 = 0.9279$ (Using Table II, Appendix B)

Since the p-value is so large, H_0 would not be rejected for any reasonable value of α. There is no evidence to indicate the mean is greater than 50.

8.27 First, find the value of the test statistic $z = \dfrac{\bar{x} - \mu_0}{\sigma_{\bar{x}}} \approx \dfrac{10.7 - 10}{3.1/\sqrt{50}} = 1.60$

p-value $= p = P(z \leq -1.60 \text{ or } z \geq 1.60) = 2P(z \geq 1.60) = 2(0.5 - 0.4452) = 2(0.0548) = 0.1096$
(using Table II, Appendix B)

There is no evidence to reject H_0 for $\alpha \leq 0.10$.

8.29 a. The p-value reported by SPSS is for a two-tailed test. Thus, $P(z \leq -1.63) + P(z \geq 1.63) = 0.1032$. For this one-tailed test, the p-value $= p = P(z \leq -1.63) = 0.1032/2 = 0.0516$.

Since the p-value $= 0.0516 > \alpha = 0.05$, H_0 is not rejected. There is insufficient evidence to indicate $\mu < 75$ at $\alpha = 0.05$.

b. For this one-tailed test, the p-value $= p = P(z \leq 1.63)$. Since $P(z \leq -1.63) = 0.1032/2 = 0.0516$, $P(z \leq 1.63) = 1 - 0.0516 = 0.9484$.

Since the p-value $= p = 0.9484 > \alpha = 0.10$, H_0 is not rejected. There is insufficient evidence to indicate $\mu < 75$ at $\alpha = 0.10$.

c. For this one-tailed test, the p-value $= p = P(z \geq 1.63) = 0.1032/2 = 0.0516$.

Since the p-value $= p = 0.0516 < \alpha = 0.10$, H_0 is rejected. There is sufficient evidence to indicate $\mu > 75$ at $\alpha = 0.10$.

d. For this two-tailed test, the p-value $= p = 0.1032$.

Since the p-value $= 0.1032 > \alpha = 0.01$, H_0 is not rejected. There is insufficient evidence to indicate $\mu \neq 75$ at $\alpha = 0.01$.

8.31 There are 2 conditions required for a valid large-sample hypothesis test for μ. They are:

1. A random sample is selected from a target population.
2. The sample size n is large, i.e., $n \geq 30$. (Due to the Central Limit Theorem, this condition guarantees that the test statistic will be approximately normal regardless of the shape of the underlying probability distribution of the population.)

8.33 a. $H_0: \mu = 100$
$H_a: \mu > 100$

The test statistic is $z = \frac{\bar{x} - \mu_0}{\sigma_{\bar{x}}} = \frac{\bar{x} - \mu_0}{\sigma / \sqrt{n}} = \frac{110 - 100}{60 / \sqrt{100}} = 1.67$

The rejection region requires $\alpha = 0.05$ in the upper tail of the z distribution. From Table II, Appendix B, $z_{0.05} = 1.645$. The rejection region is $z > 1.645$.

Since the observed value of the test statistic falls in the rejection region, $(z = 1.67 > 1.645)$, H_0 is rejected. There is sufficient evidence to indicate the true population mean is greater than 100 at $\alpha = 0.05$.

b. $H_0: \mu = 100$
$H_a: \mu \neq 100$

The test statistic is $z = \frac{\bar{x} - \mu_0}{\sigma_{\bar{x}}} = \frac{110 - 100}{60 / \sqrt{100}} = 1.67$

The rejection region requires $\alpha / 2 = 0.05 / 2 = 0.025$ in each tail of the z distribution. From Table II, Appendix B, $z_{0.025} = 1.96$. The rejection region is $z < -1.96$ or $z > 1.96$.

Since the observed value of the test statistic does not fall in the rejection region, $(z = 1.67 \not> 1.96)$, H_0 is not rejected. There is insufficient evidence to indicate μ differs from 100 at $\alpha = 0.05$.

c. In part **a**, $p - \text{value} = p = P(z > 1.67) = 0.5 - 0.4525 = 0.0475$. In part **b**, $p - \text{value} = p = P(z < -1.67) + P(z > 1.67) = (0.5 - 0.4525) + (0.5 - 0.4525) = 0.095$. In part **a**, we have a one-tailed test. In part **b**, we have a two-tailed test. Because the alternative hypothesis in part **a** is more specific than the one in **b**, it is easier to reject H_0.

8.35 a. $z = \frac{\bar{x} - \mu}{\sigma_{\bar{x}}} = \frac{215 - 200}{80 / \sqrt{100}} = 1.88$

The decision rule is: Reject H_0 if $z > 1.88$.

b. $\alpha = P(z > 1.88) = 0.5 - 0.4699 = 0.0301$ (using Table II, Appendix B)

8.37 a. Let μ = mean level of support for corporate sustainability. To see if the mean level of support differs from 75, we test:

$H_0: \mu = 75$
$H_a: \mu \neq 75$

b. A Type I error would be concluding the true mean level of support is different from 75 when, in fact, the mean level of support equals 75.

A Type II error would be concluding the true mean level of support is equal to 75 when, in fact, the mean level of support is different from 75.

c. From the printout, the *p*-value is $p = 0.000$.

d. Because $p = 0.000 < \alpha = 0.05$, we reject H_0. There is sufficient evidence to conclude that the true mean level of support for corporate sustainability is different from 75.

e. Because the sample size is so large, we do not have to make any assumptions about the distribution of support levels. The sample must be random and the observations must be independent.

8.39 Let μ = mean emotional empathy score for females.

a. To test whether female college students score higher than 3.0 on the emotional empathy scale, we test:

$H_0: \mu = 3.0$
$H_a: \mu > 3.0$

b. The test statistic is $z = \dfrac{\bar{x} - \mu_o}{\sigma_{\bar{x}}} = \dfrac{3.28 - 3.0}{\frac{0.5}{\sqrt{30}}} = 3.07$

c. The *p*-value is $p = P(z \geq 3.07) = 0.5 - 0.4989 = 0.0011$.

d. Since the *p*-value is less than $\alpha = 0.01$ $(p = 0.0011 < 0.01)$, H_0 is rejected. There is sufficient evidence to indicate female college students score higher than 3.0 on the emotional empathy scale at $\alpha = 0.01$.

e. The value of α can be just greater than 0.0011 and we could still reject H_0.

8.41 Let μ = mean facial WHR of CEOs. To see if the mean facial WHR of CEOs differs from 2.2, we test:

$H_0: \mu = 2.2$
$H_a: \mu \neq 2.2$

The test statistic is $z = \dfrac{\bar{x} - \mu_0}{\sigma_{\bar{x}}} = \dfrac{1.96 - 2.2}{\dfrac{0.15}{\sqrt{55}}} = -11.87$

Since no α is given, we will use $\alpha = 0.05$. The rejection region requires $\alpha / 2 = 0.05 / 2 = 0.025$ in each tail of the z distribution. From Table II, Appendix B, $z_{0.025} = 1.96$. The rejection region is $z < -1.96$ or $z > 1.96$.

Since the observed value of the test statistic falls in the rejection region $(z = -11.87 < -1.96)$, H_0 is rejected. There is sufficient evidence to indicate the mean facial WHR of CEOs is different from 2.2 at $\alpha = 0.05$. Using 2.2 as the mean facial WHR is not recommended.

The *p*-value is $p = P(z < -11.87) + P(z > 11.87) \approx 0$.

8.43 Let $\mu =$ mean disclosure score. To determine if the true mean disclosure score exceeds 3, we test:

$$H_0: \mu = 3$$
$$H_a: \mu > 3$$

The test statistic is $z = \dfrac{\bar{x} - \mu_o}{\sigma_{\bar{x}}} = \dfrac{3.26 - 3}{\dfrac{0.93}{\sqrt{222}}} = 4.17$

The rejection region requires $\alpha = 0.01$ in the upper tail of the z distribution. From Table II, Appendix B, $z_{0.01} = 2.33$. The rejection region is $z > 2.33$.

Since the observed value of the test statistic falls in the rejection region $(z = 4.17 > 2.33)$, H_0 is rejected. There is sufficient evidence to indicate the true mean disclosure score exceeds 3 at $\alpha = 0.01$.

8.45 a. Using MINITAB, the descriptive statistics are:

Descriptive Statistics: Bones

```
Variable   N  Mean     StDev  Minimum     Q1  Median     Q3  Maximum
Bones     41  9.258 8  1.204    6.230  8.615   9.200  9.935   12.000
```

To determine if the population mean ratio of all bones of this particular species differs from 8.5, we test:

$$H_0: \mu = 8.5$$
$$H_a: \mu \neq 8.5$$

The test statistic is $z = \dfrac{\bar{x} - \mu_0}{\sigma_{\bar{x}}} = \dfrac{9.258 - 8.5}{1.204 / \sqrt{41}} = 4.03$

The rejection region requires $\alpha / 2 = 0.01 / 2 = 0.005$ in each tail of the z distribution. From Table II, Appendix B, $z_{0.005} = 2.58$. The rejection region is $z < -2.58$ or $z > 2.58$.

Since the observed value of the test statistic falls in the rejection region $(z = 4.03 > 2.58)$, H_0 is rejected. There is sufficient evidence to indicate that the true population mean ratio of all bones of this particular species differs from 8.5 at $\alpha = 0.01$.

b. The practical implications of the test in part **a** is that the species from which these particular bones came from is probably not species A.

8.47 Let μ = mean difference in error rates for all subjects who perform the memorization tasks. To see if the mean difference in error rates for all subjects who perform the memorization tasks exceeds 75, we test:

$H_0: \mu = 75$
$H_a: \mu > 75$

The test statistic is $z = \dfrac{\bar{x} - \mu_0}{\sigma_{\bar{x}}} = \dfrac{78.1885 - 75}{\dfrac{63.2429}{\sqrt{71}}} = 0.42$

The rejection region requires $\alpha = 0.01$ in the upper tail of the z distribution. From Table II, Appendix B, $z_{0.01} = 2.33$. The rejection region is $z > 2.33$.

Since the observed value of the test statistic does not fall in the rejection region $(z = 0.42 \not> 2.33)$, H_0 is not rejected. There is insufficient evidence to indicate the mean difference in error rates for all subjects who perform the memorization tasks exceeds 75 at $\alpha = 0.01$.

8.49 a. To determine if the mean social interaction score of all Connecticut mental health patients differs from 3, we test:

$H_0: \mu = 3$
$H_a: \mu \neq 3$

The test statistic is $z = \dfrac{\bar{x} - \mu_0}{\sigma_{\bar{x}}} = \dfrac{2.95 - 3}{1.10 / \sqrt{6,681}} = -3.72$

The rejection region requires $\alpha / 2 = 0.01 / 2 = 0.005$ in each tail of the z distribution. From Table IV, Appendix B, $z_{0.005} = 2.58$. The rejection region is $z < -2.58$ or $z > 2.58$.

Since the observed value of the test statistic falls in the rejection region $(z = -3.72 < -2.58)$, H_0 is rejected. There is sufficient evidence to indicate that the mean social interaction score of all Connecticut mental health patients differs from 3 at $\alpha = 0.01$.

b. From the test in part a, we found that the mean social interaction score was statistically different from 3. However, the sample mean score was 2.95. Practically speaking, 2.95 is very similar to 3.0. The very large sample size, $n = 6,681$, makes it very easy to find statistical significance, even when no practical significance exists.

c. Because the variable of interest is measured on a 5-point scale, it is very unlikely that the population of the ratings will be normal. However, because the sample size was extremely large, $(n = 6,681)$, the Central Limit Theorem will apply. Thus, the distribution of $\bar{x}$ will be normal, regardless of the distribution of *x*. Thus, the analysis used above is appropriate.

8.51 The *z* and *t* distributions are alike in that they are both symmetric, bell-shaped distributions centered at 0. They differ in variability, though. The *z* distribution always has $\sigma^2 = 1$, while the variance of the *t* distribution is always larger and depends on the degrees of freedom.

8.53 a. $P(t > 1.440) = 0.10$
(Using Table III, Appendix B, with $df = 6$)

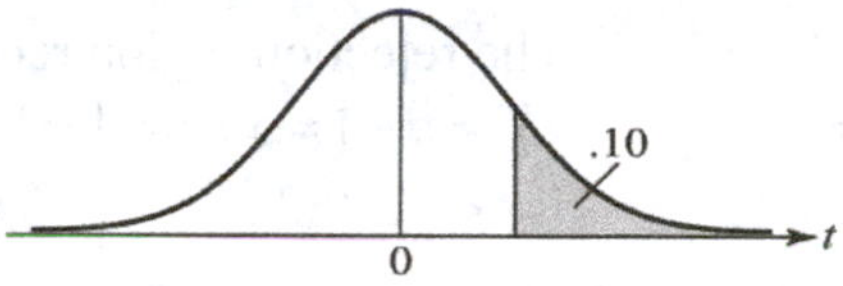

b. $P(t < -1.782) = 0.05$
(Using Table III, Appendix B, with $df = 12$)

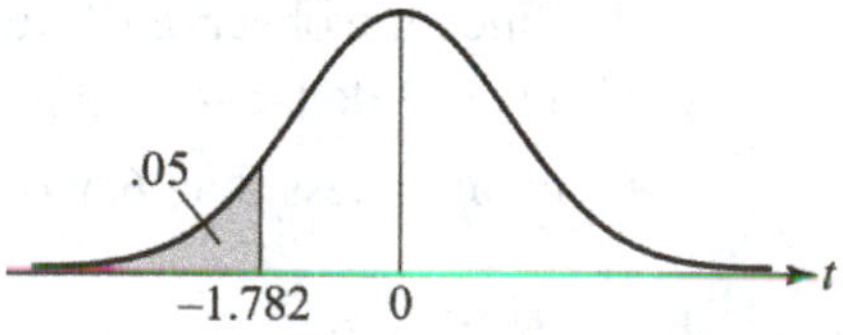

c. $P(t < -2.060) = P(t > 2.060) = 0.025$
(Using Table III, Appendix B, with $df = 25$)

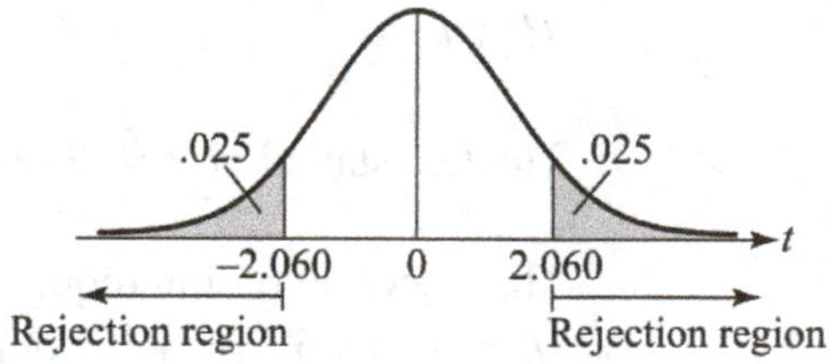

8.55 a. The rejection region requires $\alpha / 2 = 0.05 / 2 = 0.025$ in each tail of the *t* distribution with $df = n - 1 = 14 - 1 = 13$. From Table III, Appendix B, $t_{0.025} = 2.160$. The rejection region is $t < -2.160$ or $t > 2.160$.

b. The rejection region requires $\alpha = 0.01$ in the upper tail of the *t* distribution with $df = n - 1 = 24 - 1 = 23$. From Table III, Appendix B, $t_{0.01} = 2.500$. The rejection region is $t > 2.500$.

c. The rejection region requires $\alpha = 0.10$ in the upper tail of the *t* distribution with $df = n - 1 = 9 - 1 = 8$. From Table III, Appendix B, $t_{0.10} = 1.397$. The rejection region is $t > 1.397$.

d. The rejection region requires $\alpha = 0.01$ in the lower tail of the t distribution with $df = n-1 = 12-1 = 11$. From Table III, Appendix B, $t_{0.01} = 2.718$. The rejection region is $t < -2.718$.

e. The rejection region requires $\alpha/2 = 0.10/2 = 0.05$ in each tail of the t distribution with $df = n-1 = 20-1 = 19$. From Table III, Appendix B, $t_{0.05} = 1.729$. The rejection region is $t < -1.729$ or $t > 1.729$.

f. The rejection region requires $\alpha = 0.05$ in the lower tail of the t distribution with $df = n-1 = 4-1 = 3$. From Table III, Appendix B, $t_{0.05} = 2.353$. The rejection region is $t < -2.353$.

8.57 a. $H_0: \mu = 6$
$H_a: \mu < 6$

The test statistic is $t = \dfrac{\bar{x} - \mu_0}{s/\sqrt{n}} = \dfrac{4.8 - 6}{1.3/\sqrt{5}} = -2.064$.

The rejection region requires $\alpha = 0.05$ in the lower tail of the t distribution with $df = n-1 = 5-1 = 4$. From Table III, Appendix B, $t_{0.05} = 2.132$. The rejection region is $t < -2.132$.

Since the observed value of the test statistic does not fall in the rejection region $(t = -2.064 \not< -2.132)$, H_0 is not rejected. There is insufficient evidence to indicate the mean is less than 6 at $\alpha = 0.05$.

b. $H_0: \mu = 6$
$H_a: \mu \neq 6$

The test statistic is $t = -2.064$ (from **a**).

The rejection region requires $\alpha/2 = 0.05/2 = 0.025$ in each tail of the t distribution with $df = n-1 = 5-1 = 4$. From Table III, Appendix B, $t_{0.025} = 2.776$. The rejection region is $t < -2.776$ or $t > 2.776$.

Since the observed value of the test statistic does not fall in the rejection region $(t = -2.064 \not< -2.776)$, H_0 is not rejected. There is insufficient evidence to indicate the mean is different from 6 at $\alpha = 0.05$.

c. For part **a**, the p-value $= P(t \le -2.064)$.

From Table III, with $df = 4$, $0.05 < P(t \le -2.064) < 0.10$ or $0.05 < p-\text{value} < 0.10$. Using MINITAB, the p-value is $p = 0.0539809$.

For part **b**, the p-value $= P(t \le -2.064) + P(t \ge 2.064)$.

From Table III, with $df = 4$, $2(0.05) < p-\text{value} < 2(0.10)$ or $0.10 < p-\text{value} < 0.20$.
Using MINITAB, the *p*-value is $p = 2(0.0539809) = 0.1079618$.

8.59 a. Let μ = average trap spacing measurement. To determine if the true average trap spacing measurement is different from 95, we test:

$$H_0: \mu = 95$$
$$H_a: \mu \neq 95$$

b. For this example, the sample mean is 89.9. If another sample of size 7 was selected, chances are, the sample mean would not be 89.9. The next sample could possibly have a sample mean greater than 95. We need to determine how unusual a value 89.9 is for $\bar{x}$ if the true mean is 95. Also, we need to take into account the variability in the data.

c. The test statistic is $t = \dfrac{\bar{x}-\mu_0}{\frac{s}{\sqrt{n}}} = \dfrac{89.9-95}{\frac{11.6}{\sqrt{7}}} = -1.163$

d. To find the *p*-value, we need to find $p = P(t \leq -1.163) + P(t \geq 1.163)$. Using MINITAB, with $df = n-1 = 7-1 = 6$, $p = P(t \leq -1.163) + P(t \geq 1.163) = 0.1445 + 0.1445 = 0.289$.

e. Let $\alpha = 0.10$. The probability of concluding that the average strap spacing measurement is different from 95 when, in fact, it is equal to 95 is 0.10.

f. Since the *p*-value is greater than $\alpha = 0.10$ $(p = 0.289 > 0.10)$, H_0 is not rejected. There is insufficient evidence to indicate the average trap spacing measurement is different from 95 at $\alpha = 0.10$.

g. We must assume that a random sample was drawn from a normal population of trap spacing measurements.

h. From Exercise 7.41, the 95% confidence interval is $(79.095, 100.625)$. Since 95 falls in the interval, it is a likely value for μ. We would not reject it. This agrees with the results in part f.

8.61 To determine if the mean Dental Anxiety Scale score differs from 11, we test:

$$H_0: \mu = 11$$
$$H_a: \mu \neq 11$$

The test statistic is $t = \dfrac{\bar{x}-\mu_0}{s/\sqrt{n}} = \dfrac{10.7-11}{3.6/\sqrt{15}} = -0.32$

The rejection region requires $\alpha/2 = 0.05/2 = 0.025$ in each tail of the t distribution with $df = n-1 = 15-1 = 14$. From Table III, Appendix B, $t_{0.025} = 2.145$. The rejection region is

$t < -2.145$ or $t > 2.145$.

Since the observed value of the test statistic does not fall in the rejection region $(t = -0.32 \not< -2.145)$, H_0 is not rejected. There is insufficient evidence to indicate the mean Dental Anxiety Scale score differs from 11 at $\alpha = 0.05$.

8.63 Let μ = mean dentary depth of molars. To determine if the sample of 18 cheek teeth come from some other extinct primate species, we test:

$$H_0: \mu = 15$$
$$H_a: \mu \neq 15$$

The test statistic is $t = \dfrac{\bar{x} - \mu_0}{\frac{s}{\sqrt{n}}} = 3.229$ (From the printout.)

From the printout, the p-value is $p = 0.005$. Since the p-value is so small, we would reject H_0 for any reasonable value of α. There is sufficient evidence to indicate that the sample of 18 cheek teeth come from some other extinct primate species other than species A.

8.65 Using MINITAB, the results are:

One-Sample T: Strength

```
Test of μ = 300 vs > 300

Variable  N     Mean   StDev  SE Mean  99% Lower Bound      T      P
Strength  3   361.00    8.19     4.73           328.09  12.91  0.003
```

Let μ = mean yield strength of the steel bars. To determine if the true mean yield strength of the steel bars exceeds 300, we test:

$$H_0: \mu = 300$$
$$H_a: \mu > 300$$

The test statistic is $t = 12.91$ and the p-value is $p = 0.003$.

Since the p-value is less than $\alpha = 0.01$ $(p = 0.003 < 0.01)$, H_0 is rejected. There is sufficient evidence to indicate the mean yield strength of the steel bars exceeds 300 at $\alpha = 0.01$.

8.67 Using MINITAB, the results are:

One-Sample T: Ratio

```
Test of μ = 0.5 vs ≠ 0.5

Variable  N    Mean   StDev  SE Mean         95% CI              T      P
Ratio     8  0.4225  0.1219   0.0431  (0.3206, 0.5244)   -1.80  0.115
```

We will choose $\alpha = 0.05$. Let μ = average ratio of repetition for all participants in a similar memory study. To determine if the true average ratio of repetition for all participants in a similar memory study differs from 0.5, we test:

$$H_0: \mu = 0.5$$
$$H_a: \mu \neq 0.5$$

The test statistic is $t = -1.80$ and the *p*-value is $p = 0.115$.

Since the *p*-value is not less than $\alpha = 0.05$ $(p = 0.115 \nless 0.05)$, H_0 is not rejected. There is insufficient evidence to indicate the average ratio of repetition for all participants in a similar memory study differs from 0.5at $\alpha = 0.05$.

8.69 From Exercise 7.48, $\bar{x} = 358.45$ and $s = 117.8172$. To determine if the mean skidding distance is less than 425 meters, we test:

$$H_0: \mu = 425$$
$$H_a: \mu < 425$$

The test statistic is $t = \dfrac{\bar{x} - \mu_o}{s/\sqrt{n}} = \dfrac{358.45 - 425}{\dfrac{117.8172}{\sqrt{20}}} = -2.53$

The rejection region requires $\alpha = 0.10$ in the lower tail of the *t*-distribution with $df = n - 1 = 20 - 1 = 19$.
From Table III, Appendix B, $t_{0.10} = 1.328$. The rejection region is $t < -1.328$.

Since the observed value of the test statistic falls in the rejection region $(t = -2.53 < -1.328)$, H_0 is rejected. There is sufficient evidence to indicate the mean skidding distance is less than 425 meters at $\alpha = 0.10$. Thus, there is enough evidence to refute the claim.

8.71 a. To determine if the true mean bias differs from 0, we test:

$$H_0: \mu = 0$$
$$H_a: \mu \neq 0$$

From the printout, the test statistics is $t = 0.21$ and the *p*-value is $p = 0.840$.

Since the *p*-value is larger than $\alpha = 0.10$, $(p = 0.840 > 0.10)$, H_0 is not rejected. There is insufficient evidence to indicate that the true mean bias differs from 0.

b. The scores given are the average scores of different people. There are several people with averages scores that are positive and several with average scores which are negative. These scores will tend to cancel each other out. Also, each individual could travel in circles to the right and circles to the left. Again, these would tend to cancel each other out.

8.73 Typically, qualitative data are associated with making inferences about a population proportion.

8.75 The sample size is large enough if both np_o and nq_o are greater than or equal to 15.

a. $np_o = 500(0.05) = 25$ and $nq_o = 500(1-0.05) = 500(0.95) = 475$. Since both of these values are greater than 15, the sample size is large enough to use the normal approximation.

b. $np_o = 100(0.99) = 99$ and $nq_o = 100(1-0.99) = 100(0.01) = 1$. Since the second number is less than 15, the sample size is not large enough to use the normal approximation.

c. $np_o = 50(0.2) = 10$ and $nq_o = 50(1-0.2) = 50(0.8) = 40$. Since the first number is less than 15, the sample size is not large enough to use the normal approximation.

d. $np_o = 20(0.2) = 4$ and $nq_o = 20(1-0.2) = 20(0.8) = 16$. Since the first number is less than 15, the sample size is not large enough to use the normal approximation.

e. $np_o = 10(0.4) = 4$ and $nq_o = 10(1-0.4) = 10(0.6) = 6$. Since both numbers are less than 15, the sample size is not large enough to use the normal approximation.

8.77 a. $$z = \frac{\hat{p} - p_0}{\sqrt{\frac{p_0 q_0}{n}}} = \frac{0.84 - 0.9}{\sqrt{\frac{0.9(0.1)}{100}}} = -2.00$$

b. The denominator in Exercise 8.760 is $\sqrt{\frac{0.75(0.25)}{100}} = 0.0433$ as compared to $\sqrt{\frac{0.9(0.1)}{100}} = 0.03$ in part **a**. Since the denominator in this problem is smaller, the absolute value of z is larger.

c. The rejection region requires $\alpha = 0.05$ in the lower tail of the z distribution. From Table II, Appendix B, $z_{0.05} = 1.645$. The rejection region is $z < -1.645$.

Since the observed value of the test statistic falls in the rejection region $(z = -2.00 < -1.645)$, H_0 is rejected. There is sufficient evidence to indicate the population proportion is less than 0.9 at $\alpha = 0.05$.

d. The p-value $= P(z \le -2.00) = 0.5 - 0.4772 = 0.0228$ (from Table II, Appendix B).

8.79 a. From Exercise 7.58, $n = 50$ and since p is the proportion of consumers who do not like the snack food, $\hat{p}$ will be:

$$\hat{p} = \frac{\text{Number of 0's in sample}}{n} = \frac{29}{50} = 0.58$$

In order for the inference to be valid, the sample size must be large enough. The sample size is large enough if both np_o and nq_o are greater than or equal to 15.

$np_o = 50(0.5) = 25$ and $nq_o = 50(1-0.5) = 50(0.5) = 25$. Since both of these values are greater than 15, the sample size is large enough to use the normal approximation.

$H_0: p = 0.5$
$H_a: p > 0.5$

The test statistic is $z = \dfrac{\hat{p} - p_0}{\sigma_{\hat{p}}} = \dfrac{\hat{p} - p_0}{\sqrt{\dfrac{p_0 q_0}{n}}} = \dfrac{0.58 - 0.5}{\sqrt{\dfrac{0.5(1-0.5)}{50}}} = 1.13$

The rejection region requires $\alpha = 0.10$ in the upper tail of the z distribution. From Table II, Appendix B, $z_{0.10} = 1.28$. The rejection region is $z > 1.28$.

Since the observed value of the test statistic does not fall in the rejection region $(z = 1.13 \not> 1.28)$, H_0 is not rejected. There is insufficient evidence to indicate the proportion of customers who do not like the snack food is greater than 0.5 at $\alpha = 0.10$.

b. p-value $= P(z \geq 1.13) = 0.5 - 0.3708 = 0.1292$ (Using Table II, Appendix B.)

8.81 a. Let p = proportion of alligators that move toward the underwater sound. To determine if the true proportion is higher than expected by chance, we test:

$H_0: p = 0.5$
$H_a: p > 0.5$

b. $\hat{p} = \dfrac{x}{n} = \dfrac{42}{50} = 0.84$

c. The test statistic is $z = \dfrac{\hat{p} - p_0}{\sigma_{\hat{p}}} = \dfrac{\hat{p} - p_0}{\sqrt{\dfrac{p_0 q_0}{n}}} = \dfrac{0.84 - 0.5}{\sqrt{\dfrac{0.5(1-0.5)}{50}}} = 4.81$

d. The p-value of the test is $p = P(z > 4.81) \approx 0$.

e. Since the p-value is so small, H_o is rejected for any reasonable value of α. There is sufficient evidence to indicate the true proportion of alligators that move toward the underwater sound is greater than 0.5.

8.83 Let p = proportion of calves that are dehorned. To determine if the proportion of calves that are dehorned is different from 0.80, we test:

$H_0: p = 0.8$
$H_a: p \neq 0.8$

$$\hat{p} = \frac{x}{n} = \frac{515}{639} = 0.806$$

The test statistic is $z = \frac{\hat{p} - p_0}{\sigma_{\hat{p}}} = \frac{\hat{p} - p_0}{\sqrt{\frac{p_0 q_0}{n}}} = \frac{0.806 - 0.8}{\sqrt{\frac{0.8(1-0.8)}{639}}} = 0.38$

The p-value is $p = P(z > 0.38) + P(z < -0.38) = (0.5 - .1480) + (0.5 - .1480) = 0.704$.

Since the p-value is not small, H_o is not rejected for any reasonable value of α. There is insufficient evidence to indicate the true proportion of calves that are dehorned is different from 0.8. Thus, the results support the claim.

8.85 Let p = true fraction of adult TV viewers who are subscription streamers. To determine if the proportion of adult TV viewers who are subscription streamers differs from 1/3, we test:

$$H_0: p = 1/3$$
$$H_a: p \neq 1/3$$

$$\hat{p} = \frac{x}{n} = \frac{785}{2,242} = 0.35$$

The test statistic is $z = \frac{\hat{p} - p_0}{\sigma_{\hat{p}}} = \frac{\hat{p} - p_0}{\sqrt{\frac{p_0 q_0}{n}}} = \frac{0.35 - 0.333}{\sqrt{\frac{0.333(1-0.333)}{2,242}}} = 1.71$

The rejection region requires $\alpha / 2 = 0.10 / 2 = 0.05$ in each tail of the z distribution. From Table II, Appendix B, $z_{0.05} = 1.645$. The rejection region is $z < -1.645$ or $z > 1.645$.

Since the observed value of the test statistic falls in the rejection region $(z = 1.71 > 1.645)$, H_o is rejected There is sufficient evidence to indicate the true proportion of adult TV viewers who are subscription streamers differs from 1/3 at $\alpha = 0.10$.

8.87 Let p = true proportion of students who prefer the new, computerized method of identifying common conifers.

The sample proportion is $\hat{p} = \frac{138}{171} = 0.81$.

In order for the inference to be valid, the sample size must be large enough. The sample size is large enough if both np_o and nq_o are greater than or equal to 15.

$np_o = 171(0.7) = 119.7$ and $nq_o = 171(1-0.7) = 171(0.3) = 51.3$. Since both of these values are greater than 15, the sample size is large enough to use the normal approximation.

To determine if more than 70% of all students prefer the new, computerized method of identifying common conifers, we test:

$H_0: p = 0.70$
$H_a: p > 0.70$

The test statistic is $z = \frac{\hat{p} - p_0}{\sigma_{\hat{p}}} = \frac{\hat{p} - p_0}{\sqrt{\frac{p_0 q_0}{n}}} = \frac{0.81 - 0.70}{\sqrt{\frac{0.70(1-0.70)}{171}}} = 3.14$

The *p*-value for the test is $p = P(z \geq 3.14) \approx 0.5 - 0.5 = 0$. (From Table II, Appendix B.) Since the *p*-value is so small, we would reject H_0 for any reasonable value of α. There is sufficient evidence to indicate that more than 70% of all students prefer the new, computerized method of identifying common conifers. Thus, Confir ID should be added to the curriculum at SRU.

8.89 Let p = true detection rate. To determine if the true detection rate is different from 0.70, we test:

$H_0: p = 0.70$
$H_a: p \neq 0.70$

$\hat{p} = \frac{x}{n} = \frac{1554}{2376} = 0.654$

The test statistic is $z = \frac{\hat{p} - p_0}{\sigma_{\hat{p}}} = \frac{\hat{p} - p_0}{\sqrt{\frac{p_0 q_0}{n}}} = \frac{0.654 - 0.70}{\sqrt{\frac{0.70(1-0.70)}{2376}}} = -4.89$

The rejection region requires $\alpha / 2 = 0.10 / 2 = 0.05$ in each tail of the z distribution. From Table II, Appendix B, $z_{0.05} = 1.645$. The rejection region is $z < -1.645$ or $z > 1.645$.

Since the observed value of the test statistic falls in the rejection region $(z = -4.89 < -1.645)$, H_o is rejected. There is sufficient evidence to indicate the true detection rate is different from 0.70 at $\alpha = 0.10$.

8.91 First, we calculate the estimate for p: $\hat{p} = \frac{x}{n} = \frac{37}{148} = 0.25$

In order for the inference to be valid, the sample size must be large enough. The sample size is large enough if both np_o and nq_o are greater than or equal to 15.

$np_o = 148(0.2) = 29.6$ and $nq_o = 148(1-0.2) = 148(0.8) = 118.4$. Since both of these values are greater than 15, the sample size is large enough to use the normal approximation.

To determine if more than 20% of all freshman college students believe in the Big Bang Theory, we test:

$H_0: p = 0.20$
$H_a: p > 0.20$

The test statistic is $z = \dfrac{\hat{p} - p_o}{\sqrt{\dfrac{p_o q_o}{n}}} = \dfrac{0.25 - 0.20}{\sqrt{\dfrac{0.20(0.80)}{148}}} = 1.52$

Since no value of α was given, we will choose $\alpha = 0.05$. The rejection region requires $\alpha = .05$ in the upper tail of the *z*-distribution. From Table II, Appendix B, $z_{0.05} = 1.645$. The rejection region is $z > 1.645$.

Since the observed value of the test statistic does not fall in the rejection region $(z = 1.52 \not> 1.645)$, H_0 is not rejected. There is insufficient evidence to indicate that more than 20% of all freshman college students believe in the Big Bang Theory at $\alpha = 0.05$.

Thus, we would not be willing to state that more than 20% of all freshman college students believe in the Big Bang Theory. We are 95% confident of this decision.

8.93 We know that the miscarriage rate for pregnant women expecting a boy in the United States is 3 in 1,000 or 0.003. Let p = miscarriage rate for pregnant women expecting a boy in the United States during September 2001.

To determine if the male fetal death rate in September, 2001 was higher than expected, we would test:

$H_0: p = 0.003$
$H_a: p > 0.003$

The rejection region requires $\alpha = 0.05$ in the upper tail of the *z* distribution. From Table II, Appendix B, $z_{0.05} = 1.645$. The rejection region is $z > 1.645$.

The test statistic would be $z = \dfrac{\hat{p} - p_o}{\sqrt{\dfrac{p_o q_o}{n}}} = \dfrac{\hat{p} - 0.003}{\sqrt{\dfrac{0.003(0.997)}{2000}}} = 1.645$.

We now need to solve this for $\hat{p}$.

$$\frac{\hat{p} - 0.003}{\sqrt{\dfrac{0.003(0.997)}{2000}}} = 1.645 \Rightarrow \frac{\hat{p} - 0.003}{0.001223} = 1.645 \Rightarrow \hat{p} - 0.003 = 0.002 \Rightarrow \hat{p} = 0.005$$

Thus, there would need to be $2000(0.005) = 10$ miscarriages in the 2,000 pregnant women to support the claim.

8.95 The power of a test is equal to $1-\beta$. As β increases, the power decreases.

8.97 a.

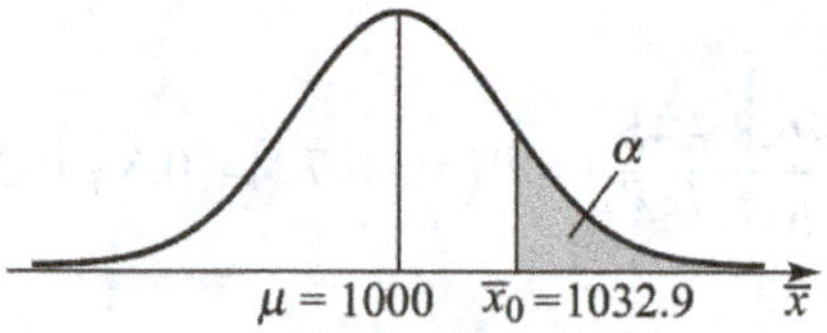

b. $z=\dfrac{\bar{x}_0-\mu_0}{\sigma_{\bar{x}}} \Rightarrow \bar{x}_0=\mu_0+z_\alpha\sigma_{\bar{x}}=\mu_0+z_\alpha\dfrac{\sigma}{\sqrt{n}}$ where $z_\alpha=z_{.05}=1.645$ from Table II, Appendix B.

Thus, $\bar{x}_0=1000+1.645\dfrac{120}{\sqrt{36}}=1032.9$

c.

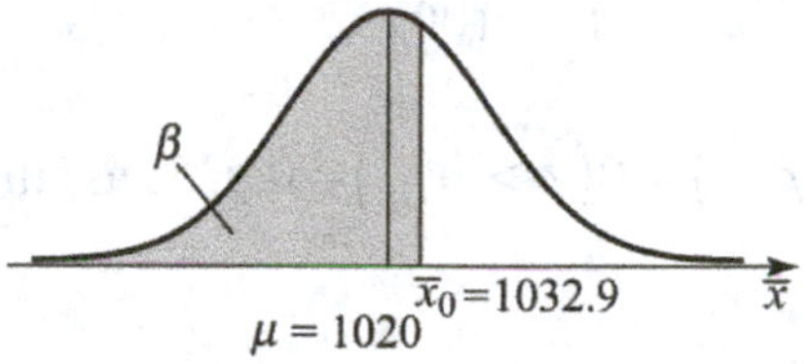

d. $\beta=P(\bar{x}_0<1032.9 \text{ when } \mu=1020)=P\left(z<\dfrac{1032.9-1020}{120/\sqrt{36}}\right)$

$=P(z<0.65)=0.5+0.2422=0.7422$

e. $Power=1-\beta=1-0.7422=0.2578$

8.99 a. The sampling distribution of $\bar{x}$ will be approximately normal (by the Central Limit Theorem) with $\mu_{\bar{x}}=\mu=50$ and $\sigma_{\bar{x}}=\dfrac{\sigma}{\sqrt{n}}=\dfrac{20}{\sqrt{64}}=2.5$.

b. The sampling distribution of $\bar{x}$ will be approximately normal (by the Central Limit Theorem) with $\mu_{\bar{x}}=\mu=45$ and $\sigma_{\bar{x}}=\dfrac{\sigma}{\sqrt{n}}=\dfrac{20}{\sqrt{64}}=2.5$.

c. First, find $\bar{x}_0$ so that $P(\bar{x}<\bar{x}_0)=0.10$ under H_0.

$$P(\bar{x}<\bar{x}_0)=P\left(z<\frac{\bar{x}_0-50}{20/\sqrt{64}}\right)=P\left(z<\frac{\bar{x}_0-50}{2.5}\right)=P(z<z_0)=0.10$$

From Table II, Appendix B, $z_{0.10}=-1.28$

$$z_0 = \frac{\bar{x}_0 - 50}{2.5} \Rightarrow -1.28 = \frac{\bar{x}_0 - 50}{2.5} \Rightarrow \bar{x}_0 = 46.8$$

Now, find

$$\beta = P(\bar{x} > 46.8 \text{ when } \mu = 45) = P\left(z > \frac{46.8 - 45}{20/\sqrt{64}}\right) = P(z > 0.72) = 0.5 - 0.2642 = 0.2358$$

d. $Power = 1 - \beta = 1 - 0.2358 = 0.7642$

8.101 a. The sampling distribution of $\hat{p}$ will be approximately normal (by the Central Limit Theorem) with $\mu_{\hat{p}} = p_o = 0.7$ and $\sigma_{\hat{p}} = \frac{p_o q_o}{n} = \sqrt{\frac{0.7(0.3)}{100}} = 0.0458$.

b. The sampling distribution of $\hat{p}$ will be approximately normal (by the Central Limit Theorem) with $\mu_{\hat{p}} = p_o = 0.65$ and $\sigma_{\hat{p}} = \frac{p_o q_o}{n} = \sqrt{\frac{0.65(0.35)}{100}} = 0.0477$.

c. First, find $\hat{p}_{o,L}$ and $\hat{p}_{o,U}$ such that $P(\hat{p} < \hat{p}_{o,L}) = P(\hat{p} > \hat{p}_{o,U}) = 0.025$, assuming H_0 is true.

$$P(\hat{p} < \hat{p}_{o,L}) = P\left(z < \frac{\hat{p}_{o,L} - 0.7}{0.0458}\right) = P(z < z_{o,L}) = 0.025$$

From Table II, Appendix B, $z_{o,L} = -1.96$

Thus, $z_{o,L} = \frac{\hat{p}_{o,L} - 0.7}{0.0458} \Rightarrow \hat{p}_{o,L} = -1.96(0.0458) + 0.7 = 0.6102$

$$P(\hat{p} > \hat{p}_{o,U}) = P\left(z > \frac{\hat{p}_{o,U} - 0.7}{0.0458}\right) = P(z > z_{o,U}) = 0.025$$

From Table II, Appendix B, $z_{o,U} = 1.96$

Thus, $z_{o,U} = \frac{\hat{p}_{o,U} - 0.7}{0.0458} \Rightarrow \hat{p}_{o,U} = 1.96(0.0458) + 0.7 = 0.7898$

Now, find

$$\beta = P(0.6102 < \hat{p} < 0.7898 \text{ when } \mu_{\hat{p}} = 0.65) = P\left(\frac{0.6102 - 0.65}{0.0477} < z < \frac{0.7898 - 0.65}{0.0477}\right)$$
$$= P(-0.83 < z < 2.93) = 0.2967 + 0.4983 = 0.7950$$

(Using Table II, Appendix B)

d. $\beta = P\left(0.6102 < \hat{p} < 0.7898 \text{ when } \mu_{\hat{p}} = 0.71\right) = P\left(\frac{0.6102 - 0.71}{0.0477} < z < \frac{0.7898 - 0.71}{0.0477}\right)$

$= P(-2.09 < z < 1.67) = 0.4817 + 0.4525 = 0.9342$

8.103 a. First, we find $\bar{x}_o$ such that $P(\bar{x} < \bar{x}_o) = 0.01$, under H_0.

From Exercise 8.102, we know that $z_o = -2.33$. Thus,

$$z_o = \frac{\bar{x}_o - \mu_o}{\frac{\sigma}{\sqrt{n}}} = \frac{\bar{x}_o - 20}{\frac{11.9}{\sqrt{100}}} = \frac{\bar{x}_o - 20}{1.19} = -2.33$$

$$\Rightarrow \bar{x}_o = -2.33(1.19) + 20 = 17.227$$

For $\mu = 19$,

$$Power = P(\bar{x} < 17.227 \text{ when } \mu = 19) = P\left(z < \frac{17.227 - 19}{1.19}\right)$$

$$= P(z < -1.49) = 0.5 - 0.4319 = 0.0681$$

For $\mu = 18$,

$$Power = P(\bar{x} < 17.227 \text{ when } \mu = 18) = P\left(z < \frac{17.227 - 18}{1.19}\right)$$

$$= P(z < -0.65) = 0.5 - 0.2422 = 0.2578$$

For $\mu = 16$,

$$Power = P(\bar{x} < 17.227 \text{ when } \mu = 16) = P\left(z < \frac{17.227 - 16}{1.19}\right)$$

$$= P(z < 1.03) = 0.5 + 0.3485 = 0.8485$$

For $\mu = 14$,

$$Power = P(\bar{x} < 17.227 \text{ when } \mu = 14) = P\left(z < \frac{17.227 - 14}{1.19}\right)$$

$$= P(z < 2.71) = 0.5 + 0.4966 = 0.9966$$

For $\mu = 12$,

$$Power = P(\bar{x} < 17.227 \text{ when } \mu = 12) = P\left(z < \frac{17.227 - 12}{1.19}\right)$$

$$= P(z < 4.39) \approx 0.5 + 0.5 = 1$$

Using MINITAB, the plot is:

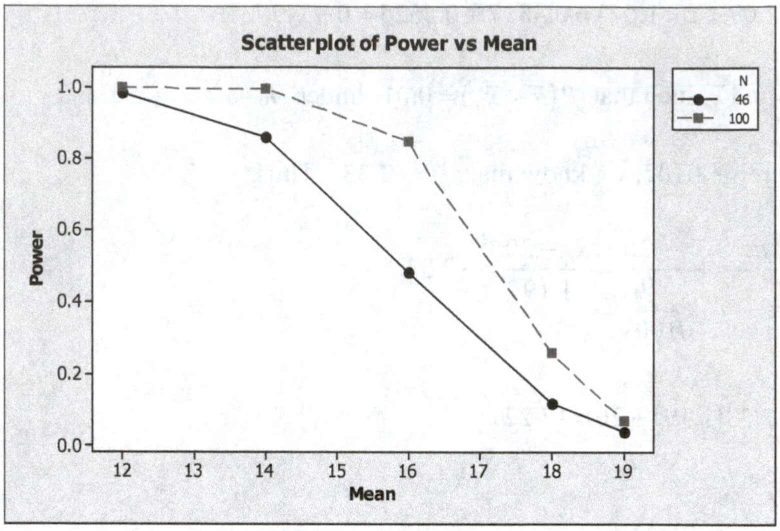

The power is much greater for $n = 100$ than for $n = 46$.

8.105 First, we must compute the value of $\bar{x}$ corresponding to the border of the rejection region. From Exercise 8.46,

$$H_0: \mu = 10{,}000$$
$$H_a: \mu > 10{,}000$$

The rejection region is $z > 1.645$ with $\alpha = 0.05$. The border value for the one-sided test is:

$$\bar{x}_0 = \mu_0 + z_\alpha \sigma_{\bar{x}} = \mu_0 + z_\alpha \frac{\sigma}{\sqrt{n}} = 10{,}000 + 1.645 \frac{1600}{\sqrt{67}} = 10{,}000 + 321.55 = 10{,}321.55$$

Next, we convert this border value to a z-score using the alternative distribution with mean $\mu_a = 10{,}500$. The z-score is:

$$z = \frac{\bar{x}_0 - \mu_a}{\sigma_{\bar{x}}} = \frac{10{,}321.55 - 10{,}500}{\frac{1{,}600}{\sqrt{67}}} = -0.91$$

The probability of making a Type II error is:

$$\beta = P\left(\bar{x} \le 10{,}321.55 \mid \mu_a = 10{,}500\right) = P\left(z \le -0.91\right) = 0.5 - 0.3186 = 0.1814$$

8.107 First, we find $\hat{p}_{oL}$ such that $P\left(\hat{p} < \hat{p}_{oL}\right) = 0.005$ and $\hat{p}_{oU}$ such that $P\left(\hat{p} > \hat{p}_{oU}\right) = 0.005$

From Exercise 8.82, we know that $z_{oL} = -2.575$ and $z_{oU} = 2.58$. Thus,

$$z_{oL} = \frac{\hat{p}_{oL} - p_o}{\sqrt{\frac{p_o q_o}{n}}} = \frac{\hat{p}_{oL} - 0.5}{\sqrt{\frac{0.5(0.5)}{121}}} = \frac{\hat{p}_{oL} - 0.5}{0.0455} = -2.58$$

$$\Rightarrow \hat{p}_{oL} = -2.58(0.0455) + 0.5 = 0.383$$

and

$$z_{oU} = \frac{\hat{p}_{oU} - p_o}{\sqrt{\frac{p_o q_o}{n}}} = \frac{\hat{p}_{oU} - 0.5}{\sqrt{\frac{0.5(0.5)}{121}}} = \frac{\hat{p}_{oU} - 0.5}{0.0455} = 2.58$$

$$\Rightarrow \hat{p}_{oU} = 2.58(0.0455) + 0.5 = 0.617$$

$$Power = P(\hat{p}_{oL} < 0.383 \text{ when } p = 0.65) + P(\hat{p}_{oH} > 0.617 \text{ when } p = 0.65)$$

$$= P\left(z < \frac{0.383 - 0.65}{\sqrt{\frac{0.65(0.35)}{121}}}\right) + P\left(z > \frac{0.617 - 0.65}{\sqrt{\frac{0.65(0.35)}{121}}}\right)$$

$$= P(z < -6.16) + P(z > -0.76) = 0 + 0.5 + 0.2764 = 0.7764$$

8.109 The conditions required for a valid test for σ^2 are:

1. A random sample is selected from the target population.
2. The population from which the sample is selected has a distribution that is approximately normal.

8.111 The statement "When the sample size n is large, no assumptions about the population are necessary to test the population variance σ^2" is false. The population from which the sample is drawn must be approximately normal.

8.113 a. $df = n - 1 = 16 - 1 = 15$; reject H_0 if $\chi^2 < 6.26214$ or $\chi^2 > 27.4884$

b. $df = n - 1 = 23 - 1 = 22$; reject H_0 if $\chi^2 > 40.2894$

c. $df = n - 1 = 15 - 1 = 14$; reject H_0 if $\chi^2 > 21.0642$

d. $df = n - 1 = 13 - 1 = 12$; reject H_0 if $\chi^2 < 3.57056$

e. $df = n - 1 = 7 - 1 = 6$; reject H_0 if $\chi^2 < 1.63539$ or $\chi^2 > 12.5916$

f. $df = n - 1 = 25 - 1 = 24$; reject H_0 if $\chi^2 < 13.8484$

8.115 a. $H_0: \sigma^2 = 1$

$H_a: \sigma^2 > 1$

The test statistic is $\chi^2 = \frac{(n-1)s^2}{\sigma_0^2} = \frac{(100-1)(1.84)}{1} = 182.16$

The rejection region requires $\alpha = 0.05$ in the upper tail of the χ^2 distribution with $df = n - 1 = 100 - 1 = 99$. From Table IV, Appendix B, $\chi^2_{0.05} \approx 124.324$. The rejection region is $\chi^2 > 124.324$.

Since the observed value of the test statistic falls in the rejection region $(\chi^2 = 182.16 > 124.324)$, H_0 is rejected. There is sufficient evidence to indicate the variance is larger than 1 at $\alpha = 0.05$.

b. In part **b** of Exercise 8.114, the test statistic was $\chi^2 = 11.04$. In Exercise 8.114, we did not reject H_0, but we did in this Exercise. As the sample size increases, it becomes easier to reject H_0.

8.117 a. To determine if the standard deviation is less than 10% or the variance is less than 100, we test:

$H_0: \sigma^2 = 100$

$H_a: \sigma^2 < 100$

b. The rejection region requires $\alpha = 0.05$ in the lower tail of the χ^2 distribution with $df = n - 1 = 200 - 1 = 199$. Using MINITAB, $\chi^2_{0.95} = 167.361$. The rejection region is $\chi^2 < 167.361$.

c. The definition of α is the probability of concluding that the standard deviation is less than 10 (or the variance is less than 100) when, in fact, the standard deviation is 10 (or the variance is 100).

d. From the printout, the test statistic is $\chi^2 = 154.81$ with a *p*-value of $p = 0.009$.

e. Since the observed value of the test statistic falls in the rejection region $(\chi^2 = 154.81 < 167.361)$, H_0 is rejected. There is sufficient evidence to conclude the true standard deviation is less than 10% (or true variance is less than 100) at $\alpha = 0.05$.

f. We must assume that a random sample was selected from the target population and that the population from which the sample was selected is approximately normal.

8.119 a. The parameter of interest is σ^2, the variance of the population of rock bounces.

b. To determine if the variance differs from 10 m^2, we test

$H_0: \sigma^2 = 10$

$H_a: \sigma^2 \neq 10$

c. The test statistic is $\chi^2 = \frac{(n-1)s^2}{\sigma_0^2} = \frac{(13-1)(4.0947291)^2}{10} = 20.12$

d. The rejection region requires $\alpha/2 = 0.10/2 = 0.05$ in each tail of the χ^2 distribution with $df = n-1 = 13-1 = 12$. From Table IV, Appendix B, $\chi^2_{0.05} = 21.0261$ and $\chi^2_{0.95} = 5.22603$. The rejection region is $\chi^2 > 21.0261$ or $\chi^2 < 5.22603$.

e. Using MINITAB, the *p*-value for the test is $p = 0.130$.

f. Since the observed value of the test statistic does not fall in the rejection region $(\chi^2 = 20.12 \not> 21.0261$ and $\chi^2 = 20.12 \not< 5.22603)$, H_0 is not rejected. There is insufficient evidence to indicate the variance differs from 10 m^2 at $\alpha = 0.10$.

g. We must assume that a random sample was selected from the population of rock bounces and that the population of rock bounces from which the sample was selected is approximately.

8.121 a. The rejection region requires $\alpha/2 = 0.01/2 = 0.005$ in each tail of the χ^2 distribution with $df = n-1 = 46-1 = 45$. From Table IV, Appendix B, $\chi^2_{0.005} \approx 66.7659$ and $\chi^2_{0.995} \approx 20.7065$. The rejection region is $\chi^2 < 20.7065$ or $\chi^2 > 66.7659$.

b. The test statistic is $\chi^2 = \frac{(n-1)s^2}{\sigma_o^2} = \frac{(46-1)(11.9)^2}{100} = 63.7245$

c. Since the observed value of the test statistic does not fall in the rejection region $(\chi^2 = 63.7245 \not> 66.7659$ and $\chi^2 = 63.7245 \not< 20.7065)$, H_0 is not rejected. There is insufficient evidence to indicate the variance is different from 100 at $\alpha = 0.01$.

8.123 Using MINITAB, the descriptive statistics are:

Descriptive Statistics: Force

```
Variable   N    Mean   StDev
Force     12  163.22    4.99
```

To determine if the true standard deviation of maximum strand force is less than 5 (variance is less than 25), we test:

$H_o: \sigma^2 = 25$
$H_a: \sigma^2 < 25$

The test statistic is $\chi^2 = \frac{(n-1)s^2}{\sigma_0^2} = \frac{(12-1)(4.99)^2}{5^2} = 10.96$

The rejection region requires $\alpha/2 = 0.10/2 = 0.05$ in each tail of the χ^2 distribution with $df = n-1 = 12-1 = 11$. From Table IV, Appendix B, $\chi^2_{0.05} = 19.6751$ and $\chi^2_{0.95} = 4.57481$. The rejection region is $\chi^2 > 19.6751$ or $\chi^2 < 4.57481$.

Since the observed value of the test statistic does not fall in the rejection region $\left(\chi^2 = 10.96 \not< 4.57481 \text{ and } \chi^2 = 10.96 \not> 19.6751\right)$, H_0 is not rejected. There is insufficient evidence to indicate the standard deviation of the population of maximum strand force is less than 5 at $\alpha = 0.10$.

8.125 To determine if the true standard deviation of the point-spread errors exceed 15 (variance exceeds 225), we test:

$$H_0: \sigma^2 = 225$$
$$H_a: \sigma^2 > 225$$

The test statistic is $\chi^2 = \dfrac{(n-1)s^2}{\sigma_0^2} = \dfrac{(240-1)13.3^2}{225} = 187.896$

The rejection region requires $\alpha = 0.05$ in the upper tail of the χ^2 distribution with $df = n-1 = 240-1 = 239$. Using a MINITAB, $\chi^2_{0.05} = 276.062$. The rejection region is $\chi^2 > 276.062$.

Since the observed value of the test statistics does not fall in the rejection region $\left(\chi^2 = 187.896 \not> 276.062\right)$, H_0 is not rejected. There is insufficient evidence to indicate that the true standard deviation of the point-spread errors exceed 15 at $\alpha = 0.05$.

(Since the observed variance (or standard deviation) is less than the hypothesized value of the variance (or standard deviation) under H_0, there is no way H_0 will be rejected for any reasonable value of α.)

8.127 To determine if the variance of birthweights of babies delivered by cocaine-dependent women is less than 200,000, we test:

$$H_0: \sigma^2 = 200{,}000$$
$$H_a: \sigma^2 < 200{,}000$$

The test statistic is $\chi^2 = \dfrac{(n-1)s^2}{\sigma_0^2} = \dfrac{(16-1)(410)^2}{200{,}000} = 12.6075$

The rejection region requires $\alpha = 0.01$ in the lower tail of the χ^2 distribution with $df = n-1 = 16-1 = 15$. From Table IV, Appendix B, $\chi^2_{0.99} = 5.22935$. The rejection region is $\chi^2 < 5.22935$.

Since the observed value of the test statistic does not fall in the rejection region $\left(\chi^2 = 12.6075 \not< 5.22935\right)$, H_0 is not rejected. There is insufficient evidence to indicate that the variance of birthweights of babies delivered by cocaine-dependent women is less than 200,000 at $\alpha = 0.01$.

8.129 The smaller the *p*-value associated with a test of hypothesis, the stronger the support for the **alternative** hypothesis. The *p*-value is the probability of observing your test statistic or anything more unusual, given the null hypothesis is true. If this value is small, it would be very unusual to observe this test statistic if the null hypothesis were true. Thus, it would indicate the alternative hypothesis is true.

8.131 The elements of the test of hypothesis that should be specified prior to analyzing the data are: null hypothesis, alternative hypothesis, and significance level.

8.133 The larger the *p*-value associated with a test of hypothesis, the stronger the support for the **null** hypothesis. The *p*-value is the probability of observing your test statistic or anything more unusual, given the null hypothesis is true. If this value is large, it would not be very unusual to observe this test statistic if the null hypothesis were true. Thus, it would lend support that the null hypothesis is true.

8.135 a. $H_0: p = 0.35$
$H_a: p > 0.35$

The test statistic is $z = \dfrac{\hat{p} - p_0}{\sqrt{\dfrac{p_0 q_0}{n}}} = \dfrac{0.29 - 0.35}{\sqrt{\dfrac{0.35(0.65)}{200}}} = -1.78$

The rejection region requires $\alpha = 0.05$ in the lower tail of the *z* distribution. From Table II, Appendix B, $z_{0.05} = 1.645$. The rejection region is $z < -1.645$.

Since the observed value of the test statistic falls in the rejection region $(z = -1.78 < -1.645)$, H_0 is rejected. There is sufficient evidence to indicate $p < 0.35$ at $\alpha = 0.05$.

b. $H_0: p = 0.35$
$H_a: p \neq 0.35$

The test statistic is $z < -1.78$ (from **a**).

The rejection region requires $\alpha / 2 = 0.05 / 2 = 0.025$ in each tail of the *z* distribution. From Table II, Appendix B, $z_{0.025} = 1.96$. The rejection region is $z < -1.96$ or $z > 1.96$.

Since the observed value of the test statistic does not fall in the rejection region $(z = -1.78 \not< -1.96)$, H_0 is not rejected. There is insufficient evidence to indicate p is different from .35 at $\alpha = 0.05$.

c. For confidence coefficient 0.95, $\alpha = 0.05$ and $\alpha/2 = 0.05/2 = 0.025$. From Table II, Appendix B, $z_{0.025} = 1.96$. The confidence interval is:

$$\hat{p} \pm z_{0.025}\sqrt{\frac{\hat{p}\hat{q}}{n}} \Rightarrow 0.29 \pm 1.96\sqrt{\frac{0.29(0.71)}{200}} \Rightarrow 0.29 \pm 0.063 \Rightarrow (0.227, 0.353)$$

d. For confidence coefficient 0.99, $\alpha = 0.01$ and $\alpha/2 = 0.01/2 = 0.005$. From Table II, Appendix B, $z_{0.005} = 2.58$. The confidence interval is:

$$\hat{p} \pm z_{0.005}\sqrt{\frac{\hat{p}\hat{q}}{n}} \Rightarrow 0.29 \pm 2.58\sqrt{\frac{0.29(0.71)}{200}} \Rightarrow 0.29 \pm 0.083 \Rightarrow (0.207, 0.373)$$

e. $$n = \frac{(z_{\alpha/2})^2 pq}{B^2} = \frac{2.58^2(0.29)(0.71)}{(0.05)^2} = 548.2 \approx 549$$

(We used $\hat{p} = 0.29$ to estimate the value of p.)

8.137 a. $H_0: \sigma^2 = 30$
$H_a: \sigma^2 > 30$

The test statistic is $\chi^2 = \frac{(n-1)s^2}{\sigma_0^2} = \frac{(41-1)(6.9)^2}{30} = 63.48$

The rejection region requires $\alpha = 0.05$ in the upper tail of the χ^2 distribution with $df = n - 1 = 41 - 1 = 40$. From Table IV, Appendix B, $\chi^2_{0.05} = 55.7585$. The rejection region is $\chi^2 > 55.7585$.

Since the observed value of the test statistic falls in the rejection region $(\chi^2 = 63.48 > 55.7585)$, H_0 is rejected. There is sufficient evidence to indicate the variance is larger than 30 at $\alpha = 0.05$.

b. $H_0: \sigma^2 = 30$
$H_a: \sigma^2 \neq 30$

The test statistic is $\chi^2 = 63.48$ (from part **a**).

The rejection region requires $\alpha/2 = 0.05/2 = 0.025$ in each tail of the χ^2 distribution with $df = n - 1 = 41 - 1 = 40$. From Table IV, Appendix B, $\chi^2_{0.025} = 59.3417$ and $\chi^2_{0.975} = 24.4331$. The rejection region is $\chi^2 < 24.4331$ or $\chi^2 > 59.3417$.

Since the observed value of the test statistic falls in the rejection region $\left(\chi^2 = 63.48 > 59.3417\right)$, H_0 is rejected. There is sufficient evidence to indicate the variance is not 30 at $\alpha = 0.05$.

8.139 a. The null hypothesis would be: $H_0: p = 0.45$

b. The null hypothesis would be: $H_0: \mu = 2.5$

8.141 a. To determine if the true mean score of all sleep-deprived subjects is less than 80, we test:

$$H_0: \mu = 80$$
$$H_a: \mu < 80$$

The test statistic is $t = \dfrac{\bar{x} - \mu_0}{s / \sqrt{n}} = \dfrac{63 - 80}{17 / \sqrt{12}} = -3.46$

The rejection region requires $\alpha = 0.05$ in the lower tail of the t distribution with $df = n - 1 = 12 - 1 = 11$. From Table III, Appendix B, $t_{0.05} = 1.796$. The rejection region is $t < -1.796$.

Since the observed value of the test statistic falls in the rejection region $\left(t = -3.46 < -1.796\right)$, H_0 is rejected. There is sufficient evidence to indicate that the true mean score of all sleep-deprived subjects is less than 80 at $\alpha = 0.05$.

b. We must assume that the overall test scores of sleep-deprived students are normally distributed.

8.143 Let $\mu =$ mean Mach rating score of all purchasing managers.

a. To determine if the true mean Mach rating score of all purchasing managers is different from 85, we test,

$$H_0: \mu = 85$$
$$H_a: \mu \neq 85$$

b. A Type I error would be concluding that the true mean Mach rating score is different from 85 when, in fact, it is equal to 85.

c. By definition, $\alpha =$ probability of committing a Type I error. With a value of $\alpha = 0.10$, this indicates that in repeated sampling, we will reject H_0 when it is true about 10% of the time.

d. The rejection region requires $\alpha / 2 = 0.10 / 2 = 0.05$ in each tail of the z distribution. From Table II, Appendix B, $z_{0.05} = 1.645$. The rejection region is $z < -1.645$ or $z > 1.645$.

e. The test statistic is $z = \dfrac{\bar{x} - \mu}{\sigma_{\bar{x}}} \approx \dfrac{99.6 - 85}{12.6 / \sqrt{122}} = 12.80$.

Since the observed value of the test statistic falls in the rejection region $(z = 12.80 > 1.645)$, H_0 is rejected. There is sufficient evidence to indicate the true mean Mach rating score is different from 85 at $\alpha = 0.10$.

f. No. Because the sample size is sufficiently large $(n = 122)$, the Central Limit Theorem says that the distribution of $\bar{x}$ will be approximately normal.

8.145 a. Let p = proportion of the subjects who felt that the masculinization of face shape decreased attractiveness of the male face. If there is no preference for the unaltered or morphed male face, then $p = 0.5$.

For this problem, $\hat{p} = x / n = 58 / 67 = 0.866$.

In order for the inference to be valid, the sample size must be large enough. The sample size is large enough if both np_o and nq_o are greater than or equal to 15.

$np_o = 67(0.5) = 33.5$ and $nq_o = 67(1 - 0.5) = 67(0.5) = 33.5$. Since both of these values are greater than 15, the sample size is large enough to use the normal approximation.

To determine if the subjects showed a preference for either the unaltered or morphed face, we test:

$H_0: p = 0.5$
$H_a: p \neq 0.5$

b. The test statistic is $z = \dfrac{\hat{p} - p_0}{\sqrt{\dfrac{p_0 q_0}{n}}} = \dfrac{0.866 - 0.50}{\sqrt{\dfrac{0.50(0.50)}{67}}} = 5.99$

c. By definition, the p-value is the probability of observing our test statistic or anything more unusual, given H_0 is true. For this problem, $p = P(z \leq -5.99) + P(z \geq 5.99) \approx 0.5 - 0.5 + 0.5 - 0.5 = 0$. This corresponds to what is listed in the Exercise.

d. The rejection region requires $\alpha / 2 = 0.01 / 2 = 0.005$ in each tail of the z distribution. From Table II, Appendix B, $z_{0.005} = 2.58$. The rejection region is $z < -2.58$ or $z > 2.58$.

Since the observed value of the test statistic falls in the rejection region $(z = 5.99 > 2.58)$, H_0 is rejected. There is sufficient evidence to indicate that the subjects showed a preference for either the unaltered or morphed face at $\alpha = 0.01$.

8.147 To determine if the mean alkalinity level of water in the tributary exceeds 50 mpl, we test:

$H_0: \mu = 50$
$H_a: \mu > 50$

The test statistic is $z = \frac{\bar{x} - \mu_0}{\sigma_{\bar{x}}} = \frac{67.8 - 50}{14.4 / \sqrt{100}} = 12.36$

The rejection region requires $\alpha = 0.01$ in the upper tail of the z distribution. From Table II, Appendix B, $z_{0.01} = 2.33$. The rejection region is $z > 2.33$.

Since the observed value of the test statistic falls in the rejection region $(z = 12.36 > 2.33)$, H_0 is rejected. There is sufficient evidence to indicate that the mean alkalinity level of water in the tributary exceeds 50 mpl at $\alpha = 0.01$.

8.149 a. A Type I error would be to conclude that the mean response for the population of all New York City public school children is not 3 when, in fact, the mean response is equal to 3.

A Type II error would be to conclude that the mean response for the population of all New York City public school children is equal to 3 when, in fact, the mean response is not equal to 3.

b. To determine if the mean response for the population of all New York City public school children is different from 3, we test:

$H_0: \mu = 3$
$H_a: \mu \neq 3$

The test statistic is $z = \frac{\bar{x} - \mu_o}{\sigma_{\bar{x}}} = \frac{2.15 - 3}{\frac{1.05}{\sqrt{11,160}}} = -85.52$

The rejection region requires $\alpha / 2 = 0.05 / 2 = 0.025$ in each tail of the z distribution. From Table II, Appendix B, $z_{0.025} = 1.96$. The rejection region is $z < -1.96$ or $z > 1.96$.

Since the observed value of the test statistic falls in the rejection region $(z = -85.52 < -1.96)$, H_0 is rejected. There is sufficient evidence to indicate the mean response for the population of all New York City public school children different from 3 at $\alpha = 0.05$.

c. To determine if the mean response for the population of all New York City public school children is different from 3, we test:

$H_0: \mu = 3$
$H_a: \mu \neq 3$

The test statistic is $z = \dfrac{\bar{x} - \mu_o}{\sigma_{\bar{x}}} = \dfrac{2.15 - 3}{\dfrac{1.05}{\sqrt{11{,}160}}} = -85.52$

The rejection region requires $\alpha / 2 = 0.10 / 2 = 0.05$ in each tail of the z distribution. From Table II, Appendix B, $z_{0.05} = 1.645$. The rejection region is $z < -1.645$ or $z > 1.645$.

Since the observed value of the test statistic falls in the rejection region $(z = -85.52 < -1.645)$, H_0 is rejected. There is sufficient evidence to indicate the mean response for the population of all New York City public school children different from 3 at $\alpha = .10$.

8.151 To determine if the true mean PTSD score of all World War II aviator POWs is less than 16, we test:

$$H_0: \mu = 16$$
$$H_a: \mu < 16$$

The test statistic is $z = \dfrac{\bar{x} - \mu_o}{\sigma_{\bar{x}}} = \dfrac{9.00 - 16}{9.32/\sqrt{33}} = -4.31$

The rejection region requires $\alpha = 0.10$ in the lower tail of the z distribution. From Table II, Appendix B, $z_{0.10} = 1.28$. The rejection region is $z < -1.28$.

Since the observed value of the test statistic falls in the rejection region $(z = -4.31 < -1.28)$, H_0 is rejected. There is sufficient evidence to indicate that the true mean PTSD score of all World War II aviator POWs is less than 16 at $\alpha = 0.10$.

8.153 a. To determine if the true mean inbreeding coefficient μ for this species of wasp exceeds 0, we test:

$$H_0: \mu = 0$$
$$H_a: \mu > 0$$

The test statistic is $z = \dfrac{\bar{x} - \mu_0}{\sigma_{\bar{x}}} = \dfrac{0.044 - 0}{0.884 / \sqrt{197}} = 0.70$

The rejection region requires $\alpha = 0.05$ in the upper tail of the z distribution. From Table II, Appendix B, $z_{0.05} = 1.645$. The rejection region is $z > 1.645$.

Since the observed value of the test statistic does not fall in the rejection region $(z = 0.70 \not> 1.645)$, H_0 is not rejected. There is insufficient evidence to indicate that the true mean inbreeding coefficient μ for this species of wasp exceeds 0 at $\alpha = 0.05$.

b. This result agrees with that of Exercise 7.139. The confidence interval in Exercise 7.139 was $(-0.118,\ 0.206)$. Since this interval contains 0, there is no evidence to indicate that the mean is different from 0. This is the same conclusion that was reached in part **a**.

8.155 Some preliminary calculations: $\hat{p} = \frac{x}{n} = \frac{401}{835} = 0.48$

In order for the inference to be valid, the sample size must be large enough. The sample size is large enough if both np_o and nq_o are greater than or equal to 15.

$np_o = 835(0.45) = 375.75$ and $nq_o = 835(1-0.45) = 835(0.45) = 459.25$. Since both of these values are greater than 15, the sample size is large enough to use the normal approximation.

To determine if more than 45% of male youths are raised in a single-parent family, we test:

$$H_0: p = 0.45$$
$$H_a: p > 0.45$$

The test statistic is $z = \dfrac{\hat{p} - p_0}{\sqrt{\dfrac{p_0 q_0}{n}}} = \dfrac{0.48 - 0.45}{\sqrt{\dfrac{0.45(0.55)}{835}}} = 1.74$

The rejection region requires $\alpha = 0.05$ in the upper tail of the z distribution. From Table II, Appendix B, $z_{0.05} = 1.645$. The rejection region is $z > 1.645$.

Since the observed value of the test statistic falls in the rejection region $(z = 1.74 > 1.645)$, H_0 is rejected. There is sufficient evidence that more than 45% of male youths are raised in a single-parent family at $\alpha = 0.05$.

8.157 a. To determine if the production process should be halted, we test:

$$H_0: \mu = 3$$
$$H_a: \mu > 3 \qquad \text{where } \mu = \text{mean amount of PCB in the effluent.}$$

The test statistic is $z = \dfrac{\bar{x} - \mu_0}{\sigma_{\bar{x}}} = \dfrac{3.1 - 3}{0.5/\sqrt{50}} = 1.41$

The rejection region requires $\alpha = 0.01$ in the upper tail of the z distribution. From Table II, Appendix B, $z_{0.01} = 2.33$. The rejection region is $z > 2.33$.

Since the observed value of the test statistic does not fall in the rejection region, $(z = 1.41 \not> 2.33)$, H_0 is not rejected. There is insufficient evidence to indicate the mean amount of PCB in the effluent is more than 3 parts per million at $\alpha = 0.01$. Do not halt the manufacturing process.

b. As plant manager, I do not want to shut down the plant unnecessarily. Therefore, I want $\alpha = P$(shut down plant when $\mu = 3$) to be small.

8.159 a. No, it increases the risk of falsely rejecting H_0, i.e., closing the plant unnecessarily.

b. First, find $\bar{x}_0$ such that $P(\bar{x} > \bar{x}_0) = P(z > z_0) = 0.05$.

From Table II, Appendix B, $z_0 = 1.645$

$$z = \frac{\bar{x} - \mu_0}{\sigma / \sqrt{n}} \Rightarrow 1.645 = \frac{\bar{x}_0 - 3}{0.5 / \sqrt{50}} \Rightarrow \bar{x}_0 = 3.116$$

Then, compute

$$\beta = P(\bar{x}_0 \le 3.116 \text{ when } \mu = 3.1) = P\left(z \le \frac{3.116 - 3.1}{0.5/\sqrt{50}}\right) = P(z \le 0.23) = 0.5 + 0.0910 = 0.5910$$

$$Power = 1 - \beta = 1 - 0.5910 = 0.4090$$

c. The power of the test increases as α increases.

8.161 The point estimate for p is $\hat{p} = \frac{x}{n} = \frac{85}{124} = 0.685$

In order for the inference to be valid, the sample size must be large enough. The sample size is large enough if both np_o and nq_o are greater than or equal to 15.

$np_o = 124(0.167) = 20.7$ and $nq_o = 124(1 - 0.167) = 124(0.833) = 103.3$. Since both of these values are greater than 15, the sample size is large enough to use the normal approximation.

To determine if the students will tend to choose the three-grill display so that Grill #2 is a compromise between a more and a less desirable grill $(p > 0.167)$, we test:

$$H_0: p = 0.167$$
$$H_a: p > 0.167$$

The test statistic is $z = \frac{\hat{p} - p_0}{\sqrt{\frac{p_0 q_0}{n}}} = \frac{0.685 - 0.167}{\sqrt{\frac{0.167(0.822)}{124}}} = 15.46$

The rejection region requires $\alpha = 0.05$ in the upper tail of the z distribution. From Table II, Appendix B, $z_{0.05} = 1.645$. The rejection region is $z > 1.645$.

Since the observed value of the test statistic falls in the rejection region $(z = 15.46 > 1.645)$, H_0 is rejected. There is sufficient evidence to indicate the students will tend to choose the three-grill display so that Grill #2 is a compromise between a more and a less desirable grill $(p > 0.167)$ at $\alpha = 0.05$.

8.163 To determine if the mean amount of cesium in lichen specimens differs from .003, we test:

$$H_0: \mu = 0.003$$
$$H_a: \mu \neq 0.003$$

From the printout, the test statistic is $t = 3.725$ and the *p*-value is $p = 0.0058$.

Since the p-value is less than α $(p = 0.0058 < 0.10)$, H_0 is rejected. There is sufficient evidence to indicate the mean amount of cesium in lichen specimens differs from .003 at $\alpha = 0.10$.

8.165 a. A Type I error is rejecting the null hypothesis when it is true. In this problem, we would be concluding that the individual is a liar when, in fact, the individual is telling the truth.

A Type II error is accepting the null hypothesis when it is false. In this problem, we would be concluding that the individual is telling the truth when, in fact, the individual is a liar.

b. The probability of a Type I error would be the probability of concluding the individual is a liar when he/she is telling the truth. From the problem, it stated that the polygraph would indicate that of 500 individuals who were telling the truth, 185 would be liars. Thus, an estimate of the probability of a Type I error would be $185/500 = 0.37$.

The probability of a Type II error would be the probability of concluding the individual is telling the truth when he/she is a liar. From the problem, it stated that the polygraph would indicate that of 500 individuals who were liars, 120 would be telling the truth. Thus, an estimate of the probability of a Type II error would be $120/500 = 0.24$.

8.167 a. Using MINITAB, the descriptive statistics are:

Descriptive Statistics: PCB

```
Variable   N  Mean    StDev      Minimum     Q1  Median      Q3  Maximum
PCB       48 9.292    2.103  0.000000000  9.000   9.000  10.000   13.000
```

To determine if the data can refute the manufacturer's claim, we test:

$$H_0: \mu = 10$$
$$H_a: \mu < 10$$

The test statistic is $z = \dfrac{\bar{x} - \mu_0}{\sigma_{\bar{x}}} = \dfrac{9.292 - 10}{2.103/\sqrt{48}} = -2.33$

The rejection region requires $\alpha = 0.05$ in the lower tail of the z distribution. From Table II, Appendix B, $z_{0.05} = 1.645$. The rejection region is $z < -1.645$.

Since the observed value of the test statistic falls in the rejection region $(z = -2.33 < -1.645)$, H_0 is rejected. There is sufficient evidence to indicate that the true average number of inspections is less than 10 at $\alpha = 0.05$.

We agree with the potential buyer who doubts the manufacturer's claim.

b. First, we must compute the value of $\bar{x}$ corresponding to the border of the rejection region. The rejection region is $z < -1.645$ with $\alpha = 0.05$. The border value for the one-sided test is:

$$\bar{x}_0 = \mu_0 - z_\alpha \sigma_{\bar{x}} = \mu_0 - z_\alpha \frac{\sigma}{\sqrt{n}} = 10 - 1.645 \frac{1.2}{\sqrt{48}} = 10 - 0.285 = 9.715$$

Next, we convert this border value to a z-score using the alternative distribution with mean $\mu_a = 9.5$. The z-score is $z = \frac{\bar{x}_0 - \mu_a}{\sigma_{\bar{x}}} = \frac{9.715 - 9.5}{\frac{1.2}{\sqrt{48}}} = 1.24$

The probability of rejecting H_0 is:

$$Power = P(\bar{x} \leq 9.715 \mid \mu_a = 9.5) = P(z \leq 1.24) = 0.5 + 0.3925 = 0.8925$$

8.169 a. We want to show that more than 17,000 of the 18,200 signatures are valid. Thus, we want to show that the proportion of valid signatures is greater than 17,000/18,200 = .934.

From the Exercise, $\hat{p} = 98/100 = 0.98$.

In order for the inference to be valid, the sample size must be large enough. The sample size is large enough if both np_o and nq_o are greater than or equal to 15.

$np_o = 100(0.934) = 93.4$ and $nq_o = 100(1 - 0.934) = 100(0.066) = 6.6$. Since the second value is not greater than 15, the sample size may not be large enough to use the normal approximation.

To determine if more than 0.934 of the signatures are valid, we test:

$H_0: p = 0.934$
$H_a: p > 0.934$

The test statistic is $z = \frac{\hat{p} - p_0}{\sqrt{\frac{p_0 q_0}{n}}} = \frac{0.98 - 0.934}{\sqrt{\frac{0.934(0.066)}{100}}} = 1.85$

Since no α is given in the problem, we will choose $\alpha = 0.05$. The rejection region requires $\alpha = 0.05$ in the upper tail of the z distribution. From Table II, Appendix B, $z_{0.05} = 1.645$. The rejection region is $z > 1.645$.

Since the observed value of the test statistic falls in the rejection region $(z = 1.85 > 1.645)$, H_0 is rejected. There is sufficient evidence to indicate that the true proportion of valid signatures is greater than 0.934 at $\alpha = .05$. This indicates that more than 17,000 of the 18,200 signatures are valid.

b. We want to show that more than 16,000 of the 18,200 signatures are valid. Thus, we want to show that the proportion of valid signatures is greater than $16,000/18,200 = 0.879$.

Again, from the Exercise, $\hat{p} = 98/100 = 0.98$.

In order for the inference to be valid, the sample size must be large enough. The sample size is large enough if both np_o and nq_o are greater than or equal to 15.

$np_o = 100(0.879) = 87.9$ and $nq_o = 10(1-0.879) = 100(0.121) = 12.1$. Since the second value is not greater than 15, the sample size may not be large enough to use the normal approximation.

To determine if more than 0.879 of the signatures are valid, we test:

$H_0: p = 0.879$
$H_a: p > 0.879$

The test statistic is $z = \frac{\hat{p} - p_0}{\sqrt{\frac{p_0 q_0}{n}}} = \frac{0.98 - 0.879}{\sqrt{\frac{0.879(0.121)}{100}}} = 3.10$

Since no α is given in the problem, we will choose $\alpha = 0.05$. The rejection region requires $\alpha = 0.05$ in the upper tail of the z distribution. From Table II, Appendix B, $z_{0.05} = 1.645$. The rejection region is $z > 1.645$.

Since the observed value of the test statistic falls in the rejection region $(z = 3.10 > 1.645)$, H_0 is rejected. There is sufficient evidence to indicate that the true proportion of valid signatures is greater than .879 at $\alpha = 0.05$. This indicates that more than 16,000 of the 18,200 signatures are valid.

Chapter 9

Inferences Based on Two Samples

Confidence Intervals and Tests of Hypotheses

9.1 When the samples are large, the sampling distribution of $(\bar{x}_1 - \bar{x}_2)$ is approximately normal by the Central Limit Theorem. The mean of the sampling distribution is $(\mu_1 - \mu_2)$ and the standard deviation is $\sqrt{\frac{\sigma_1^2}{n_1} + \frac{\sigma_2^2}{n_2}}$.

9.3 a. No. Both populations must be normal.

b. No. Both population variances must be equal.

c. No. Both populations must be normal.

d. Yes.

e. No. Both populations must be normal.

9.5 The confidence interval for $(\mu_1 - \mu_2)$ is $(-10, -4)$. The correct inference is **b:** $\mu_1 < \mu_2$. Since all the values in the interval are negative, there is evidence that $\mu_1 < \mu_2$.

9.7 a. $\mu_{\bar{x}_1} = \mu_1 = 14$; $\sigma_{\bar{x}} = \frac{\sigma_1}{\sqrt{n_1}} = \frac{4}{\sqrt{100}} = .4$

b. $\mu_{\bar{x}_2} = \mu_2 = 10$; $\sigma_{\bar{x}_2} = \frac{\sigma_2}{\sqrt{n_2}} = \frac{3}{\sqrt{100}} = .3$

c. $\mu_{\bar{x}_1 - \bar{x}_2} = \mu_1 - \mu_2 = 14 - 10 = 4$; $\sigma_{\bar{x}_1 - \bar{x}_2} = \sqrt{\frac{\sigma_1^2}{n_1} + \frac{\sigma_2^2}{n_2}} = \sqrt{\frac{4^2}{100} + \frac{3^2}{100}} = \sqrt{\frac{25}{100}} = 0.5$

d. Since $n_1 \geq 30$ and $n_2 \geq 30$, the sampling distribution of $\bar{x}_1 - \bar{x}_2$ is approximately normal by the Central Limit Theorem.

9.9 Some preliminary calculations are:

$$\bar{x}_1 = \frac{\sum x_1}{n_1} = \frac{11.8}{5} = 2.36 \qquad s_1^2 = \frac{\sum x_1^2 - \frac{\left(\sum x_1\right)^2}{n_1}}{n_1 - 1} = \frac{30.78 - \frac{(11.8)^2}{5}}{5 - 1} = 0.733$$

$$\bar{x}_2 = \frac{\sum x_2}{n_2} = \frac{14.4}{4} = 3.6 \qquad s_2^2 = \frac{\sum x_2^2 - \frac{\left(\sum x_2\right)^2}{n_2}}{n_2 - 1} = \frac{53.1 - \frac{(14.4)^2}{4}}{4 - 1} = 0.42$$

a. $$s_p^2 = \frac{(n_1-1)s_1^2+(n_2-1)s_2^2}{n_1+n_2-2} = \frac{(5-1)(0.773)+(4-1)(0.42)}{5+4-2} = \frac{4.192}{7} = 0.5989$$

b. $H_0: \mu_1 - \mu_2 = 0$
$H_a: \mu_1 - \mu_2 < 0$

The test statistic is $t = \frac{(\bar{x}_1-\bar{x}_2)-D_0}{\sqrt{s_p^2\left(\frac{1}{n_1}+\frac{1}{n_2}\right)}} = \frac{(2.36-3.6)-0}{\sqrt{0.5989\left(\frac{1}{5}+\frac{1}{4}\right)}} = \frac{-1.24}{0.5191} = -2.39$

The rejection region requires $\alpha = 0.10$ in the lower tail of the t distribution with $df = n_1 + n_2 - 2 = 5 + 4 - 2 = 7$. From Table III, Appendix B, $t_{0.10} = 1.415$. The rejection region is $t < -1.415$.

Since the test statistic falls in the rejection region $(t = -2.39 < -1.415)$, H_0 is rejected. There is sufficient evidence to indicate that $\mu_2 > \mu_1$ at $\alpha = 0.10$.

c. A small sample confidence interval is needed because $n_1 = 5 < 30$ and $n_2 = 4 < 30$.

For confidence coefficient 0.90, $\alpha = 0.10$ and $\alpha / 2 = 0.05$. From Table III, Appendix B, with $df = n_1 + n_2 - 2 = 5 + 4 - 2 = 7$, $t_{0.05} = 1.895$. The 90% confidence interval for $(\mu_1 - \mu_2)$ is:

$$(\bar{x}_1-\bar{x}_2) \pm t_{0.05}\sqrt{s_p^2\left(\frac{1}{n_1}+\frac{1}{n_2}\right)} \Rightarrow (2.36-3.6) \pm 1.895\sqrt{0.5989\left(\frac{1}{5}+\frac{1}{4}\right)}$$

$$\Rightarrow -1.24 \pm 0.98 \Rightarrow (-2.22, -0.26)$$

d. The confidence interval in part **c** provides more information about $(\mu_1 - \mu_2)$ than the test of hypothesis in part **b**. The test in part **b** only tells us that μ_2 is greater than μ_1. However, the confidence interval estimates the actual difference between μ_1 and μ_2.

9.11 a. The p-value is $p = 0.115$.

Since the p-value is not small, there is no evidence to reject H_0. There is no evidence to indicate that the population means are different for $\alpha \le 0.10$.

b. The p-value would be half of the p-value in part **a**. The p-value is $p = 0.1150 / 2 = 0.0575$.

There is no evidence to reject H_0 for $\alpha = 0.05$. There is no evidence to indicate that the mean for population 1 is less than the mean for population 2 for $\alpha = 0.05$.

There is evidence to reject H_0 for $\alpha > 0.0575$. There is evidence to indicate that the mean for population 1 is less than the mean for population 2 for $\alpha > 0.0575$.

9.13 a. The p-value is $p = 0.000$. Because this p-value is so small, we would reject H_0 and conclude that there was a difference in mean leadership values between captains from successful and unsuccessful teams.

b. The p-value is $p = 0.907$. Because this p-value is so large, we would not reject H_0 and conclude that there was insufficient evidence to conclude a difference in mean leadership values between flight attendants from successful and unsuccessful teams.

9.15 a. Let μ_1 = mean test score of students in classrooms using software products and μ_2 = mean test score of students in classrooms not using software products. The parameter of interest is $\mu_1 - \mu_2$.

b. To determine if the mean test score of students in classrooms using software products is higher than the mean test score of students in classrooms not using software products, we test:

$$H_0: \mu_1 - \mu_2 = 0$$
$$H_a: \mu_1 - \mu_2 > 0$$

c. Since the p-value is so large $(p = 0.62)$, H_0 is not rejected for any reasonable value of α. There is insufficient evidence to indicate the mean test score of students in classrooms using software products is higher than the mean test score of students in classrooms not using software products. This agrees with the conclusion of the DOE.

9.17 a. Let μ_1 = mean recall of those who received only visual aspects of the ad and μ_2 = mean recall of those who received an audiovisual presentation. To determine if the mean recall of ad information by those who receive an audiovisual presentation is different from those who receive only the visual aspects of the ad, we test:

$$H_0: \mu_1 - \mu_2 = 0$$
$$H_a: \mu_1 - \mu_2 \neq 0$$

b. Some preliminary calculations are:

$$s_p^2 = \frac{(n_1 - 1)s_1^2 + (n_2 - 1)s_2^2}{n_1 + n_2 - 2} = \frac{(20-1)1.98^2 + (20-1)2.13^2}{20 + 20 - 2} = 4.22865$$

The test statistic is $t = \dfrac{\bar{x}_1 - \bar{x}_2}{\sqrt{s_p^2\left(\dfrac{1}{n_1} + \dfrac{1}{n_2}\right)}} = \dfrac{3.70 - 3.30}{\sqrt{4.22865\left(\dfrac{1}{20} + \dfrac{1}{20}\right)}} = 0.62$

c. The rejection region requires $\alpha / 2 = 0.10 / 2 = 0.05$ in each tail of the t distribution with $df = n_1 + n_2 - 2 = 20 + 20 - 2 = 38$. From Table III, Appendix B, $t_{0.05} \approx 1.684$. The rejection region is $t < -1.684$ or $t > 1.684$.

d. Since the observed value of the test statistic does not fall in the rejection region

$(t = 0.62 \not> 1.684)$, H_0 is not rejected. There is insufficient evidence to indicate the mean recall of ad information by those who receive an audiovisual presentation is different from those who receive only the visual aspects of the ad at $\alpha = 0.10$.

Although we cannot prove the researchers' theory is correct (Accept H_o), there is evidence that supports the theory.

e. A *p*-value of 0.62 is the probability of observing our test statistic or anything more unusual. For this problem, the *p*-value $= P(t > 0.62) + P(t < -0.62) = 0.62$. Since the *p*-value is so large, H_o will not be rejected for any reasonable value of α.

f. In order for the inference to be valid, the following conditions must be met:

1. Both populations being sampled from are normal
2. Both population variances must be the same
3. The samples must be random and independent.

9.19 a. From the printout, the 95% confidence interval is $(-0.60,\ 7.95)$. We are 95% confident that the difference in mean FNE scores for bulimic and normal female students is between -0.60 and 7.95.

b. We must assume that the distribution of FNE scores for bulimic female students and the distribution of the FNE scores for the normal female students are normally distributed. We must also assume that the variances of the two populations are equal and the samples are random and independent.

Both sample distributions look somewhat mound-shaped and the sample variances are fairly close in value. Thus, both assumptions appear to be reasonably satisfied.

9.21 a. Let μ_1 = mean choice score for consumers with a flexed arm and μ_2 = mean choice score for consumers with an extended arm.

$$s_p^2 = \frac{(n_1-1)s_1^2 + (n_2-1)s_2^2}{n_1+n_2-2} = \frac{(11-1)4^2 + (11-1)2^2}{11+11-2} = 10$$

To determine if the mean choice score for consumers with flexed arms is greater than the mean choice score for consumers with extended arms, we test:

$$H_0: \mu_1 - \mu_2 = 0$$
$$H_a: \mu_1 - \mu_2 > 0$$

The test statistic is $t = \dfrac{(\bar{x}_1 - \bar{x}_2) - D_o}{\sqrt{s_p^2\left(\dfrac{1}{n_1} + \dfrac{1}{n_2}\right)}} = \dfrac{(59-43)-0}{\sqrt{10\left(\dfrac{1}{11} + \dfrac{1}{11}\right)}} = 11.87$

The rejection region requires $\alpha = 0.05$ in the upper tail of the t-distribution with $df = n_1 + n_2 - 2 = 11 + 11 - 2 = 20$. From Table III, Appendix B, $t_{0.05} = 1.725$. The rejection region is $t > 1.725$.

Since the observed value of the test statistic falls in the rejection region $(t = 11.87 > 1.725)$, H_0 is rejected. There is sufficient evidence to indicate the mean choice score for consumers with flexed arms is greater than the mean choice score for consumers with extended arms at $\alpha = 0.05$.

b. $$s_p^2 = \frac{(n_1-1)s_1^2 + (n_2-1)s_2^2}{n_1+n_2-2} = \frac{(11-1)10^2 + (11-1)15^2}{11+11-2} = 162.5$$

To determine if the mean choice score for consumers with flexed arms is greater than the mean choice score for consumers with extended arms, we test:

$$H_0: \mu_1 - \mu_2 = 0$$
$$H_a: \mu_1 - \mu_2 > 0$$

The test statistic is $t = \dfrac{(\bar{x}_1 - \bar{x}_2) - D_o}{\sqrt{s_p^2\left(\dfrac{1}{n_1} + \dfrac{1}{n_2}\right)}} = \dfrac{(59-43)-0}{\sqrt{162.5\left(\dfrac{1}{11} + \dfrac{1}{11}\right)}} = 2.94$

From part a, the rejection region is $t > 1.725$.

Since the observed value of the test statistic falls in the rejection region $(t = 2.94 > 1.725)$, H_0 is rejected. There is sufficient evidence to indicate the mean choice score for consumers with flexed arms is greater than the mean choice score for consumers with extended arms at $\alpha = 0.05$.

9.23 Let μ_1 = mean drug concentration for site 1 and μ_2 = mean drug concentration for site 2. To determine if there is a difference in the mean drug concentration of the 2 sites, we test:

$$H_0: \mu_1 - \mu_2 = 0$$
$$H_a: \mu_1 - \mu_2 \neq 0$$

From the printout, the test statistic is $t = 0.57$ and the p-value is $p = 0.573$. Since the p-value is so large, H_0 is not rejected. There is insufficient evidence to indicate a difference in the mean drug concentration of the 2 sites for any reasonable value of α.

9.25 Let μ_1 = mean total rating for group 1 (strengthen favored position) and μ_2 = mean total rating for group 2 (weaken opposing position).

Some preliminary calculations are:

$$s_p^2 = \frac{(n_1-1)s_1^2+(n_2-1)s_2^2}{n_1+n_2-2} = \frac{(26-1)12.5^2+(26-1)12.2^2}{26+26-2} = \frac{7,627.25}{50} = 152.545$$

To determine if the difference in the mean total ratings for the two groups differ, we test:

$H_0: \mu_1 - \mu_2 = 0$
$H_a: \mu_1 - \mu_2 \neq 0$

The test statistic is $t = \dfrac{(\bar{x}_1-\bar{x}_2)-0}{\sqrt{s_p^2\left(\dfrac{1}{n_1}+\dfrac{1}{n_2}\right)}} = \dfrac{(28.6-24.9)-0}{\sqrt{152.545\left(\dfrac{1}{26}+\dfrac{1}{26}\right)}} = 1.08$

The rejection region requires $\alpha/2 = 0.05/2 = 0.025$ in each tail of the t distribution. From Table III, Appendix B, with $df = n_1 + n_2 - 2 = 26 + 26 - 2 = 50$, $t_{0.025} \approx 2.010$. The rejection region is $t < -2.010$ or $t > 2.010$.

Since the observed value of the test statistic does not fall in the rejection region $(t = 1.08 \not> 2.010)$, H_0 is not rejected. There is insufficient evidence to indicate there is a difference in the mean total ratings for the two groups at $\alpha = 0.05$.

9.27 a. Let μ_1 = mean MFS score for the violent event group and μ_2 = mean MFS score for the avoided-violent event group. The target parameter of interest is $\mu_1 - \mu_2$ = difference in the mean MFS score between the 2 groups.

b. We are given the sample mean scores for the two groups. However, in order to determine if there is a difference in the 2 population means, we also need to know the population standard deviations for the 2 groups or the variances for the 2 groups. This information was not provided.

c. The p-value of the test is

$$p = P(z \leq -1.21) + P(z \geq 1.21) = (0.5 - 0.3869) + (0.5 - 0.3869)$$
$$= 0.1131 + 0.1131 = 0.2262$$

d. Since the observed p-value is greater than α $(p = 0.2262 > \alpha = 0.10)$, H_0 is not rejected. There is insufficient evidence to indicate a difference in the mean MFS score between the violent event group and the avoided-violent event group at $\alpha = 0.10$.

9.29 a. From the information given, we have no idea what the standard deviations are. Whether the population means are different or not depends on how variable the data are.

b. Let μ_1 = mean pain intensity rating for blacks and μ_2 = mean pain intensity rating for whites.

To determine if blacks, on average, have a higher pain intensity rating than whites, we test:

$$H_0: \mu_1 - \mu_2 = 0$$
$$H_a: \mu_1 - \mu_2 > 0$$

The rejection region requires $\alpha = 0.05$ in the upper tail of the z distribution. From Table II, Appendix B, $z_{0.05} = 1.645$. Thus, in order to reject H_0, the test statistic would have to be greater than 1.645.

Substituting known values into the test statistic, we can solve for s_p.

$$z = \frac{\bar{x}_1 - \bar{x}_2}{\sqrt{s_p^2\left(\frac{1}{n_1} + \frac{1}{n_2}\right)}} > 1.645 \quad \Rightarrow \frac{8.2 - 6.9}{\sqrt{s_p^2\left(\frac{1}{55} + \frac{1}{159}\right)}} > 1.645$$

$$\Rightarrow 1.3 > 1.645(s_p)\sqrt{\left(\frac{1}{55} + \frac{1}{159}\right)} \quad \Rightarrow \frac{1.3}{1.645(0.1564)} > s_p \quad \Rightarrow 5.053 > s_p$$

Thus, if both s_1 and s_2 were less than 5.053, then we would conclude that blacks, on average, would have a higher pain intensity than whites. (It is possible that either s_1 or s_2 could be greater than 5.053 and we would still reject H_0. As long as $s_p < 5.053$, we would reject H_0.)

c. From part **b**, we know we would reject H_0 if $s_p < 5.053$. Thus, we would not reject H_0 if $s_p \geq 5.053$. This could happen if both s_1 and s_2 are greater than 5.053. (It is possible that either s_1 or s_2 could be less than 5.053 and we would still reject H_0. As long as $s_p \geq 5.053$, we would reject H_0.)

9.31 In a paired difference experiment, the observations should be paired before collecting the data.

9.33 True.

9.35 a. The rejection region requires $\alpha = 0.05$ in the upper tail of the t distribution with $df = n_d - 1 = 10 - 1 = 9$. From Table III, Appendix B, $t_{0.05} = 1.833$. The rejection region is $t > 1.833$.

b. From Table III, with $df = n_d - 1 = 20 - 1 = 19$, $t_{0.10} = 1.328$. The rejection region is $t > 1.328$.

c. From Table III, with $df = n_d - 1 = 5 - 1 = 4$, $t_{0.025} = 2.776$. The rejection region is $t > 2.776$.

d. From Table III, with $df = n_d - 1 = 9 - 1 = 8$, $t_{0.01} = 2.896$. The rejection region is $t > 2.896$.

9.37 a.

Pair	Difference
1	3
2	2
3	2
4	4
5	0
6	1

$$\bar{x}_d = \frac{\sum x_d}{n_d} = \frac{12}{6} = 2 \qquad s_d^2 = \frac{\sum x_d^2 - \frac{\left(\sum x_d\right)^2}{n_d}}{n_d - 1} = \frac{\left(34 - \frac{(12)^2}{6}\right)}{5} = 2$$

b. $\mu_d = \mu_1 - \mu_2$

c. For confidence coefficient 0.95, $\alpha = 0.05$ and $\alpha/2 = 0.025$. From Table III, Appendix B, with $df = n_d - 1 = 6 - 1 = 5$, $t_{0.025} = 2.571$. The confidence interval is:

$$\bar{x}_d \pm t_{\alpha/2}\frac{s_d}{\sqrt{n_d}} \Rightarrow 2 \pm 2.571\frac{\sqrt{2}}{\sqrt{6}} \Rightarrow 2 \pm 1.484 \Rightarrow (0.516, 3.484)$$

d. $H_0: \mu_d = 0$
$H_a: \mu_d \neq 0$

The test statistic is $t = \frac{\bar{x}_d - 0}{s_d/\sqrt{n_d}} = \frac{2}{\sqrt{2}/\sqrt{6}} = 3.46$

The rejection region requires $\alpha/2 = 0.05/2 = 0.025$ in each tail of the t distribution with $df = n_d - 1 = 6 - 1 = 5$. From Table III, Appendix B, $t_{0.025} = 2.571$. The rejection region is $t < -2.571$ or $t > 2.571$.

Since the observed value of the test statistic falls in the rejection region $(t = 3.46 > 2.571)$, H_0 is rejected. There is sufficient evidence to indicate that the mean difference is different from 0 at $\alpha = 0.05$.

9.39 a. $H_0: \mu_d = 10$
$H_a: \mu_d \neq 10$

The test statistic is $z = \frac{\bar{x}_d - 10}{s_d/\sqrt{n_d}} = \frac{11.7 - 10}{6/\sqrt{40}} = 1.79$

The rejection region requires $\alpha / 2 = 0.05 / 2 = 0.025$ in each tail of the z distribution. From Table II, Appendix B, $z_{0.025} = 1.96$. The rejection region is $z < -1.96$ or $z > 1.96$.

Since the observed value of the test statistic does not fall in the rejection region $(z = 1.79 \not> 1.96)$, H_0 is not rejected. There is insufficient evidence to indicate the difference in the population means is different from 10 at $\alpha = 0.05$.

b. The p-value is

$$p = P(z \le -1.79) + P(z \ge 1.79) = (0.5 - 0.4633) + (0.5 - 0.4633)$$
$$= 0.0367 + 0.0367 = 0.0734$$

9.41 a. To determine if the mean score for the fictitious brand is greater than the mean for the commercially available brand, we test:

$$H_0: \mu_d = 0$$
$$H_a: \mu_d > 0$$

b. Because the data were collected as paired data (each student rated each brand), it has to be analyzed as paired data using a paired-difference test.

c. The sample size was 408. Thus, we can use the z distribution. The test statistic is $z = 5.71$.

The rejection region requires $\alpha = 0.05$ in the upper tail of the z distribution. From Table II, Appendix B, $z_{0.05} = 1.645$. The rejection region is $z > 1.645$.

Since the test statistic falls in the rejection region $(z = 5.71 > 1.645)$, H_0 is rejected. There is sufficient evidence to indicate the mean score for the fictitious brand is greater than the mean for the commercially available brand at $\alpha = 0.05$.

d. The p-value is $p = P(z \ge 5.71) \approx 0.5 - 0.5 = 0$.

e. Yes. Because the p-value is smaller than $\alpha = 0.01$ $(p = 0 < 0.01)$, H_0 is rejected.

9.43 a. The health status is measured on all patients two times, both before and after handling museum objects. Thus, the two sets of measurements are not independent. The data need to be analyzed as paired differences.

b. The differences between the "before" and "after" measurements are:

Session	Difference	Session	Difference
1	-7	17	0
2	-12	18	-11
3	-9	19	-11
4	-9	20	-10
5	1	21	-12
6	-13	22	-7
7	-16	23	-4
8	-3	24	-2
9	-7	25	-8
10	-2	26	-1
11	-6	27	4
12	-10	28	-17
13	-9	29	-11
14	-9	30	-12
15	1	31	-5
16	-14	32	-13

c. Using MINITAB, the descriptive statistics are:

Descriptive Statistics: Before, After, Diff

```
Variable   N    Mean  StDev  Minimum      Q1  Median      Q3  Maximum
Before    32   51.66  10.09    30.00   43.00   51.00   59.00    73.00
After     32   59.28   9.60    42.00   52.25   59.00   64.75    83.00
Diff      32  -7.625  5.272  -17.000 -11.750  -9.000  -3.250    4.000
```

The mean of the differences is $\mu_d = -7.625$ and the standard deviation of the differences is $s_d = 5.272$.

d. For confidence coefficient 0.90, $\alpha = 0.10$ and $\alpha/2 = 0.05$. From Table II, Appendix B, $z_{0.05} = 1.645$. The confidence interval is:

$$\bar{x}_d \pm z_{0.05}\frac{s_d}{\sqrt{n_d}} \Rightarrow -7.625 \pm 1.645\frac{5.272}{\sqrt{32}} \Rightarrow -7.625 \pm 1.533 \Rightarrow (-9.158, -6.092)$$

e. We are 90% confident that the true difference in mean health status before and after handling museum objects is between -9.158 and -6.092. Since the interval is entirely negative, there is evidence to indicate the mean after handling the museum objects is higher than the mean before handling the museum objects. Thus, handling museum objects has a positive impact on a sick patient's well-being.

9.45 a. Let μ_1 = mean BOLD score for the first session and μ_2 = mean BOLD score when the placebo was applied. The target parameter is $\mu_d = \mu_1 - \mu_2$, the difference in the mean BOLD scores between the first session and the session involving the placebo.

b. A paired difference design was used to collect the data. Each experimental person had two observations measured on it.

c. To determine if there was a placebo effect, we test:

$H_o: \mu_d = 0$
$H_a: \mu_d > 0$

d. The test statistic is $t = \frac{\bar{x}_d - 0}{s_d / \sqrt{n_d}} = \frac{0.21 - 0}{.47 / \sqrt{24}} = 2.19$

Since no α value was given, we will use $\alpha = 0.05$. The rejection region requires $\alpha = 0.05$ in the upper tail of the t distribution with $df = n_d - 1 = 24 - 1 = 23$. From Table III, Appendix B, $t_{0.05} = 1.714$. The rejection region is $t > 1.714$.

Since the observed value of the test statistic falls in the rejection region $(t = 2.19 > 1.714)$, H_0 is rejected. There is sufficient evidence to indicate that there is a placebo effect at $\alpha = 0.05$.

e. Since the p-value is less than α $(p = 0.02 < \alpha = 0.05)$, H_0 is rejected. There is sufficient evidence to indicate that there is a placebo effect at $\alpha = 0.05$.

9.47 Let μ_1 = mean energy supplied by east-west panels, μ_2 = mean energy supplied by north-south panels and $\mu_d = \mu_1 - \mu_2$. This is a paired difference problem. The differences are:

Month	*East-West*	*North-South*	*diff*
February	8658	8921	-263
April	7930	8317	-387
July	5120	5274	-154
September	6862	7148	-286
October	8608	8936	-328

Using MINITAB, the descriptive statistics are:

Descriptive Statistics: EastWest, NorthSouth, diff

```
Variable     N    Mean  StDev
EastWest     5    7436   1484
NorthSouth   5    7719   1548
diff         5  -283.6   86.4
```

To determine if the mean energy supplied by north-south panels is greater than that of east-west panels, we test:

$H_0: \mu_d = 0$
$H_a: \mu_d > 0$

The test statistic is $t = \frac{\bar{x}_d - 0}{\frac{s_d}{\sqrt{n_d}}} = \frac{-283.6 - 0}{\frac{86.4}{\sqrt{5}}} = -7.34$.

No α was given so we will use $\alpha = 0.05$. The rejection region requires $\alpha = 0.05$ in the lower tail of the t distribution. From Table III, Appendix B, with $df = n_d - 1 = 5 - 1 = 4$, $t_{0.05} = 2.132$. The rejection region is $t < -2.132$.

Since the observed value of the test statistic falls in the rejection region $(t = -7.34 < -2.132)$, H_0 is rejected. There is sufficient evidence to indicate the mean energy supplied by north-south panels is greater than that of east-west panels at $\alpha = 0.05$.

9.49 Using MINITAB, the descriptive statistics are:

Descriptive Statistics: Initial, Final, Diff

Variable	N	Mean	StDev
Initial	3	5.640	1.075
Final	3	5.453	1.125
Diff	3	0.1867	0.1106

Let μ_1 = mean initial pH and μ_2 = mean final pH. The parameter of interest is $\mu_d = \mu_1 - \mu_2$. To determine if the mean initial pH differs significantly from the mean pH after 30 days, we test:

$$H_0: \mu_d = 0$$
$$H_a: \mu_d \neq 0$$

The test statistics is $t = \dfrac{\bar{x}_d - 0}{\frac{s_d}{\sqrt{n_d}}} = \dfrac{0.1867}{\frac{0.1106}{\sqrt{3}}} = 2.92$

The rejection region requires $\alpha / 2 = 0.05 / 2 = 0.025$ in each tail of the t distribution. From Table III, Appendix B, with $df = n_d - 1 = 3 - 1 = 2$, $t_{0.025} = 4.303$. The rejection region is $t < -4.303$ or $t > 4.303$.

Since the observed value of the test statistic does not fall in the rejection region $(t = 2.92 \not> 4.303)$, H_0 is not rejected. There is insufficient evidence to indicate the mean initial pH differs significantly from the mean pH after 30 days at $\alpha = 0.05$.

9.51 a. Since all the data were collected from twins, the observations from each pair are not independent of each other. Thus, the data were collected as paired data. If the data are collected as paired data, they must be analyzed as paired data.

b. Let μ_1 = mean level of overall leisure activity for the control group and μ_2 = mean level of overall leisure activity for the demented group. The target parameter is $\mu_d = \mu_1 - \mu_2$, the difference in the mean level of overall leisure activity for the control group and the demented group.

Using MINITAB, the descriptive statistics are:

Descriptive Statistics: Diff

```
Variable     N  Mean   StDev
Diff       107 1.953 10.082
```

We will use a 95% confidence interval to estimate the difference in the mean level of leisure activity between the control group and the demented group. For confidence coefficient 0.95, $\alpha = 0.05$ and $\alpha/2 = 0.05/2 = 0.025$. From Table II, Appendix B, $z_{0.025} = 1.96$. The 95% confidence interval is:

$$\bar{x}_d \pm z_{.025}\frac{s_d}{\sqrt{n_d}} \Rightarrow 1.953 \pm 1.96\frac{10.082}{\sqrt{107}} \Rightarrow 1.953 \pm 1.910 \Rightarrow (0.043,\ 3.863)$$

We are 95% confident that the difference in the mean level of leisure activity between the control group and the demented group is between 0.043 and 3.863. Since 0 is not contained in this interval, there is evidence of a difference in the mean level of leisure activity between the control group and the demented group at $\alpha = .05$. Since all the values are positive, there is evidence that the mean level of leisure activity for the control group is larger than that for the demented group.

9.53 Using MINITAB, the descriptive statistics are:

Descriptive Statistics: Before, After, Diff

```
Variable   N   Mean   StDev
Before    13  2.513  1.976
After     13  1.506  1.448
Diff      13  1.007  1.209
```

Let μ_1 = mean number of crashes before the camera and μ_2 = mean number of crashes after the camera. The parameter of interest is $\mu_d = \mu_1 - \mu_2$. To determine if the mean number of crashes after the camera is smaller than before the camera, we test:

$H_0: \mu_d = 0$
$H_a: \mu_d > 0$

The test statistics is $t = \dfrac{\bar{x}_d - 0}{\frac{s_d}{\sqrt{n_d}}} = \dfrac{1.007}{\frac{1.209}{\sqrt{13}}} = 3.00$

Since no α is given, we will use $\alpha = 0.05$. The rejection region requires $\alpha = 0.05$ in the upper tail of the t distribution. From Table III, Appendix B, with $df = n_d - 1 = 13 - 1 = 12$, $t_{0.05} = 1.782$. The rejection region is $t > 1.782$.

Since the observed value of the test statistic falls in the rejection region $(t = 3.00 > 1.782)$, H_0 is rejected. There is sufficient evidence to indicate the mean number of crashes after the camera is smaller than before the camera at $\alpha = 0.05$.

9.55 The conditions required for a large-sample inference about $(p_1 - p_2)$ are:

1. The two samples are randomly selected in an independent manner from the two target populations.
2. The sample sizes, n_1 and n_2, are both large so that the sampling distribution of $(\hat{p}_1 - \hat{p}_2)$ will be approximately normal. (This condition will be satisfied if all of the following are true: $n_1\hat{p}_1 \geq 15$, $n_1\hat{q}_1 \geq 15$ and $n_2\hat{p}_2 \geq 15$, $n_2\hat{q}_2 \geq 15$.)

9.57 a. The distribution of x_1 is binomial with n_1 trials and the probability of success, p_1. The distribution of x_2 is binomial with n_2 trials and the probability of success, p_2.

b. For large samples, the sampling distribution of $(\hat{p}_1 - \hat{p}_2)$ is approximately normal with mean $(p_1 - p_2)$, and standard deviation $\sigma_{\hat{p}_1 - \hat{p}_2} = \sqrt{\frac{p_1 q_1}{n_1} + \frac{p_2 q_2}{n_2}}$.

9.59 For confidence coefficient 0.95, $\alpha = 1 - 0.95 = 0.05$ and $\alpha/2 = 0.05/2 = 0.025$. From Table II, Appendix B, $z_{0.025} = 1.96$. The 95% confidence interval for $p_1 - p_2$ is approximately:

a. $(\hat{p}_1 - \hat{p}_2) \pm z_{\alpha/2}\sqrt{\frac{\hat{p}_1\hat{q}_1}{n_1} + \frac{\hat{p}_2\hat{q}_2}{n_2}}$

$$\Rightarrow (0.65 - 0.58) \pm 1.96\sqrt{\frac{0.65(1-0.65)}{400} + \frac{0.58(1-0.58)}{400}} \Rightarrow 0.07 \pm 0.067 \Rightarrow (0.003,\ 0.137)$$

b. $(\hat{p}_1 - \hat{p}_2) \pm z_{\alpha/2}\sqrt{\frac{\hat{p}_1\hat{q}_1}{n_1} + \frac{\hat{p}_2\hat{q}_2}{n_2}}$

$$\Rightarrow (0.31 - 0.25) \pm 1.96\sqrt{\frac{0.31(1-0.31)}{180} + \frac{0.25(1-0.25)}{250}} \Rightarrow 0.06 \pm 0.086 \Rightarrow (-0.026,\ 0.146)$$

c. $(\hat{p}_1 - \hat{p}_2) \pm z_{\alpha/2}\sqrt{\frac{\hat{p}_1\hat{q}_1}{n_1} + \frac{\hat{p}_2\hat{q}_2}{n_2}}$

$$\Rightarrow (0.46 - 0.61) \pm 1.96\sqrt{\frac{0.46(1-0.46)}{100} + \frac{0.61(1-0.61)}{120}} \Rightarrow -0.15 \pm 0.131 \Rightarrow (-0.281,\ -0.019)$$

9.61 $H_0: p_1 - p_2 = 0.1$

$H_a: p_1 - p_2 > 0.1$

Since D_0 is not equal to 0, the test statistic is:

$$z=\frac{(\hat{p}_1-\hat{p}_2)-D_0}{\sqrt{\frac{p_1q_1}{n_1}+\frac{p_2q_2}{n_2}}}=\frac{(\hat{p}_1-\hat{p}_2)-D_0}{\sqrt{\frac{\hat{p}_1\hat{q}_1}{n_1}+\frac{\hat{p}_2\hat{q}_2}{n_2}}}=\frac{(0.4-0.2)-0.1}{\sqrt{\frac{0.4(1-0.4)}{50}+\frac{0.2(1-0.2)}{60}}}=\frac{1}{0.0864}=1.16$$

The rejection region requires $\alpha = 0.05$ in the upper tail of the z distribution. From Table II, Appendix B, $z_{0.05} = 1.645$. The rejection region is $z > 1.645$.

Since the observed value of the test statistic does not fall in the rejection region $(z = 1.16 \not> 1.645)$, H_0 is not rejected. There is insufficient evidence to show $(p_1 - p_2) > 0.1$ at $\alpha = 0.05$.

9.63 a. $\hat{p}_1 = \frac{x_1}{n_1} = \frac{746}{1358} = 0.549$

b. $\hat{p}_2 = \frac{x_2}{n_2} = \frac{967}{1379} = 0.701$

c. For confidence coefficient 0.90, $\alpha = 0.10$ and $\alpha/2 = 0.10/2 = 0.05$. From Table II, Appendix B, $z_{0.05} = 1.645$. The 90% confidence interval is:

$$(\hat{p}_1-\hat{p}_2)\pm z_{0.05}\sqrt{\frac{\hat{p}_1\hat{q}_1}{n_1}+\frac{\hat{p}_2\hat{q}_2}{n_2}}\Rightarrow(0.549-0.701)\pm 1.645\sqrt{\frac{0.549(0.451)}{1358}+\frac{0.701(0.299)}{1379}}$$

$$\Rightarrow -0.152\pm 0.030\Rightarrow(-0.182,\ -0.122)$$

d. We are 90% confident that the true value of the difference in the proportion of Dutch boys and Dutch girls who have never bullied another student is between –0.182 and –0.122.

e. Since the 90% confidence interval contains only negative numbers, the proportion of girls who do not bully is larger than that for boys. Thus, the proportion of Dutch boys who bully is significantly greater than the proportion of Dutch girls who bully.

9.65 a. One population of interest is all hospital patients in India admitted in January and the other is all hospital patients in India admitted in May.

b. Let p_1 = proportion of all hospital patients in India admitted in January who were treated for malaria and p_2 = proportion of all hospital patients in India admitted in May who were treated for malaria.

$\hat{p}_1 = \frac{x_1}{n_1} = \frac{32}{192} = 0.167$ $\qquad \hat{p}_2 = \frac{x_2}{n_2} = \frac{34}{403} = 0.084$ $\qquad \hat{p}_1 - \hat{p}_2 = 0.167 - 0.084 = 0.083$

c. For confidence coefficient 0.90, $\alpha = 0.10$ and $\alpha/2 = 0.10/2 = 0.05$. From Table II, Appendix B, $z_{0.05} = 1.645$. The 90% confidence interval is:

$$(0.167-0.084)\pm 1.645\sqrt{\frac{0.167(0.833)}{192}+\frac{0.084(0.916)}{403}}$$

$$\Rightarrow 0.083\pm 0.050\Rightarrow (0.033,\ 0.133)$$

d. Because 0 is not contained in the confidence interval, there is sufficient evidence to indicate a difference exists in the true malaria admission rate in January and May at $\alpha = 0.10$

9.67 a. Let p_1 = proportion of web users using white welcome screens who broke off the survey and p_2 = proportion of web users using red welcome screens who broke off the survey.

$$\hat{p}_1=\frac{x_1}{n_1}=\frac{49}{190}=0.258 \qquad \hat{p}_2=\frac{x_2}{n_2}=\frac{37}{183}=0.202$$

b. To determine if the break-off rate for users of red welcome screens is lower than the break-off rate for users for white welcome screens, we test:

$$H_0: p_1 - p_2 = 0$$
$$H_a: p_1 - p_2 > 0$$

c. Some preliminary calculations are: $\hat{p}=\frac{x_1+x_2}{n_1+n_2}=\frac{49+37}{190+183}=0.2306$

The test statistic is $z=\frac{\hat{p}_1-\hat{p}_2}{\sqrt{\hat{p}\hat{q}\left(\frac{1}{n_1}+\frac{1}{n_2}\right)}}=\frac{0.258-0.202}{\sqrt{0.2306(0.7694)\left(\frac{1}{190}+\frac{1}{183}\right)}}=1.28$

The rejection region requires $\alpha = 0.10$ in the upper tail of the z-distribution. From Table II, Appendix B, $z_{0.10} = 1.28$. The rejection region is $z > 1.28$.

Since the observed value of the test statistic does not fall in the rejection region $(z = 1.28 \not> 1.28)$, H_0 is not rejected. There is insufficient evidence to indicate the break-off rate for users of red welcome screens is lower than the break-off rate for users for white welcome screens at $\alpha = 0.10$.

9.69 Let p_1 = proportion of water cultures that test positive for Salmonella and p_2 = proportion of wildlife cultures that test positive for Salmonella.

$$\hat{p}_1=\frac{x_1}{n_1}=\frac{18}{252}=0.0714 \qquad \hat{p}_2=\frac{x_2}{n_2}=\frac{20}{476}=0.0420$$

$$\hat{p}=\frac{x_1+x_2}{n_1+n_2}=\frac{18+20}{252+476}=0.0522$$

To determine if the proportion of water cultures that contain Salmonella differs from the proportion of wildlife cultures that contain Salmonella, we test:

$$H_0: p_1 - p_2 = 0$$
$$H_a: p_1 - p_2 \neq 0$$

The test statistic is $z = \dfrac{\hat{p}_1 - \hat{p}_2}{\sqrt{\hat{p}\hat{q}\left(\dfrac{1}{n_1} + \dfrac{1}{n_2}\right)}} = \dfrac{0.0714 - 0.0420}{\sqrt{0.0522(0.9478)\left(\dfrac{1}{252} + \dfrac{1}{476}\right)}} = 1.70$

The rejection region requires $\alpha / 2 = 0.01/2 = 0.005$ in each tail of the z-distribution. From Table II, Appendix B, $z_{0.005} = 2.58$. The rejection region is $z < -2.58$ or $z > 2.58$.

Since the observed value of the test statistic does not fall in the rejection region $(z = 1.70 \not> 2.58)$, H_0 is not rejected. There is insufficient evidence to indicate the proportion of water cultures that contain Salmonella differs from the proportion of wildlife cultures that contain Salmonella at $\alpha = 0.01$.

9.71 Let p_1 = proportion of traffic signs that fail the minimum FHWA retroreflectivity requirements for signs maintained by NCDOT and p_2 = proportion of traffic signs that fail the minimum FHWA retroreflectivity requirements for signs maintained by the county.

$$\hat{p}_1 = \frac{x_1}{n_1} = \frac{512}{1000} = 0.512 \qquad \hat{p}_2 = \frac{x_2}{n_2} = \frac{328}{1000} = 0.328$$

$$\hat{p} = \frac{x_1 + x_2}{n_1 + n_2} = \frac{512 + 328}{1000 + 1000} = 0.420$$

To determine if the proportion of traffic signs that fail the minimum FHWA retroreflectivity requirements differ for signs maintained by NCDOT and signs maintained by the county, we test:

$$H_0: p_1 - p_2 = 0$$
$$H_a: p_1 - p_2 \neq 0$$

The test statistic is $z = \dfrac{\hat{p}_1 - \hat{p}_2}{\sqrt{\hat{p}\hat{q}\left(\dfrac{1}{n_1} + \dfrac{1}{n_2}\right)}} = \dfrac{0.512 - 0.328}{\sqrt{0.420(0.580)\left(\dfrac{1}{1000} + \dfrac{1}{1000}\right)}} = 8.34$

The rejection region requires $\alpha / 2 = 0.05/2 = 0.025$ in each tail of the z-distribution. From Table II, Appendix B, $z_{0.025} = 1.96$. The rejection region is $z < -1.96$ or $z > 1.96$.

Since the observed value of the test statistic falls in the rejection region $(z = 8.34 > 1.96)$, H_0 is rejected. There is sufficient evidence to indicate the proportion of traffic signs that fail the minimum FHWA retroreflectivity requirements differ for signs maintained by NCDOT and signs maintained by the county at $\alpha = 0.05$.

9.73 Let p_1 = recall rate of the 60 to 72 year-old seniors and p_2 = recall rate of the 73 to 83 year-old seniors.

Some preliminary calculations are:

$$\hat{p}_1 = \frac{x_1}{n_1} = \frac{31}{40} = 0.775 \qquad \hat{p}_2 = \frac{x_2}{n_2} = \frac{22}{40} = 0.55 \qquad \hat{p} = \frac{x_1 + x_2}{n_1 + n_2} = \frac{31+22}{40+40} = 0.6625$$

To determine if the recall rates of the two groups differ, we test:

$$H_0: p_1 - p_2 = 0$$
$$H_a: p_1 - p_2 \neq 0$$

The test statistic is $z = \frac{(\hat{p}_1 - \hat{p}_2) - 0}{\sqrt{\hat{p}\hat{q}\left(\frac{1}{n_1} + \frac{1}{n_2}\right)}} = \frac{(0.775 - 0.55) - 0}{\sqrt{0.6625(0.3375)\left(\frac{1}{40} + \frac{1}{40}\right)}} = 2.13$

The rejection region requires $\alpha / 2 = 0.05 / 2 = 0.025$ in each tail of the z distribution. From Table II, Appendix B, $z_{0.025} = 1.96$. The rejection region is $z < -1.96$ or $z > 1.96$.

Since the observed value of the test statistic falls in the rejection region $(z = 2.13 > 1.96)$, H_0 is rejected. There is sufficient evidence to indicate the recall rates of the two groups differ at $\alpha = 0.05$.

9.75 Let p_1 = proportion of volunteers who figured out the third rule in the group that slept and p_2 = proportion of volunteers who figured out the third rule in the group that stayed awake all night. The parameter of interest is $p_1 - p_2$, or the difference in the proportions of volunteers who figured out the third rule between the 2 groups.

Some preliminary calculations are:

$$\hat{p}_1 = \frac{x_1}{n_1} = \frac{39}{50} = 0.78 \qquad \hat{p}_2 = \frac{x_2}{n_2} = \frac{15}{50} = 0.30$$

For confidence coefficient 0.90, $\alpha = 0.10$ and $\alpha / 2 = 0.10 / 2 = 0.05$. From Table II, Appendix B, $z_{0.05} = 1.645$. The 90% confidence interval is:

$$(0.78 - 0.30) \pm 1.645\sqrt{\frac{0.78(0.22)}{50} + \frac{0.30(0.70)}{50}} \Rightarrow 0.48 \pm 0.144 \Rightarrow (0.336,\ 0.624)$$

We are 90% confident that the difference in the proportions of volunteers who figured out the third rule between the 2 groups is between 0.336 and 0.624. Since both end points are greater than 0, there is evidence that the proportion of those who slept who figured out the third rule is greater than the proportion of those who stayed awake all night who figured out the third rule.

9.77 Let p_1 = proportion of Yayoi farmers with LEH defects, p_2 = proportion of Eastern Jomon foragers with LEH defects, and p_3 = proportion of Western Jomon foragers with LEH defects.

First, we will test Theory 1.

Some preliminary calculations are: $\hat{p} = \frac{n_1\hat{p}_1 + n_2\hat{p}_2}{n_1 + n_2} = \frac{182(0.631) + 164(0.482)}{182 + 164} = 0.560$

To determine if foragers with a broad-based economy (Eastern Jomon) have a lower prevalence of LEH defects than early agriculturists (Yayoi), we test:

$$H_0: p_1 - p_2 = 0$$
$$H_a: p_1 - p_2 > 0$$

The test statistic is $z = \frac{\hat{p}_1 - \hat{p}_2}{\sqrt{\hat{p}\hat{q}\left(\frac{1}{n_1} + \frac{1}{n_2}\right)}} = \frac{0.631 - 0.482}{\sqrt{0.560(0.440)\left(\frac{1}{182} + \frac{1}{164}\right)}} = 2.79$

The rejection region requires $\alpha = 0.01$ in the upper tail of the z-distribution. From Table II, Appendix B, $z_{0.01} = 2.33$. The rejection region is $z > 2.33$.

Since the observed value of the test statistic falls in the rejection region $(z = 2.79 > 2.33)$, H_0 is rejected. There is sufficient evidence to indicate that foragers with a broad-based economy (Eastern Jomon) have a lower prevalence of LEH defects than early agriculturists (Yayoi) at $\alpha = 0.01$.

Next, we test Theory 2.

Some preliminary calculations are: $\hat{p} = \frac{n_1\hat{p}_1 + n_3\hat{p}_3}{n_1 + n_3} = \frac{182(0.631) + 122(0.648)}{182 + 122} = 0.638$

To determine if foragers with a wet rice economy (Western Jomon) have a different prevalence of LEH defects than early agriculturists (Yayoi), we test:

$$H_0: p_1 - p_2 = 0$$
$$H_a: p_1 - p_2 \neq 0$$

The test statistic is $z = \frac{\hat{p}_1 - \hat{p}_2}{\sqrt{\hat{p}\hat{q}\left(\frac{1}{n_1} + \frac{1}{n_2}\right)}} = \frac{0.631 - 0.648}{\sqrt{0.638(0.362)\left(\frac{1}{182} + \frac{1}{122}\right)}} = -0.30$

The rejection region requires $\alpha = 0.01/2 = 0.005$ in each tail of the z-distribution. From Table II, Appendix B, $z_{0.005} = 2.58$. The rejection region is $z < -2.58$ or $z > 2.58$.

Since the observed value of the test statistic does not fall in the rejection region $(z=-0.30 \not< -2.58)$, H_0 is not rejected. There is insufficient evidence to indicate that foragers with a wet rice economy (Western Jomon) have a different prevalence of LEH defects than early agriculturists (Yayoi) at $\alpha = 0.01$.

9.79 We can estimate the values of p_1 and p_2 based on prior samples, by using an educated guess of the values, or most conservatively, estimating p_1 and p_2 with 0.5.

9.81 $$n_1 = n_2 = \frac{(z_{\alpha/2})^2(\sigma_1^2 + \sigma_2^2)}{(SE)^2}$$

For confidence coefficient 0.95, $\alpha = 1 - 0.95 = 0.05$ and $\alpha/2 = 0.05/2 = 0.025$. From Table II, Appendix B, $z_{0.025} = 1.96$.

$$n_1 = n_2 = \frac{1.96^2(15+15)}{2.2^2} = 23.8 \approx 24$$

9.83 a. For confidence coefficient 0.99, $\alpha = 1 - 0.99 = 0.01$ and $\alpha/2 = 0.01/2 = 0.005$. From Table II, Appendix B, $z_{0.005} = 2.58$.

$$n_1 = n_2 = \frac{(z_{\alpha/2})^2(p_1q_1 + p_2q_2)}{(SE)^2} = \frac{2.58^2(0.4(1-0.4) + 0.7(1-0.7))}{0.01^2}$$

$$= \frac{2.99538}{0.0001} = 29{,}953.8 \approx 29{,}954$$

b. For confidence coefficient 0.90, $\alpha = 1 - 0.90 = 0.10$ and $\alpha/2 = 0.10/2 = 0.05$. From Table II, Appendix B, $z_{0.05} = 1.645$. Since we have no prior information about the proportions, we use $p_1 = p_2 = 0.5$ to get a conservative estimate. For a width of 0.05, the standard error is 0.5.

$$n_1 = n_2 = \frac{(z_{\alpha/2})^2(p_1q_1 + p_2q_2)}{(SE)^2} = \frac{(1.645)^2(0.5(1-0.5) + 0.5(1-0.5))}{0.025^2} = 2{,}164.82 \approx 2{,}165$$

c. From part **b**, $z_{0.05} = 1.645$.

$$n_1 = n_2 = \frac{(z_{\alpha/2})^2(p_1q_1 + p_2q_2)}{(SE)^2} = \frac{1.645^2(0.2(1-0.2) + 0.3(1-0.3))}{0.03^2}$$

$$= \frac{1.00123}{0.0009} = 1{,}112.48 \approx 1{,}113$$

9.85 a. The parameter of interest is the difference in two population means, $\mu_1 - \mu_2$.

b. For 95% confidence, $\alpha = 0.05$ and $\alpha/2 = 0.05/2 = 0.025$. From Table II, Appendix B, $z_{0.025} = 1.96$.

c. For this problem, $SE = 10$.

d. From Exercise 9.24, $\hat{\sigma}_1^2 = s_1^2 = 615.3$ for handshakes. For fist bumps, $\hat{\sigma}_2^2 = s_2^2 = 36$.

e. $$n_1 = n_2 = \frac{(z_{\alpha/2})^2(\sigma_1^2 + \sigma_2^2)}{(SE)^2} = \frac{(1.96)^2(615.3 + 36)}{10^2} = 25.02 \approx 26$$

9.87 For confidence coefficient 0.95, $\alpha = 0.05$ and $\alpha/2 = 0.05/2 = 0.025$. From Table II, Appendix B, $z_{0.025} = 1.96$.

$$n_1 = n_2 = \frac{(z_{\alpha/2})^2(\sigma_B^2 + \sigma_N^2)}{(SE)^2} = \frac{1.96^2(25 + 25)}{2^2} = 48.02 \approx 49$$

9.89 a. For confidence coefficient 0.95, $\alpha = 0.05$ and $\alpha/2 = 0.05/2 = 0.025$. From Table II, Appendix B, $z_{0.025} = 1.96$. The standard error is $SE = 0.015$. From Exercise 9.72, $\hat{p}_1 = 0.184$ and $\hat{p}_2 = 0.177$.

$$n_1 = n_2 = \frac{(z_{\alpha/2})^2(p_1q_1 + p_2q_2)}{(SE)^2} = \frac{(1.96)^2((0.184)(0.816) + (0.177)(0.823))}{(0.015)^2} = 5050.7 \approx 5051$$

In order to estimate the difference in proportions using a 95% confidence interval to within 0.015, we would need to sample 5,051 observations from each population.

b. It would be extremely difficult to obtain information on over 10,000 patients. In addition, the cost would be extremely high.

c. A difference of 0.015 is very small. This difference would be a difference of only 15 out of every 1,000 patients. In practice, this difference would be almost meaningless.

9.91 For confidence coefficient 0.90, $\alpha = 0.10$ and $\alpha/2 = 0.10/2 = 0.05$. From Table II, Appendix B, $z_{0.05} = 1.645$. From Exercise 9.47, $s_d = 86.4$. The standard error is $SE = 25$.

$$n_d = \frac{(z_{\alpha/2})^2(\sigma_d^2)}{(SE)^2} = \frac{1.645^2(86.4)^2}{25^2} = 32.3 \approx 33$$

We would need to sample a total of 33 months. Therefore, we would need to sample 28 additional months.

9.93 For confidence coefficient 0.95, $\alpha = 1 - 0.95 = 0.05$ and $\alpha/2 = 0.05/2 = 0.025$. From Table II, Appendix B, $z_{0.025} = 1.96$. The standard error is $SE = 5$.

$$n_1 = n_2 = \frac{(z_{\alpha/2})^2(\sigma_1^2 + \sigma_2^2)}{(SE)^2} = \frac{(1.96)^2(9^2 + 9^2)}{5^2} = 24.9 \approx 25$$

We would need to sample 25 subjects from each group.

9.95 For confidence coefficient 0.90, $\alpha = 1 - 0.90 = 0.10$ and $\alpha/2 = 0.10/2 = 0.05$. From Table II, Appendix B, $z_{0.05} = 1.645$.

$$n_1 = n_2 = \frac{(z_{\alpha/2})^2(\sigma_1^2 + \sigma_2^2)}{(SE)^2} = \frac{1.645^2(5^2 + 5^2)}{1^2} = 135.3 \Rightarrow 136$$

9.97 The conditions required for valid inferences about σ_1^2 / σ_2^2 are:

1. The two sampled populations are normally distributed.
2. The samples are randomly and independently selected from their respective populations.

9.99 The statement "$H_0: \sigma_1^2 = \sigma_2^2$ is equivalent to $H_0: \sigma_1^2 / \sigma_2^2 = 0$" is false.

The correct statement is: $H_0: \sigma_1^2 = \sigma_2^2$ is equivalent to $H_0: \sigma_1^2 / \sigma_2^2 = 1$.

9.101 a. With $\nu_1 = 2$ and $\nu_2 = 30$, $P(F \geq 4.18) = 0.025$ (Table VII, Appendix B)

b. With $\nu_1 = 24$ and $\nu_2 = 10$, $P(F \geq 1.94) = 0.10$ (Table V, Appendix B)

Thus, $P(F < 1.94) = 1 - P(F \geq 1.94) = 1 - 0.10 = 0.90$

c. With $\nu_1 = 9$ and $\nu_2 = 1$, $P(F \geq 6022) = 0.01$ (Table VIII, Appendix B)

Thus, $P(F < 6022) = 1 - P(F \geq 6022) = 1 - 0.01 = 0.99$

d. With $\nu_1 = 30$ and $\nu_2 = 30$, $P(F \geq 1.847) = 0.05$ (Table VI, Appendix B)

9.103 The test statistic for each of these is:

$$F = \frac{\text{Larger sample variance}}{\text{Smaller sample variance}}$$

a. The rejection region requires $\alpha/2 = 0.20/2 = 0.10$ in the upper tail of the F distribution. If $s_1^2 > s_2^2$, $\nu_1 = 8$ and $\nu_2 = 40$. From Table V, Appendix B, $F_{0.10} = 1.83$. The rejection region is $F > 1.83$. If $s_1^2 < s_2^2$, $\nu_1 = 40$ and $\nu_2 = 8$. From Table V, Appendix B, $F_{0.10} = 2.36$. The rejection region is $F > 2.36$.

b. The rejection region requires $\alpha/2 = 0.10/2 = 0.05$ in the upper tail of the F distribution. If $s_1^2 > s_2^2$, $\nu_1 = 8$ and $\nu_2 = 40$. From Table VI, Appendix B, $F_{0.05} = 2.18$. The rejection region is $F > 2.18$. If $s_1^2 < s_2^2$, $\nu_1 = 40$ and $\nu_2 = 8$. From Table VI, Appendix B, $F_{0.05} = 3.04$. The rejection region is $F_{0.05} = 3.04$.

c. The rejection region requires $\alpha/2 = 0.05/2 = 0.025$ in the upper tail of the F distribution. If $s_1^2 > s_2^2$, $\nu_1 = 8$ and $\nu_2 = 40$. From Table VII, Appendix B, $F_{0.025} = 2.53$. The rejection region is $F > 2.53$. If $s_1^2 < s_2^2$, $\nu_1 = 40$ and $\nu_2 = 8$. From Table VII, Appendix B, $F_{0.025} = 3.84$. The rejection region is $F > 3.84$.

d. The rejection region requires $\alpha/2 = 0.02/2 = 0.01$ in the upper tail of the F distribution. If $s_1^2 > s_2^2$, $\nu_1 = 8$ and $\nu_2 = 40$. From Table VIII, Appendix B, $F_{0.01} = 2.99$. The rejection region is $F > 2.99$. If $s_1^2 < s_2^2$, $\nu_1 = 40$ and $\nu_2 = 8$. From Table VIII, Appendix B, $F_{0.01} = 5.12$. The rejection region is $F > 5.12$.

9.105 a. $H_0: \sigma_1^2 = \sigma_2^2$
$H_a: \sigma_1^2 \neq \sigma_2^2$

The test statistic is $F = \dfrac{\text{Larger sample variance}}{\text{Smaller sample variance}} = \dfrac{s_2^2}{s_1^2} = \dfrac{9.85}{2.87} = 3.43$

The rejection region requires $\alpha/2 = 0.05/2 = 0.025$ in the upper tail of the F distribution with $\nu_1 = n_2 - 1 = 25 - 1 = 24$ and $\nu_2 = n_1 - 1 = 16 - 1 = 15$. From Table VII, Appendix B, $F_{0.025} = 2.70$. The rejection region is $F > 2.70$.

Since the observed value of the test statistic falls in the rejection region $(F = 3.43 > 2.70)$, H_0 is rejected. There is sufficient evidence to indicate $\sigma_1^2 \neq \sigma_2^2$ at $\alpha = 0.05$.

b. The p-value for the test is $p = 2P(F \geq 3.43)$. Using MINITAB, with $\nu_1 = n_2 - 1 = 25 - 1 = 24$ and $\nu_2 = n_1 - 1 = 16 - 1 = 15$, $P(F > 3.29) = 0.0068$. Thus, the p-value is $p = 0.0068$.

9.107 a. For level of confidence 0.90, $\alpha = 0.05$ and $\alpha/2 = 0.10/2 = 0.05$. Using MINITAB, with $\nu_1 = n_1 - 1 = 406 - 1 = 405$ and $\nu_2 = n_2 - 1 = 230 - 1 = 229$, $F_{L,0.05} = 1.21610$. Using MINITAB, with $\nu_1 = n_2 - 1 = 230 - 1 = 229$ and $\nu_2 = n_1 - 1 = 406 - 1 = 405$, $F_{U,0.05} = 1.20886$. The 90% confidence interval is

$$\left(\frac{s_1^2}{s_2^2}\right)\left(\frac{1}{F_{L,0.05}}\right) < \frac{\sigma_1^2}{\sigma_2^2} < \left(\frac{s_1^2}{s_2^2}\right)F_{U,0.05} \Rightarrow \left(\frac{0.40^2}{0.30^2}\right)\left(\frac{1}{1.21610}\right) < \frac{\sigma_1^2}{\sigma_2^2} < \left(\frac{0.40^2}{0.30^2}\right)1.20886$$

$$\Rightarrow 1.462 < \frac{\sigma_1^2}{\sigma_2^2} < 2.149$$

b. Yes. 90% confidence interval does not contain 1. Thus, there is evidence that the variance of the number of hippo trails in the National Reserve is different from the variance of the number of hippo trails in the pastoral ranch.

c. The test of hypothesis will yield the same results. If the two population variances are the same, then the ratio would be 1. From the above interval, 1 is not a likely value for the value of the ratio. Thus, we would reject it.

To determine if the variance of the number of hippo trails in the National Reserve is different from the variance of the number of hippo trails in the pastoral ranch, we test.

$$H_0: \sigma_1^2 = \sigma_2^2$$
$$H_a: \sigma_1^2 \neq \sigma_2^2$$

The test statistic is $F = \frac{s_1^2}{s_2^1} = \frac{0.4^2}{0.3^2} = 1.778$

The rejection region requires $\alpha / 2 = 0.10 / 2 = 0.05$ in the upper tail of the F distribution with $\nu_1 = n_1 - 1 = 406 - 1 = 405$ and $\nu_2 = n_2 - 1 = 230 - 1 = 229$. Using MINITAB, $F_{0.05} = 1.2161$. The rejection region is $F > 1.2161$.

Since the observed value of the test statistic falls in the rejection region $(F = 1.778 > 1.2161)$, H_0 is rejected. There is sufficient evidence to conclude that the variance of the number of hippo trails in the National Reserve is different from the variance of the number of hippo trails in the pastoral ranch at $\alpha = 0.10$.

9.109 a. To determine if the variances of the two groups are different, we test:

$$H_0: \sigma_1^2 = \sigma_2^2$$
$$H_a: \sigma_1^2 \neq \sigma_2^2$$

b. The test statistic is $F = \frac{\text{Larger sample variance}}{\text{Smaller sample variance}} = \frac{s_2^2}{s_1^2} = \frac{2.13^2}{1.98^2} = 1.157$

c. The rejection region requires $\alpha / 2 = 0.10 / 2 = 0.05$ in the upper tail of the F distribution with $\nu_1 = n_2 - 1 = 20 - 1 = 19$ and $\nu_2 = n_1 - 1 = 20 - 1 = 19$. Using MINITAB, $F_{0.05} = 2.17$. The rejection region is $F > 2.17$.

d. Since the observed value of the test statistic does not fall in the rejection region $(F = 1.157 \ngtr 2.17)$, H_0 is not rejected. There is insufficient evidence to indicate that the two variances are different at $\alpha = 0.10$. It appears that the assumption of equal variances is valid.

e. It appears that the inference derived in Exercise 9.17 is valid.

9.111 To determine if the variances of the 2 groups are different, we test:

$$H_0: \sigma_1^2 = \sigma_2^2$$
$$H_a: \sigma_1^2 \neq \sigma_2^2$$

The test statistic is $F = \dfrac{\text{Larger sample variance}}{\text{Smaller sample variance}} = \dfrac{s_1^2}{s_2^2} = \dfrac{12.5^2}{12.2^2} = 1.050$

The rejection region requires $\alpha/2 = 0.05/2 = 0.025$ in the upper tail of the F distribution with $\nu_1 = n_1 - 1 = 26 - 1 = 25$ and $\nu_2 = n_2 - 1 = 26 - 1 = 25$. Using MINITAB, $F_{0.025} = 2.23$. The rejection region is $F > 2.23$.

Since the observed value of the test statistic does not fall in the rejection region $(F = 1.050 \ngtr 2.23)$, H_0 is not rejected. There is insufficient evidence to indicate a difference in the population variances between the two groups at $\alpha = 0.05$. Thus, the assumption appears to be valid.

9.113 Let σ_1^2 = variance of the time for the Trail Making Test for schizophrenics and σ_2^2 = variance of the time for the Trail Making Test for normal individuals.

To determine if schizophrenics have a wider range in time on the Trail Making Test than normal subjects, we test:

$$H_0: \sigma_1^2 = \sigma_2^2$$
$$H_a: \sigma_1^2 > \sigma_2^2$$

The test statistic is $F = \dfrac{\text{Larger sample variance}}{\text{Smaller sample variance}} = \dfrac{s_1^2}{s_2^2} = \dfrac{45.45^2}{16.34^2} = 7.737$

The rejection region requires $\alpha = 0.01$ in the upper tail of the F distribution with $\nu_1 = n_1 - 1 = 41 - 1 = 40$ and $\nu_2 = n_2 - 1 = 49 - 1 = 48$. Using MINITAB, $F_{0.01} = 2.02$. The rejection region is $F > 2.02$.

Since the observed value of the test statistic falls in the rejection region $(F = 7.737 > 2.02)$, H_0 is rejected. There is sufficient evidence to indicate that schizophrenics have a wider range in time on the Trail Making Test than normal subjects at $\alpha = 0.01$.

9.115 Using MINITAB, the descriptive statistics are:

Descriptive Statistics: Honey, DM

Variable	N	Mean	StDev	Variance	Minimum	Median	Maximum
Honey	35	10.714	2.855	8.151	4.000	11.000	16.000
DM	33	8.333	3.256	10.604	3.000	9.000	15.000

Let σ_1^2 = variance of the coughing improvement scores for DM dosage group and σ_2^2 = variance of the coughing improvement scores for Honey dosage group.

To determine if the variances of the coughing improvement scores differ for the two groups, we test:

$$H_0: \sigma_1^2 = \sigma_2^2$$
$$H_a: \sigma_1^2 \neq \sigma_2^2$$

The test statistic is $F = \dfrac{\text{Larger sample variance}}{\text{Smaller sample variance}} = \dfrac{s_1^2}{s_2^2} = \dfrac{10.604}{8.151} = 1.301$.

The rejection region requires $\alpha/2 = 0.10/2 = 0.05$ in the upper tail of the F distribution with $\nu_1 = n_1 - 1 = 33 - 1 = 32$ and $\nu_2 = n_2 - 1 = 35 - 1 = 34$. Using MINITAB, $F_{0.05} = 1.78$. The rejection region is $F > 1.78$.

Since the observed value of the test statistic does not fall in the rejection region $(F = 1.301 \not> 1.78)$, H_0 is not rejected. There is insufficient evidence to indicate that the variances of the coughing improvement scores differ for the two groups at $\alpha = 0.10$. Since there is no difference in the variances for the two groups, neither treatment is preferred over the other.

9.117 Some preliminary calculations:

$$s_1^2 = \frac{\sum x_1^2 - \frac{\left(\sum x_1\right)^2}{n_1}}{n_1 - 1} = \frac{6620.96 - \frac{227^2}{9}}{9-1} = 110.6778$$

$$s_2^2 = \frac{\sum x_2^2 - \frac{\left(\sum x_2\right)^2}{n_2}}{n_2 - 1} = \frac{596.68 - \frac{66.4^2}{9}}{9-1} = 13.3494$$

a. Let σ_1^2 = variance of the order-to-delivery times for the Persian Gulf and σ_2^2 = variance of the order-to-delivery times for Bosnia. To determine if the variances of the order-to-delivery times for the Persian Gulf and Bosnia differ, we test:

$$H_0: \sigma_1^2 = \sigma_2^2$$
$$H_a: \sigma_1^2 \neq \sigma_2^2$$

The test statistic is $F = \frac{\text{Larger sample variance}}{\text{Smaller sample variance}} = \frac{s_1^2}{s_2^2} = \frac{110.6778}{13.3494} = 8.29$

The rejection region requires $\alpha / 2 = 0.05 / 2 = 0.025$ in the upper tail of the F distribution with $\nu_1 = n_1 - 1 = 9 - 1 = 8$ and $\nu_2 = n_2 - 1 = 9 - 1 = 8$. From Table VII, Appendix B, $F_{0.025} = 4.03$. The rejection region is $F > 4.03$.

Since the observed value of the test statistic falls in the rejection region $(F = 8.29 > 4.03)$, H_0 is rejected. There is sufficient evidence to indicate the variances of the order-to-delivery times for Persian Gulf and Bosnia differ at $\alpha = 0.05$.

b. No. Since both sample sizes are less than 30, we must use the two-sample t-test to test for differences in means. However, one of the assumptions for the two-sample t-test is that the variances of the two populations being sampled from are equal. Since we rejected H_0 in part a, there is evidence to indicate the variances of the two populations are not the same.

9.119 a. To compare average SAT scores of males and females, the target parameter is $\mu_1 - \mu_2$.

b. To compare the difference between mean waiting times at two supermarket checkout lines, the target parameter is $\mu_1 - \mu_2$.

c. To compare proportions of Democrats and Republicans who favor legalization of marijuana, the target parameter is $p_1 - p_2$.

d. To compare variation in salaries of NBA players picked in the 1st round and the 2nd round, the target parameter is σ_1^2 / σ_2^2.

e. To compare the difference in dropout rates of college student-athletes and regular students, the target parameter is $p_1 - p_2$.

9.121 a. $s_p^2 = \frac{(n_1 - 1)s_1^2 + (n_1 - 1)s_2^2}{n_1 + n_2 - 2} = \frac{11(74.2) + 13(60.5)}{12 + 14 - 2} = 66.7792$

$H_0: \mu_1 - \mu_2 = 0$

$H_a: \mu_1 - \mu_2 > 0$

The test statistic is $t = \frac{(\bar{x}_1 - \bar{x}_2) - 0}{\sqrt{s_p^2\left(\frac{1}{n_1} + \frac{1}{n_2}\right)}} = \frac{(17.8 - 15.3) - 0}{\sqrt{66.7792\left(\frac{1}{12} + \frac{1}{14}\right)}} = 0.78$

The rejection region requires $\alpha = 0.05$ in the upper tail of the t distribution with $df = n_1 + n_2 - 2 = 12 + 14 - 2 = 24$. From Table III, Appendix B, $t_{0.05} = 1.711$. The rejection region is $t > 1.711$.

Since the observed value of the test statistic does not fall in the rejection region $(t = 0.78 \not> 7.711)$, H_0 is not rejected. There is insufficient evidence to indicate that $\mu_1 > \mu_2$ at $\alpha = 0.05$.

b. For confidence coefficient 0.99, $\alpha = 0.01$ and $\alpha/2 = 0.01/2 = 0.005$. From Table III, Appendix B, with $df = n_1 + n_2 - 2 = 12 + 14 - 2 = 24$, $t_{0.005} = 2.797$. The confidence interval is:

$$(\bar{x}_1 - \bar{x}_2) \pm t_{0.005}\sqrt{s_p^2\left(\frac{1}{n_1}+\frac{1}{n_2}\right)} \Rightarrow (17.8 - 15.3) \pm 2.797\sqrt{66.7792\left(\frac{1}{12}+\frac{1}{14}\right)}$$

$$\Rightarrow 2.50 \pm 8.99 \Rightarrow (-6.49, 11.49)$$

c. For confidence coefficient 0.99, $\alpha = 0.01$ and $\alpha/2 = 0.01/2 = 0.005$. From Table II, Appendix B, $z_{0.005} = 2.58$.

$$n_1 = n_2 = \frac{(z_{\alpha/2})^2(\sigma_1^2 + \sigma_2^2)}{(SE)^2} = \frac{(2.58)^2(74.2 + 60.5)}{2^2} = 224.15 \approx 225$$

9.123 a. For confidence coefficient 0.90, $\alpha = 0.10$ and $\alpha/2 = 0.05$. From Table II, Appendix B, $z_{0.05} = 1.645$. The confidence interval is:

$$(\bar{x}_1 - \bar{x}_2) \pm z_{0.05}\sqrt{\frac{s_1^2}{n_1}+\frac{s_2^2}{n_2}} \Rightarrow (12.2 - 8.3) \pm 1.645\sqrt{\frac{2.1}{135}+\frac{3.0}{148}} \Rightarrow 3.90 \pm 0.31 \Rightarrow (3.59, 4.21)$$

b. $H_0: \mu_1 - \mu_2 = 0$
$H_a: \mu_1 - \mu_2 \neq 0$

The test statistic is $z = \dfrac{(\bar{x}_1 - \bar{x}_2) - 0}{\sqrt{\dfrac{s_1^2}{n_1}+\dfrac{s_2^2}{n_2}}} = \dfrac{(12.2 - 8.3) - 0}{\sqrt{\dfrac{2.1}{135}+\dfrac{3.0}{148}}} = 20.60$

The rejection region requires $\alpha/2 = 0.01/2 = 0.005$ in each tail of the z distribution. From Table II, Appendix B, $z_{0.005} = 2.58$. The rejection region is $z < -2.58$ or $z > 2.58$.

Since the observed value of the test statistic falls in the rejection region $(z = 20.60 > 2.58)$, H_0 is rejected. There is sufficient evidence to indicate that $\mu_1 \neq \mu_2$ at $\alpha = 0.01$.

c. For confidence coefficient 0.90, $\alpha = 0.10$ and $\alpha/2 = 0.10/2 = 0.05$. From Table II, Appendix B, $z_{0.05} = 1.645$.

$$n_1 = n_2 = \frac{(z_{\alpha/2})^2(\sigma_1^2 + \sigma_2^2)}{(SE)^2} = \frac{(1.645)^2(2.1 + 3.0)}{(0.2)^2} = 345.02 \approx 346$$

9.125 a. The observed proportion of St. John's wort patients who were in remission is:

$$\hat{p}_1 = \frac{x_1}{n_1} = \frac{14}{98} = 0.143$$

b. The observed proportion of placebo patients who were in remission is:

$$\hat{p}_2 = \frac{x_2}{n_2} = \frac{5}{102} = 0.049$$

c. Some preliminary calculations are:

$$\hat{p} = \frac{x_1 + x_2}{n_1 + n_2} = \frac{14+5}{98+102} = \frac{19}{200} = 0.095 \qquad \hat{q} = 1 - \hat{p} = 1 - 0.095 = 0.905$$

To determine if the proportion of St. John's wort patients in remission exceeds the proportion of placebo patients in remission, we test:

$$H_0: p_1 - p_2 = 0$$
$$H_a: p_1 - p_2 > 0$$

The test statistic is $z = \frac{\hat{p}_1 - \hat{p}_2}{\sqrt{\hat{p}\hat{q}\left(\frac{1}{n_1} + \frac{1}{n_2}\right)}} = \frac{0.143 - 0.049}{\sqrt{0.095(0.905)\left(\frac{1}{98} + \frac{1}{102}\right)}} = 2.27$

The rejection region requires $\alpha = 0.01$ in the upper tail of the z distribution. From Table II, Appendix B, $z_{0.01} = 2.33$. The rejection region is $z > 2.33$.

Since the observed value of the test statistic does not fall in the rejection region $(z = 2.27 \not> 2.33)$, H_0 is not rejected. There is insufficient evidence to indicate the proportion of St. John's wort patients in remission exceeds the proportion of placebo patients in remission at $\alpha = 0.01$.

d. The hypotheses and test statistic are the same as in part **c**.

The rejection region requires $\alpha = 0.10$ in the upper tail of the z distribution. From Table II, Appendix B, $z_{0.10} = 1.28$. The rejection region is $z > 1.28$.

Since the observed value of the test statistic falls in the rejection region $(z = 2.27 > 1.28)$, H_0 is rejected. There is sufficient evidence to indicate the proportion of St. John's wort patients in remission exceeds the proportion of placebo patients in remission $\alpha = 0.10$.

e. By changing the value of α in this problem, the conclusion changes. When α is small

$(\alpha = 0.01)$, we did not reject H_0. When α increases $(\alpha = 0.10)$, we rejected H_0. As α increases, the chance of making a Type I error increases, but the chance of a Type II error decreases.

9.127 From the printout, the 95% confidence interval for the difference in mean seabird densities of oiled and unoiled transects is $(-2.93,\ 2.49)$. We are 95% confident that the true difference in mean seabird densities of oiled and unoiled transects is between −2.93 and 2.49. Since 0 is contained in the interval, there is no evidence to indicate that the mean seabird densities are different for the oiled and unoiled transects.

9.129 a. The variable measured for this experiment is the time needed to match the picture with the sentence.

b. The experimental units are the climbers.

c. Since the same climbers were timed at the base camp and at a camp 5 miles above sea level, the data should be analyzed as a paired difference experiment.

9.131 a. Let p_1 = proportion of African-American drivers who were searched and p_2 = proportion of White drivers who were searched. The parameter of interest is $p_1 - p_2$, or the difference in the proportions of drivers searched between the two ethnic groups.

Some preliminary calculations are:

$$\hat{p}_1 = \frac{x_1}{n_1} = \frac{12{,}016}{61{,}688} = 0.195 \qquad \hat{p}_2 = \frac{x_2}{n_2} = \frac{5{,}312}{106{,}892} = 0.050$$

$$\hat{p} = \frac{x_1 + x_2}{n_1 + n_2} = \frac{12{,}016 + 5{,}312}{61{,}688 + 106{,}892} = 0.103$$

To determine if there is a difference in the proportions of African-American and White drivers who are searched by the LA police, we test:

$$H_0: p_1 - p_2 = 0$$
$$H_a: p_1 - p_2 \neq 0$$

The test statistic is $z = \dfrac{(\hat{p}_1 - \hat{p}_2) - 0}{\sqrt{\hat{p}\hat{q}\left(\dfrac{1}{n_1} + \dfrac{1}{n_2}\right)}} = \dfrac{(0.195 - 0.050) - 0}{\sqrt{0.103(0.897)\left(\dfrac{1}{61{,}688} + \dfrac{1}{106{,}892}\right)}} = 94.35$

The rejection region requires $\alpha / 2 = 0.05 / 2 = 0.025$ in each tail of the z distribution. From Table II, Appendix B, $z_{0.025} = 1.96$. The rejection region is $z < -1.96$ or $z > 1.96$.

Since the observed value of the test statistic falls in the rejection region $(z = 94.35 > 1.96)$, H_0 is rejected. There is sufficient evidence to indicate a difference in the proportions of African-American and White drivers who are searched by the LA police at $\alpha = 0.05$.

b. Let p_1 = "hit rate" of African-American drivers and p_2 = "hit rate" of White drivers. The parameter of interest is $p_1 - p_2$, or the difference in the "hit rates" for the two ethnic groups.

Some preliminary calculations are:

$$\hat{p}_1 = \frac{x_1}{n_1} = \frac{5{,}134}{12{,}016} = 0.427 \qquad \hat{p}_2 = \frac{x_2}{n_2} = \frac{3{,}006}{5{,}312} = 0.566$$

For confidence coefficient 0.95, $\alpha = 0.05$ and $\alpha / 2 = 0.05 / 2 = 0.025$. From Table II, Appendix B, $z_{0.025} = 1.96$. The 95% confidence interval is:

$$(0.427 - 0.566) \pm 1.96\sqrt{\frac{0.427(0.573)}{12{,}016} + \frac{0.566(0.434)}{5{,}312}} \Rightarrow -0.139 \pm 0.016 \Rightarrow (-0.155,\ -0.123)$$

We are 95% confident that the difference in "hit rates" between African-American and White drivers searched by the LA police is between −0.155 and −0.123. Since both end points of the interval are negative, there is evidence that the "hit rate" for African-American drives is less than that for White drivers.

9.133 a. $\hat{p}_1 = \frac{x_1}{n_1} = \frac{29}{189} = 0.153$

b. $\hat{p}_2 = \frac{x_2}{n_2} = \frac{32}{149} = 0.215$

c. For confidence coefficient 0.90, $\alpha = 0.10$ and $\alpha / 2 = 0.10 / 2 = 0.05$. From Table II, Appendix B, $z_{0.05} = 1.645$. The 90% confidence interval is:

$$(\hat{p}_1 - \hat{p}_2) \pm z_{.05}\sqrt{\frac{\hat{p}_1\hat{q}_1}{n_1} + \frac{\hat{p}_2\hat{q}_2}{n_2}} \Rightarrow (0.153 - 0.215) \pm 1.645\sqrt{\frac{0.153(0.847)}{189} + \frac{0.215(0.785)}{149}}$$
$$\Rightarrow -0.062 \pm 0.070 \Rightarrow (-0.132,\ 0.008)$$

d. We are 90% confident that the difference in the true proportion of super-experienced bidders who fall prey to the winner's curse and the true proportion of less-experienced bidders who fall prey to the winner's curse is between −0.132 and 0.008. Since 0 is contained in the interval, there is no evidence to indicate that experience in bidding affects the likelihood of the winner's curse occurring.

9.135 For confidence coefficient 0.95, $\alpha = 0.05$ and $\alpha / 2 = 0.05 / 2 = 0.025$. From Table III, Appendix B, with $df = n_d - 1 = 26 - 1 = 25$, $t_{0.025} = 2.060$. The 95% confidence interval is:

$$\bar{x}_d \pm t_{0.025}\frac{s_d}{\sqrt{n_d}} \Rightarrow 10.5 \pm 2.060\frac{7.6}{\sqrt{26}} \Rightarrow 10.5 \pm 3.07 \Rightarrow (7.43,\ 13.57)$$

We are 95% confident that the difference in the mean anxiety levels between before and after the visit is between 7.43 and 13.57. Since all of these values are positive, we can conclude that mean anxiety level after the visit is significantly less than the anxiety level before the visit.

9.137 a. Let μ_1 = mean reading response times for the tongue-twister list and μ_2 = mean reading response times for the control list. Then $\mu_d = \mu_1 - \mu_2$.

To compare the mean reading response times for the tongue-twister and control lists, we test:

$$H_0: \mu_d = 0$$
$$H_a: \mu_d \neq 0$$

b. The test statistic is $z = \dfrac{\overline{x}_d - 0}{s_d / \sqrt{n_d}} = \dfrac{.25 - 0}{0.78 / \sqrt{42}} = 2.08$

The p-value is $P(z \le -2.08) + P(z \ge 2.08) = 2P(z \ge 2.08) = 2(0.5000 - 0.4812)$ $= 2(0.0188) = 0.0376$. (Table II, Appendix B.)

c. Since the observed p-value is less than α $(p = 0.0376 < 0.05)$, H_0 is rejected. There is sufficient evidence to conclude that the mean reading response times differ for the tongue-twister and control lists at $\alpha = 0.05$.

9.139 a. Using MINITAB, the descriptive statistics are:

Descriptive Statistics: Species

Variable	Region	N	Mean	StDev	Variance
Species	Dry Steppe	5	14.00	21.31	454.00
	Gobi Desert	6	11.83	18.21	331.77

Let μ_1 = average number of ant species found in the Dry Steppe and μ_2 = average number of ant species found in the Gobi Desert.

$$s_p^2 = \frac{(n_1 - 1)s_1^2 + (n_2 - 1)s_2^2}{n_1 + n_2 - 2} = \frac{(5-1)21.31^2 + (6-1)18.21^2}{5 + 6 - 2} = 386.0539$$

To determine if there is a difference in the average number of ant species found at sites in the two regions, we test:

$$H_0: \mu_1 - \mu_2 = 0$$
$$H_a: \mu_1 - \mu_2 \neq 0$$

The test statistic is $t = \dfrac{(\overline{x}_1 - \overline{x}_2) - D_o}{\sqrt{s_p^2\left(\dfrac{1}{n_1} + \dfrac{1}{n_2}\right)}} = \dfrac{(14 - 11.83) - 0}{\sqrt{386.0539\left(\dfrac{1}{5} + \dfrac{1}{6}\right)}} = 0.18$

The rejection region requires $\alpha / 2 = 0.05/2 = 0.025$ in each tail of the *t*-distribution with $df = n_1 + n_2 - 2 = 5 + 6 - 2 = 9$. From Table III, Appendix B, $t_{0.025} = 2.262$. The rejection region is $t < -2.262$ or $t > 2.262$.

Since the observed value of the test statistic does not fall in the rejection region $(t = 0.18 \not> 2.262)$, H_0 is not rejected. There is insufficient evidence to indicate a difference in the average number of ant species found at sites in the two regions at $\alpha = 0.05$.

b. To determine if the variances of the two populations are equal, we test:

$$H_0: \sigma_1^2 = \sigma_2^2$$
$$H_a: \sigma_1^2 \neq \sigma_2^2$$

The test statistic is $F = \dfrac{\text{Larger sample variance}}{\text{Smaller sample variance}} = \dfrac{s_1^2}{s_2^2} = \dfrac{21.31^2}{18.21^2} = 1.37$

The rejection region requires $\alpha / 2 = 0.05/2 = 0.025$ in the upper tail of the *F* distribution with $\nu_1 = n_1 - 1 = 5 - 1 = 4$ and $\nu_2 = n_2 - 1 = 6 - 1 = 5$. Using Table VII, Appendix B, $F_{0.025} = 7.39$. The rejection region is $F > 7.39$.

Since the observed value of the test statistic does not fall in the rejection region $(F = 1.37 \not> 7.39)$, H_0 is not rejected. There is insufficient evidence to indicate a difference in the two population variances at $\alpha = 0.05$.

9.141 Let p_1 = proportion of rice weevils found dead after 4 days of exposure to nitrogen gas and p_2 = proportion of rice weevils found dead after 3.5 days of exposure to nitrogen gas. The parameter of interest is $p_1 - p_2$, or the difference in the proportions of rice weevils found dead between the 2 different exposures to nitrogen gas.

Some preliminary calculations are:

$$\hat{p}_1 = \frac{x_1}{n_1} = \frac{31{,}386}{31{,}421} = 0.999 \qquad \hat{p}_2 = \frac{x_2}{n_2} = \frac{23{,}516}{23{,}676} = 0.993$$

$$\hat{p} = \frac{x_1 + x_2}{n_1 + n_2} = \frac{31{,}386 + 23{,}516}{31{,}421 + 23{,}676} = 0.996$$

To determine if there is a difference in the proportions of rice weevils found dead between the 2 exposure times, we test:

$$H_o: p_1 - p_2 = 0$$
$$H_a: p_1 - p_2 \neq 0$$

The test statistic is $z = \dfrac{(\hat{p}_1 - \hat{p}_2) - 0}{\sqrt{\hat{p}\hat{q}\left(\dfrac{1}{n_1} + \dfrac{1}{n_2}\right)}} = \dfrac{(0.999 - 0.993) - 0}{\sqrt{0.996(0.004)\left(\dfrac{1}{31{,}421} + \dfrac{1}{23{,}676}\right)}} = 11.05$

The rejection region requires $\alpha / 2 = 0.10 / 2 = 0.05$ in each tail of the z distribution. From Table II, Appendix B, $z_{0.05} = 1.645$. The rejection region is $z < -1.645$ or $z > 1.645$.

Since the observed value of the test statistic falls in the rejection region $(z = 11.05 > 1.645)$, H_o is rejected. There is sufficient evidence to indicate a difference in the proportions of rice weevils found dead between the 2 exposure times at $\alpha = 0.10$.

9.143 Let p_1 = proportion of world-class athletes not competing in the 1999 World Championships and p_2 = proportion of world-class athletes not competing in the 2000 Olympic Games. If the new EPO test is effective, then the proportion of nonparticipating athletes will increase.

Some preliminary calculations are:

$$\hat{p}_1 = \frac{x_1}{n_1} = \frac{159}{830} = 0.1916 \qquad \hat{p}_2 = \frac{x_2}{n_2} = \frac{133}{825} = 0.1612 \qquad \hat{p} = \frac{x_1 + x_2}{n_1 + n_2} = \frac{159 + 133}{833 + 825} = 0.1764$$

To determine if the new test for EPO was effective in deterring an athlete's participation in the 2000 Olympics, we test:

$$H_o : p_1 - p_2 = 0$$
$$H_a : p_1 - p_2 < 0$$

The test statistic is $z = \dfrac{\hat{p}_1 - \hat{p}_2}{\sqrt{\hat{p}\hat{q}\left(\dfrac{1}{n_1} + \dfrac{1}{n_2}\right)}} = \dfrac{(0.1916 - 0.1612)}{\sqrt{0.1764(0.8236)\left(\dfrac{1}{830} + \dfrac{1}{825}\right)}} = 1.62$

The rejection region requires $\alpha = 0.10$ in the lower tail of the z distribution. From Table II, Appendix B, $z_{0.10} = 1.28$. The rejection region is $z < -1.28$.

Since the observed value of the test statistic does not fall in the rejection region $(z = 1.62 \nless -1.28)$, H_0 is not rejected. There is insufficient evidence to indicate the new test was effective in deterring an athlete's participation in the Olympics at $\alpha = 0.10$.

9.145 Let μ_1 = mean number of high frequency vocal responses for piglets castrated using Method 1 and μ_2 = mean number of high frequency vocal responses for piglets castrated using Method 2.

Some preliminary calculations are:

$$s_p^2 = \frac{(n_1 - 1)s_1^2 + (n_2 - 1)s_2^2}{n_1 + n_2 - 2} = \frac{(24 - 1)(0.09)^2 + (25 - 1)(0.09)^2}{24 + 25 - 2} = 0.0081$$

To determine if the mean number of high frequency vocal responses differ for piglets castrated by the 2 methods, we test:

$H_0: \mu_1 - \mu_2 = 0$
$H_a: \mu_1 - \mu_2 \neq 0$

The test statistic is $t = \dfrac{\bar{x}_1 - \bar{x}_2}{\sqrt{s_p^2\left(\dfrac{1}{n_1} + \dfrac{1}{n_2}\right)}} = \dfrac{0.74 - 0.70}{\sqrt{0.0081\left(\dfrac{1}{24} + \dfrac{1}{25}\right)}} = 1.56$

The rejection region requires $\alpha / 2 = 0.05 / 2 = 0.025$ in each tail of the t distribution. Using MINITAB, with $df = n_1 + n_2 - 2 = 24 + 25 - 2 = 47$, $t_{0.025} = 2.012$. The rejection region is $t < -2.012$ or $t > 2.012$.

Since the observed value of the test statistic does not fall in the rejection region $(t = 1.56 \not> 2.012)$, H_0 is not rejected. There is insufficient evidence to indicate that the mean number of high frequency vocal responses differ for piglets castrated by the 2 methods at $\alpha = 0.05$.

9.147 a. Some preliminary calculations are:

$$s_1^2 = \frac{\sum x_1^2 - \frac{\left(\sum x_1\right)^2}{n_1}}{n_1 - 1} = \frac{4209 - \frac{(145)^2}{5}}{5 - 1} = 1$$

$$s_2^2 = \frac{\sum x_2^2 - \frac{\left(\sum x_2\right)^2}{n_2}}{n_2 - 1} = \frac{4540 - \frac{(150)^2}{5}}{5 - 1} = 10$$

Let σ_1^2 = variance of measurements for instrument 1 and σ_2^2 = variance of measurements for instrument 2.

To determine if the variances of the measurements for the two instruments differ, we test:

$H_0: \sigma_1^2 = \sigma_2^2$
$H_a: \sigma_1^2 \neq \sigma_2^2$

The test statistic is $F = \dfrac{\text{Larger sample variance}}{\text{Smaller sample variance}} = \dfrac{s_2^2}{s_1^2} = \dfrac{10}{1} = 10$

Since no α was given, we will use $\alpha = 0.05$. The rejection region requires $\alpha / 2 = 0.05 / 2 = 0.025$ in the upper tail of the F distribution with numerator

$\nu_1 = n_2 - 1 = 5 - 1 = 4$ and $\nu_2 = n_1 - 1 = 5 - 1 = 4$. From Table VII, Appendix B, $F_{0.025} = 9.60$. The rejection region is $F > 9.60$.

Since the observed value of the test statistic falls in the rejection region $(F = 10 > 9.60)$, H_0 is rejected. There is sufficient evidence to indicate the variances of the measurements for the two instruments differ at $\alpha = 0.05$.

b. We must assume that the samples are randomly and independently selected from populations that are normally distributed.

9.149 a. Let p_1 = proportion of 9th grade boys who gambled weekly or daily on any game in year 1 and p_2 = proportion of 9th grade boys who gambled weekly or daily on any game in year 2.

Some preliminary calculations are:

$$\hat{p}_1 = \frac{x_1}{n_1} = \frac{4,684}{21,484} = 0.218 \qquad \hat{p}_2 = \frac{x_2}{n_2} = \frac{5,313}{23,199} = 0.229$$

$$\hat{p} = \frac{x_1 + x_2}{n_1 + n_2} = \frac{4,684 + 5,313}{21,484 + 23,199} = 0.224$$

To determine if the proportions of 9th grade boys who gambled weekly or daily on any game in year 1 and year 2 differ, we test:

$H_0: p_1 - p_2 = 0$
$H_a: p_1 - p_2 \neq 0$

The test statistic is $z = \dfrac{(\hat{p}_1 - \hat{p}_2) - 0}{\sqrt{\hat{p}\hat{q}\left(\dfrac{1}{n_1} + \dfrac{1}{n_2}\right)}} = \dfrac{(0.218 - 0.229) - 0}{\sqrt{0.224(0.776)\left(\dfrac{1}{21,484} + \dfrac{1}{23,199}\right)}} = -2.79$

The rejection region requires $\alpha / 2 = 0.01/2 = 0.005$ in each tail of the z distribution. From Table II, Appendix B, $z_{0.005} = 2.58$. The rejection region is $z < -2.58$ or $z > 2.58$.

Since the observed value of the test statistic falls in the rejection region $(z = -2.79 < -2.58)$, H_0 is rejected. There is sufficient evidence to indicate the proportions of 9th grade boys who gambled weekly or daily on any game in year 1 and year 2 differ at $\alpha = 0.01$.

b. Yes. If the sample sizes are large enough, any difference can be found to be statistically different.

To measure the magnitude of the difference in the proportions of 9th grade boys who gambled weekly or daily on any game between year 1 and year 2, we will construct a 95% confidence interval. For confidence coefficient 0.95, $\alpha = 0.05$ and

$\alpha/2 = 0.05/2 = 0.025$. From Table II, Appendix B, $z_{0.025} = 1.96$. The 95% confidence interval is:

$$(\hat{p}_1 - \hat{p}_2) \pm z_{0.025}\sqrt{\left(\frac{\hat{p}_1\hat{q}_1}{n_1} + \frac{\hat{p}_2\hat{q}_2}{n_2}\right)} \Rightarrow (0.218 - 0.229) \pm 1.96\sqrt{\frac{0.218(0.782)}{21{,}484} + \frac{0.229(0.771)}{23{,}199}}$$
$$\Rightarrow -0.011 \pm 0.0077 \Rightarrow (-0.0187,\ -0.0033)$$

We are 95% confident that the difference in the proportions of 9th grade boys who gambled weekly or daily on any game between year 1 and year 2 is between −0.0187 and −0.0033.

9.151 Using MINITAB, the descriptive statistics are:

Descriptive Statistics: CREATIVE, INFO, DECPERS, SKILLS, TASKID, AGE, EDYRS

Variable	GROUP	N	Mean	StDev	Minimum	Q1	Median	Q3	Maximum
CREATIVE	NOSPILL	67	4.4478	0.5304	3.0000	4.0000	4.0000	5.0000	5.0000
	SPILLOV	47	5.2553	0.4408	5.0000	5.0000	5.0000	6.0000	6.0000
INFO	NOSPILL	67	4.657	2.226	1.000	3.000	5.000	7.000	7.000
	SPILLOV	47	5.213	1.719	1.000	4.000	5.000	7.000	7.000
DECPERS	NOSPILL	67	2.746	1.964	1.000	1.000	2.000	5.000	7.000
	SPILLOV	47	3.319	2.023	1.000	1.000	3.000	5.000	7.000
SKILLS	NOSPILL	67	4.821	1.100	2.000	4.000	5.000	5.000	7.000
	SPILLOV	47	5.851	1.161	2.000	5.000	6.000	7.000	7.000
TASKID	NOSPILL	67	4.239	1.156	1.000	3.000	4.000	5.000	7.000
	SPILLOV	47	4.809	2.028	1.000	3.000	5.000	7.000	7.000
AGE	NOSPILL	67	45.343	7.971	21.000	41.000	45.000	50.000	63.000
	SPILLOV	47	46.298	5.767	30.000	44.000	47.000	51.000	54.000
EDYRS	NOSPILL	67	13.224	1.312	12.000	12.000	13.000	14.000	18.000
	SPILLOV	47	13.085	1.060	12.000	12.000	13.000	14.000	16.000

Let μ_1 = mean characteristic for the workers with positive spillover and μ_2 = mean characteristic for the workers with no spillover. To determine if there is a difference in the mean characteristic between the positive spillover group and the no spillover group, we test:

$H_0: \mu_1 - \mu_2 = 0$
$H_a: \mu_1 - \mu_2 \neq 0$

The test statistic is $z = \frac{\bar{x}_1 - \bar{x}_2}{\sqrt{\left(\frac{s_1^2}{n_1} + \frac{s_2^2}{n_2}\right)}}$

Since no α was given, we will run all tests at $\alpha = 0.05$. The rejection region requires $\alpha/2 = 0.05/2 = 0.025$ in each tail of the z distribution. From Table II, Appendix B, $z_{0.025} = 1.96$. The rejection region is $z < -1.96$ or $z > 1.96$.

Use of Creative Ideas:

The test statistic is $z = \dfrac{\bar{x}_1 - \bar{x}_2}{\sqrt{\left(\dfrac{s_1^2}{n_1} + \dfrac{s_2^2}{n_2}\right)}} = \dfrac{5.2553 - 4.4478}{\sqrt{\dfrac{(0.4408)^2}{47} + \dfrac{(0.5304)^2}{67}}} = 8.85$

Since the observed value of the test statistic falls in the rejection region $(z = 8.85 > 1.96)$, H_0 is rejected. There is sufficient evidence to indicate a difference in the mean use of creative ideas between the positive spillover group and the no spillover group at $\alpha = 0.05$.

Utilization of Information:

The test statistic is $z = \dfrac{\bar{x}_1 - \bar{x}_2}{\sqrt{\left(\dfrac{s_1^2}{n_1} + \dfrac{s_2^2}{n_2}\right)}} = \dfrac{5.213 - 4.657}{\sqrt{\dfrac{1.719^2}{47} + \dfrac{2.226^2}{67}}} = 1.50$

Since the observed value of the test statistic does not fall in the rejection region $(z = 1.50 \not> 1.96)$, H_0 is not rejected. There is insufficient evidence to indicate a difference in the mean utilization of information between the positive spillover group and the no spillover group at $\alpha = 0.05$.

Decisions Regarding Personnel Matters:

The test statistic is $z = \dfrac{\bar{x}_1 - \bar{x}_2}{\sqrt{\left(\dfrac{s_1^2}{n_1} + \dfrac{s_2^2}{n_2}\right)}} = \dfrac{3.319 - 2.746}{\sqrt{\dfrac{2.023^2}{47} + \dfrac{1.964^2}{67}}} = 1.51$

Since the observed value of the test statistic does not fall in the rejection region $(z = 1.51 \not> 1.96)$, H_0 is not rejected. There is insufficient evidence to indicate a difference in the mean participation in decisions regarding personnel matters between the positive spillover group and the no spillover group at $\alpha = 0.05$.

Good Use of Job Skills:

The test statistic is $z = \dfrac{\bar{x}_1 - \bar{x}_2}{\sqrt{\left(\dfrac{s_1^2}{n_1} + \dfrac{s_2^2}{n_2}\right)}} = \dfrac{5.851 - 4.821}{\sqrt{\dfrac{1.161^2}{47} + \dfrac{1.100^2}{67}}} = 4.76$

Since the observed value of the test statistic falls in the rejection region $(z = 4.76 > 1.96)$, H_0 is rejected. There is sufficient evidence to indicate a difference in the mean good use of skills between the positive spillover group and the no spillover group at $\alpha = 0.05$.

Task Identity:

The test statistic is $z = \dfrac{\bar{x}_1 - \bar{x}_2}{\sqrt{\left(\dfrac{s_1^2}{n_1} + \dfrac{s_2^2}{n_2}\right)}} = \dfrac{4.809 - 4.239}{\sqrt{\dfrac{2.028^2}{47} + \dfrac{1.156^2}{67}}} = 1.74$

Since the observed value of the test statistic does not fall in the rejection region $(z = 1.74 \not> 1.96)$, H_0 is not rejected. There is insufficient evidence to indicate a difference in the mean task identity between the positive spillover group and the no spillover group at $\alpha = 0.05$.

Of the 5 job-related characteristics, there are only 2 that have significantly different means for the 2 groups. They are "use of creative ideas" and "good use of skills". For both of these characteristics, the mean for the positive spillover group was significantly higher than the mean for the no spillover group.

Age:

The test statistic is $z = \dfrac{\bar{x}_1 - \bar{x}_2}{\sqrt{\left(\dfrac{s_1^2}{n_1} + \dfrac{s_2^2}{n_2}\right)}} = \dfrac{46.298 - 45.343}{\sqrt{\dfrac{5.767^2}{47} + \dfrac{7.971^2}{67}}} = 0.74$

Since the observed value of the test statistic does not fall in the rejection region $(z = 0.74 \not> 1.96)$, H_0 is not rejected. There is insufficient evidence to indicate a difference in the mean age between the positive spillover group and the no spillover group at $\alpha = 0.05$.

Education:

The test statistic is $z = \dfrac{\bar{x}_1 - \bar{x}_2}{\sqrt{\left(\dfrac{s_1^2}{n_1} + \dfrac{s_2^2}{n_2}\right)}} = \dfrac{13.085 - 13.224}{\sqrt{\dfrac{1.060^2}{47} + \dfrac{1.312^2}{67}}} = -0.62$

Since the observed value of the test statistic does not fall in the rejection region $(z = -0.62 \not< 1.96)$, H_0 is not rejected. There is insufficient evidence to indicate a difference in the mean years of education between the positive spillover group and the no spillover group at $\alpha = 0.05$.

Chapter 10

Analysis of Variance
Comparing More than Two Means

10.1 Since only one factor is utilized, the treatments are the four levels (A, B, C, D) of the qualitative factor.

10.3 One has no control over the levels of the factors in an observational experiment. One does have control of the levels of the factors in a designed experiment.

10.5 a. This is an observational study. The economist has no control over the factor levels or unemployment rates.

b. This is a designed study. The psychologist selects the feedback programs of interest and randomly assigns five rats to each program.

c. This is a designed study. The marketer has control of the selection of the national publications.

d. This is an observational study. The load on the generators is not controlled by the utility.

e. This is an observational study. One has no control over the distance of the haul, the goods hauled, or the price of diesel fuel.

f. This is an observational study. The student does not control which state is assigned to a portion of the country.

10.7 a. The experimental units are the Boer goats.

b. The response variable is the weight loss of the goats.

c. The factor is the group to which the goats were assigned.

d. The factor levels are the 4 different groups – Group A (goats administered AA 30 minutes prior to transportation), Group B (goats administered AA 30 minutes following transportation), Group C (goats not administered AA prior to or following transportation), and Group D (goats not administered AA nor transported).

10.9 a. The experimental unit is a patient.

b. The response variable is the score on the Hamilton Depression Rating Scale.

c. The factor is the drug combination group.

d. There are 4 factor levels. Group 1 received daily doses of the antidepressant drug floxetine and a placebo; group 2 received the antidepressant drug mirtazapine plus floxetine; group 3 received mirtazapine plus veniafaxine; and group 4 received mirtazapine plus bupropion.

10.11 a. The experimental unit is a healthy adult.

b. The response variable is the postural index.

c. There are 2 factors – gender and strength knowledge.

d. There are 2 levels of gender – male and female. There are 2 levels of strength knowledge – either provided or not provided.

e. There are a total of $2\times 2=4$ treatments: male, knowledge provided; male, knowledge not provided; female knowledge provided; and female, knowledge not provided.

10.13 a. The experimental units are the cockatiels.

b. This experiment is a designed experiment. The birds were randomly divided into 3 groups and each group received a different treatment.

c. There is one factor in this study. The factor is the group.

d. There are three levels of the group variable – Group 1 received purified water, Group 2 received purified water and liquid sucrose, and Group 3 received purified water and liquid sodium chloride.

e. There are 3 treatments in the study. Because there is only one factor, the treatments are the same as the factor levels.

f. The response variable is the liquid consumption.

10.15 a. There are 2 factors in this experiment – Temperature and Type of yeast. Temperature has 4 levels – 45, 48, 51, and 54°C. Type of yeast has 2 levels – Baker's and Brewer's.

b. The response variable is the autolysis yield (recorded as a percentage).

c. There are a total of $4\times 2=8$ treatments for this experiment.

d. This is a designed experiment. The levels of temperature and type of yeast are controlled by the researcher.

10.17 For a completely randomized design, independent random samples are selected for each treatment, or treatments are randomly assigned to experimental units.

10.19 The conditions required for a valid ANOVA F-test in a completely randomized design are:

1. The samples are randomly selected in an independent manner from *k* treatment populations.
2. All *k* sampled populations have distributions that are approximately normal.
3. The *k* population variances are equal (i.e., $\sigma_1^2=\sigma_2^2=\ldots=\sigma_k^2$)

10.21 a. Using Table VI, $F_{0.05} = 6.59$ with $\nu_1 = 3$, $\nu_2 = 4$.

b. Using Table VIII, $F_{0.01} = 16.69$ with $\nu_1 = 3$, $\nu_2 = 4$.

c. Using Table V, $F_{0.10} = 1.61$ with $\nu_1 = 20$, $\nu_2 = 40$.

d. Using Table VII, $F_{0.025} = 3.87$ with $\nu_1 = 12$, $\nu_2 = 9$.

10.23 a. In the second dot diagram **B**, the difference between the sample means is small relative to the variability within the sample observations. In the first dot diagram **A**, the values in each of the samples are grouped together with a range of 4, while in the second diagram **B**, the range of values is 8.

b. For diagram **A**,

$$\bar{x}_1 = \frac{\sum x_1}{n} = \frac{7+8+9+9+10+11}{6} = \frac{54}{6} = 9$$

$$\bar{x}_2 = \frac{\sum x_2}{n} = \frac{12+13+14+14+15+16}{6} = \frac{84}{6} = 14$$

For diagram **B**,

$$\bar{x}_1 = \frac{\sum x_1}{n} = \frac{5+5+7+11+13+13}{6} = \frac{54}{6} = 9$$

$$\bar{x}_2 = \frac{\sum x_2}{n} = \frac{10+10+12+16+18+18}{6} = \frac{84}{6} = 14$$

c. For diagram **A**,

$$\text{SST} = \sum_{i=1}^{2} n_i(\bar{x}_i - \bar{x})^2 = 6(9-11.5)^2 + 6(14-11.5)^2 = 75$$

$$\left(\bar{x} = \frac{\sum x}{n} = \frac{54+84}{12} = 11.5\right)$$

For diagram **B**,

$$\text{SST} = \sum_{i=1}^{2} n_i(\bar{x}_i - \bar{x})^2 = 6(6-11.5)^2 + 6(14-11.5)^2 = 75$$

d. For diagram **A**,

$$s_1^2 = \frac{\sum x_1^2 - \frac{\left(\sum x_1\right)^2}{n_1}}{n_1 - 1} = \frac{496 - \frac{54^2}{6}}{6-1} = 2$$

$$s_2^2 = \frac{\sum x_2^2 - \frac{\left(\sum x_2\right)^2}{n_2}}{n_2 - 1} = \frac{1186 - \frac{84^2}{6}}{6-1} = 2$$

$$SSE = (n_1 - 1)s_1^2 + (n_2 - 1)s_2^2 = (6-1)^2(2) + (6-1)^2(2) = 20$$

For diagram **B**,

$$s_1^2 = \frac{\sum x_1^2 - \frac{\left(\sum x_1\right)^2}{n_1}}{n_1 - 1} = \frac{558 - \frac{54^2}{6}}{6-1} = 14.4$$

$$s_2^2 = \frac{\sum x_2^2 - \frac{\left(\sum x_2\right)^2}{n_2}}{n_2 - 1} = \frac{1248 - \frac{84^2}{6}}{6-1} = 14.4$$

$$SSE = (n_1 - 1)s_1^2 + (n_2 - 1)s_2^2 = (6-1)(14.4) + (6-1)(14.4) = 144$$

e. For diagram **A**, $SS(\text{Total}) = SST + SSE = 75 + 20 = 95$

$$\text{SST is } \frac{SST}{SS(\text{Total})} \times 100\% = \frac{75}{95} \times 100\% = 78.95\% \text{ of SS(Total)}$$

For diagram **B**, $SS(\text{Total}) = SST + SSE = 75 + 144 = 219$

$$\text{SST is } \frac{SST}{SS(\text{Total})} \times 100\% = \frac{75}{219} \times 100\% = 34.25\% \text{ of SS(Total)}$$

f. For diagram **A**, $MST = \frac{SST}{k-1} = \frac{75}{2-1} = 75$

$$MSE = \frac{SSE}{n-k} = \frac{20}{12-2} = 2 \qquad F = \frac{MST}{MSE} = \frac{75}{2} = 37.5$$

For diagram **B**, $MST = \frac{SST}{k-1} = \frac{75}{2-1} = 75$

$$MSE = \frac{SSE}{n-k} = \frac{144}{12-2} = 14.4 \qquad F = \frac{MST}{MSE} = \frac{75}{14.4} = 5.21$$

g. The rejection region for both diagrams requires $\alpha = 0.05$ in the upper tail of the F distribution with $\nu_1 = k - 1 = 2 - 1 = 1$ and $\nu_2 = n - k = 12 - 2 = 10$. From Table VI, Appendix B, $F_{0.05} = 4.96$. The rejection region is $F > 4.96$.

For diagram **A**, since the observed value of the test statistic falls in the rejection region $(F = 37.5 > 4.96)$, H_0 is rejected. There is sufficient evidence to indicate the samples were drawn from populations with different means at $\alpha = 0.05$.

For diagram **B**, since the observed value of the test statistic falls in the rejection region $(F = 5.21 > 4.96)$, H_0 is rejected. There is sufficient evidence to indicate the samples were drawn from populations with different means at $\alpha = 0.05$.

h. We must assume both populations are normally distributed with common variances.

10.25 Refer to Exercises 10.23 and 10.24, the ANOVA table is:

For diagram **A**:

Source	df	SS	MS	F
Treatment	1	75	75	37.5
Error	10	20	2	
Total	11	95		

For diagram **B**:

Source	df	SS	MS	F
Treatment	1	75	75	5.21
Error	10	144	14.4	
Total	11	219		

10.27 For all parts, the hypotheses are:

H_0: $H_0: \mu_1 = \mu_2 = \mu_3 = \mu_4 = \mu_5$
H_a: At least two treatment means differ

The rejection region for all parts is the same.

The rejection region requires $\alpha = 0.10$ in the upper tail of the F distribution with $\nu_1 = k - 1 = 5 - 1 = 4$ and $\nu_2 = n - k = 30 - 5 = 25$. From Table V, Appendix B, $F_{0.10} = 2.18$. The rejection region is $F > 2.18$.

a. $\text{SST} = 0.2(500) = 100$ $\quad\quad$ $\text{SSE} = \text{SS(Total)} - \text{SST} = 500 - 100 = 400$

$$\text{MST} = \frac{\text{SST}}{k-1} = \frac{100}{5-1} = 25 \quad\quad \text{MSE} = \frac{\text{SSE}}{n-k} = \frac{400}{30-5} = 16$$

$$F = \frac{\text{MST}}{\text{MSE}} = \frac{25}{16} = 1.5625$$

Since the observed value of the test statistic does not fall in the rejection region $(F = 1.5625 \not> 2.18)$, H_0 is not rejected. There is insufficient evidence to indicate differences among the treatment means at $\alpha = 0.10$.

b. $\text{SST} = 0.5(500) = 250$ $\quad\quad$ $\text{SSE} = \text{SS(Total)} - \text{SST} = 500 - 250 = 250$

$$\text{MST} = \frac{\text{SST}}{k-1} = \frac{250}{5-1} = 62.5 \quad\quad \text{MSE} = \frac{\text{SSE}}{n-k} = \frac{250}{30-5} = 10$$

$$F = \frac{\text{MST}}{\text{MSE}} = \frac{62.5}{10} = 6.25$$

Since the observed value of the test statistic falls in the rejection region $(F = 6.25 > 2.18)$, H_0 is rejected. There is sufficient evidence to indicate differences among the treatment means at $\alpha = 0.10$.

c. $SST = 0.8(500) = 400$ $SSE = SS(Total) - SST = 500 - 400 = 100$

$MST = \frac{SST}{k-1} = \frac{400}{5-1} = 100$ $MSE = \frac{SSE}{n-k} = \frac{100}{30-5} = 4$

$F = \frac{MST}{MSE} = \frac{100}{4} = 25$

Since the observed value of the test statistic falls in the rejection region, $(F = 25 > 2.18)$, H_0 is rejected. There is sufficient evidence to indicate differences among the treatment means at $\alpha = 0.10$.

d. The F ratio increases as the treatment sum of squares increases.

10.29 a. Answers will vary. Using MINITAB, the numbers selected to receive the usual follow-up care are:

2, 3, 7, 9, 11, 12, 13, 14, 15, 16, 17, 18, 19, 20, 22, 23, 25, 26, 27, 31, 32, 33, 35, 42, 44, 45, 47, 48, 49, 51, 52, 56, 57, 63, 64, 65, 66, 67, 69, 70, 71, 72, 75, 78, 79, 80, 82, 83, 86, 87, 88, 89, 93, 94, 95, 96, 99, 101, 102, 103, 104, 105, 106, 107, 108, 110, 111, 112, 114, 115, 116, 117, 118, 120, 121, 123, 125, 126, 128, 130, 138, 140, 142, 143, 144, 145, 147, 152, 154, 155, 156, 159, 161, 163, 165, 166, 168, 170, 171, 173, 174, 175, 176, 177, 179, 181, 182, 183, 186, 188, 190, 191, 192, 193, 194, 196, 198, 199, 202, 204, 207, 208, 244, 212, 218, 219, 221, 224, 228, 229, 231, 232, 233, 234, 235, 240, 243, 244, 246, 252, 259, 262, 264, 267, 269, 270, 273, 274, 275, 276, 277, 278, 279, 283, 285, 286, 288, 292, 298, 300, 302, 306, 311, 314, 315, 317, 321, 323, 329, 330, 332, 334, 335, 338, 340, 342, 344, 345, 347, 354, 356, 359, 360, 362, 363, 369, 370, 371, 372, 373, 375, 376, 377, 378, 381, 383, 384, 385, 386, 391, 392, 394, 401, 402, 405

All other numbers will be assigned to the group attending yoga classes.

b. The results from this assignment would not be valid. Those patients with the most severe cancer would probably be more fatigued and sleepy than those with less severe cancer, regardless of which group they were assigned. To assign all the most severe cancer patients to one group will skew the results.

10.31 a. This is a completely randomized design.

b. The dependent variable is the decrease in the number of cards sold.

c. The factor in the experiment is the type of pay-cut. There are three factor levels: unilateral wage cut, general wage cut, and baseline.

d. The null hypothesis would be that there is no difference in the mean decrease in cards sold for the three groups or:

$H_0: \mu_1 = \mu_2 = \mu_3$

e. Since the *p*-value is less than $\alpha\left(p = 0.001 < 0.01\right)$, H_o is rejected. There is sufficient evidence to indicate that there is a difference in mean decrease in cards sold among the three groups at $\alpha = 0.01$.

10.33 a. This is a completely randomized design.

b. The experimental units are the 168 college students. The dependent variable is the tanning attitude. The treatments are the three groups – tanned models, models with no tan, and no models.

c. To determine if there is a difference in mean tanning attitude among the three groups, we test:

$$H_0 : \mu_1 = \mu_2 = \mu_3$$

d. The sample means are random variables. If the experiment were repeated and new sample means computed, they would be different from the means in this experiment. We need to determine if these sample means would be unusual if there really is no difference in the three population means.

e. Since the *p*-value is less than $\alpha\left(p = 0.03 < 0.05\right)$, H_o is rejected. There is sufficient evidence to indicate that there is a difference in mean tanning attitudes among the three groups at $\alpha = 0.05$.

f. The conditions required for a valid ANOVA *F*-test in a completely randomized design are:

1. The samples are randomly selected in an independent manner from 3 treatment populations.
2. All 3 sampled populations have distributions that are approximately normal.
3. The 3 population variances are equal (i.e., $\sigma_1^2 = \sigma_2^2 = \sigma_3^2$)

10.35 a. The experimental design used is a completely randomized design.

b. There are 4 treatments in this experiment. The four treatments are the 4 "colonies" to which the robots were assigned – 3, 6, 9, or 12 robots per colony. The dependent variable is the energy expended per robot.

c. To determine if the mean energy expended (per robot) of the four different colony sizes differed, we test:

$H_0 : \mu_1 = \mu_2 = \mu_3 = \mu_4$
H_a: At least two treatment means differ

d. The test statistic is $F = 7.70$ and the *p*-value is $p < 0.001$. Since the *p*-value is less than $\alpha = 0.05$, H_0 is rejected. There is sufficient evidence to indicate that mean energy expended (per robot) in the four different colony sizes differed at $\alpha = 0.05$.

10.37 a. The experimental units are the participants in the study.

b. The dependent variable is the brand recall score.

c. There is one factor in this study – TV viewing group. Since there is only one factor, the treatments correspond to the factor levels of this variable. Thus, the treatments are the same as the three levels of TV viewer group. These 3 levels are violent content code, sex content code, and neutral TV.

d. The means given are only sample means. If new samples were selected and sample means computed, the values and order of the sample means could change. In addition, the variances are not taken into account.

e. The test statistic is $F = 20.45$ and the *p*-value is $p = 0.000$.

f. Since the *p*-value is less than α $(p = 0.000 < 0.01)$, H_o is rejected. There is sufficient evidence to indicate differences in the mean recall scores among the three viewing groups at $\alpha = 0.01$. The researchers can conclude that the content of the TV show affects the recall of imbedded commercials.

10.39 Using MINITAB, the ANOVA is:

One-way ANOVA: 30, 35, 40, 45

```
Source  DF   Adj SS   Adj MS  F-Value  P-Value
Factor   3  4589.47  1529.82  2026.26    0.000
Error    8     6.04     0.76
Total   11  4595.51

Model Summary

       S    R-sq  R-sq(adj)  R-sq(pred)
0.868907  99.87%     99.82%      99.70%

Means

Factor  N     Mean  StDev        90% CI
30      3  102.567  1.358  (101.634, 103.500)
35      3  101.600  0.458  (100.667, 102.533)
40      3   96.767  0.513  ( 95.834,  97.700)
45      3   55.433  0.839  ( 54.500,  56.366)

Pooled StDev = 0.868907
```

To determine if high temperatures inhibit mean concentration of ethanol, we test:

$H_0: \mu_1 = \mu_2 = \mu_3 = \mu_4$
H_a: At least one mean differs

The test statistic is $F = 2026.26$ and the *p*-value is $p = 0.000$.

Since the *p*-value is less than $\alpha(p = 0.000 < 0.10)$, H_o is rejected. There is sufficient evidence to indicate that high temperatures inhibit mean concentration of ethanol among the four temperatures at $\alpha = 0.10$.

10.41 a. This experiment is a completely randomized design. The 3 treatments are honey dosage, DM dosage, and no dosage.

b. Using MINITAB, the ANOVA is:

One-way ANOVA: Honey, DM, Control

```
Source   DF      SS      MS      F      P
Factor    2   318.51  159.25  17.51  0.000
Error   102   927.72    9.10
Total   104  1246.23

S = 3.016   R-Sq = 25.56%    R-Sq(adj) = 24.10%

                                Individual 95% CIs For Mean Based on
                                Pooled StDev
Level     N    Mean   StDev  -----+---------+---------+---------+----
Honey    35  10.714   2.855                         (-----*-----)
DM       33   8.333   3.256              (-----*------)
Control  37   6.514   2.940  (-----*-----)
                             -----+---------+---------+---------+----
                                6.4       8.0       9.6      11.2
Pooled StDev = 3.016
```

To determine if there are differences in the mean improvement scores among the 3 groups, we test:

$H_0: \mu_1 = \mu_2 = \mu_3$
H_a: At least two treatment means differ

From the printout, the test statistic is $F = 17.51$ and the p-value is $p = 0.000$.

Since the p-value is so small $(p = 0.000)$, H_0 is rejected. There is sufficient evidence to indicate a difference in mean improvement scores among the 3 groups for any reasonable value of α.

10.43 a. Some preliminary calculations are:

$$\bar{x} = \frac{n_1\bar{x}_1 + n_2\bar{x}_2 + n_3\bar{x}_3}{n_1 + n_2 + n_3} = \frac{26(10.5) + 25(3.9) + 25(1.4)}{26 + 25 + 25} = \frac{405.5}{76} = 5.34$$

$$\text{SST} = \sum n_i(\bar{x}_i - \bar{x})^2 = 26(10.5 - 5.34)^2 + 25(3.9 - 5.34)^2 + 25(1.4 - 5.34)^2 = 1{,}132.1956$$

b. $$\text{SSE} = (n_1 - 1)s_1^2 + (n_2 - 1)s_2^2 + (n_3 - 1)s_3^2$$
$$= (26 - 1)7.6^2 + (25 - 1)7.5^2 + (25 - 1)7.5^2 = 4{,}144$$

c.

Source	df	SS	MS	F
Treatment	3 – 1 = 2	1,132.1956	566.0978	9.972
Error	76 – 3 = 73	4,144.0000	56.7671	
Total	76 – 1 = 75			

$$\text{MST} = \frac{\text{SST}}{k-1} = \frac{1{,}132.1956}{3-1} = 566.0978 \qquad \text{MSE} = \frac{\text{SSE}}{n-k} = \frac{4{,}144}{76-3} = 56.7671$$

$$F = \frac{\text{MST}}{\text{MSE}} = \frac{566.0978}{56.7671} = 9.972$$

d. To determine if there are differences in the mean drops among the three groups, we test:

$H_0: \mu_1 = \mu_2 = \mu_3$
H_a: At least 2 means differ

The test statistic is $F = 9.972$.

The rejection region requires $\alpha = 0.01$ in the upper tail of the F distribution with $\nu_1 = k-1 = 3-1 = 2$ and $\nu_2 = n-k = 26+25+25-3 = 73$. Using MINITAB, $F_{0.01} = 4.91$. The rejection region is $F > 4.91$.

Since the observed value of the test statistic falls in the rejection region $(F = 9.972 > 4.91)$, H_0 is rejected. There is sufficient evidence to indicate a difference in mean drops in anxiety levels among the 3 groups at $\alpha = 0.01$.

e. 1. We must assume that the probability distributions for the 3 groups are normal. This assumption seems reasonable, but we do not have the actual data to check it.

2. We must assume that the variances for the three groups are equal. Since the standard deviations for the 3 groups are almost the same, this assumption seems to be valid.
3. We must assume that the observations are random and independent. This also seems to be a reasonable assumption because the problem states that the patients were randomly assigned to a group.

Since all assumptions seem valid, the results of the analysis should be valid.

10.45 The experimentwise error rate is the probability of declaring at least one pair of means different, given all of the means are the same.

10.47 a. The confidence interval $(-10, 5)$ for $(\mu_1 - \mu_2)$ means that we are confident that the difference between $\mu_1 - \mu_2$ is between -10 and 5. Since this interval spans both positive and negative numbers, we cannot conclude that either mean is significantly larger.
b. The confidence interval $(-10, -5)$ for $(\mu_1 - \mu_2)$ means that we are confident that the difference between $\mu_1 - \mu_2$ is between -10 and -5. Since this interval spans only negative numbers, we can conclude that μ_2 is significantly larger than μ_1.

c. The confidence interval $(5, 10)$ for $(\mu_1 - \mu_2)$ means that we are confident that the difference between $\mu_1 - \mu_2$ is between 5 and 10. Since this interval spans only positive numbers, we can conclude that μ_1 is significantly larger than μ_2.

10.49 $(\mu_1 - \mu_2)$: $(2, 15)$ Since all values in the interval are positive, μ_1 is significantly greater than μ_2.

$(\mu_1 - \mu_3)$: $(4, 7)$ Since all values in the interval are positive, μ_1 is significantly greater than μ_3.

$(\mu_1 - \mu_4)$: $(-10, 3)$ Since 0 is in the interval, μ_1 is not significantly different from μ_4. However, since the center of the interval is less than 0, μ_4 is larger than μ_1.

$(\mu_2 - \mu_3)$: $(-5, 11)$ Since 0 is in the interval, μ_2 is not significantly different from μ_3. However, since the center of the interval is greater than 0, μ_2 is larger than μ_3.

$(\mu_2 - \mu_4)$: $(-12, -6)$ Since all values in the interval are negative, μ_4 is significantly greater than μ_2.

$(\mu_3 - \mu_4)$: $(-8, -5)$ Since all values in the interval are negative, μ_4 is significantly greater than μ_3.

Thus, the largest mean is μ_4 followed by μ_1, μ_2, and μ_3.

10.51 a. The total number of pairwise comparisons is $c = k(k-1)/2 = 3(3-1)/2 = 3$. The pairs are:

Unilateral wage cut and General wage cut; Unilateral wage cut and Baseline;

General wage cut and Baseline

b. Tukey's multiple comparison procedure controls the experimentwise error rate at the selected value of α.

c. Since the confidence interval contains only positive values, it indicates that the mean decrease in promotional cards sold for baseline is greater than the mean decrease in promotional cards sold for the general wage cut.

10.53 The multiple comparison of means indicate that the mean consumer attitude towards tanning for those in the Models/No Tan group is significantly less than the mean consumer attitude towards tanning for those in both the Tanned Models and the No Models group. There is no significant difference in the mean consumer attitude towards tanning between those in the Tanned Models group and those in the No Models group. It does appear that the type of product advertisement can influence a consumer's attitude towards tanning.

10.55 a. The total number of pairwise comparisons made in the Bonferroni analysis is $c = k(k-1)/2 = 4(4-1)/2 = 6$.

b. The Sourdough treatment yielded the significantly highest mean soluble magnesium level. There is no significant difference in the mean soluble magnesium level between the Control and the Yeast groups. These two treatments yielded the lowest mean soluble magnesium level.

c. The experimentwise error rate is 0.05. This means that the probability of declaring at least one pair significantly different when, in fact, none are different is 0.05.

10.57 a. This ANOVA table is used to test whether the mean number of alternatives listed differs among the 3 groups. Since the *p*-value is so small ($p = .001$), H_0 is rejected. There is sufficient evidence of a difference in the mean number of alternatives listed among the 3 groups.

b. The .05 means that the probability of concluding at least two means differ when, in fact, none are different is .05.

c. From the results, we can conclude that the mean number of alternatives listed for the guilt group is significantly greater than the mean number of alternatives listed for the other two groups. There is no significant difference in the mean number of alternatives listed between the angry group and the neutral group.

10.59 a. The treatments for this experiment are the 7 different cheeses.

b. The dependent variable is the color change index (CCI).

c. Since the *p*-value is less than 0.05, reject H_0. There is evidence to indicate a difference in mean CCI among the 7 cheeses at $\alpha = 0.05$.

d. The total number of pairwise comparisons is $c = k(k-1)/2 = 7(7-1)/2 = 21$.

e. The mean CCI for Mozzarella cheese is significantly higher than the mean CCI for all the other cheeses. The mean CCI for Provolone cheese is significantly higher than the mean CCI for all the other cheeses except for Mozzarella. There is no significant difference in the mean CCI between Gruyere cheese and Cheddar cheese, and these means are significantly greater than the mean CCI for Colby cheese, Edam cheese, and Emmental cheese. There is no significant difference in the mean CCI between Edam cheese and Emmental cheese and these means are both significantly higher than the mean CCI of Colby cheese.

10.61 Using MINITAB, the ANOVA results and the results from Tukey's multiple comparison procedure are:

One-way ANOVA: Honey, DM, Control

```
Source   DF       SS      MS      F      P
Factor    2   318.51  159.25  17.51  0.000
Error   102   927.72    9.10
Total   104  1246.23
```

```
S = 3.016   R-Sq = 25.56%   R-Sq(adj) = 24.10%

                                 Individual 95% CIs For Mean Based on
                                 Pooled StDev
Level     N    Mean   StDev  -----+---------+---------+---------+----
Honey    35  10.714   2.855                             (-----*-----)
DM       33   8.333   3.256              (-----*------)
Control  37   6.514   2.940  (-----*-----)
                             -----+---------+---------+---------+----
                                6.4       8.0       9.6      11.2

Pooled StDev = 3.016

Tukey 95% Simultaneous Confidence Intervals
All Pairwise Comparisons

Individual confidence level = 98.06%
Honey subtracted from:

           Lower   Center   Upper   ----+---------+---------+---------+-----
DM        -4.120   -2.381  -0.642            (-----*------)
Control   -5.890   -4.201  -2.511   (------*------)
                                    ----+---------+---------+---------+-----
                                      -5.0      -2.5       0.0       2.5


DM subtracted from:

           Lower   Center   Upper   ----+---------+---------+---------+-----
Control   -3.535   -1.820  -0.104              (------*------)
                                    ----+---------+---------+---------+-----
                                      -5.0      -2.5       0.0       2.5
```

The means for the three groups in order are 10.714 (honey dosage), 8.333 (DM dosage), and 6.514 (no dosage). The confidence interval for the difference in mean improvement between the no dosage and the honey dosage is $(-5.890,\ -2.511)$. Since both endpoints are negative, the mean improvement for the honey dosage is significantly greater than that for the no dosage or control. The confidence interval for the difference in mean improvement between the DM dosage and the honey dosage is $(-4.120,\ -0.642)$. Since both endpoints are negative, the mean improvement for the honey dosage is significantly greater than that for the DM dosage. The confidence interval for the difference in mean improvement between the no dosage and the DM dosage is $(-3.535,\ -0.104)$. Since both endpoints are negative, the mean improvement for the DM dosage is significantly greater than that for the no dosage or control.

This agrees with the statement in the article.

10.63 Using MINITAB, the results of Tukey's multiple comparison procedure are:

```
Tukey 95% Simultaneous Confidence Intervals
All Pairwise Comparisons

Individual confidence level = 98.94%
```

```
AR subtracted from:

     Lower    Center     Upper   ---------+---------+---------+---------+
AC  -0.3751  -0.1745   0.0260          (------*------)
A   -0.5769  -0.3764  -0.1758   (-----*------)
P   -0.2405  -0.0400   0.1605             (------*-----)
                                 ---------+---------+---------+---------+
                                      -0.30      0.00      0.30      0.60

AC subtracted from:

    Lower    Center     Upper   ---------+---------+---------+---------+
A  -0.4023  -0.2018  -0.0013         (-----*------)
P  -0.0660   0.1345   0.3351                    (-----*------)
                                ---------+---------+---------+---------+
                                     -0.30      0.00      0.30      0.60

A subtracted from:

   Lower   Center   Upper   ---------+---------+---------+---------+
P  0.1358  0.3364  0.5369                         (-----*------)
                            ---------+---------+---------+---------+
                                 -0.30      0.00      0.30      0.60
```

Mean:	0.4400	0.4000	0.2655	0.0636
Group:	AR	P	AC	A

(A line is drawn under the means for AR, P, and AC.)

The mean for the Alcohol only group is significantly less than the means for the other 3 groups. No other differences exist. The researchers' theories were supported. The Alcohol only group performed the worst. There were no significant differences between the Placebo group and the Alcohol plus caffeine group, between the Placebo group and the Alcohol plus reward group, and between the alcohol plus caffeine group and the alcohol plus reward group.

10.65 A paired difference experiment is a randomized block design with only 2 experimental units per block. In the paired difference design, the pairs are the blocks. However, a randomized block design is not limited to 2 experimental units per block.

10.67 The conditions required for a valid ANOVA F-test in a randomized block design are:

1. The b blocks are randomly selected and all k treatments are applied (in random order) to each block.
2. The distributions of observations corresponding to all bk block-treatment combinations are approximately normal.
3. The bk block-treatment distributions have equal variances.

10.69 a.

Treatment	$df = k - 1 = 3 - 1 = 2$
Block	$df = b - 1 = 3 - 1 = 2$
Error	$df = n - k - b + 1 = 9 - 3 - 3 + 1 = 4$
Total	$df = n - 1 = 9 - 1 = 8$

$\text{SSB} = \sum_{i=1}^{b} \frac{B_i^2}{k} - \text{CM}$ from Appendix C where $\text{CM} = \frac{\left(\sum x_i\right)^2}{n} = \frac{49^2}{9} = 266.7778$

$$\text{SSB} = \frac{17^2}{3} + \frac{15^2}{3} + \frac{17^2}{3} - 266.7778 = .8889$$

$$\text{SSE} = \text{SS(Total)} - \text{SST} - \text{SSB} = 30.2222 - 21.5555 - 0.8889 = 7.7778$$

$$\text{MST} = \frac{\text{SST}}{k-1} = \frac{21.5555}{2} = 10.7778 \qquad \text{MSB} = \frac{\text{SSB}}{b-1} = \frac{0.8889}{2} = 0.4445$$

$$\text{MSE} = \frac{\text{SSE}}{n-k-b+1} = \frac{7.7778}{4} = 1.9445$$

$$F_T = \frac{\text{MST}}{\text{MSE}} = \frac{10.7778}{1.9445} = 5.54 \qquad F_B = \frac{\text{MSB}}{\text{MSE}} = \frac{0.4445}{1.9445} = 0.23$$

The ANOVA table is:

Source	**df**	**SS**	**MS**	***F***
Treatment	2	21.5555	10.7778	5.54
Block	2	0.8889	0.4445	0.23
Error	4	7.7778	1.9445	
Total	8	30.2222		

b. To determine if differences exist among the treatment means, we test:

$H_0: \mu_1 = \mu_2 = \mu_3$
H_a: At least two treatment means differ

c. The test statistic is $F = \frac{\text{MST}}{\text{MSE}} = 5.54$

d. A Type I error would be concluding at least two treatment means differ when they do not.

A Type II error would be concluding all the treatment means are the same when at least two differ.

e. The rejection region requires $\alpha = 0.05$ in the upper tail of the F distribution with $\nu_1 = k - 1 = 3 - 1 = 2$ and $\nu_2 = n - k - b + 1 = 9 - 3 - 3 + 1 = 4$. From Table VI, Appendix B, $F_{0.05} = 6.94$. The rejection region is $F > 6.94$.

Since the observed value of the test statistic does not fall in the rejection region $(F = 5.54 \not> 6.94)$, H_0 is not rejected. There is insufficient evidence to indicate at least two of the treatment means differ at $\alpha = 0.05$.

10.71 a. The ANOVA Table is as follows:

Source	**df**	**SS**	**MS**	***F***	***p***
Treatment	2	12.032	6.016	50.96	0.000
Block	3	71.749	23.916	202.59	0.000
Error	6	0.708	0.118		
Total	11	84.489			

b. To determine if the treatment means differ, we test:

$H_0: \mu_A = \mu_B = \mu_C$
H_a: At least two treatment means differ

The test statistic is $F = \dfrac{\text{MST}}{\text{MSE}} = 50.96$

The rejection region requires $\alpha = 0.05$ in the upper tail of the F distribution with $\nu_1 = k-1 = 3-1 = 2$ and $\nu_2 = n-k-b+1 = 12-3-4+1 = 6$. From Table VI, Appendix B, $F_{0.05} = 5.14$. The rejection region is $F > 5.14$.

Since the observed value of the test statistic falls in the rejection region $(F = 50.96 > 5.14)$, H_0 is rejected. There is sufficient evidence to indicate that the treatment means differ at $\alpha = 0.05$.

c. To see if the blocking was effective, we test:

$H_0: \mu_1 = \mu_2 = \mu_3 = \mu_4$
H_a: At least two block means differ

The test statistic is $F = \dfrac{\text{MSB}}{\text{MSE}} = 202.59$

The rejection region requires $\alpha = 0.05$ in the upper tail of the F distribution with $\nu_1 = b-1 = 4-1 = 3$ and $\nu_2 = n-k-b+1 = 12-3-4+1 = 6$. From Table VI, Appendix B, $F_{0.05} = 4.76$. The rejection region is $F > 4.76$.

Since the observed value of the test statistic falls in the rejection region $(F = 202.59 > 4.76)$, H_0 is rejected. There is sufficient evidence to indicate that blocking was effective in reducing the experimental error at $\alpha = 0.05$.

d. From the printout, we are given the sample means: $\bar{x}_B = 5.7$, $\bar{x}_A = 4.5$, and $\bar{x}_C = 3.2$. The confidence interval comparing the means for treatments C and B is $(-3.196, -1.704)$. Since the endpoints are both negative, it indicates that the mean for treatment B is significantly greater than the mean from treatment C.

The confidence interval comparing the means for treatments B and A is $(0.379, 1.8706)$. Since the endpoints are both positive, it indicates that the mean for treatment B is significantly greater than the mean from treatment A.

The confidence interval comparing the means for treatments C and A is $(-2.071, -0.5794)$. Since the endpoints are both negative, it indicates that the mean for treatment A is significantly greater than the mean from treatment C.

Thus, the rankings of the treatment means are: $\mu_B > \mu_A > \mu_C$

e. The assumptions necessary to assure the validity of the inferences above are:

1. The probability distributions of observations corresponding to all *bk* block-treatment combinations are normal.
2. The variances of all the *bk* probability distributions are equal.
3. The *b* blocks are randomly selected and all *k* treatments are applied (in random order) to each block.

10.73 a. The experimental design for this experiment is a randomized complete block design because each subject responded to each treatment. Here, the subjects are the blocks.

b. The response variable is the WTP amount for insuring the art work. The treatments are the two conditions under which the subjects provided the WTP amount for insuring the art work – insuring the painting with no probabilities of the hazards and insuring the sculpture with probabilities of the hazards. The blocks are the subjects.

c. To determine if the mean WTP amounts differ for the painting and the sculpture, we test:

$H_0: \mu_P = \mu_S$
$H_a: \mu_P \neq \mu_S$

To determine if differences exist among the block means, we test:

$H_0: \mu_1 = \mu_2 = \cdots = \mu_{84}$
H_a: At least two block means differ

10.75 a. To determine if there is a significant difference in the mean TT index across the 15 weeks of the college football season, we test:

$H_0: \mu_1 = \mu_2 = \mu_3 = \cdots = \mu_{15}$
H_a: At least two treatment means differ

The test statistic is $F = 2.57$ and the p-value is $p = 0.0044$. Since the p-value is less than $\alpha = 0.01$ $(p = 0.0044 < 0.01)$, H_0 is rejected. There is sufficient evidence of a difference in the mean TT index among the 15 weeks of the college football season at $\alpha = 0.01$.

b. To determine if blocking on seasons was effective in removing an extraneous source of variation in the data, we test:

$H_0: \mu_1 = \mu_2 = \mu_3 = \cdots = \mu_6$
H_a: At least two blocking means differ

The test statistic is $F = 5.94$ and the p-value is $p = 0.0001$. Since the p-value is less than $\alpha = 0.01$ $(p = 0.0001 < 0.01)$, H_0 is rejected. There is sufficient evidence that blocking on seasons was effective in removing an extraneous source of variation in the data at $\alpha = 0.01$.

c. The number of pairwise comparisons is $c = k(k-1)/2 = 15(15-1)/2 = 105$.

d. There is only one significant difference in means among the 105 pairs compared. This difference is between the means of weeks 6 and 14. Thus, the mean TT score for week 6 is not significantly greater than the mean TT scores for all weeks except week 14. Similarly, the mean TT score for week 14 is not significantly less than the mean TT scores for all weeks except week 6. We can only say that week 6 is more topsy-turvy than week 14.

10.77 a. The treatments are the 8 activities.

b. The blocks in this experiment are the 15 adults.

c. Since the p-value is less than α $(p = 0.001 < 0.01)$, H_o is rejected. There is sufficient evidence to indicate that the mean heart rates differ among the 8 activities at $\alpha = 0.01$.

d. The mean heart rate for Treadmill jogging was significantly higher than all others. The mean heart rate for Brisk treadmill walking had the next highest mean. The mean heart rates for Wii aerobics, Wii muscle conditioning and Wii yoga were not significantly different from each other and were in the next highest group. There was no significant difference in the mean heart rates for Handheld gaming and Rest. These means were significantly lower than all the other means.

10.79 a. The treatments are the 4 pre-slaughter phases and the blocks are the cows.

b. Using MINITAB, the ANOVA is:

Two-way ANOVA: Y versus Phase, Cow

```
Source  DF       SS       MS     F      P
Phase    3   521.13  173.708  3.63  0.030
Cow      7  1922.88  274.696  5.74  0.001
Error   21  1004.88   47.851
Total   31  3448.88

S = 6.917   R-Sq = 70.86%   R-Sq(adj) = 56.99%


                    Individual 95% CIs For Mean Based on
                    Pooled StDev
Phase      Mean     -------+---------+---------+---------+--
1       108.000                        (-------*-------)
2        97.000     (---------*-------)
3       103.125               (---------*-------)
4       105.125                   (-------*--------)
                    -------+---------+---------+---------+--
                        96.0     102.0     108.0     114.0
```

c. To determine if there are differences in mean heart rates among the four pre-slaughter phases, we test:

$H_0: \mu_1 = \mu_2 = \mu_3 = \mu_4$
H_a: At least two blocking means differ

The test statistic is $F = 3.63$ and the p-value is $p = 0.030$. Since the p-value is less than $\alpha = 0.05$ $(p = 0.030 < 0.05)$, H_0 is rejected. There is sufficient evidence to indicate differences in mean heart rates exist among the four pre-slaughter phases at $\alpha = 0.05$.

d. Using MINITAB, the results of Tukey's multiple comparison procedure are:

```
Tukey 95.0% Simultaneous Confidence Intervals
Response Variable Y
All Pairwise Comparisons among Levels of Phase
Phase = 1  subtracted from:

Phase    Lower   Center    Upper   -------+---------+---------+--------
2       -20.64   -11.00   -1.364   (-------*-------)
3       -14.51    -4.87    4.761        (-------*-------)
4       -12.51    -2.88    6.761          (-------*-------)
                                   -------+---------+---------+--------
                                        -12         0        12

Phase = 2  subtracted from:

Phase    Lower   Center   Upper   -------+---------+---------+--------
3       -3.511    6.125   15.76                (-------*-------)
4       -1.511    8.125   17.76                  (-------*-------)
                                  -------+---------+---------+--------
                                       -12         0        12

Phase = 3  subtracted from:

Phase    Lower   Center   Upper   -------+---------+---------+--------
4       -7.636    2.000   11.64             (-------*-------)
                                  -------+---------+---------+--------
                                       -12         0        12
```

If we rank the means in order from the smallest to the larges we get:

	Phase 1	Phase 4	Phase 3	Phase 2
Means:	Phase 1	Phase 4	Phase 3	Phase 2
Treatments:	108.000	105.125	103.125	97.000

Comparing phase 1 mean to phase 2 mean, the confidence interval is $(-20.64, -1.364)$. Since both end points are negative, the mean heart rate at phase 1 is significantly greater than that at phase 2.

Comparing phase 1 mean to phase 3 mean, the confidence interval is $(-14.51, 4.761)$. Since 0 is in the confidence interval, there is no evidence that the mean heart rate at phase 1 is greater than that at phase 3.

Since there is no difference in the means for phases 1 and 3, there will not be a difference in the means for phases 1 and 4.

Comparing phase 4 mean to phase 2 mean, the confidence interval is $(-1.511, 17.76)$. Since 0 is in the confidence interval, there is no evidence that the mean heart rate at phase 4 is greater than that at phase 2.

Since there is no difference in the means for phases 4 and 2, there will not be a difference in the means for phases 4 and 3. There will also not be a difference in the means for phases 2 and 3.

Thus, the only difference in mean heart rates is between phases 1 and 2, with the mean heart rate for phase 1 significantly greater than that for phase 2.

10.81 Using SAS, the results of the randomized complete block experiment are:

```
                              The GLM Procedure

Dependent Variable: RATING

                                        Sum of
  Source                      DF       Squares     Mean Square    F Value    Pr > F

  Model                       11   5596.900000      508.809091      46.40    <.0001

  Error                       18    197.400000       10.966667

  Corrected Total             29   5794.300000


                 R-Square     Coeff Var      Root MSE    RATING Mean

                 0.965932      6.331923      3.311596       52.30000


  Source                      DF     Type I SS     Mean Square    F Value    Pr > F

  CANDIDATE                    2    110.600000       55.300000       5.04    0.0182
  SUBJECT                      9   5486.300000      609.588889      55.59    <.0001


  Source                      DF   Type III SS     Mean Square    F Value    Pr > F

  CANDIDATE                    2    110.600000       55.300000       5.04    0.0182
  SUBJECT                      9   5486.300000      609.588889      55.59    <.0001


                              The GLM Procedure

                  Tukey's Studentized Range (HSD) Test for RATING

   NOTE: This test controls the Type I experimentwise error rate, but it generally has a
                       higher Type II error rate than REGWQ.

                      Alpha                                    0.05
                      Error Degrees of Freedom                   18
                      Error Mean Square                    10.96667
                      Critical Value of Studentized Range   3.60930
                      Minimum Significant Difference         3.7797

              Means with the same letter are not significantly different.

               Tukey Grouping          Mean       N     CANDIDATE

                           A         54.600      10     HIGH
                           A
                      B    A         52.400      10     NEUTRAL
                      B
                      B              49.900      10     LOW
```

To determine if there are differences in the mean ratings among the three candidates, we test:

$H_0: \mu_1 = \mu_2 = \mu_3$
H_a: At least one mean differs

The test statistic is $F = 5.04$ and the p-value is $p = 0.0182$.

There was no α given so we will use $\alpha = 0.05$. Since the p-value is less than $\alpha\left(p = 0.0182 < 0.05\right)$, H_0 is rejected. There is sufficient evidence to indicate there are differences in the mean ratings among the three types of candidates at $\alpha = 0.05$.

From the Tukey's multiple comparison, the mean rating for the High candidate is sufficiently greater than the mean rating for the Low candidate. There is no significant difference in the mean ratings between candidates in the High and Neutral categories and between candidates in the Neutral and Low categories.

10.83 Using MINITAB, the ANOVA results are:

Two-way ANOVA: Temp versus Plant, Student

```
Source    DF      SS       MS       F      P
Plant      2   0.122  0.06100   0.02  0.981
Student    9  18.415  2.04611   0.63  0.754
Error     18  58.038  3.22433
Total     29  76.575

S = 1.796   R-Sq = 24.21%   R-Sq(adj) = 0.00%

                Individual 95% CIs For Mean Based on
                Pooled StDev
Plant   Mean    -------+---------+---------+---------+--
Live   95.21    (----------------*----------------)
None   95.20    (----------------*----------------)
Photo  95.34       (----------------*----------------)
                -------+---------+---------+---------+--
                    94.50     95.20     95.90     96.60
```

To determine if there are differences among the mean temperatures among the three treatments, we test:

$H_0: \mu_1 = \mu_2 = \mu_3$
H_a: At least two treatment means differ

The test statistic is $F = 0.02$ and the associated p-value is $p = 0.981$. Since the p-value is very large, H_0 is not rejected. There is no evidence of a difference in mean temperature among the three treatments for any reasonable value of α. Since there is no difference, we do not need to compare the means. It appears that the presence of plants or pictures of plants does not reduce stress.

10.85 In a complete factorial experiment, the treatments are formed by taking every combination of all levels of both factors. The total number of treatments is equal to the total number of factor-level combinations.

10.87 The conditions that are required for valid inferences for a factorial ANOVA are:

1. The response distribution for each factor-level combination (treatment) is normal.
2. The response variance is constant for all treatments.
3. Random and independent samples of experimental units are associated with each treatment.

10.89 a. There are two factors.

b. No, we cannot tell whether the factors are qualitative or quantitative.

c. Yes. There are 3 levels of factor *A* and 5 levels of factor *B*.

d. A treatment would consist of a combination of one level of factor *A* and one level of factor *B*. There are a total of $3\times 5=15$ treatments.

e. One problem with only one replicate is there are no degrees of freedom for error. This is overcome by having at least two replicates.

10.91 Some preliminary calculations are:

df for *A* is $a-1=3-1=2$
df for *B* is $b-1=4-1=3$
df for *AB* is $(a-1)(b-1)=(3-1)(4-1)=2(3)=6$
df for Error is $n-ab=24-(3)(4)=24-12=12$
df for Total is $n-1=24-1=23$

$$\text{SSE}=\text{SS(Total)}-\text{SSA}-\text{SSB}-\text{SSAB}=18.1-0.8-5.3-9.6=2.4$$

$$\text{MSA}=\frac{\text{SSA}}{a-1}=\frac{0.8}{3-1}=0.40 \qquad \text{MSB}=\frac{\text{SSB}}{b-1}=\frac{5.3}{4-1}=1.7667$$

$$\text{MSAB}=\frac{\text{SSAB}}{(a-1)(b-1)}=\frac{9.6}{(3-1)(4-1)}=1.60$$

$$\text{MSE}=\frac{\text{MSE}}{n-ab}=\frac{2.4}{24-3(4)}=0.20$$

$$F_A=\frac{\text{MSA}}{\text{MSE}}=\frac{0.40}{0.20}=2.00 \qquad F_B=\frac{\text{MSB}}{\text{MSE}}=\frac{1.7667}{0.20}=8.83$$

$$F_{AB}=\frac{\text{MSAB}}{\text{MSE}}=\frac{1.60}{0.20}=8.00$$

The ANOVA table is:

Source	df	SS	MS	*F*
A	2	0.8	0.4000	2.00
B	3	5.3	1.7667	8.83
AB	6	9.6	1.6000	8.00
Error	12	2.4	0.2000	
Total	23	18.1		

b. $SSTrt = SSA + SSB + SSAB = 0.8 + 5.3 + 9.6 = 15.7$.

$$MST = \frac{SST}{ab-1} = \frac{15.7}{3(4)-1} = 1.4273 \qquad F_T = \frac{MST}{MSE} = \frac{1.4273}{0.20} = 7.14$$

To determine if the treatment means differ, we test:

$H_0: \mu_1 = \mu_2 = \cdots = \mu_{12}$
H_a: At least two treatment means differ

The test statistic is $F = 7.14$.

The rejection region requires $\alpha = 0.05$ in the upper tail of the F distribution with $\nu_1 = ab - 1 = 3(4) - 1 = 11$ and $\nu_2 = n - ab = 24 - 3(4) = 12$. Using MINITAB, $F_{0.05} = 2.72$. The rejection region is $F > 2.72$.

Since the observed value of the test statistic falls in the rejection region $(F = 7.14 > 2.72)$, H_0 is rejected. There is sufficient evidence to indicate the treatment means differ at $\alpha = 0.05$.

c. Yes. We need to partition the Treatment Sum of Squares into the Main Effects and Interaction Sum of Squares. Then we test whether factors *A* and *B* interact. Depending on the conclusion of the test for interaction, we either test for main effects or compare the treatment means.

d. Two factors are said to interact if the effects of one factor on the dependent variable are not the same at different levels of the second factor. If the factors interact, then tests for main effects are not necessary. We need to compare the treatment means for one factor at each level of the second.

e. To determine if the factors interact, we test:

H_0: Factors *A* and *B* do not interact to affect the response mean
H_a: Factors *A* and *B* do interact to affect the response mean

The test statistic is $F = \frac{MSAB}{MSE} = 8.00$

The rejection region requires $\alpha = 0.05$ in the upper tail of the F distribution with $\nu_1 = (a-1)(b-1) = (3-1)(4-1) = 6$ and $\nu_2 = n - ab = 24 - 3(4) = 12$. From Table VI, Appendix B, $F_{0.05} = 3.00$. The rejection region is $F > 3.00$.

Since the observed value of the test statistic falls in the rejection region $(F = 8.00 > 3.00)$, H_0 is rejected. There is sufficient evidence to indicate the two factors interact to affect the response mean at $\alpha = 0.05$.

f. No. Testing for main effects is not warranted because the interaction was significant. Instead, we compare the treatment means of one factor at each level of the second factor.

10.93 a. $\text{SSA} = 0.2(1000) = 200$, $\text{SSB} = 0.1(1000) = 100$, $\text{SSAB} = 0.1(1000) = 100$

$$\text{SSE} = \text{SS(Total)} - \text{SSA} - \text{SSB} - \text{SSAB} = 1000 - 200 - 100 - 100 = 600$$

$$\text{SST} = \text{SSA} + \text{SSB} + \text{SSAB} = 200 + 100 + 100 = 400$$

$$\text{MSA} = \frac{\text{SSA}}{a-1} = \frac{200}{3-1} = 100 \qquad \text{MSB} = \frac{\text{SSB}}{b-1} = \frac{100}{3-1} = 50$$

$$\text{MSAB} = \frac{\text{SSAB}}{(a-1)(b-1)} = \frac{100}{(3-1)(3-1)} = 25 \qquad \text{MSE} = \frac{\text{SSE}}{n-ab} = \frac{600}{27-3(3)} = 33.333$$

$$\text{MST} = \frac{\text{SST}}{ab-1} = \frac{400}{3(3)-1} = 50$$

$$F_A = \frac{\text{MSA}}{\text{MSE}} = \frac{100}{33.333} = 3.00 \qquad F_B = \frac{\text{MSB}}{\text{MSE}} = \frac{50}{33.333} = 1.50$$

$$F_{AB} = \frac{\text{MSAB}}{\text{MSE}} = \frac{25}{33.333} = 0.75 \qquad F_T = \frac{\text{MST}}{\text{MSE}} = \frac{50}{33.333} = 1.50$$

Source	**df**	**SS**	**MS**	***F***
A	2	200	100	3.00
B	2	100	50	1.50
AB	4	100	25	0.75
Error	18	600	33.333	
Total	26	1000		

To determine whether the treatment means differ, we test:

$H_0: \mu_1 = \mu_2 = \cdots = \mu_9$
H_a: At least two treatment means differ

The test statistic is $F_T = \dfrac{\text{MST}}{\text{MSE}} = 1.50$

Suppose $\alpha = 0.05$. The rejection region requires $\alpha = 0.05$ in the upper tail of the F distribution with $\nu_1 = ab - 1 = 3(3) - 1 = 8$ and $\nu_2 = n - ab = 27 - 3(3) = 18$. From Table VI, Appendix B, $F_{0.05} = 2.51$. The rejection region is $F > 2.51$.

Since the observed value of the test statistic does not fall in the rejection region $(F = 1.50 \not> 2.51)$, H_0 is not rejected. There is insufficient evidence to indicate the treatment means differ at $\alpha = 0.05$. Since there are no treatment mean differences, we have nothing more to do.

b. $SSA = 0.1(1000) = 100$, $SSB = 0.1(1000) = 100$, $SSAB = 0.5(1000) = 500$

$$SSE = SS(Total) - SSA - SSB - SSAB = 1000 - 100 - 100 - 500 = 300$$

$$SST = SSA + SSB + SSAB = 100 + 100 + 500 = 700$$

$$MSA = \frac{SSA}{a-1} = \frac{100}{3-1} = 50 \qquad MSB = \frac{SSB}{b-1} = \frac{100}{3-1} = 50$$

$$MSAB = \frac{SSAB}{(a-1)(b-1)} = \frac{500}{(3-1)(3-1)} = 125 \qquad MSE = \frac{SSE}{n-ab} = \frac{300}{27-3(3)} = 16.667$$

$$MST = \frac{SST}{ab-1} = \frac{700}{9-1} = 87.5$$

$$F_A = \frac{MSA}{MSE} = \frac{50}{16.667} = 3.00 \qquad F_B = \frac{MSB}{MSE} = \frac{50}{16.667} = 3.00$$

$$F_{AB} = \frac{MSAB}{MSE} = \frac{125}{16.667} = 7.50 \qquad F_T = \frac{MST}{MSE} = \frac{87.5}{16.667} = 5.25$$

Source	**df**	**SS**	**MS**	***F***
A	2	100	50	3.00
B	2	100	50	3.00
AB	4	500	125	7.50
Error	18	300	16.667	
Total	26	1000		

To determine if the treatment means differ, we test:

$H_0: \mu_1 = \mu_2 = \cdots = \mu_9$
H_a: At least two treatment means differ

The test statistic is $F = \frac{MST}{MSE} = 5.25$

The rejection region requires $\alpha = 0.05$ in the upper tail of the F distribution with $\nu_1 = ab - 1 = 3(3) - 1 = 8$ and $\nu_2 = n - ab = 27 - 3(3) = 18$. From Table VI, Appendix B, $F_{0.05} = 2.51$. The rejection region is $F > 2.51$.

Since the observed value of the test statistic falls in the rejection region $(F = 5.25 > 2.51)$, H_0 is rejected. There is sufficient evidence to indicate the treatment means differ at $\alpha = 0.05$.

Since the treatment means differ, we next test for interaction between factors *A* and *B*. To determine if factors *A* and *B* interact, we test:

H_0: Factors *A* and *B* do not interact to affect the mean response
H_a: Factors *A* and *B* do interact to affect the mean response

The test statistic is $F = \dfrac{\text{MSAB}}{\text{MSE}} = 7.50$

The rejection region requires $\alpha = 0.05$ in the upper tail of the *F* distribution with $\nu_1 = (a-1)(b-1) = (3-1)(3-1) = 4$ and $\nu_2 = n - ab = 27 - 3(3) = 18$. From Table VI, Appendix B, $F_{0.05} = 2.93$. The rejection region is $F > 2.93$.

Since the observed value of the test statistic falls in the rejection region $(F = 7.50 > 2.93)$, H_0 is rejected. There is sufficient evidence to indicate the factors *A* and *B* interact at $\alpha = 0.05$. Since interaction is present, no tests for main effects are necessary.

c. $\text{SSA} = 0.4(1000) = 400$, $\text{SSB} = 0.1(1000) = 100$, $\text{SSAB} = 0.2(1000) = 200$

$\text{SSE} = \text{SS(Total)} - \text{SSA} - \text{SSB} - \text{SSAB} = 1000 - 400 - 100 - 200 = 300$

$\text{SST} = \text{SSA} + \text{SSB} + \text{SSAB} = 400 + 100 + 200 = 700$

$\text{MSA} = \dfrac{\text{SSA}}{a-1} = \dfrac{400}{3-1} = 200$ $\quad\quad$ $\text{MSB} = \dfrac{\text{SSB}}{b-1} = \dfrac{100}{3-1} = 50$

$\text{MSAB} = \dfrac{\text{SSAB}}{(a-1)(b-1)} = \dfrac{200}{(3-1)(3-1)} = 50$

$\text{MSE} = \dfrac{\text{SSE}}{n-ab} = \dfrac{300}{27-3(3)} = 16.667$ $\quad\quad$ $\text{MST} = \dfrac{\text{SST}}{ab-1} = \dfrac{700}{3(3)-1} = 87.5$

$F_A = \dfrac{\text{MSA}}{\text{MSE}} = \dfrac{200}{16.667} = 12.00$ $\quad\quad$ $F_B = \dfrac{\text{MSB}}{\text{MSE}} = \dfrac{50}{16.667} = 3.00$

$F_{AB} = \dfrac{\text{MS}AB}{\text{MSE}} = \dfrac{50}{16.667} = 3.00$ $\quad\quad$ $F_T = \dfrac{\text{MST}}{\text{MSE}} = \dfrac{87.5}{16.667} = 5.25$

Source	**df**	**SS**	**MS**	***F***
A	2	400	200	12.00
B	2	100	50	3.00
AB	4	200	50	3.00
Error	18	300	16.667	
Total	26	1000		

To determine if the treatment means differ, we test:

$H_0: \mu_1 = \mu_2 = \cdots = \mu_9$
H_a: At least two treatment means differ

The test statistic is $F = \dfrac{\text{MST}}{\text{MSE}} = 5.25$

The rejection region requires $\alpha = 0.05$ in the upper tail of the F distribution with $\nu_1 = ab - 1 = 3(3) - 1 = 8$ and $\nu_2 = n - ab = 27 - 3(3) = 18$. From Table VI, Appendix B, $F_{0.05} = 2.51$. The rejection region is $F > 2.51$.

Since the observed value of the test statistic falls in the rejection region $(F = 5.25 > 2.51)$, H_0 is rejected. There is sufficient evidence to indicate the treatment means differ at $\alpha = 0.05$.

Since the treatment means differ, we next test for interaction between factors A and B. To determine if factors A and B interact, we test:

H_0: Factors A and B do not interact to affect the mean response
H_a: Factors A and B do interact to affect the mean response

The test statistic is $F = \dfrac{\text{MS}AB}{\text{MSE}} = 3.00$

The rejection region requires $\alpha = 0.05$ in the upper tail of the F distribution with $\nu_1 = (a-1)(b-1) = (3-1)(3-1) = 4$ and $\nu_2 = n - ab = 27 - 3(3) = 18$. From Table VI, Appendix B, $F_{0.05} = 2.93$. The rejection region is $F > 2.93$.

Since the observed value of the test statistic falls in the rejection region $(F = 3.00 > 2.93)$, H_0 is rejected. There is sufficient evidence to indicate the factors A and B interact at $\alpha = 0.05$. Since interaction is present, no tests for main effects are necessary.

d. $\text{SSA} = 0.4(1000) = 400$, $\quad \text{SSB} = 0.4(1000) = 400$, $\quad \text{SSAB} = 0.1(1000) = 100$

$\text{SSE} = \text{SS(Total)} - \text{SSA} - \text{SSB} - \text{SSAB} = 1000 - 400 - 400 - 100 = 100$

$\text{SST} = \text{SSA} + \text{SSB} + \text{SSAB} = 400 + 400 + 100 = 900$

$$\text{MSA} = \frac{\text{SSA}}{a-1} = \frac{400}{3-1} = 200 \qquad \text{MSB} = \frac{\text{SSB}}{b-1} = \frac{400}{3-1} = 200$$

$$\text{MSAB} = \frac{\text{SSAB}}{(a-1)(b-1)} = \frac{100}{(3-1)(3-1)} = 25 \qquad \text{MSE} = \frac{\text{SSE}}{n-ab} = \frac{100}{27-3(3)} = 5.556$$

$$\text{MST} = \frac{\text{SST}}{ab-1} = \frac{900}{3(3)-1} = 112.5$$

$$F_A = \frac{\text{MSA}}{\text{MSE}} = \frac{200}{5.556} = 36.00 \qquad F_B = \frac{\text{MSB}}{\text{MSE}} = \frac{200}{5.556} = 36.00$$

$$F_{AB} = \frac{\text{MSAB}}{\text{MSE}} = \frac{25}{5.556} = 4.50 \qquad F_T = \frac{\text{MST}}{\text{MSE}} = \frac{112.5}{5.556} = 20.25$$

Source	df	SS	MS	*F*
A	2	400	200	36.00
B	2	400	200	36.00
AB	4	100	25	4.50
Error	18	100	5.556	
Total	26	1000		

To determine if the treatment means differ, we test:

$H_0: \mu_1 = \mu_2 = \cdots = \mu_9$
H_a: At least two treatment means differ

The test statistic is $F = \dfrac{\text{MST}}{\text{MSE}} = 20.25$

The rejection region requires $\alpha = 0.05$ in the upper tail of the F distribution with $\nu_1 = ab - 1 = 3(3) - 1 = 8$ and $\nu_2 = n - ab = 27 - 3(3) = 18$. From Table VI, Appendix B, $F_{0.05} = 2.51$. The rejection region is $F > 2.51$.

Since the observed value of the test statistic falls in the rejection region $(F = 20.25 > 2.51)$, H_0 is rejected. There is sufficient evidence to indicate the treatment means differ at $\alpha = 0.05$.

Since the treatment means differ, we next test for interaction between factors *A* and *B*. To determine if factors *A* and *B* interact, we test:

H_0: Factors *A* and *B* do not interact to affect the mean response
H_a: Factors *A* and *B* do interact to affect the mean response

The test statistic is $F = \dfrac{\text{MSAB}}{\text{MSE}} = 4.50$

The rejection region requires $\alpha = 0.05$ in the upper tail of the F distribution with $\nu_1 = (a-1)(b-1) = (3-1)(3-1) = 4$ and $\nu_2 = n - ab = 27 - 3(3) = 18$. From Table VI, Appendix B, $F_{0.05} = 2.93$. The rejection region is $F > 2.93$.

Since the observed value of the test statistic falls in the rejection region $(F = 4.50 > 2.93)$, H_0 is rejected. There is sufficient evidence to indicate the factors *A* and *B* interact at $\alpha = 0.05$. Since interaction is present, no tests for main effects are necessary.

10.95 a. This is a complete factorial experiment.

b. There are 2 factors – age and diet. Age has 2 levels – young and old. Diet has 2 levels – fine limestone and coarse limestone. There are a total of $2 \times 2 = 4$ treatments – young, fine limestone; young, coarse limestone; old, fine limestone; and old, coarse limestone.

c. The experimental unit is the hen.

d. The dependent variable is shell thickness.

e. The factors do not interact. This means that the effect of diet on shell thickness is the same for each age group of hens.

f. The shell thickness is not affected by the age of the hens.

g. The shell thickness is affected by the diet. The mean shell thickness of eggs produced by hens on the coarse limestone diet is significantly greater than the shell thickness of eggs produced by hens on the fine limestone diet.

10.97 a. The ANOVA table is:

Source	*df*	*F-value*	*p-value*
Order (vice/virtue)	1		
Menu (homogeneous/mixed)	1		
Order x Menu	1	11.25	<0.001
Error	176		
Total	179		

b. To determine if Order and Menu interact, we test:

H_0: Order and Menu do not interact
H_a: Order and Menu do interact

The test statistic is $F = 11.25$ and the *p*-value is $p < 0.001$. Since the *p*-value is so small $(p < 0.001)$, H_0 is rejected. There is sufficient evidence to indicate that Order and Menu interact to affect the mean amount customer would be willing to pay at any value of $\alpha > 0.001$.

c. No. Since the interaction is significant, there is no need to test for the main effects.

d. Using MINITAB, a graph of the means is:

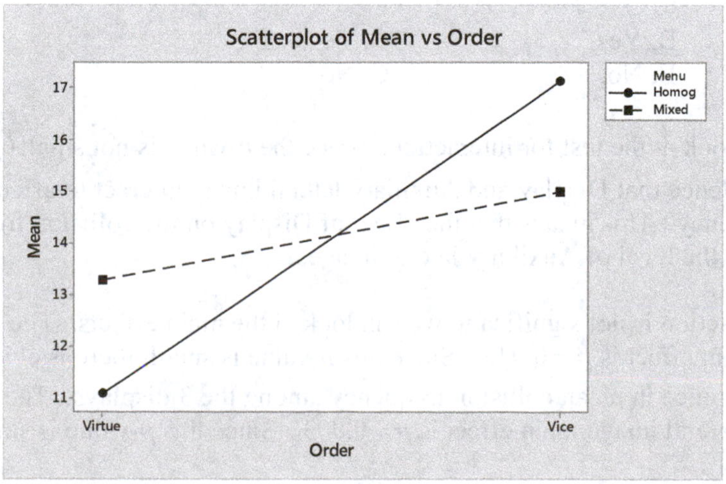

10.99 a. If the two factors, justice reparation potential and producer need, interact, then the effect of producer need on the mean intention to purchase depends on the value of producer need.

b. To determine if justice reparation potential and producer need interact to affect the mean intention to purchase, we test:

H_0: Justice reparation potential and producer need do not interact
H_a: Justice reparation potential and producer need do interact

The test statistic is $F = 20.55$ and the p-value is $p = 0.000$. Since the p-value is less than $\alpha\,(p = 0.000 < 0.01)$, H_0 is rejected. There is sufficient evidence to indicate justice reparation potential and producer need interact to affect the mean intention to purchase at $\alpha = 0.01$.

c. No. Because the interaction is significant, the main effect tests are not necessary.

d. Because the lines are not parallel, the graph indicates that interaction between justice reparation potential and producer need exists.

e. Yes. When Justice Restoration Potential is low, then mean purchase intention is lower when Need is high than when Need is moderate. When Justice Restoration Potential is High, then mean purchase intention is lower when Need is moderate than when Need is high.

10.101 a. The summary ANOVA table is:

Source	**df**
VR display	$a-1=3-1=2$
Lateral Image	$b-1=2-1=1$
VR x Lat	$(a-1)(b-1)=2(1)=2$
Error	$nab-ab=12-6=6$
Total	$nab-1=12-1=11$

b. There are a total of $3\times 2=6$ treatments in this experiment. They are:

A, Yes	B, Yes	C, Yes
A, No	B, No	C, No

c. We must first look at the test for interaction. Since the p-value is not small $(p = 0.411)$, there is no evidence that Display and Auxiliary lateral image interact to affect the collision frequency. This means that the effect of Display on the collision frequency is not affected by the level of Auxiliary lateral image.

Since the interaction is not significant, we can look at the main effects. The p-value for the Display main effect is $p = 0.045$. Since this p-value is small, there is evidence to indicate a difference in mean collision frequency among the 3 displays. The p-value for the Auxiliary lateral image main effect is $p = 0.003$. Since this p-value is small, there is evidence to indicate a difference in mean collision frequency among the 2 Auxiliary lateral images.

10.103 Using SAS, the results of the analysis are:

```
                          The ANOVA Procedure

Dependent Variable: VEG

                                     Sum of
  Source                   DF        Squares     Mean Square    F Value    Pr > F

  Model                     8    556.1072222      69.5134028      31.36    <.0001

  Error                    27     59.8425000       2.2163889

  Corrected Total          35    615.9497222


              R-Square     Coeff Var      Root MSE      VEG Mean

              0.902845      7.124172      1.488754      20.89722


  Source                   DF       Anova SS     Mean Square    F Value    Pr > F

  HEIGHT                    2     77.9022222      38.9511111      17.57    <.0001
  FREQ                      2    387.9772222     193.9886111      87.52    <.0001
  HEIGHT*FREQ               4     90.2277778      22.5569444      10.18    <.0001
```

To determine if Mow height and Mow frequency interact to affect the mean vegetation height, we test:

> H_0: Mow height and Mow frequency do not interact to affect the mean vegetation height
> H_a: Mow height and Mow frequency interact to affect the mean vegetation height

The test statistic is $F = 10.18$ and the *p*-value is $p < 0.0001$. Because the *p*-value is very small $(p < 0.0001)$, H_0 is rejected. There is sufficient evidence to indicate Mow height and Mow frequency interact to affect the mean vegetation height for any reasonable value of α.

Since interaction exists, no main effect tests are run. We now run multiple comparisons on the mean vegetation heights for mow frequencies at each level of mow height. The results are:

```
------------------------------------- HEIGHT=5 -----------------------------------

                          The ANOVA Procedure

               Tukey's Studentized Range (HSD) Test for VEG

 NOTE: This test controls the Type I experimentwise error rate, but it generally has a
                    higher Type II error rate than REGWQ.


               Alpha                                        0.05
               Error Degrees of Freedom                        9
               Error Mean Square                          2.5375
               Critical Value of Studentized Range       3.94840
               Minimum Significant Difference             3.1448
```

```
             Means with the same letter are not significantly different.

              Tukey Grouping          Mean      N    FREQ

                           A        22.350      4    2

                           B        18.000      4    3
                           B
                           B        17.075      4    1

------------------------------------ HEIGHT=10 ------------------------------------

                                The ANOVA Procedure

                   Tukey's Studentized Range (HSD) Test for VEG

     NOTE: This test controls the Type I experimentwise error rate, but it generally has a
                          higher Type II error rate than REGWQ.

                   Alpha                                        0.05
                   Error Degrees of Freedom                        9
                   Error Mean Square                        1.555278
                   Critical Value of Studentized Range       3.94840
                   Minimum Significant Difference              2.462

             Means with the same letter are not significantly different.

              Tukey Grouping          Mean      N    FREQ

                           A       23.3000      4    3
                           A
                           A       23.2500      4    2

                           B       15.8750      4    1

------------------------------------ HEIGHT=20 ------------------------------------
                                The ANOVA Procedure

                   Tukey's Studentized Range (HSD) Test for VEG

     NOTE: This test controls the Type I experimentwise error rate, but it generally has a
                          higher Type II error rate than REGWQ.

                   Alpha                                        0.05
                   Error Degrees of Freedom                        9
                   Error Mean Square                        2.556389
                   Critical Value of Studentized Range       3.94840
                   Minimum Significant Difference             3.1565

             Means with the same letter are not significantly different.

              Tukey Grouping          Mean      N    FREQ

                           A        26.150      4    2
                           A
                           A        26.025      4    3

                           B        16.050      4    1
```

For mowing height 5, the mean vegetation height for mowing frequency 2 is significantly higher than the mean vegetation height for mowing frequencies 1 and 3. There is no significant difference in the mean vegetation height between mowing frequencies 1 and 3.

For mowing height 10, the mean vegetation height for mowing frequencies 2 and 3 are significantly higher than the mean vegetation height for mowing frequency 1. There is no significant difference in the mean vegetation height between mowing frequencies 2 and 3.

For mowing height 20, the mean vegetation height for mowing frequencies 2 and 3 are significantly higher than the mean vegetation height for mowing frequency 1. There is no significant difference in the mean vegetation height between mowing frequencies 2 and 3.

10.105 a. The ANOVA table is:

Source	**df**	**F**	***p*-value**
Age	$3-1=2$	11.93	<0.001
Book	$3-1=2$	23.64	<0.001
Age x Book	$(3-1)(3-1)=4$	2.99	<0.05
Error	$108-3(3)=99$		
Total	107		

b. The total number of treatments is $3\times 3=9$. The treatments are:

18, Photos	24, Photos	30, Photos
18, Drawings	24, Drawings	30, Drawings
18, Control	24, Control	30, Control

c. To determine if the interaction between Age and Book exists, we test:

H_0: Age and Book do not interact to affect reenactment scores
H_a: Age and Book do interact to affect reenactment scores

The test statistic is $F=2.99$ and the *p*-value is $p<0.05$. Since the *p*-value is less than $\alpha=0.05$, H_0 is rejected. There is sufficient evidence to indicate that Age and Book interact to affect target action scores at $\alpha=0.05$. This means that the effect of Age on the reenactment scores depends on the level of Book.

d. We do not need to conduct the main effect tests because the interaction between Book and Age is significant. If the interaction is significant, it is possible that it could cover up any main effects.

e. Using MINITAB, a graph of the results is:

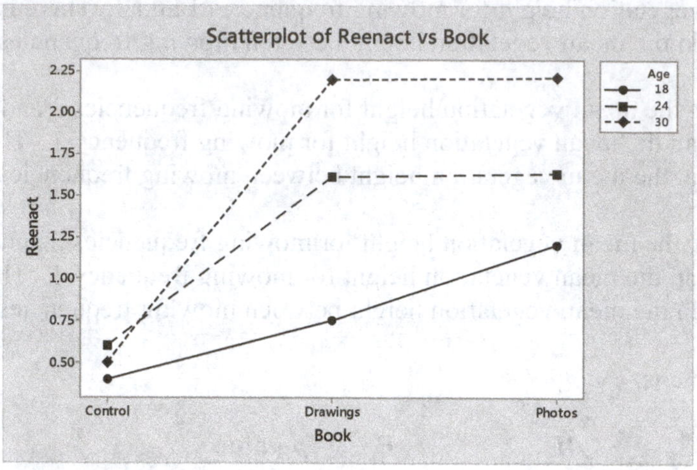

From the multiple comparisons, we can conclude:

Children at age 18 months: The mean reenactment score for children using books with Photos is significantly greater than the mean reenactment score for children using books with no photos or drawings. There are no other differences.

Children at age 24 months: The mean reenactment scores for children using books with Photos and books with Drawings are significantly greater than the mean reenactment score for children using books with no photos or drawings. There is no difference between the mean reenactment scores for children using books with Photos and children using books with Drawings.

Children at age 30 months: The mean reenactment scores for children using books with Photos and books with Drawings are significantly greater than the mean reenactment score for children using books with no photos or drawings. There is no difference between the mean reenactment scores for children using books with Photos and children using books with Drawings.

10.107 Using MINITAB, the results of the ANOVA are:

General Linear Model: NUMBER versus GROUP, SET

```
Factor  Type   Levels  Values
GROUP   fixed       3  3, 6, 12
SET     fixed       3  FIRST, LAST, MIDDLE


Analysis of Variance for NUMBER, using Adjusted SS for Tests

Source      DF   Seq SS   Adj SS   Adj MS      F      P
GROUP        2   15.267   15.267    7.633   7.59  0.001
SET          2   62.600   62.600   31.300  31.11  0.000
GROUP*SET    4    7.133    7.133    1.783   1.77  0.142
Error       81   81.500   81.500    1.006
Total       89  166.500


S = 1.00308   R-Sq = 51.05%   R-Sq(adj) = 46.22%
```

```
Unusual Observations for NUMBER

Obs    NUMBER       Fit    SE Fit   Residual   St Resid
 37  0.00000  2.30000  0.31720  -2.30000     -2.42 R
 66  5.00000  2.60000  0.31720   2.40000      2.52 R

R denotes an observation with a large standardized residual.

             Means

SET        N  NUMBER
FIRST    30  3.0000
LAST     30  1.1000
MIDDLE  30  1.4000


GROUP    N  NUMBER
 3      30  2.4000
 6      30  1.4333
12      30  1.6667

Tukey 95.0% Simultaneous Confidence Intervals
Response Variable NUMBER
All Pairwise Comparisons among Levels of GROUP
GROUP =  3  subtracted from:

GROUP    Lower    Center     Upper  ---+---------+---------+---------+---
 6      -1.586  -0.9667  -0.3477   (---------*--------)
12      -1.352  -0.7333  -0.1143       (--------*--------)
                                    ---+---------+---------+---------+---
                                    -1.40     -0.70      0.00      0.70


GROUP =  6  subtracted from:

GROUP     Lower  Center   Upper   ---+---------+---------+---------+---
12      -0.3857  0.2333  0.8523                  (--------*--------)
                                   ---+---------+---------+---------+---
                                   -1.40     -0.70      0.00      0.70
Tukey 95.0% Simultaneous Confidence Intervals
Response Variable NUMBER
All Pairwise Comparisons among Levels of SET
SET = FIRST  subtracted from:

SET      Lower  Center   Upper  -----+---------+---------+---------+-
LAST    -2.519  -1.900  -1.281  (-----*-----)
MIDDLE  -2.219  -1.600  -0.981     (-----*-----)
                                -----+---------+---------+---------+-
                                  -2.0      -1.0       0.0       1.0

SET = LAST  subtracted from:

SET        Lower  Center   Upper  -----+---------+---------+---------+-
MIDDLE  -0.3190  0.3000  0.9190                          (-----*-----)
                                  -----+---------+---------+---------+-
                                    -2.0      -1.0       0.0       1.0
```

To determine if group size and photo set interact to affect the number of selections, we test:

H_0: Group size and Photo set do not interact to affect the number of selections
H_a: Group size and Photo set interact to affect the number of selections

The test statistic is $F = 1.77$ and the *p*-value is $p = 0.142$. Since the *p*-value is not small, H_0 is not rejected. There is insufficient evidence to indicate that group size and photo set interact to affect the number of selections for any reasonable level of α.

Since there is no evidence of an interaction, we will next test for the main effects. To determine if group size had an effect on the mean number of selections, we test:

$H_0 : \mu_1 = \mu_2 = \mu_3$
H_a: At least two group size means differ

The test statistic is $F = 7.59$ and the *p*-value is $p = 0.001$. Since the *p*-value is so small, H_0 is rejected. There is sufficient evidence to indicate that group size has an effect on the mean number of selections for any level of α greater than 0.001.

To determine if photo set had an affect on the mean number of selections, we test:

$H_0 : \mu_1 = \mu_2 = \mu_3$
H_a: At least two photo set means differ

The test statistic is $F = 31.11$ and the *p*-value is $p = 0.000$. Since the *p*-value is so small, H_0 is rejected. There is sufficient evidence to indicate that photo set has an affect the mean number of selections for any level of α greater than 0.000.

Since both main effects are significant, we will run Tukey's multiple comparison procedure on each main effect to find where the differences exist.

The mean number of selections made for the different group sizes are:

Means:	2.400	1.433	1.667
Groups:	3	6	12

The confidence interval comparing size 3 to size 6 is $(-1.586, -0.3477)$. Since both endpoints of the interval are negative, the mean number of selections for size 3 is significantly greater than the mean number of selections for size 6. The confidence interval comparing size 3 to size 12 is $(-1.352, -0.1143)$. Since both endpoints of the interval are negative, the mean number of selections for size 3 is significantly greater than the mean number of selections for size 12. The confidence interval comparing size 6 to size 12 is $(-0.3857, 0.8523)$. Since 0 is contained in the interval, there is no difference in the mean number of selections between sizes 6 and 12. Thus, there are significantly more selections made for group size 3 than for the other two sizes.

The mean number of selections made for the different photo sets are:

Means:	3.00	1.40	1.10
Groups:	First	Middle	Last

The confidence interval comparing the first photo set to the last photo set is $(-2.519, -1.281)$. Since both endpoints of the interval are negative, the mean number of selections for the first photo set is significantly greater than the mean number of selections for the last photo set. The confidence interval comparing the first photo set to the middle photo set is $(-2.219, -0.981)$. Since both endpoints of the interval are negative, the mean number of selections for the first photo set is significantly greater than the mean number of selections for the middle photo set. The confidence interval comparing the middle photo set to the last photo set is $(-0.3190, 0.9190)$. Since 0 is contained in the interval, there is no difference in the mean number of selections between the last photo set and the middle photo set. Thus, there are significantly more selections made for the first photo set than for the other two photo sets.

10.109 a. The treatment totals for the 4 treatments are:

Low Load, Ambiguous: $25(18.0) = 450$

Low Load, Common: $25(7.8) = 195$

High Load, Ambiguous: $25(6.1) = 152.5$

High Load, Common: $25(6.3) = 157.5$

b. $$CM = \frac{\left(\sum y_i\right)^2}{n} = \frac{(450+152.5+195+157.5)^2}{25+25+25+25} = \frac{955^2}{100} = 9{,}120.25$$

c. $$\text{SS(Load)} = \frac{\sum A_i^2}{br} - CM = \frac{(450+195)^2 + (152.5+157.5)^2}{2(25)} - 9{,}120.25$$
$$= \frac{645^2 + 310^2}{50} - 9{,}120.25 = \frac{512{,}125}{50} - 9{,}120.25$$
$$= 10{,}242.5 - 9{,}120.25 = 1{,}122.25$$

$$\text{SS(Name)} = \frac{\sum B_i^2}{ar} - CM = \frac{(450+152.5)^2 + (195+157.5)^2}{2(25)} - 9{,}120.25$$
$$= \frac{602.5^2 + 352.5^2}{50} - 9{,}120.25 = \frac{487{,}262.5}{50} - 9{,}120.25$$
$$= 9{,}745.25 - 9{,}120.25 = 625$$

$$SS(\text{Load}\times\text{Name})=\frac{\sum\sum AB_{ij}^2}{r}-SS(\text{Name})-SS(\text{Load})-CM$$
$$=\frac{450^2+195^2+152.5^2+157.5^2}{25}-625-1{,}122.25-9{,}120.25$$
$$=11{,}543.5-625-1{,}122.25-9{,}120.25=676$$

d. The treatment variances and sums of squares are:

Low Load, Ambiguous: $s_{LA}^2=15^2=225$,
$\sum(x_i-\bar{x}_{LA})^2=(n_{LA}-1)s_{LA}^2=(25-1)225=5{,}400$

Low Load, Common: $s_{LC}^2=9.5^2=90.25$,
$\sum(x_i-\bar{x}_{LC})^2=(n_{LC}-1)s_{LC}^2=(25-1)90.25=2{,}166$

High Load, Ambiguous: $s_{HA}^2=9.5^2=90.25$,
$\sum(x_i-\bar{x}_{HA})^2=(n_{HA}-1)s_{HA}^2=(25-1)90.25=2{,}166$

High Load, Common: $s_{HC}^2=10^2=100$,
$\sum(x_i-\bar{x}_{HC})^2=(n_{HC}-1)s_{HC}^2=(25-1)100=2{,}400$

e. $SSE=5{,}400+2{,}166+2{,}166+2{,}400=12{,}132$

f. $SS(\text{Total})=SS(\text{Load})+SS(\text{Name})+SS(\text{Load x Name})+SSE$
$=1{,}122.25+625+676+12{,}132=14{,}555.25$

g. The ANOVA table is:

Source	**df**	**SS**	**MSE**	**F**
Load	$2-1=1$	1,122.25	1,122.25	8.88
Name	$2-1=1$	625	625	4.95
Load x Name	$1\times1=1$	676	676	5.35
Error	$100-4=96$	12,132	126.375	
Total	99	14,555.25		

h. The researchers calculated the F value to be $F=5.34$. We calculated the F to be $F=5.35$. The difference could be due to some round-off error.

i. To determine if Load and Name interact to affect the number of jelly beans taken, we test:

H_0: Load and Name do not interact to affect the number of jelly beans taken
H_a: Load and Name do interact to affect the number of jelly beans taken

The test statistic is $F = 5.35$.

The rejection region requires $\alpha = 0.05$ in the upper tail of the F distribution $\nu_1 = (a-1)(b-1) = (2-1)(2-1) = 1$ and $\nu_2 = n - ab = 100 - 2(2) = 96$. Using MINITAB, $F_{0.05} = 3.94$. The rejection region is $F > 3.94$.

Since the observed value of the test statistic falls in the rejection region $(F = 5.35 > 3.94)$, H_0 is rejected. There is sufficient evidence to indicate that Load and Name interact to affect the number of jelly beans taken at $\alpha = 0.05$.

Since the interaction is significant, there is no need to run the main effects test. Using MINITAB, a graphical display of the results is:

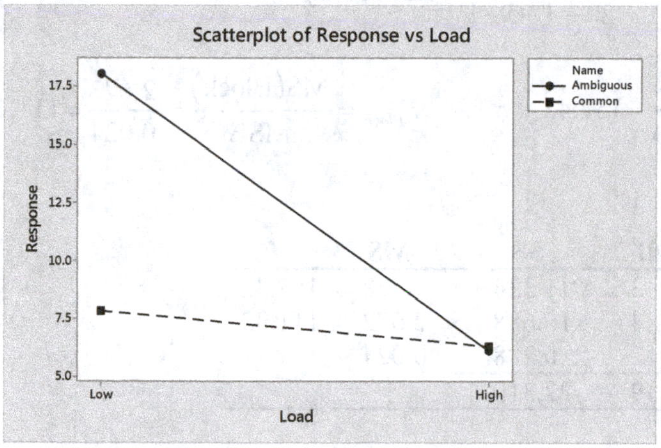

From the graph, it appears that when the Name of the jelly beans is Ambiguous, there is a significant difference in the mean number of jelly beans taken for the two loads. When Load is Low, there is a significantly greater mean number of jelly beans taken than when Load is high. When the Name of the jelly beans is Common, there appears to be no significant difference in the mean number of jelly beans taken between the 2 levels of load.

j. In order for the techniques to be valid, the following assumptions are necessary:

1. We must assume that the response distribution for each factor-level combination (treatment) is normal
2. We must assume that the response variance is constant for all treatments.
3. We must assume that we have random and independent samples of experimental units associated with each treatment.

10.111 In a completely randomized design, independent random selection of treatments to be assigned to experimental units is required; in a randomized block design, sets (blocks) of matched experimental units are employed, with an experimental unit in each block randomly assigned a treatment.

10.113 When the overall level of significance of a multiple comparisons procedure is α, the level of significance for each comparison is less than α.

10.115 a. $\text{SST} = \text{SS(Total)} - \text{SS(Block)} - \text{SSE} = 22.31 - 10.688 - 0.288 = 11.334$

$$\text{MST} = \frac{\text{SST}}{k-1} = \frac{11.334}{4-1} = 3.778, \quad df = k-1 = 4-1 = 3$$

$$\text{MS(Block)} = \frac{\text{SS(Block)}}{b-1} = \frac{10.688}{5-1} = 2.672, \quad df = b-1 = 5-1 = 4$$

$$\text{MSE} = \frac{\text{SSE}}{n-k-b+1} = \frac{.288}{20-4-5+1} = 0.024, \quad df = n-k-b+1 = 20-4-5+1 = 12$$

$$F_T = \frac{\text{MST}}{\text{MSE}} = \frac{3.778}{0.024} = 157.42 \qquad F_{Block} = \frac{\text{MS(Block)}}{\text{MSE}} = \frac{2.672}{0.024} = 111.33$$

The ANOVA Table is:

Source	**df**	**SS**	**MS**	***F***
Treatment	3	11.334	3.778	157.42
Block	4	10.688	2.672	111.33
Error	12	0.288	0.024	
Total	19	22.310		

b. To determine if there is a difference among the treatment means, we test:

$H_0: \mu_A = \mu_B = \mu_C = \mu_D$
H_a: At least two treatment means differ

The test statistic is $F = \frac{\text{MST}}{\text{MSE}} = 157.42$

The rejection region requires $\alpha = 0.05$ in the upper tail of the F distribution with $\nu_1 = k-1 = 4-1 = 3$ and $\nu_2 = n-k-b+1 = 20-4-5+1 = 12$. From Table VI, Appendix B, $F_{0.05} = 3.49$. The rejection region is $F > 3.49$.

Since the observed value of the test statistic falls in the rejection region $(F = 157.42 > 3.49)$, H_0 is rejected. There is sufficient evidence to indicate a difference among the treatment means at $\alpha = 0.05$.

c. Since there is evidence of differences among the treatment means, we need to compare the treatment means. The number of pairwise comparisons is $c = k(k-1)/2 = 4(4-1)/2 = 6$.

d. To determine if there are differences among the block means, we test:

H_0: All the block means are the same
H_a: At least two block means differ

The test statistic is $F = \dfrac{\text{MS(Block)}}{\text{MSE}} = 111.33$

The rejection region requires $\alpha = 0.05$ in the upper tail of the F distribution with $\nu_1 = b - 1 = 5 - 1 = 4$ and $\nu_2 = n - k - b + 1 = 20 - 4 - 5 + 1 = 12$. From Table VI, Appendix B, $F_{0.05} = 3.26$. The rejection region is $F > 3.26$.

Since the observed value of the test statistic falls in the rejection region $(F = 111.33 > 3.26)$, H_0 is rejected. There is sufficient evidence that the block means differ at $\alpha = 0.05$.

10.117 a. For this study, the experimental units are accountants.

b. The response variable is income.

c. There are 2 factors for this study: Mach rating and gender.

d. There are 3 levels of Mach rating: high, moderate, and low. There are 2 levels od gender: male and female.

e. There are a total of $3 \times 2 = 6$ treatments. They are: high male; high female; moderate male; moderate female; low male; and low female.

10.119 a. A 2×2 factorial design was used for this study. The two factors are color and type of questions, each at 2 levels. There are $2 \times 2 = 4$ treatments. The 4 treatments are "blue, difficult", "blue, simple", "red, difficult", and "red, simple".

b. Since the p-value is small $(p < 0.03)$, H_0 is rejected. There is sufficient evidence to indicate that color and type of questions interact to affect mean exam scores. This means that the effect of color on the mean exam scores depends on the difficulty of the questions.

c. Using MINITAB, the graph is:

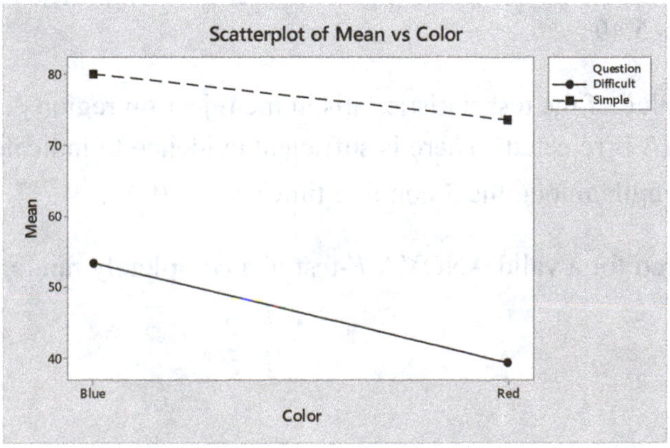

Looking at the graph, the lines are not parallel. The difference between the Blue and Red "simple" questions is not as great as the difference between the Blue and Red "difficult" questions. The effect of color on the exam scores depends on the level of difficulty.

10.121 a. The experimental units are the NCAA tennis coaches. The dependent variable is the rating on a 7-point scale of how important the coaches think the web site was. There is one factor and it is the division of the college/university that employs the tennis coach. Since there is only one factor, the treatments are the different levels of the factor. The factor has 3 levels and thus, 3 treatments which are Division I, Division II and Division III.

b. To determine if the mean ratings of the web sites are different for the different levels of coaches, we test:

$H_0: \mu_{I} = \mu_{II} = \mu_{III}$
H_a: At least one mean differs

c. Since the *p*-value is less than $\alpha\left(p < 0.003 < 0.05\right)$, H_0 is rejected. There is sufficient evidence to indicate a difference in mean responses among the different divisions of the colleges and universities at $\alpha = 0.05$.

d. From the results, there is no significant difference in the mean response between Division II and Division III coaches for the question concerning the usefulness of the Prospective Student-Athlete Form on the web site. The mean response for the Division I coaches is significantly larger than the mean responses for the Division II and Division III coaches. Thus, Division I coaches think the web site is significantly more useful than do the Division II and Division III coaches.

10.123 a. The experimental units are the teeth. The treatments are the 3 different bonding times: 1 hour, 24 hours, and 48 hours. The response variable is the breaking strength (in Mpa).

b. To determine if there is a difference in the mean breaking strength among the 3 bonding times, we test:

$H_0: \mu_1 = \mu_2 = \mu_3$
H_a: At least two treatment means differ

c. The rejection region requires $\alpha = 0.01$ in the upper tail of the *F* distribution with $\nu_1 = k - 1 = 3 - 1 = 2$ and $\nu_2 = n - k = 30 - 3 = 27$. Using MINITAB, $F_{0.01} = 5.49$. The rejection region is $F > 5.49$.

d. Since the observed value of the test statistic falls in the rejection region $\left(F = 61.62 > 5.49\right)$, H_0 is rejected. There is sufficient evidence to indicate a difference in mean breaking strength among the 3 bonding times at $\alpha = 0.01$.

e. The conditions required for a valid ANOVA *F*-test in a completely randomized design are:

1. The samples are randomly selected in an independent manner from k treatment populations.
2. All k sampled populations have distributions that are approximately normal.
3. The k population variances are equal (i.e., $\sigma_1^2 = \sigma_2^2 = ... = \sigma_k^2$).

f. To indicate that the means for 24 and 48 hours were not different, we would order the sample means and connect the means for 24 and 48 hours.

1 hour	48 hours	24 hours
3.32	5.03	5.07

The practical conclusions that we could draw from this experiment is that to get the strongest bond, we should use 24 hours. The mean strength for 24 hours is significantly greater than the mean strength for 1 hour. However, there is no difference in the mean strength of the bond between 24 and 48 hours. There is no need to wait the 48 hours.

10.125 a. There are two factors. The factor measuring luckiness has three levels: lucky, unlucky, and uncertain. The factor measuring competition has two levels: competitive and noncompetitive.

b. There are three tests:

To determine if interaction between Luckiness and Competition exists, we test:

H_0: The factors Luckiness and Competition do not
H_a: Luckiness and Competition interact to affect mean response

The test statistic is $F = 0.72$ and the p-value is $p = 0.72$. Since the p-value is so large, H_0 is not rejected. There is insufficient evidence to indicate Luckiness and Competition interact for any reasonable value of α.

To determine if the mean percentage guessed correctly differs among the three levels of luckiness, we test:

$H_0: \mu_1 = \mu_2 = \mu_3$
H_a: At least two Luckiness mean percentages differ

The test statistic is $F = 1.39$ and the p-value is $p = 0.26$. Since the p-value is so large, H_0 is not rejected. There is insufficient evidence to indicate differences in the mean percentages guessed correctly among the three levels of Luckiness for any level of $\alpha < 0.26$.

To determine if differences exist in the mean percentages guessed correctly between the two competition levels, we test:

$H_0: \mu_1 = \mu_2$
H_a: The two competition means differ.

The test statistic is $F = 2.84$ and the p-value is $p = 0.10$. Since the p-value is not small, H_0 is not rejected. There is insufficient evidence to indicate the mean percentages guessed correctly differ between the two levels of Competition for $\alpha < 0.10$.

10.127 a. This is a randomized block design. The blocks are the 12 plots of land. The treatments are the three methods used on the shrubs: fire, clipping, and control. The response variable is the mean number of flowers produced. The experimental units are the 36 shrubs.

b.

		Treatment	
	Fire	Clipping	Control
Plot 1	Shrub 2	Shrub 1	Shrub 3
2	Shrub 3	Shrub 2	Shrub 1
. . .	. . .	. . .	. . .
12	Shrub 1	Shrub 3	Shrub 2

c. To determine if there is a difference in the mean number of flowers produced among the three treatments, we test:

$H_0: \mu_1 = \mu_2 = \mu_3$

H_a: The mean number of flowers produced differ for at least two of the methods.

The test statistic is $F = 5.42$ and $p = 0.009$. We can reject the null hypothesis at the $\alpha > .009$ level of significance. At least two of the methods differ with respect to the mean number of flowers produced by pawpaws for $\alpha > 0.009$.

d. The mean number of flowers produced by the Control is significantly less than the mean number of flowers produced by the Clipping method and the Burning method. The mean number of flowers produced by the Clipping method does not differ from the mean number of flowers produced by the Burning method.

10.129 Using MINITAB, the descriptive statistics are:

Descriptive Statistics: THICKNESS, OVERRUN, STRENGTH

Variable	HOUSING	N	Mean	StDev	Minimum	Maximum
THICKNESS	BARN	6	0.50000	0.01414	0.48000	0.52000
	CAGE	10	0.4230	0.0350	0.3700	0.4700
	FREE	6	0.5017	0.0279	0.4700	0.5500
	ORGANIC	6	0.4817	0.0387	0.4300	0.5200

OVERRUN	BARN	6	513.33	8.38	501.00	526.00
	CAGE	10	480.60	12.91	462.00	502.00
	FREE	6	517.50	8.17	510.00	531.00
	ORGANIC	6	529.17	10.65	511.00	544.00
STRENGTH	BARN	6	39.333	1.120	37.600	40.300
	CAGE	10	37.320	2.127	33.000	40.200
	FREE	6	37.17	3.79	31.50	40.60
	ORGANIC	6	35.97	3.04	32.60	40.20

Using MINITAB, the ANOVAs are:

One-way ANOVA: THICKNESS versus HOUSING

Source	DF	Adj SS	Adj MS	F-Value	P-Value
HOUSING	3	0.03429	0.011430	11.74	0.000
Error	24	0.02338	0.000974		
Total	27	0.05767			

One-way ANOVA: OVERRUN versus HOUSING

Source	DF	Adj SS	Adj MS	F-Value	P-Value
HOUSING	3	10788	3595.9	31.36	0.000
Error	24	2752	114.7		
Total	27	13540			

One-way ANOVA: STRENGTH versus HOUSING

Source	DF	Adj SS	Adj MS	F-Value	P-Value
HOUSING	3	35.12	11.707	1.70	0.193
Error	24	164.82	6.867		
Total	27	199.94			

To determine if there are differences in mean shell thicknesses among the four housing systems, we test:

$H_o: \mu_1 = \mu_2 = \mu_3 = \mu_4$
H_a: At least two means differ

The test statistics is $F = 11.74$ and the p-value is $p = 0.000$. Since the p-value is so small, H_0 is rejected. There is sufficient evidence to indicate there are differences in mean shell thicknesses among the four housing systems for any reasonable value of α.

To determine if there are differences in mean percent overruns among the four housing systems, we test:

$H_o: \mu_1 = \mu_2 = \mu_3 = \mu_4$
H_a: At least two means differ

The test statistics is $F = 31.36$ and the p-value is $p = 0.000$. Since the p-value is so small, H_0 is rejected. There is sufficient evidence to indicate there are differences in mean percent overruns among the four housing systems for any reasonable value of α.

To determine if there are differences in mean penetration strengths among the four housing systems, we test:

$H_o: \mu_1 = \mu_2 = \mu_3 = \mu_4$
H_a: At least two means differ

The test statistics is $F = 1.70$ and the p-value is $p = 0.193$. Since the p-value is not small, H_0 is not rejected. There is insufficient evidence to indicate there are differences in mean penetration strengths among the four housing systems for any $\alpha < 0.193$.

10.131 a. Using MINITAB, the ANOVA table is:

One-way ANOVA: y versus FACES

```
Source  DF     SS    MS     F      P
FACES    5  23.09  4.62  3.96  0.007
Error   30  34.99  1.17
Total   35  58.07

S = 1.080   R-Sq = 39.75%   R-Sq(adj) = 29.71%
```

To determine if differences exist among the mean dominance ratings of the six facial expressions, we test:

$H_0: \mu_1 = \mu_2 = \mu_3 = \mu_4 = \mu_5 = \mu_6$
H_a: At least two treatment means differ

where μ_i represents the mean dominance rating for facial expression *i*.

The test statistic is $F = 3.96$.

The rejection region requires $\alpha = 0.10$ in the upper tail of the F distribution with $\nu_1 = k - 1 = 6 - 1 = 5$ and $\nu_2 = n - k = 36 - 6 = 30$. Using Table V, Appendix B, $F_{0.10} = 2.05$. The rejection region is $F > 2.05$.

Since the observed value of the test statistic falls in the rejection region $(F = 3.96 > 2.05)$, H_0 is rejected. There is sufficient evidence to indicate that differences exist among the mean dominance ratings among the facial expressions at $\alpha = 0.10$.

b. Using MINITAB, the results of Tukey's multiple comparison procedure using $\alpha = 0.05$ is:

```
Tukey 95% Simultaneous Confidence Intervals
All Pairwise Comparisons among Levels of FACES

Individual confidence level = 99.51%


FACES = 1 subtracted from:

FACES    Lower   Center   Upper      +---------+---------+---------+---------
2       -2.144   -0.248   1.647               (---------*--------)
3       -3.487   -1.592   0.304         (--------*---------)
4       -1.731    0.165   2.061                 (---------*--------)
5       -3.794   -1.898  -0.003       (---------*--------)
6       -3.069   -1.173   0.722           (--------*---------)
                                     +---------+---------+---------+---------
                                      -4.0      -2.0       0.0       2.0


FACES = 2 subtracted from:

FACES    Lower   Center  Upper      +---------+---------+---------+---------
3       -3.239   -1.343  0.552          (--------*---------)
4       -1.482    0.413  2.309                   (--------*---------)
5       -3.546   -1.650  0.246        (---------*--------)
6       -2.821   -0.925  0.971            (--------*---------)
                                    +---------+---------+---------+---------
                                  -4.0      -2.0       0.0       2.0


FACES = 3 subtracted from:

FACES    Lower   Center  Upper      +---------+---------+---------+---------
4       -0.139    1.757  3.652                         (---------*--------)
5       -2.202   -0.307  1.589               (--------*---------)
6       -1.477    0.418  2.314                   (--------*---------)
                                    +---------+---------+---------+---------
                                  -4.0      -2.0       0.0       2.0


FACES = 4 subtracted from:

FACES    Lower   Center   Upper      +---------+---------+---------+---------
5       -3.959   -2.063  -0.168      (---------*--------)
6       -3.234   -1.338   0.557          (--------*---------)
                                     +---------+---------+---------+---------
                                   -4.0      -2.0       0.0       2.0


FACES = 5 subtracted from:

FACES    Lower   Center  Upper      +---------+---------+---------+---------
6       -1.171    0.725  2.621                (---------*--------)
                                    +---------+---------+---------+---------
                                  -4.0      -2.0       0.0       2.0
```

A convenient summary of the results is below, where the treatment means are listed from highest to lowest with a solid line connecting those means that are not significantly different.

Means:	1.018	0.853	0.605	-0.320	-0.738	-1.045
Groups:	Happy	Angry	Disgusted	Neutral	Fearful	Sad

The mean dominance rating for the Sad face is significantly lower than the mean dominance ratings for the Happy and Angry faces. There are no other significant differences.

10.133 a. Using MINITAB, the ANOVA results are:

ANOVA: Weight versus Diet, Size

```
Factor  Type    Levels  Values
Diet    fixed        2  1, 2
Size    fixed        2  1, 2


Analysis of Variance for Weight

Source      DF        SS        MS       F      P
Diet         1    0.0124    0.0124    0.22  0.645
Size         1    8.0679    8.0679  141.18  0.000
Diet*Size    1    0.0364    0.0364    0.64  0.432
Error       24    1.3715    0.0571
Total       27    9.4883
```

Source	**df**	**SS**	**MS**	***F***	***P***
Diet	1	0.0124	0.0124	0.22	0.645
Size	1	8.0679	8.0679	141.18	0.000
Interaction	1	0.0364	0.0364	0.64	0.432
Error	24	1.3715	0.0571		
Total	27	9.4883			

b. To determine if Diet and Size interact, we test:

H_0: The factors Diet and Size do not interact
H_a: Diet and Size interact

The test statistic is $F = 0.64$ and *p*-value is $p = 0.432$. Since the *p*-value is so large, H_0 is not rejected. There is insufficient evidence to indicate Diet and Size interact for any reasonable value of α.

To determine if the mean weight differs for the two diets, we test:

$$H_0: \mu_{Regular} = \mu_{Supplement}$$
$$H_a: \mu_{Regular} \neq \mu_{Supplement}$$

The test statistic is $F = 0.22$ and *p*-value is $p = 0.645$. Since the *p*-value is so large, H_0 is not rejected. There is insufficient evidence to indicate the mean weight of kidneys differed for the two diets for any reasonable value of α.

To determine if the mean weight differs between the two sizes, we test:

$$H_0: \mu_{Lean} = \mu_{Obese}$$
$$H_a: \mu_{Lean} \neq \mu_{Obese}$$

The test statistic is $F = 141.18$ and the p-value is $p = 0.000$. Since the p-value is so small, H_0 is rejected. There is sufficient evidence to indicate the mean weight of kidneys differs for the two sizes for any reasonable value of α.

10.135 a. Using MINITAB, the plot is:

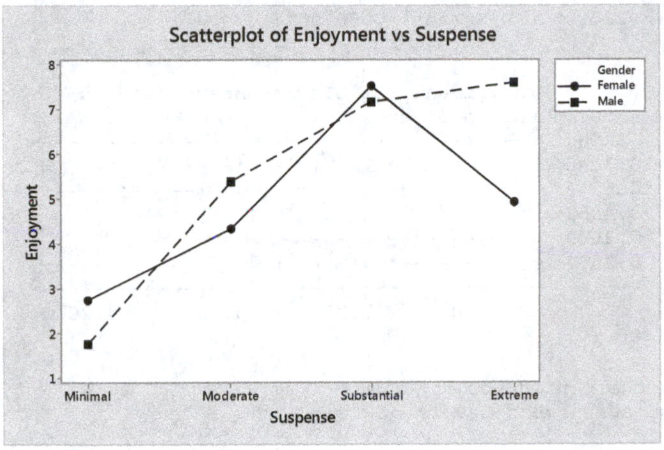

The pattern in the graph suggests interaction between suspense and gender. The difference between the mean ratings of the two genders varies depending on the suspense category. Notice that the lines cross three times.

b. To determine if interaction exists between suspense and gender, we test:

H_0: Gender and Suspense do not interact
H_a: Gender and Suspense interact

The test statistic is $F = 4.42$ and the p-value is $p = 0.007$. Since the p-value is so small, we will reject H_0. There is sufficient evidence to conclude that Gender and Suspense interact to affect the mean rating for any value of $\alpha > 0.007$.

c. Because the factors interact, there is evidence to indicate the difference between the mean enjoyment levels of males and females are different for different suspense levels.

10.137 Using MINITAB, the results of the analysis are:

One-way ANOVA: UMRB-1, UMRB-2, UMBR-3, SWRA, SD

```
Source  DF      SS     MS     F      P
Factor   4   5.836  1.459  7.25  0.001
Error   21   4.225  0.201
Total   25  10.061

S = 0.4486   R-Sq = 58.00%   R-Sq(adj) = 50.00%

                               Individual 95% CIs For Mean Based on
                               Pooled StDev
Level   N    Mean   StDev   -----+---------+---------+---------+----
UMRB-1  7  3.4971  0.3640                  (-----*-----)
UMRB-2  6  4.0167  0.5726                           (-----*-----)
UMRB-3  7  3.7614  0.4832                       (-----*-----)
SWRA    3  2.6433  0.3580   (--------*--------)
SD      3  2.7800  0.2587     (---------*--------)
                            -----+---------+---------+---------+----
                              2.40      3.00      3.60      4.20
Pooled StDev = 0.4486

Tukey 90% Simultaneous Confidence Intervals
All Pairwise Comparisons among Levels of T

Individual confidence level = 98.44%

T = SD subtracted from:

T          Lower   Center   Upper   --------+---------+---------+---------+-
SWRA     -1.1001  -0.1367  0.8267          (--------*-------)
UMRB-1   -0.0971   0.7171  1.5314                (------*------)
UMRB-2    0.4023   1.2367  2.0710                    (------*------)
UMRB-3    0.1672   0.9814  1.7957                  (------*------)
                                    --------+---------+---------+---------+-
                                         -1.2       0.0       1.2       2.4

T = SWRA subtracted from:

T         Lower  Center   Upper   --------+---------+---------+---------+-
UMRB-1  0.0396  0.8538  1.6680                     (------*------)
UMRB-2  0.5390  1.3733  2.2077                         (------*------)
UMRB-3  0.3039  1.1181  1.9323                       (-----*------)
                                  --------+---------+---------+---------+-
                                       -1.2       0.0       1.2       2.4

T = UMRB-1 subtracted from:

T          Lower  Center   Upper  --------+---------+---------+---------+-
UMRB-2   -0.1369  0.5195  1.1760                    (----*-----)
UMRB-3   -0.3664  0.2643  0.8950                  (----*----)
                                  --------+---------+---------+---------+-
                                       -1.2       0.0       1.2       2.4

T = UMRB-2 subtracted from:

T          Lower   Center   Upper  --------+---------+---------+---------+-
UMRB-3   -0.9117  -0.2552  0.4012           (-----*----)
                                   --------+---------+---------+---------+-
                                        -1.2       0.0       1.2       2.4
```

To determine if there are differences among the mean Al/Be ratios for the 5 boreholes, we test:

$H_0: \mu_1 = \mu_2 = \mu_3 = \mu_4 = \mu_5$
H_a: At least 2 means differ

From the printout, the test statistic is $F = 7.25$ and the p-value is $p = 0.001$. Since the p-value is less than $\alpha = 0.10$ $(p = 0.001 < 0.10)$, H_0 is rejected. There is sufficient evidence to indicate a difference in the mean Al/Be ratios among the 5 boreholes at $\alpha = 0.10$.

A convenient summary of the results is below, where the treatment means are listed from highest to lowest with a solid line connecting those means that are not significantly different.

Means:	4.0167	3.7614	3.4971	2.7800	2.6433
Groups:	UMRB-2	UMRB-3	UMRB-1	SD	SWRA

The mean ratio for UMRB-2 is significantly higher than the mean ratio for SD and SWRA. The mean ratio for UMRB-3 is significantly higher than the mean ratio for SD and SWRA. The mean ratio for UMRB-1 is significantly higher than the mean ratio for SWRA. There are no other significant differences. Therefore, the borehole with the highest mean ratio is either UMRB-2 or UMRB-3.

10.139 Using MINITAB, the ANOVA results are:

General Linear Model: Deviation versus Group, Trail

```
Factor     Type Levels Values
Group     fixed      4 F G M N
Trail     fixed      2 C E

Analysis of Variance for Deviation, using Adjusted SS for Tests

Source        DF     Seq SS     Adj SS     Adj MS      F      P
Group          3    16271.2    13000.6     4333.5   5.91  0.001
Trail          1    46445.5    46445.5    46445.5  63.34  0.000
Group*Trail    3     2245.2     2245.2      748.4   1.02  0.386
Error        112    82131.7    82131.7      733.3
Total        119   147093.6
```

First, we must test for treatment effects.

$SST = SS(Group) + SS(Trail) + SS(GxT) = 16,271.2 + 46,445.5 + 2,245.2 = 64,961.9$.

$df = 3 + 1 + 3 = 7$.

$$MST = \frac{SST}{ab-1} = \frac{64,961.9}{4(2)-1} = 9,280.2714 \qquad F = \frac{MST}{MSE} = \frac{9,280.2714}{733.3} = 12.66$$

To determine if there are differences in mean ratings among the 8 treatments, we test:

H_0: All treatment means are the same
H_a: At least two treatment means differ

The test statistic is $F = 12.66$.

Since no α was given, we will use $\alpha = 0.05$. The rejection region requires $\alpha = 0.05$ in the upper tail of the F distribution with $\nu_1 = ab - 1 = 4(2) - 1 = 7$ and $\nu_2 = n - ab = 120 - 4(2) = 112$. Using MINITAB, $F_{0.05} = 2.09$. The rejection region is $F > 2.09$.

Since the observed value of the test statistic falls in the rejection region $(F = 12.66 > 2.09)$, H_0 is rejected. There is sufficient evidence that differences exist among the treatment means at $\alpha = 0.05$. Since differences exist, we now test for the interaction effect between Trail and Group.

To determine if Trail and Group interact, we test:

H_0: Trail and Group do not interact
H_a: Trail and Group do interact

The test statistic is $F = 1.02$ and the p-value is $p = 0.386$.

Since the p-value is greater than α $(p = 0.386 > 0.05)$, H_0 is not rejected. There is insufficient evidence that Trail and Group interact at $\alpha = 0.05$. Since the interaction does not exist, we test for the main effects of Trail and Group.

To determine if there are differences in the mean trail deviations between the two levels of Trail, we test:

$H_0: \mu_1 = \mu_2$
$H_a: \mu_1 \neq \mu_2$

The test statistics is $F = 63.34$ and the p-value is $p = 0.000$.

Since the p-value is less than α $(p = 0.000 < 0.05)$, H_0 is rejected. There is sufficient evidence that the mean trail deviations differ between the fecal extract trail and the control trail at $\alpha = 0.05$.

To determine if there are differences in the mean trail deviations among the four levels of Group, we test:

$H_0: \mu_1 = \mu_2 = \mu_3 = \mu_4$
H_a: At least two means differ

The test statistics is $F = 5.91$ and $p = 0.001$.

Since the *p*-value is less than α $(p = 0.001 < 0.05)$, H_0 is rejected. There is sufficient evidence that the mean trail deviations differ among the four groups at $\alpha = 0.05$.

Using SAS, the multiple comparsions are:

```
             Tukey's Studentized Range (HSD) Test for EXTRACT

NOTE: This test controls the Type I experimentwise error rate, but it
generally has a higher Type II error rate than REGWQ.

                  Alpha                                    0.05
                  Error Degrees of Freedom                  112
                  Error Mean Square                    733.3187
                  Critical Value of Studentized Range   3.68830
                  Minimum Significant Difference         18.235

             Means with the same letter are not significantly different.

               Tukey Grouping          Mean        N    GROUP

                          A          53.747       30    Gravid
                          A
                     B    A          37.077       30    Nymphs
                     B
                     B               35.110       30    Females
                     B
                     B               20.917       30    Males
```

```
             Tukey's Studentized Range (HSD) Test for EXTRACT

NOTE: This test controls the Type I experimentwise error rate, but it
generally has a higher Type II error rate than REGWQ.

                  Alpha                                    0.05
                  Error Degrees of Freedom                  112
                  Error Mean Square                    733.3187
                  Critical Value of Studentized Range   2.80208
                  Minimum Significant Difference          10.39
                  Harmonic Mean of Cell Sizes          53.33333

                         NOTE: Cell sizes are not equal.

             Means with the same letter are not significantly different.

             Tukey Grouping          Mean        N    TRAIL

                          A        64.535       40    Control

                          B        22.801       80    Extract
```

The mean trail deviation for the control group was significantly greater than the mean trail deviation for the Extract group. There was no significant difference in the mean trail deviation between the gravids and the nymphs. There was no significant difference in the mean trail deviation among males, females, and nymphs. However, the mean trail deviation for gravids was significantly greater than the mean trail deviations for males and females.

10.141 Using SAS, the ANOVA results are:

```
                          The SAS System

                       The ANOVA Procedure

Dependent Variable: CORROSION

                                    Sum of
   Source                DF        Squares     Mean Square    F Value    Pr > F

   Model                  5    72.68833333     14.53766667     155.30    <.0001

   Error                  6     0.56166667      0.09361111

   Corrected Total       11    73.25000000

            R-Square     Coeff Var      Root MSE    CORROSION Mean

            0.992332      3.170563      0.305959          9.650000

   Source                DF       Anova SS     Mean Square    F Value    Pr > F

   EXPOSURE               2    63.10500000     31.55250000     337.06    <.0001
   SYSTEM                 3     9.58333333      3.19444444      34.12    0.0004

                       The ANOVA Procedure

             Tukey's Studentized Range (HSD) Test for CORROSION

    NOTE: This test controls the Type I experimentwise error rate, but it generally has
    a higher Type II error rate than REGWQ.

                 Alpha                                      0.05
                 Error Degrees of Freedom                      6
                 Error Mean Square                      0.093611
                 Critical Value of Studentized Range     4.89559
                 Minimum Significant Difference           0.8648

           Means with the same letter are not significantly different.

           Tukey Grouping          Mean      N    SYSTEM

                        A       11.0667      3    3

                        B        9.7333      3    2
                        B
                   C    B        9.0667      3    1
                   C
                   C             8.7333      3    4
```

To determine if there are differences in the epoxy treatment means, we test:

$H_0: \mu_1 = \mu_2 = \mu_3 = \mu_4$
H_a: At least two treatment means differ

From the printout, the test statistic is $F = 34.12$ and the p-value is $p = 0.0004$. Since the p-value is so small, H_0 is rejected. There is sufficient evidence to indicate a difference in the epoxy treatment means among the 4 groups for any reasonable value of α.

From the Tukey multiple comparison procedure, the mean corrosion rate for System 3 is significantly greater than the mean corrosion rate for any of the other 3 Systems. The mean corrosion rate for System 2 is significantly greater than the mean corrosion rate for System 4. No other significant differences exist. The System with the lowest corrosion rate is either System 1 or System 4. There is no significant difference in the mean corrosion rate between these two systems and they are both in the lowest group.

Chapter 11

Simple Linear Regression

11.1 A deterministic model does not allow for random error or variation, whereas a probabilistic model does. An example where a deterministic model would be appropriate is:

Let y = cost of a 2×4 piece of lumber and x = length (in feet)

The model would be $y = \beta_1 x$. There should be no variation in price for the same length of wood.

An example where a probabilistic model would be appropriate is:

Let y = sales per month of a commodity and
x = amount of money spent advertising

The model would be $y = \beta_0 + \beta_1 x + \varepsilon$. The sales per month will probably vary even if the amount of money spent on advertising remains the same.

11.3 The "line of means" is the deterministic component in a probabilistic model.

11.5

a.

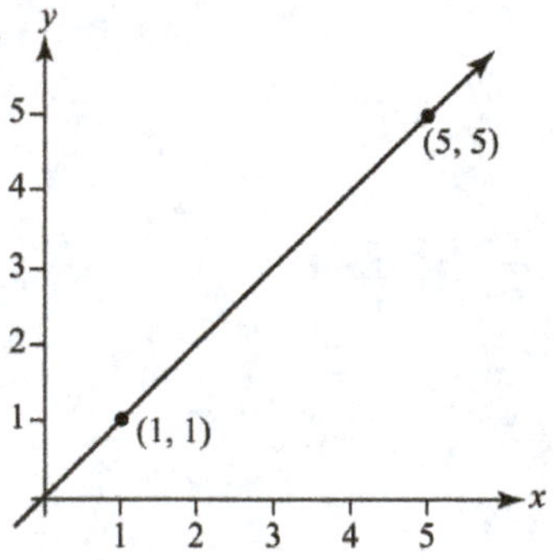

b.

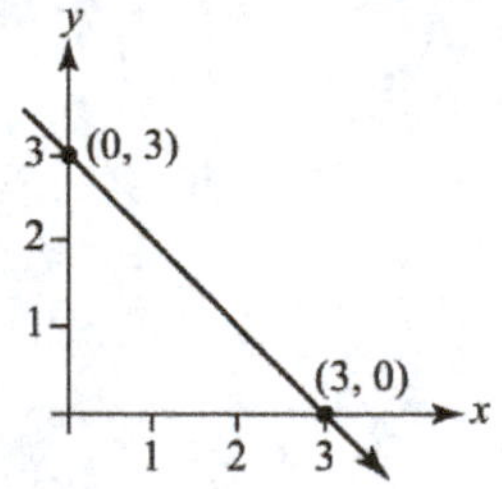

c.

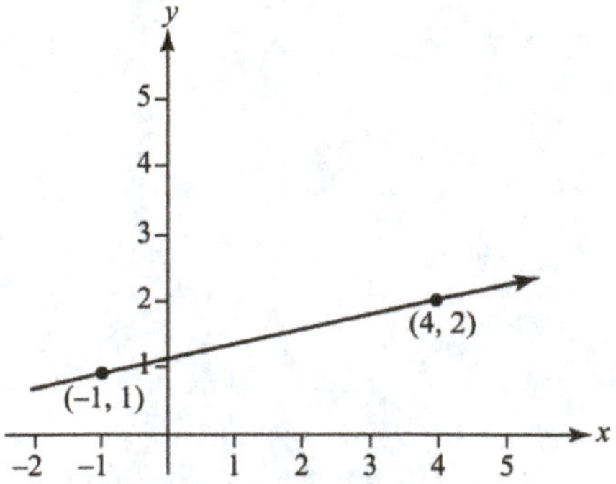

d.

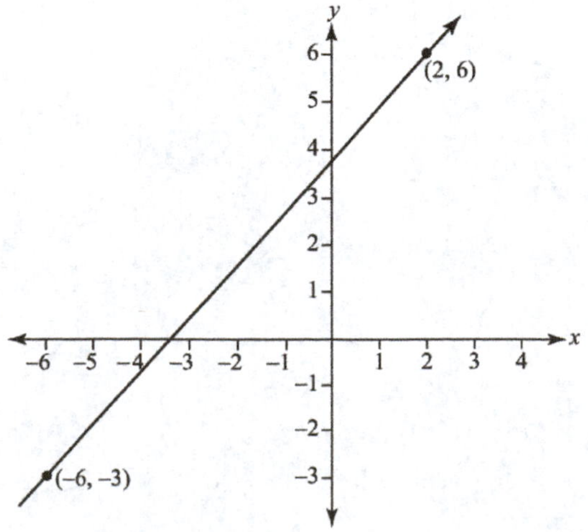

11.7 The two equations are:

$$4 = \beta_0 + \beta_1(-2) \text{ and } 6 = \beta_0 + \beta_1(4)$$

Subtracting the first equation from the second, we get

$$\begin{array}{rl} 6 = & \beta_0 + 4\beta_1 \\ -(4 = & \beta_0 - 2\beta_1) \\ \hline 2 = & 6\beta_1 \Rightarrow \beta_1 = \frac{2}{6} = \frac{1}{3} \end{array}$$

Substituting $\beta_1 = \frac{1}{3}$ into the first equation, we get:

$$4 = \beta_0 + \frac{1}{3}(-2) \Rightarrow \beta_0 = 4 + \frac{2}{3} = \frac{14}{3}$$

The equation for the line is $y = \frac{14}{3} + \frac{1}{3}x$

11.9 To graph a line, we need two points. Pick two values for x, and find the corresponding y values by substituting the values of x into the equation.

a. Let $x = 0 \Rightarrow y = 4 + (0) = 4$
and $x = 2 \Rightarrow y = 4 + (2) = 6$

b. Let $x = 0 \Rightarrow y = 5 - 2(0) = 5$
and $x = 2 \Rightarrow y = 5 - 2(2) = 1$

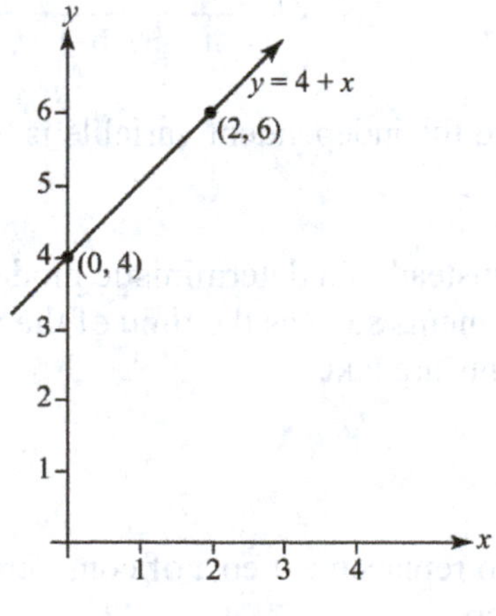

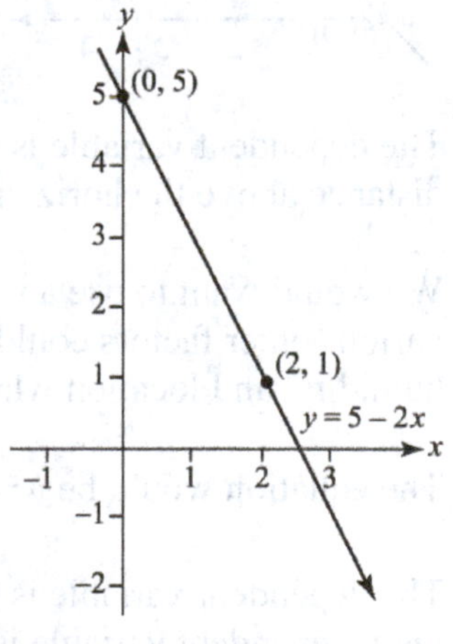

c. Let $x = 0 \Rightarrow y = -4 + 3(0) = -4$
and $x = 2 \Rightarrow y = -4 + 3(2) = 2$

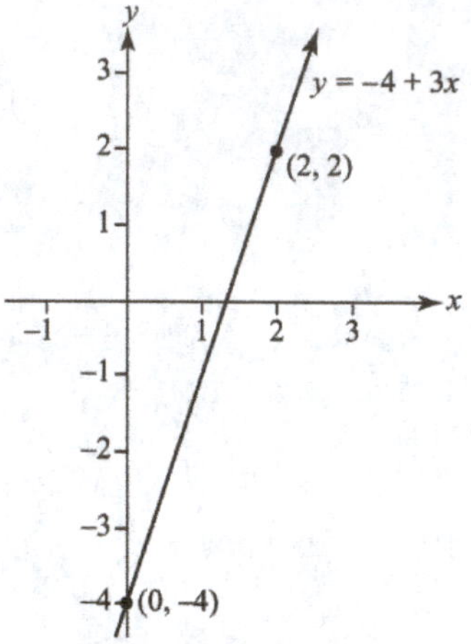

d. Let $x = 0 \Rightarrow y = -2(0) = 0$
and $x = 2 \Rightarrow y = -2(2) = -4$

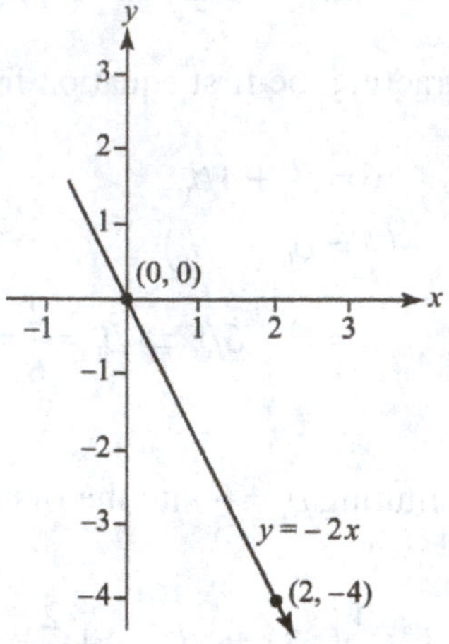

e. Let $x = 0 \Rightarrow y = 0$
and $x = 2 \Rightarrow y = 2$

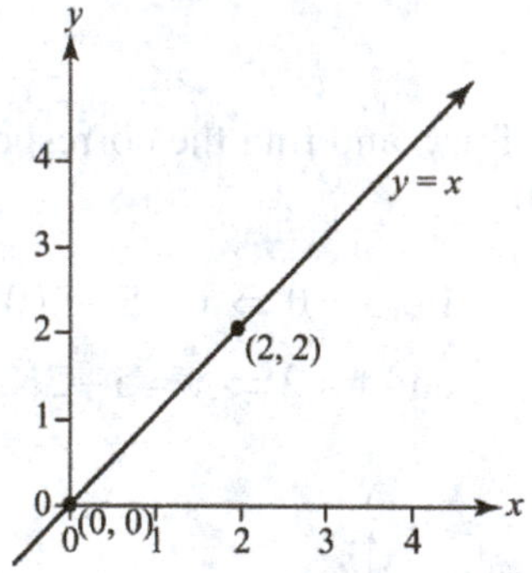

f. Let $x = 0 \Rightarrow y = 0.5 + 1.5(0) = 0.5$
and $x = 2 \Rightarrow y = 0.5 + 1.5(2) = 3.5$

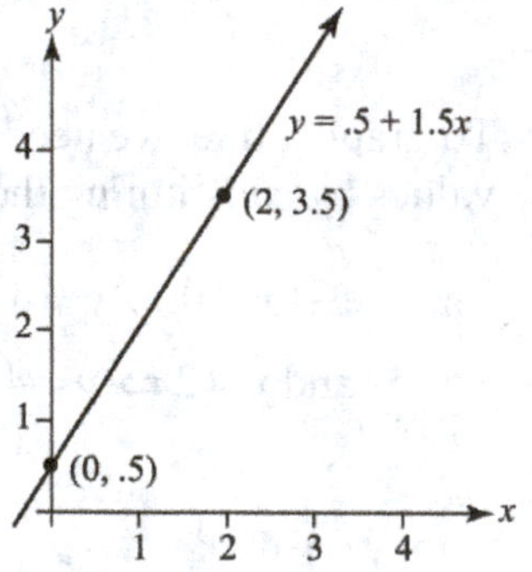

11.11 a. The dependent variable is the Moon's size and the independent variable is the Moon's distance above the horizon.

b. We would want to use a probabilistic model instead of a deterministic model because various other factors could affect the measurements such as the time of the year, humidity, and location where the measurements are taken.

c. The equation would be $y = \beta_0 + \beta_1 x + \varepsilon$.

11.13 a. The dependent variable is the ratio of repair to replacement cost of commercial pipe and the independent variable is the pipe's diameter.

b. We would want to use a probabilistic model instead of a deterministic model because various other factors could affect the measurements location of the pipe within a building and location of the building.

c. The equation would be $y = \beta_0 + \beta_1 x + \varepsilon$.

11.15 In regression, the error of prediction is the difference between the observed and predicted values of the dependent variable.

11.17 The statement "The estimates of β_0 and β_1 should be interpreted only within the sampled range of the independent variable, x." is true. We only have information about the relationship between y and x within the observed range of x.

11.19 From Exercise 11.18, $\hat{\beta}_0 = 7.10$ and $\hat{\beta}_1 = -0.78$.

The fitted line is $\hat{y} = 7.10 - 0.78x$. To obtain values for $\hat{y}$, we substitute values of x into the equation and solve for $\hat{y}$.

a.

x	y	$\hat{y} = 7.10 - 0.78x$	$(y - \hat{y})$	$(y - \hat{y})^2$
7	2	1.64	0.36	0.1296
4	4	3.98	0.02	0.0004
6	2	2.42	−0.42	0.1764
2	5	5.54	−0.54	0.2916
1	7	6.32	0.68	0.4624
1	6	6.32	−0.32	0.1024
3	5	4.76	0.24	0.0576
			$\sum(y - \hat{y}) = 0.02$	$\text{SSE} = \sum(y - \hat{y})^2 = 1.2204$

b.

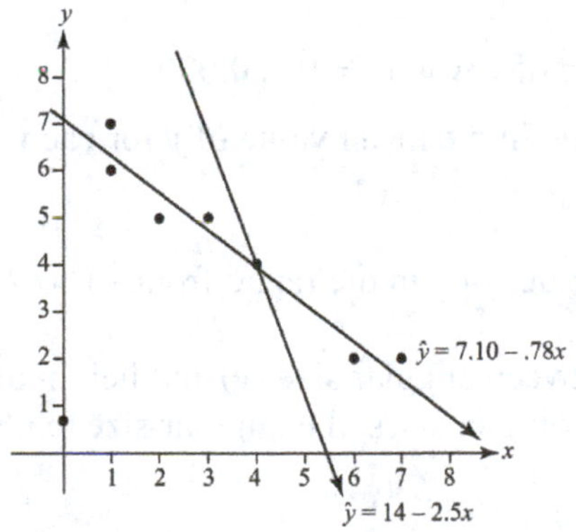

c.

x	y	$\hat{y} = 14 - 2.5x$	$(y - \hat{y})$	$(y - \hat{y})^2$
7	2	−3.5	5.5	30.25
4	4	4	0	0
6	2	−1	3	9
2	5	9	−4	16
1	7	11.5	−4.5	20.25
1	6	11.5	−5.5	30.25
3	5	6.5	−1.5	2.25
			$\sum(y - \hat{y}) = -7$	$\text{SSE} = 108.00$

11.21 a.

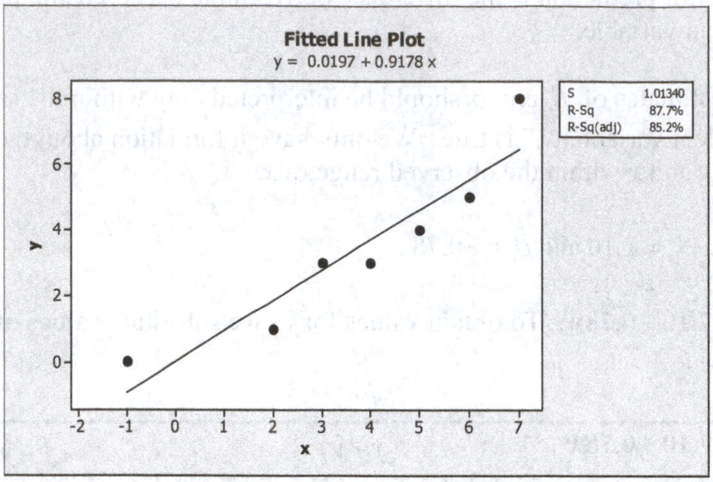

b. As x increases, y tends to increase. Thus, there appears to be a positive, linear relationship between y and x.

c. $$\hat{\beta}_1 = \frac{SS_{xy}}{SS_{xx}} = \frac{39.8571}{43.4286} = 0.9177616 \approx 0.918$$

$$\hat{\beta}_0 = \bar{y} - \hat{\beta}_1\bar{x} = 3.4286 - 0.9177616(3.7143) = 0.0197581 \approx 0.020$$

d. The line appears to fit the data quite well.

e. $\hat{\beta}_0 = 0.020$ The estimated mean value of y when $x = 0$ is 0.020.

$\hat{\beta}_1 = 0.918$ The estimated change in the mean value of y for each unit change in x is 0.918.

These interpretations are valid only for values of x in the range from −1 to 7.

11.23 a. Yes, there is evidence of a linear trend between angular size (y) and height above horizon (x). As the height above the horizon increases, the angular size tends to increase.

b. Answers will vary. Using MINITAB, a possible line is:

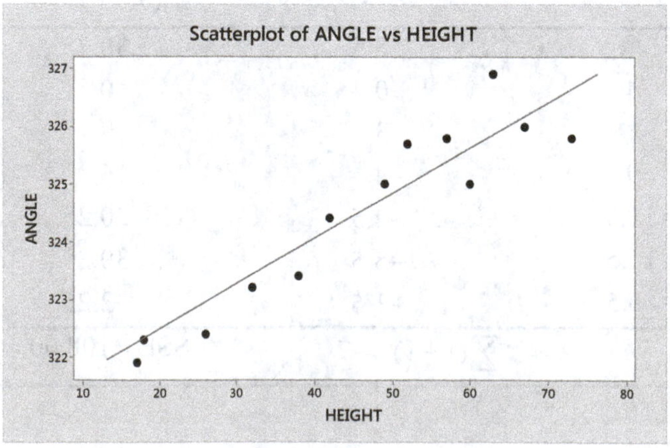

c. Answers will vary. The lines from the points to the line are as follows:

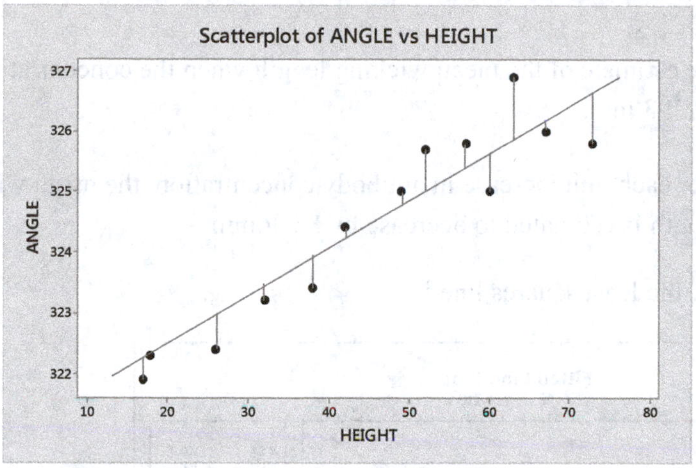

A table with the deviations and squared deviations is:

ANGLE	HEIGHT	$\hat{y}$	$(y-\hat{y})$	$(y-\hat{y})^2$
321.9	17	322.2	-0.3	0.09
322.3	18	322.3	0.0	0.00
322.4	26	323.0	-0.6	0.36
323.2	32	323.4	-0.2	0.04
323.4	38	323.9	-0.5	0.25
324.4	42	324.2	0.2	0.04
325	49	324.8	0.2	0.04
325.7	52	325.0	0.7	0.49
325.8	57	325.4	0.4	0.16
325	60	325.6	-0.6	0.36
326.9	63	325.9	1.0	1.00
326	67	326.2	-0.2	0.04
325.8	73	326.7	-0.9	0.81
			$\sum(y-\hat{y})=-0.8$	$\sum(y-\hat{y})^2=3.68$

d. From the printout, the y-intercept is $\hat{\beta}_0 = 320.63639$ and the slope is $\hat{\beta}_1 = 0.08338$. Answers will vary. From the visual line in part b, the y-intercept is $\hat{\beta}_0 \approx 320.8$ and the slope is $\hat{\beta}_1 \approx 0.077$. The values from the visual line are close to those of the least squares line.

e. From the printout, $\text{SSE} = 3.56465$. From part c, the sum of squares is $\text{SSE} = 3.68$. Although the sum of squares for the visual line is close to that of the least squares line, the sum of squares for the least squares is smaller than that for the visual line.

11.25 a. The equation for the straight-line model is $y = \beta_0 + \beta_1 x + \varepsilon$.

b. The least squares line is $\hat{y} = 19.393 - 8.036x$.

c. $\hat{\beta}_0 = 19.393$. The estimate of the mean wicking length when the concentration is 0 is 19.393 mm.

$\hat{\beta}_1 = -8.036$. For each unit increase in antibody concentration, the mean wicking length is estimated to decrease by 8.036mm.

11.27 a. Using MINITAB, the least squares line is:

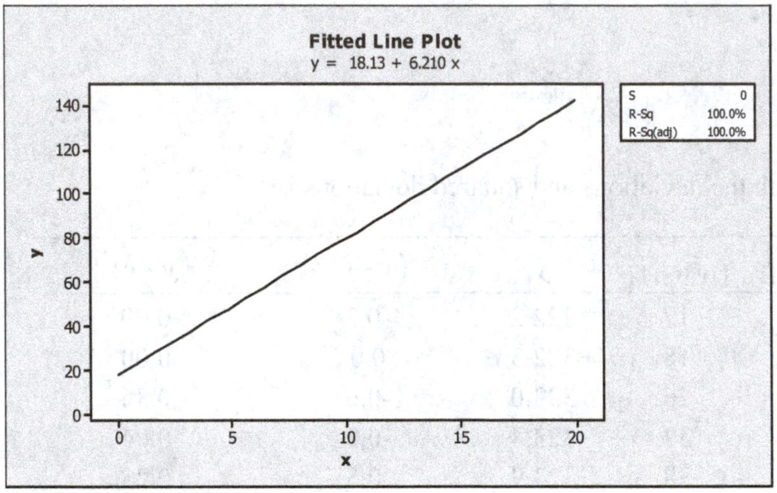

The slope of the line is positive – as x increases, y also increases.

b. $\hat{\beta}_0 = 18.13$. Since we have no information as to whether $x = 0$ is within the observed range, we will assume that it is. Then, the estimate of the mean magnitude when redshift level is 0 is 18.13.

c. $\hat{\beta}_1 = 6.21$. For each unit increase in redshift level, the mean value of magnitude is estimated to increase by 6.21.

11.29 a. The straight-line model is $y = \beta_0 + \beta_1 x + \varepsilon$.

b. Using MINITAB, the results are:

Regression Analysis: Accuracy versus Distance

```
The regression equation is
Accuracy = 250 - 0.629 Distance

Predictor      Coef   SE Coef       T      P
Constant     250.14     14.23   17.58  0.000
Distance   -0.62944   0.04759  -13.23  0.000


S = 2.23639   R-Sq = 82.2%   R-Sq(adj) = 81.7%
```

```
Analysis of Variance

Source           DF       SS       MS       F      P
Regression        1   874.99   874.99  174.95  0.000
Residual Error   38   190.06     5.00
Total            39  1065.04
```

The least squares prediction equation is: $\hat{y} = 250.14 - .6294x$

c. The estimated y-intercept is $\hat{\beta}_o = 250.14$. Since 0 is not in the observed range for values of driving distance, the y-intercept has no meaning.

d. The estimated slope of the line is $\hat{\beta}_1 = -.6294$. For each additional yard of driving distance, the mean driving accuracy value will decrease by an estimated .6294.

e. The slope will help determine if the golfer's concern is valid. If there is no significant linear relationship between driving accuracy and driving distance, then the golfer would have nothing to worry about. However, if there is a significant linear relationship between driving accuracy and driving distance and the relationship is negative, then as the driving distance increases, the mean driving accuracy will decrease.

11.31 a. The hypothesized model would be $y = \beta_0 + \beta_1 x + \varepsilon$.

b. The slope will be positive because the researcher's theory indicates a linearly increasing relationship.

c. Some preliminary calculations are:

$$\sum x_i = 300 \qquad \sum y_i = 98,494 \qquad \sum x_i y_i = 1,473,555$$
$$\sum x_i^2 = 4900 \qquad \sum y_i^2 = 456,565,950$$

$$SS_{xy} = \sum x_i y_i - \frac{\sum x_i \sum y_i}{n} = 1,473,555 - \frac{300(98,494)}{24} = 242,380$$

$$SS_{xx} = \sum x_i^2 - \frac{\left(\sum x_i\right)^2}{n} = 4900 - \frac{300^2}{24} = 1150$$

$$\hat{\beta}_1 = \frac{SS_{xy}}{SS_{xx}} = \frac{242,380}{1150} = 210.7652174$$

For each additional resonance, the mean sound wave frequency is estimated to increase by 210.765.

$$\hat{\beta}_0 = \bar{y} - \hat{\beta}_1 \bar{x} = \frac{98,494}{24} - (210.7652174)\left(\frac{300}{24}\right) = 1469.351449$$

Since 0 is not in the observed range of resonances, $\hat{\beta}_0$ has no interpretation other than being the y-intercept.

11.33 a. Using MINITAB, the results of fitting a regression model are:

Regression Analysis: IdealHt versus Height

```
The regression equation is
IdealHt = 86.0 - 0.260 Height


Predictor        Coef  SE Coef      T      P
Constant       86.018    4.786  17.97  0.000
Height       -0.26046  0.07035  -3.70  0.000


S = 3.64336   R-Sq = 8.6%   R-Sq(adj) = 8.0%


Analysis of Variance

Source           DF       SS      MS      F      P
Regression        1   181.93  181.93  13.71  0.000
Residual Error  145  1924.74   13.27
Total           146  2106.67
```

The fitted regression line is $\hat{y} = 86.0 - .260x$.

Because the parameter estimate for Height is negative (-.26046), this confirms what the researchers found.

b. Yes, this is the correct interpretation.

c. Using MINITAB, a plot of the data is:

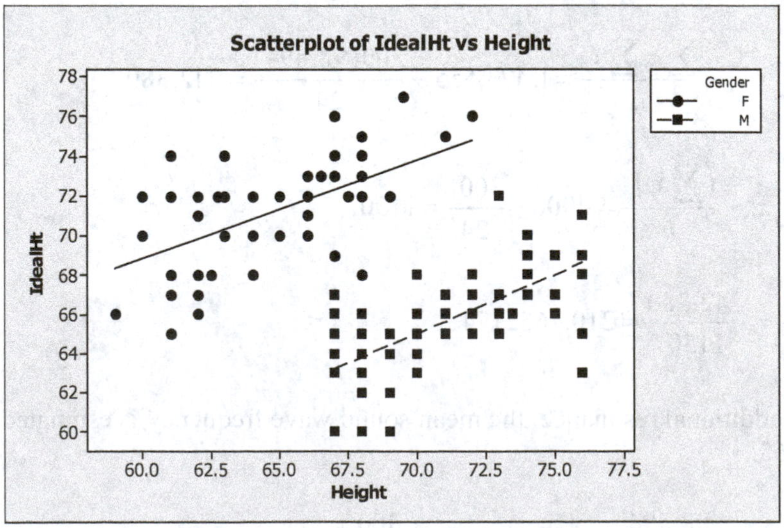

For the female students (round dots), there appears to be a positive relationship between the student's height and her ideal height. This same trend appears for the male students (squares).

d. Using MINITAB, the results of fitting the female data are:

Regression Analysis: F-IdealHt versus F-Height

```
The regression equation is
F-IdealHt = 39.3 + 0.493 F-Height


Predictor      Coef  SE Coef     T      P
Constant     39.304    6.206  6.33  0.000
F-Height    0.49261  0.09602  5.13  0.000


S = 2.32153   R-Sq = 27.3%   R-Sq(adj) = 26.3%


Analysis of Variance

Source          DF      SS      MS      F      P
Regression       1  141.84  141.84  26.32  0.000
Residual Error  70  377.27    5.39
Total           71  519.11
```

The fitted regression line is $\hat{y} = 39.3 + .493x$.

The estimated slope of the line is 0.493. For each additional inch of height for females, the mean ideal partner's height is estimated to increase by 0.493 inches.

e. Using MINITAB, the results of fitting the male data are:

Regression Analysis: M-IdealHt versus M-Height

```
The regression equation is
M-IdealHt = 23.3 + 0.596 M-Height

Predictor      Coef  SE Coef     T      P
Constant     23.274    6.333  3.67  0.000
M-Height    0.59630  0.08902  6.70  0.000


S = 2.05688   R-Sq = 38.1%   R-Sq(adj) = 37.2%


Analysis of Variance

Source          DF      SS      MS      F      P
Regression       1  189.82  189.82  44.87  0.000
Residual Error  73  308.84    4.23
Total           74  498.67
```

The fitted regression line is $\hat{y} = 23.3 + .596x$.

The estimated slope of the line is 0.596. For each additional inch of height for males, the mean ideal partner's height is estimated to increase by 0.596 inches.

f. By fitting separate lines for the males and females, the results now match the theory developed by the anthropologists.

11.35 Using MINITAB, the results are:

Regression Analysis: Mass versus Time

```
The regression equation is
Mass = 5.22 - 0.114 Time


Predictor         Coef   SE Coef       T      P
Constant        5.2207    0.2960   17.64  0.000
Time          -0.11402   0.01032  -11.05  0.000


S = 0.857257   R-Sq = 85.3%   R-Sq(adj) = 84.6%


Analysis of Variance

Source          DF       SS      MS       F      P
Regression       1   89.794  89.794  122.19  0.000
Residual Error  21   15.433   0.735
Total           22  105.227
```

The fitted regression line is $\hat{y} = 5.2207 - 0.11402x$. Since the estimate of the slope is negative, the data indicate that the mass of the spill tends to diminish as time increases. For each additional minute, the mean mass is estimated to decrease by 0.11402 pounds.

11.37 The four assumptions made about the probability distribution of ε in regression are:

1. The mean of the probability distribution of ε is 0.
2. The variance of the probability distribution of ε is constant for all settings of the independent variable x.
3. The probability distribution of ε is normal.
4. The values of ε associated with any two observed values of y are independent.

See Figure 11.9 for an illustration of these assumptions.

11.39 a. $\text{SSE} = \text{SS}_{yy} - \hat{\beta}_1 \text{SS}_{xy} = 95 - 0.75(50) = 57.5$ $\qquad s^2 = \frac{\text{SSE}}{n-2} = \frac{57.5}{20-2} = 3.19444$

b. $\text{SS}_{yy} = \sum y^2 - \frac{\left(\sum y\right)^2}{n} = 860 - \frac{50^2}{40} = 797.5$

$\text{SSE} = \text{SS}_{yy} - \hat{\beta}_1 \text{SS}_{xy} = 797.5 - 0.2(2700) = 257.5$

$s^2 = \frac{\text{SSE}}{n-2} = \frac{257.5}{40-2} = 6.776315789 \approx 6.7763$

c. $SS_{yy} = \sum(y_i - \bar{y})^2 = 58$ $\hat{\beta}_1 = \frac{SS_{xy}}{SS_{xx}} = \frac{91}{170} = 0.535294117$

$$SSE = SS_{yy} - \hat{\beta}_1 SS_{xy} = 58 - 0.535294117(91) = 9.2882353 \approx 9.288$$

$$s^2 = \frac{SSE}{n-2} = \frac{9.2882353}{10-2} = 1.161029413 \approx 1.1610$$

11.41 $SSE = SS_{yy} - \hat{\beta}_1 SS_{xy}$ where $SS_{yy} = \sum y_i^2 - \frac{\left(\sum y_i\right)^2}{n}$

From Exercise 11.18, $\sum y_i^2 = 159$, $\sum y_i = 31$,

$$SS_{yy} = \sum y^2 - \frac{\left(\sum y\right)^2}{n} = 159 - \frac{31^2}{7} = 159 - 137.2857143 = 21.7142857,$$

$SS_{xy} = -26.2857143$, $\hat{\beta}_1 = -0.779661017$

Therefore, $SSE = 21.7142857 - (-0.779661017)(-26.2857143) = 1.22033896 \approx 1.2203$

$$s^2 = \frac{SSE}{n-2} = \frac{1.22033896}{7-2} = 0.244067792, \qquad s = \sqrt{0.244067792} = 0.4940$$

We would expect most of the observations to fall within $2s = 2(0.4940) = 0.988$ units of the least squares prediction line.

For Exercise 11.21, $\sum x_i = 26$, $\sum y_i = 24$, $\sum x_i y_i = 129$, $\sum x_i^2 = 140$, $\sum y_i^2 = 124$,

$$SS_{xy} = \sum x_i y_i - \frac{\left(\sum x_i \sum y_i\right)}{n} = 129 - \frac{(26)(24)}{7} = 129 - 89.14285714 = 39.85714286$$

$$SS_{xx} = \sum x_i^2 - \frac{\left(\sum x_i\right)^2}{n} = 140 - \frac{(26)^2}{7} = 140 - 96.57142857 = 43.42857143$$

$$SS_{yy} = \sum y_i^2 - \frac{\left(\sum y_i\right)^2}{n} = 124 - \frac{(24)^2}{7} = 124 - 82.28571429 = 41.71428571$$

$$\hat{\beta}_1 = \frac{SS_{xy}}{SS_{xx}} = \frac{39.85714286}{43.42857143} = 0.917763157$$

$$SSE = SS_{yy} - \hat{\beta}_1 SS_{xy} = 41.71428571 - (0.917763157)(39.85714286)$$
$$= 41.71428571 - 36.57941726 = 5.13486841 \approx 5.1349$$

$$s^2 = \frac{\text{SSE}}{n-2} = \frac{5.13486841}{7-2} = 1.026973682 \qquad s = \sqrt{1.026973682} = 1.0134$$

We would expect most of the observations to fall within $2s = 2(1.0134) = 2.0268$ units of the least squares prediction line.

11.43 From the printout, the estimate of σ is $s = 0.56926$. We would expect most of the observations to fall within $2s = 2(0.56926) = 1.13852$ units of the least squares prediction line.

11.45 a. From the printout:

$\text{SSE} = 22.268$, $s^2 = \text{MSE} = 5.567$, and $s = \sqrt{5.567} = 2.3594$.

b. We would expect most of the observations to fall within $2s = 2(2.3594) = 4.7188$ mm of the least squares prediction line.

11.47 a. From the printout, $\text{SSE} = 2760$, $s^2 = 307.0$, and $s = 17.5111$.

b. We would expect most of the observations to be within $2s = 2(17.5111) = 35.0222$ of the least squares line.

11.49 a. From Exercise 11.30,

$$\sum y_i = 3,781.1 \qquad \sum y_i^2 = 651,612.45 \qquad \text{SS}_{xy} = -3,882.3686 \qquad \hat{\beta}_1 = -0.305444503$$

$$\text{SS}_{yy} = \sum y_i^2 - \frac{\left(\sum y_i\right)^2}{n} = 651,612.45 - \frac{3,781.1^2}{22} = 1,761.6677$$

$$\text{SSE} = \text{SS}_{yy} - \hat{\beta}_1\text{SS}_{xy} = 1,761.6677 - (-0.305444503)(-3,882.3686) = 575.8195525$$

$$s^2 = \frac{\text{SSE}}{n-2} = \frac{575.8195525}{22-2} = 28.79097763 \qquad s = \sqrt{s^2} = \sqrt{28.79099763} = 5.36572247$$

We would expect most of the observations to be within $2s = 2(5.3657) = 10.7314$ units of the least squares line.

b. From Exercise 11.30,

$$\sum y_i = 3,764.2 \qquad \sum y_i^2 = 645,221.16 \qquad \text{SS}_{xy} = -3,442.16 \qquad \hat{\beta}_1 = -0.270811187$$

$$\text{SS}_{yy} = \sum y_i^2 - \frac{\left(\sum y_i\right)^2}{n} = 645,221.16 - \frac{3,764.2^2}{22} = 1,166.54$$

$$\text{SSE} = \text{SS}_{yy} - \hat{\beta}_1\text{SS}_{xy} = 1,166.54 - (-0.270811187)(-3,442.16) = 234.3645646$$

$$s^2 = \frac{\text{SSE}}{n-2} = \frac{234.3645646}{22-2} = 11.71822823 \qquad s = \sqrt{s^2} = \sqrt{11.71822823} = 3.423189774$$

We would expect most of the observations to be within $2s = 2(3.4232) = 6.8464$ units of the least squares line.

c. Since the standard deviation $(s = 3.423)$ for the reading scores is smaller than the standard deviation for the math scores $(s = 5.366)$, the reading score can be more accurately predicted than the math score.

11.51 a. Using MINITAB, the results of fitting the male data are:

Regression Analysis: M-IdealHt versus M-Height

```
The regression equation is
M-IdealHt = 23.3 + 0.596 M-Height

Predictor      Coef  SE Coef     T      P
Constant     23.274    6.333  3.67  0.000
M-Height    0.59630  0.08902  6.70  0.000


S = 2.05688   R-Sq = 38.1%   R-Sq(adj) = 37.2%

Analysis of Variance

Source          DF      SS      MS      F      P
Regression       1  189.82  189.82  44.87  0.000
Residual Error  73  308.84    4.23
Total           74  498.67
```

The fitted regression line is $\hat{y} = 23.3 + .596x$.

The estimate of σ is $s = 2.05688$.

b. Using MINITAB, the results of fitting the female data are:

Regression Analysis: F-IdealHt versus F-Height

```
The regression equation is
F-IdealHt = 39.3 + 0.493 F-Height


Predictor      Coef  SE Coef     T      P
Constant     39.304    6.206  6.33  0.000
F-Height    0.49261  0.09602  5.13  0.000
S = 2.32153   R-Sq = 27.3%   R-Sq(adj) = 26.3%


Analysis of Variance

Source          DF      SS      MS      F      P
Regression       1  141.84  141.84  26.32  0.000
Residual Error  70  377.27    5.39
Total           71  519.11
```

The fitted regression line is $\hat{y} = 39.3 + .493x$.

The estimate of σ is $s = 2.32153$.

c. The student's height is a more accurate predictor of ideal partner's height for the males than for the females because the estimate of the standard deviation for the males is smaller than the estimate for the females.

11.53 In the equation, $E(y) = \beta_0 + \beta_1 x$, the value of β_1 is 0 if x has no linear relationship with y.

11.55 If you are running a one-tailed test of β_1 in simple linear regression, you divide the p-value from the computer printout by 2 to obtain the correct p-value of the one-tailed test.

11.57 a. For confidence coefficient 0.95, $\alpha = 0.05$ and $\alpha / 2 = 0.05 / 2 = 0.025$. From Table III, Appendix B, with $df = n - 2 = 12 - 2 = 10$, $t_{0.025} = 2.228$.

The 95% confidence interval for β_1 is:

$$\hat{\beta}_1 \pm t_{0.025} s_{\hat{\beta}_1} \Rightarrow 31 \pm 2.228(0.5071) \Rightarrow 31 \pm 1.13 \Rightarrow (29.87, 32.13)$$

$$\text{where } s_{\hat{\beta}_1} = \frac{s}{\sqrt{\text{SS}_{xx}}} = \frac{3}{\sqrt{35}} = 0.5071$$

For confidence coefficient 0.90, $\alpha = 0.10$ and $\alpha / 2 = 0.10 / 2 = 0.05$. From Table III, Appendix B, with $df = n - 2 = 12 - 2 = 10$, $t_{0.05} = 1.812$.

The 90% confidence interval for β_1 is:

$$\hat{\beta}_1 \pm t_{0.05} s_{\hat{\beta}_1} \Rightarrow 31 \pm 1.812(0.5071) \Rightarrow 31 \pm 0.92 \Rightarrow (30.08, 31.92)$$

b. $$s^2 = \frac{\text{SSE}}{n-2} = \frac{1960}{18-2} = 122.5, \qquad s = \sqrt{s^2} = 11.0680$$

For confidence coefficient, 0.95, $\alpha = 0.05$ and $\alpha / 2 = 0.05 / 2 = 0.025$. From Table III, Appendix B, with $df = n - 2 = 18 - 2 = 16$, $t_{0.025} = 2.120$. The 95% confidence interval for β_1 is:

$$\hat{\beta}_1 \pm t_{0.025} s_{\hat{\beta}_1} \Rightarrow 64 \pm 2.120(2.0207) \Rightarrow 64 \pm 4.28 \Rightarrow (59.72, 68.28)$$

$$\text{where } s_{\hat{\beta}_1} = \frac{s}{\sqrt{\text{SS}_{xx}}} = \frac{11.0680}{\sqrt{30}} = 2.0207$$

For confidence coefficient 0.90, $\alpha = 0.10$ and $\alpha / 2 = 0.10 / 2 = 0.05$. From Table III, Appendix B, with $df = n - 2 = 18 - 2 = 16$, $t_{0.05} = 1.746$.

The 90% confidence interval for β_1 is:

$$\hat{\beta}_1 \pm t_{0.05} s_{\hat{\beta}_1} \Rightarrow 64 \pm 1.746(2.0207) \Rightarrow 64 \pm 3.53 \Rightarrow (60.47, 67.53)$$

c. $s^2 = \frac{\text{SSE}}{n-2} = \frac{146}{24-2} = 6.6364$, $\quad s = \sqrt{s^2} = 2.5761$

For confidence coefficient 0.95, $\alpha = 0.05$ and $\alpha/2 = 0.05/2 = 0.025$. From Table III, Appendix B, with $df = n-2 = 24-2 = 22$, $t_{0.025} = 2.074$ The 95% confidence interval for β_1 is:

$$\hat{\beta}_1 \pm t_{0.025} s_{\hat{\beta}_1} \Rightarrow -8.4 \pm 2.074(0.322) \Rightarrow -8.4 \pm 0.67 \Rightarrow (-9.07, -7.73)$$

where $s_{\hat{\beta}_1} = \frac{s}{\sqrt{\text{SS}_{xx}}} = \frac{2.5761}{\sqrt{64}} = 0.3220$

For confidence coefficient 0.90, $\alpha = 0.10$ and $\alpha/2 = 0.10/2 = 0.05$. From Table III, Appendix B, with $df = n-2 = 24-2 = 22$, $t_{0.05} = 1.717$.

The 90% confidence interval for β_1 is:

$$\hat{\beta}_1 \pm t_{0.05} s_{\hat{\beta}_1} \Rightarrow -8.4 \pm 1.717(0.322) \Rightarrow -8.4 \pm 0.55 \Rightarrow (-8.95, -7.85)$$

11.59 a & c. Using MINITAB, a scatterplot of the data is:

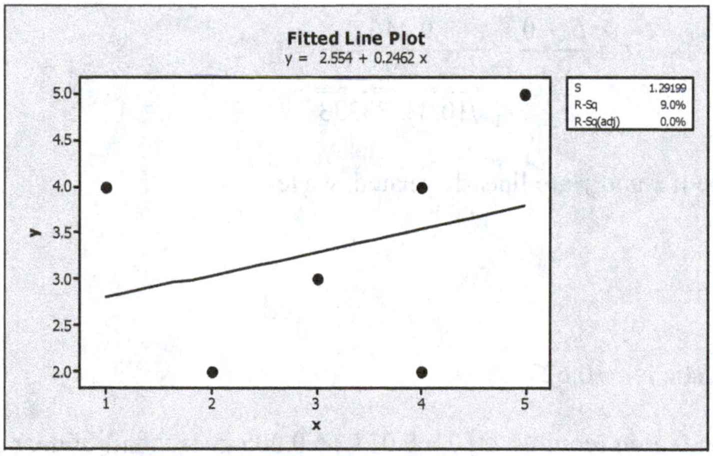

b. Some preliminary calculations are:

$\sum x_i = 19$ $\quad \sum y_i = 20$ $\quad \sum x_i y_i = 66$

$\sum x_i^2 = 71$ $\quad \sum y_i^2 = 74$

$$SS_{xy} = \sum x_i y_i - \frac{\sum x_i \sum y_i}{n} = 66 - \frac{19(20)}{6} = 2.666666667$$

$$SS_{xx} = \sum x_i^2 - \frac{\left(\sum x_i\right)^2}{n} = 71 - \frac{19^2}{6} = 10.833333333$$

$$\hat{\beta}_1 = \frac{SS_{xy}}{SS_{xx}} = \frac{2.66666667}{10.833333333} = 0.246153846$$

$$\hat{\beta}_0 = \bar{y} - \hat{\beta}_1 \bar{x} = \frac{20}{6} - (0.246153846)\left(\frac{19}{6}\right) = 2.553846155$$

The least squares prediction equation is $\hat{y} = 2.554 + 0.246x$

d. Some preliminary calculations are:

$$SS_{yy} = \sum y_i^2 - \frac{\left(\sum y_i\right)^2}{n} = 74 - \frac{20^2}{6} = 7.3333333333$$

$$SSE = SS_{yy} - \hat{\beta}_1 SS_{xy} = 7.3333333333 - (0.246153846)(2.666666667) = 6.6769$$

$$s^2 = \frac{SSE}{n-2} = \frac{6.6769}{6-2} = 1.6692 \qquad s = \sqrt{s^2} = \sqrt{1.6692} = 1.292$$

The test statistic is $t = \dfrac{\hat{\beta}_1 - 0}{s_{\hat{\beta}_1}} = \dfrac{0.246}{\dfrac{1.292}{\sqrt{10.83333333}}} = 0.627$

e. To determine if x and y are linearly related, we test:

$H_0: \beta_1 = 0$
$H_a: \beta_1 \neq 0$

The test statistic is $t = 0.627$.

The rejection region requires $\alpha / 2 = 0.01 / 2 = 0.005$ in each tail of the t distribution with $df = n - 2 = 6 - 2 = 4$. From Table III, Appendix B, $df = n - 2 = 6 - 2 = 4$. The rejection region is $t < -4.604$ or $t > 4.604$.

Since the observed value of the test statistic does not fall in the rejection region $(t = 0.627 \not> 4.604)$, H_0 is not rejected. There is insufficient evidence to indicate that x and y are linearly related at $\alpha = 0.01$.

f. For confidence coefficient 0.99, $\alpha = 0.01$ and $\alpha / 2 = 0.01/2 = 0.005$. From Table III, Appendix B, with $df = n-2 = 6-2 = 4$, $df = n-2 = 6-2 = 4$. The 99% confidence interval is:

$$\hat{\beta}_1 \pm t_{0.005} s_{\hat{\beta}_1} \Rightarrow \hat{\beta}_1 \pm t_{0.005} \frac{s}{\sqrt{SS_{xx}}} \Rightarrow 0.246 \pm 4.604 \frac{1.292}{\sqrt{10.8333333}}$$

$$\Rightarrow .246 \pm 1.807 \Rightarrow (-1.561,\ 2.053)$$

We are 99% confident that the true value of β_1 is between −1.561 and 2.053.

11.61 To determine if the engineers are able to conclude that the cost ratio increases linearly with pipe diameter, we test:

$$H_0: \beta_1 = 0$$
$$H_a: \beta_1 > 0$$

From the printout, the test statistic is $t = 14.87$.

The rejection region requires $\alpha = 0.05$ in the upper tail of the t distribution. From Table III, Appendix B, with $df = n-2 = 12-2 = 10$, $t_{0.05} = 1.812$. The rejection region is $t > 1.812$.

Since the observed test statistic falls in the rejection region $(t = 14.87 > 1.812)$, H_0 is rejected. There is sufficient evidence to indicate the engineers are able to conclude that the cost ratio increases linearly with pipe diameter at $\alpha = 0.05$.

For confidence coefficient 0.95, $\alpha = 0.05$ and $\alpha / 2 = 0.05/2 = 0.025$. From Table III, Appendix B, with $df = n-2 = 12-2 = 10$, $t_{0.025} = 2.228$. The 95% confidence interval is:

$$\hat{\beta}_1 \pm t_{0.025} s_{\hat{\beta}_1} \Rightarrow 0.00479 \pm 2.228(0.00032) \Rightarrow 0.00479 \pm 0.00071 \Rightarrow (0.00408,\ 0.00550)$$

We are 95% confident that the true value of β_1 is between 0.00408 and 0.00550.

11.63 a. Using MINITAB, a scatterplot of the data is:

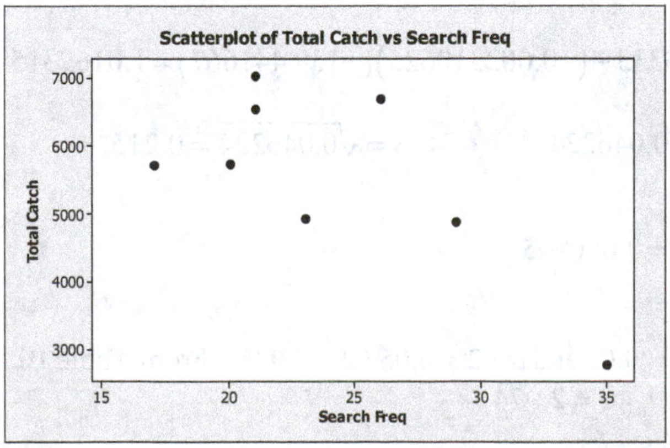

There appears to be a negative linear trend – as search frequency increases, the total catch tends to decrease.

b. From the printout, the least squares prediction equation is: $\hat{y} = 9{,}658.24 - 171.573x$. For each additional unit increase in search frequency, the mean total catch is estimated to decrease by 171.573 kilograms.

c. To determine if the total catch is negatively linearly related to search frequency, we test:

$$H_0 : \beta_1 = 0$$
$$H_a : \beta_1 < 0$$

d. From the printout, the *p*-value is $p = 0.0402/2 = 0.0201$.

e. Since the *p*-value is less than $\alpha\left(p = 0.0201 < 0.05\right)$, H_0 is rejected. There is sufficient evidence to indicate that the total catch is negatively linearly related to search frequency at $\alpha = 0.05$.

11.65 a. To determine whether driving accuracy decreases linearly as driving distance increases, we test:

$$H_0 : \beta_1 = 0$$
$$H_a : \beta_1 < 0$$

b. From the results in Exercise 11.29, the test statistic is $t = -13.23$ and the *p*-value is $p = 0.000$.

c. Since the *p*-value is less than $\alpha\left(p = 0.000 < 0.01\right)$, H_0 is rejected. There is sufficient evidence to indicate driving accuracy decreases linearly as driving distance increases at $\alpha = 0.01$.

11.67 Some preliminary calculations are:

$$\text{SS}_{yy} = \sum y_i^2 - \frac{\left(\sum y_i^{\ 2}\right)}{n} = 769.72 - \frac{(135.8)^2}{24} = 1.3183333$$

$$\text{SSE} = \text{SS}_{yy} - \hat{\beta}_1 \text{SS}_{xy} = 1.3183333 - (-0.002310625)(-130.441667) = 1.016931523$$

$$s^2 = \frac{\text{SSE}}{n-2} = \frac{1.016931523}{22} = 0.046224 \qquad s = \sqrt{0.046224} = 0.215$$

$$s_{\hat{\beta}_1} = \sqrt{\frac{s^2}{\text{SS}_{xx}}} = \sqrt{\frac{0.046224}{56{,}452.958}} = 0.000905$$

For confidence level 0.95, $\alpha = 0.05$ and $\alpha/2 = 0.05/2 = 0.025$. From Table III, Appendix B with $df = n - 2 = 24 - 2 = 22$, $t_{0.025} = 2.074$.

The confidence interval is:

$$\hat{\beta}_1 \pm t_{0.025} s_{\hat{\beta}_1} \Rightarrow -0.0023 \pm 2.074(0.000905) \Rightarrow -0.0023 \pm 0.0019 \Rightarrow (-0.0042, -0.0004)$$

We are 95% confident that the change in the mean sweetness index for each one unit change in the pectin is between −0.0042 and −0.0004.

11.69 a. The model relating Anthropogenic Index, y, to Natural Origin Index, x, is:

$$y = \beta_0 + \beta_1 x + \varepsilon$$

b. Some preliminary calculations are:

$\sum x_i = 1{,}047.22$ $\sum y_i = 1{,}283.46$ $\sum x_i y_i = 32{,}048.2$

$\sum x_i^2 = 24{,}607.5$ $\sum y_i^2 = 54{,}184.8$

$$\text{SS}_{xy} = \sum x_i y_i - \frac{\sum x_i \sum y_i}{n} = 32{,}048.2 - \frac{1{,}047.22(1{,}283.46)}{54} = 7{,}158.10776$$

$$\text{SS}_{xx} = \sum x_i^2 - \frac{\left(\sum x_i\right)^2}{n} = 24{,}607.5 - \frac{1{,}047.22^2}{54} = 4{,}298.80133$$

$$\hat{\beta}_1 = \frac{\text{SS}_{xy}}{\text{SS}_{xx}} = \frac{7{,}158.10776}{4{,}298.80133} = 1.665140399$$

$$\hat{\beta}_0 = \bar{y} - \hat{\beta}_1 \bar{x} = \frac{1{,}283.46}{54} - (1.665140399)\left(\frac{1{,}047.22}{54}\right) = -8.524228315$$

The least squares prediction equation is $\hat{y} = -8.524 + 1.665x$

c. $\hat{\beta}_1 = 1.665$. For each unit increase in the Natural Origin Index, the mean Anthropogenic Index is estimated to increase by 1.665.

$\hat{\beta}_0 = 2.522$. Since $x = 0$ is not in the observed range of the Natural Origin Index, $\hat{\beta}_0$ has no interpretation other than the y-intercept.

d. Some preliminary calculations are:

$$\text{SS}_{yy} = \sum y_i^2 - \frac{\left(\sum y_i\right)^2}{n} = 54{,}184.8 - \frac{1{,}283.46^2}{54} = 23{,}679.80793$$

$$\text{SSE} = \text{SS}_{yy} - \hat{\beta}_1 \text{SS}_{xy} = 23{,}679.80793 - (1.665140399)(7{,}158.10776) = 11{,}760.55352$$

$$s^2 = \frac{\text{SSE}}{n-2} = \frac{11{,}760.55352}{54-2} = 226.1644908$$

$$s = \sqrt{s^2} = \sqrt{226.1644908} = 15.03876627$$

To determine if the natural origin index and anthropogenic index are positively linearly related, we test:

$$H_0: \beta_1 = 0$$
$$H_a: \beta_1 > 0$$

The test statistic is $t = \dfrac{\hat{\beta}_1 - 0}{s_{\hat{\beta}_1}} = \dfrac{1.665}{\dfrac{15.0388}{\sqrt{4{,}298.80133}}} = 7.259$

The rejection region requires $\alpha = 0.05$ in the upper tail of the t distribution with $df = n-2 = 54-2 = 52$. Using MINITYAB, $t_{0.05} = 1.675$. The rejection region is $t > 1.675$.

Since the observed value of the test statistic falls in the rejection region $(t = 7.259 > 1.675)$, H_0 is rejected. There is sufficient evidence to indicate the natural origin index and anthropogenic index are positively linearly related at $\alpha = 0.05$.

e. For confidence coefficient 0.95, $\alpha = 0.05$ and $\alpha/2 = 0.05/2 = 0.025$. Using MINITAB, with $df = n-2 = 54-2 = 52$, $t_{0.025} = 2.007$. The 95% confidence interval is:

$$\hat{\beta}_1 \pm t_{0.025} s_{\hat{\beta}_1} \Rightarrow \hat{\beta}_1 \pm t_{0.025} \frac{s}{\sqrt{\text{SS}_{xx}}} \Rightarrow 1.665 \pm 2.007 \frac{15.0688}{\sqrt{4{,}298.80133}}$$
$$\Rightarrow 1.665 \pm 0.461 \Rightarrow (1.204,\ \ 2.126)$$

We are 95% confident that the change in the anthropogenic index for each unit increase in the natural origin index is between 1.204 and 2.126.

11.71 a. We are 95% confident that for each 1% increase in body mass, the mean percent increase in eye mass is between 0.25% and 0.30%.

b. We are 95% confident that for each 1% increase in body mass, the mean percent decrease in eye mass is between 0.05% and 0.14%.

11.73 Using MINITAB the results of fitting a simple regression line are:

Regression Analysis: GPa versus Fractal

```
Analysis of Variance

Source        DF   Adj SS   Adj MS   F-Value   P-Value
Regression     1   406.15   406.15     14.32     0.005
Error          8   226.86    28.36
Total          9   633.01
```

```
Model Summary

      S     R-sq  R-sq(adj)  R-sq(pred)
5.32512   64.16%     59.68%      45.91%

Coefficients

Term         Coef  SE Coef  T-Value  P-Value   VIF
Constant    272.4     67.5     4.03    0.004
Fractal    -102.0     26.9    -3.78    0.005  1.00


Regression Equation

GPa = 272.4 - 102.0 Fractal
```

For confidence coefficient 0.90, $\alpha = 0.10$ and $\alpha / 2 = 0.10 / 2 = 0.05$. From Table III, Appendix B, with $df = n - 2 = 10 - 2 = 8$, $t_{0.05} = 1.860$. The 90% confidence interval is:

$$\hat{\beta}_1 \pm t_{0.05} s_{\hat{\beta}_1} \Rightarrow -102.0 \pm 1.860(26.9) \Rightarrow -102.0 \pm 50.034 \Rightarrow (-152.034,\ -51.966)$$

We are 90% confident that the mean change in Young's Modulus for each unit change in fractal dimension is between -152.034 and -51.966.

11.75 a. From the printout, $\hat{\beta}_o = .515$, $\hat{\beta}_1 = 0.000021$, and $s = 0.0370$.

b. To determine if there is a positive linear relationship between elevation and slugging percentage, we test:

$$H_0 : \beta_1 = 0$$
$$H_a : \beta_1 > 0$$

From the printout, the test statistic is $t = 2.89$ and the p-value is $p = 0.008 / 2 = 0.004$.

Since the p-value is less than $\alpha = 0.01 (p = 0.004 < 0.01)$, H_0 is rejected. There is sufficient evidence to indicate a positive linear relationship between elevation and slugging percentage at $\alpha = 0.01$.

c. Using MINITAB, the scatterplot is:

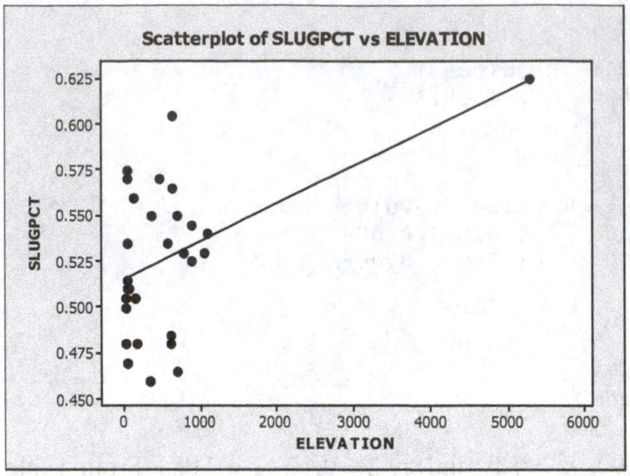

The data point corresponding to Denver is the data point furthest to the right. This point is very influential. If this point were deleted, there probably would not be a relationship between slugging percentage and elevation.

d. Using MINITAB with Denver removed, the results are:

Regression Analysis: SLUGPCT2 versus ELEVATION2

```
The regression equation is
SLUGPCT2 = 0.515 + 0.000020 ELEVATION2

Predictor          Coef      SE Coef      T      P
Constant        0.51537      0.01066  48.33  0.000
ELEVATION2   0.00002012   0.00002034   0.99  0.332

S = 0.0376839   R-Sq = 3.6%   R-Sq(adj) = 0.0%

Analysis of Variance

Source          DF        SS        MS     F      P
Regression       1  0.001389  0.001389  0.98  0.332
Residual Error  26  0.036922  0.001420
Total           27  0.038311
```

From the printout, $\hat{\beta}_o = 0.515$, $\hat{\beta}_1 = 0.000020$, and $s = 0.0377$.
To determine if there is a positive linear relationship between elevation and slugging percentage, we test:

$$H_0: \beta_1 = 0$$
$$H_a: \beta_1 > 0$$

From the printout, the test statistic is $t = 0.99$ and the *p*-value is $p = 0.332/2 = 0.166$.

Since the p-value is not less than $\alpha = 0.01 (p = 0.166 > 0.01)$, H_0 is not rejected. There is insufficient evidence to indicate a positive linear relationship between elevation and slugging percentage when Denver is deleted using $\alpha = 0.01$.

The "thin air" theory appears to be valid. No other city has an elevation that is close to that of Denver's. If Denver is not included in the analysis, there is no relationship between the slugging percentage and the elevation. Without Denver, the elevations of the cities range from 10 feet to 1082 feet. The elevation of Denver is 5277. The slugging percentage at Denver stadium is also higher than any other city. Even if we changed the value of the elevation of Denver to 1500 feet, the linear relationship between slugging percentage and elevation is still statistically significant.

11.77 The statement "The correlation coefficient is a measure of the strength of the linear relationship between x and y." is a true statement.

11.79 a. $r = 1$ implies x and y have a perfect positive linear relationship.

b. $r = -1$ implies x and y have a perfect negative linear relationship.

c. $r = 0$ implies x and y are not linearly related.

d. $r = 0.90$ implies x and y have a positive linear relationship. Since r is close to 1, the strength of the relationship is very high.

e. $r = 0.10$ implies x and y have a positive linear relationship. Since r is close to 0, the relationship is very weak.

f. $r = -0.88$ implies x and y have a negative linear relationship. Since r is close to -1, the relationship is fairly strong.

11.81 a. Using MINITAB, the scatterplot of the data is:

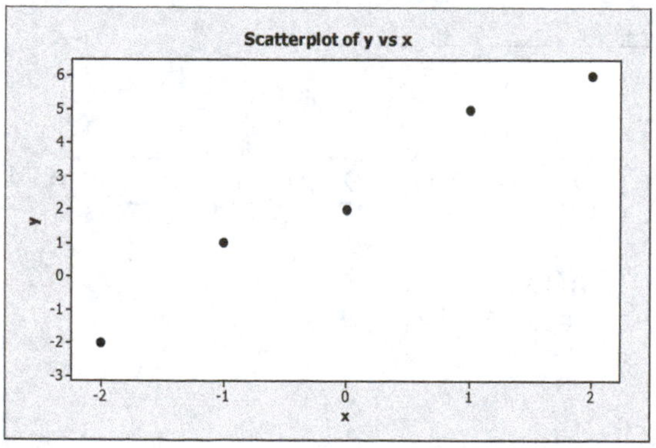

Some preliminary calculations are:

$\sum x = 0 \qquad \sum x^2 = 10 \qquad \sum xy = 20 \qquad \sum y = 12 \qquad \sum y^2 = 70$

$$SS_{xy} = \sum xy - \frac{\sum x \sum y}{n} = 20 - \frac{0(12)}{5} = 20$$

$$SS_{xx} = \sum x^2 - \frac{\left(\sum x\right)^2}{n} = 10 - \frac{0^2}{5} = 10$$

$$SS_{yy} = \sum y^2 - \frac{\left(\sum y\right)^2}{n} = 70 - \frac{12^2}{5} = 41.2$$

$$r = \frac{SS_{xy}}{\sqrt{SS_{xx}SS_{yy}}} = \frac{20}{\sqrt{10(41.2)}} = 0.9853 \qquad r^2 = 0.9853^2 = 0.9709$$

b. Using MINITAB, the scatterplot of the data is:

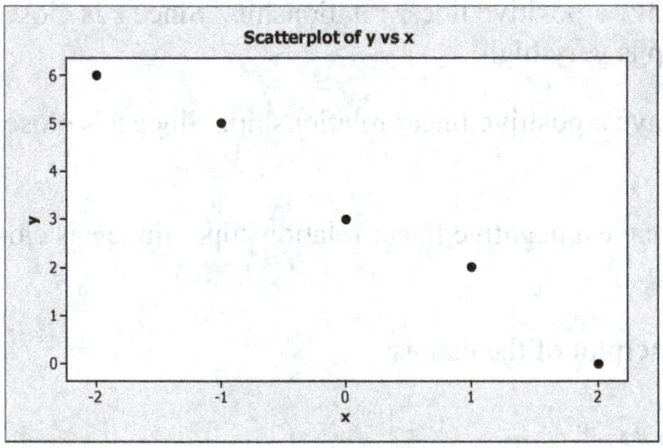

Some preliminary calculations are:

$\sum x = 0 \qquad \sum x^2 = 10 \qquad \sum xy = -15 \qquad \sum y = 16 \qquad \sum y^2 = 74$

$$SS_{xy} = \sum xy - \frac{\sum x \sum y}{n} = -15 - \frac{0(16)}{5} = -15$$

$$SS_{xx} = \sum x^2 - \frac{\left(\sum x\right)^2}{n} = 10 - \frac{0^2}{5} = 10$$

$$SS_{yy} = \sum y^2 - \frac{\left(\sum y\right)^2}{n} = 74 - \frac{16^2}{5} = 22.8$$

$$r = \frac{SS_{xy}}{\sqrt{SS_{xx}SS_{yy}}} = \frac{-15}{\sqrt{10(22.8)}} = -0.9934 \qquad r^2 = 0.9934^2 = 0.9868$$

c. Using MINITAB, the scatterplot is:

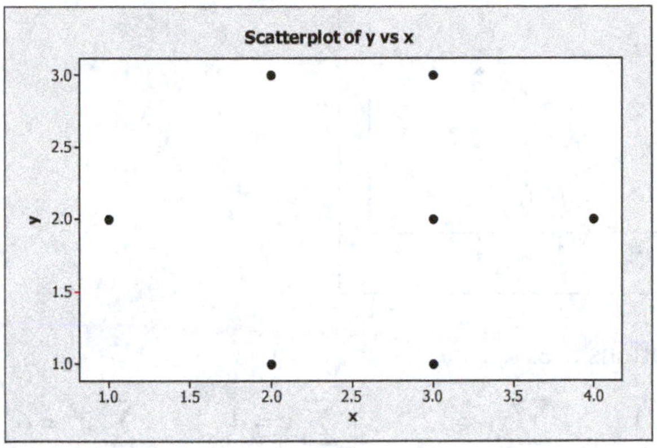

Some preliminary calculations are:

$\sum x = 18 \qquad \sum x^2 = 52 \qquad \sum xy = 36 \qquad \sum y = 14 \qquad \sum y^2 = 32$

$$SS_y = \sum xy - \frac{\sum x \sum y}{n} = 36 - \frac{18(14)}{7} = 0$$

$$SS_{xx} = \sum x^2 - \frac{\left(\sum x\right)^2}{n} = 52 - \frac{18^2}{7} = 5.71428571$$

$$SS_{yy} = \sum y^2 - \frac{\left(\sum y\right)^2}{n} = 32 - \frac{14^2}{7} = 4$$

$$r = \frac{SS_{xy}}{\sqrt{SS_{xx}SS_{yy}}} = \frac{0}{\sqrt{5.71428571(4)}} = 0 \qquad r^2 = 0^2 = 0$$

d. Using MINITAB, the scatterplot of the data is:

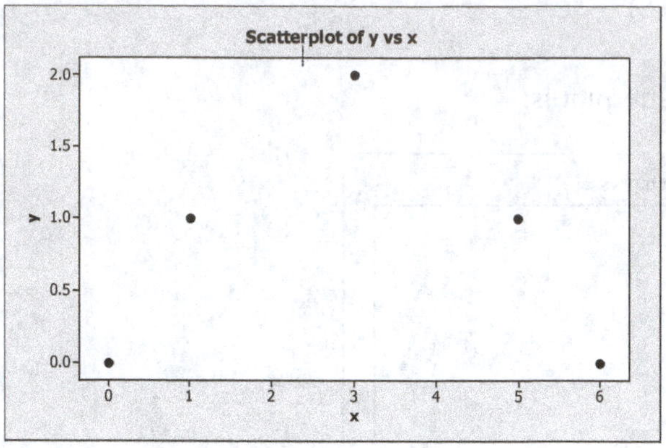

Some preliminary calculations are:

$\sum x = 15 \qquad \sum x^2 = 71 \qquad \sum xy = 12 \qquad \sum y = 4 \qquad \sum y^2 = 6$

$$SS_{xy} = \sum xy - \frac{\sum x \sum y}{n} = 12 - \frac{15(4)}{5} = 0$$

$$SS_{xx} = \sum x^2 - \frac{\left(\sum x\right)^2}{n} = 71 - \frac{15^2}{5} = 26$$

$$SS_{yy} = \sum y^2 - \frac{\left(\sum y\right)^2}{n} = 6 - \frac{4^2}{5} = 2.8$$

$$r = \frac{SS_{xy}}{\sqrt{SS_{xx}SS_{yy}}} = \frac{0}{\sqrt{26(2.8)}} = 0 \qquad r^2 = 0^2 = 0$$

11.83 From Exercises 11.21 and 11.41,

$$r^2 = 1 - \frac{SSE}{SS_{yy}} = 1 - \frac{5.13486841}{41.7142857} = 1 - 0.1231 = 0.8769$$

87.69% of the total sample variability around $\bar{y}$ is explained by the linear relationship between *y* and *x*.

11.85 a. The correlation coefficient is $r = -0.271$. There is a rather weak, negative linear relationship between the first letter of the last name and the response time.

b. Since the *p*-value is less than $\alpha\left(p = 0.018 < 0.05\right)$, H_0 is rejected. There is sufficient evidence to indicate there is a significant, negative linear relationship between the first letter of the last name and the response time.

c. Yes, the results support the researchers' *last name* effect theory. As the number of the first letter of the last name gets larger, the response time tends to decrease.

11.87 a. $r^2 = 0.18$. 18% of the sample variability around the sample mean number of points scored is explained by the linear relationship between the number of points scored and number of yards from the opposing goal line.

b. $r = -\sqrt{0.18} = -0.424$. The value of r is negative because the estimated slope of the regression line is negative.

11.89 a. The correlation coefficient of $r = 0.50$ indicates that the strength of the positive linear relationship between baseline and follow-up physical activity among the obese adults is moderate. Since the p-value is not less than $\alpha = 0.05 \left(p = 0.07 \nless 0.05\right)$, there is no evidence to indicate correlation coefficient differs from 0 at $\alpha = 0.05$.

b. Using MINITAB, a possible scatterplot with 13 data points that would yield a value of $r = 0.50$ is:

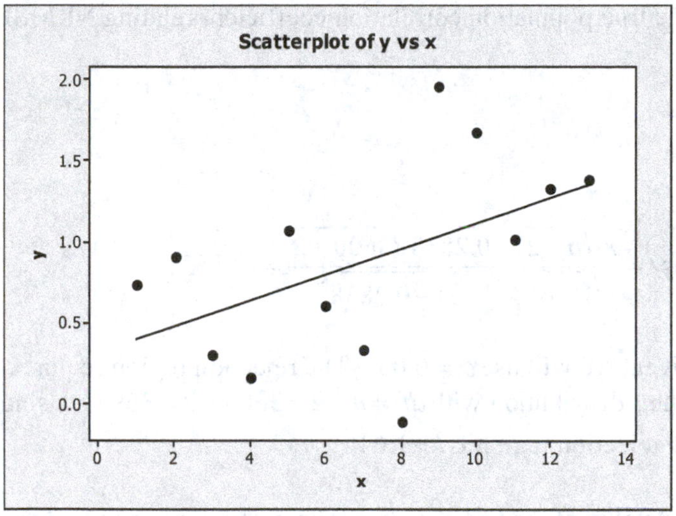

c. Since the p-value is not small $(p = 0.66)$ there is no evidence to indicate a linear relationship between baseline and follow-up physical activity among the normal adults. The correlation coefficient of $r = -0.12$ indicates that the strength of the negative linear relationship is extremely weak.

d. Using MINITAB, a possible scatterplot with 15 data points that would yield a value of $r = -0.12$ is:

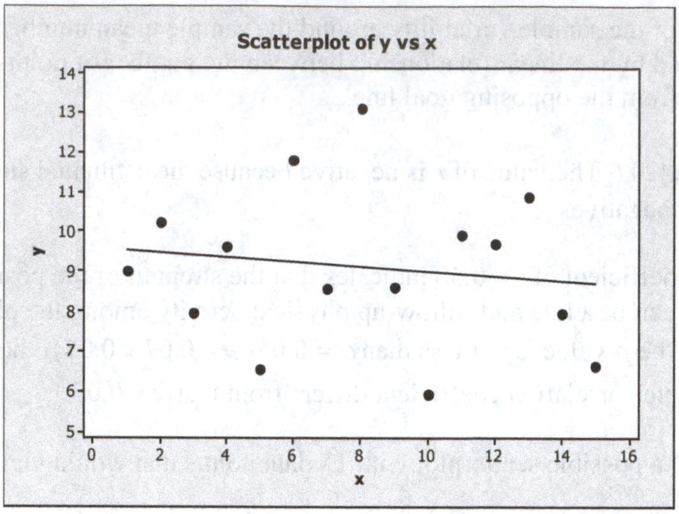

11.91 a. To determine if the true population correlation coefficient relating NRMSE and bias is positive, we test:

$H_0: \beta_1 = 0$
$H_a: \beta_1 > 0$

The test statistic is $t = \frac{r\sqrt{n-2}}{\sqrt{1-r^2}} = \frac{0.2838\sqrt{3600-2}}{\sqrt{1-0.2838^2}} = 17.753$.

Since no α was given, we will use $\alpha = 0.05$. The rejection region requires $\alpha = 0.05$ in the upper tail of the t-distribution with $df = n - 2 = 3600 - 2 = 3598$. Using MINITAB, $t_{0.05} = 1.645$. The rejection region is $t > 1.645$.

Since the observed value of the test statistics falls in the rejection region $(t = 17.753 > 1.645)$, H_0 is rejected. There is sufficient evidence to indicate the true population correlation coefficient relating NRMSE and bias is positive at $\alpha = 0.05$.

b. No. Even though there is a significant positive linear relationship between NRMSE and bias, the relationship is very weak. The coefficient of determination is $r^2 = 0.2838^2 = 0.0805$. Only 8.05% of the variation in the bias values is explained by the linear relationship between bias and NRMSE.

11.93 a. **Piano**: $r = 0.447$
Because this value is near 0.5, there is s slight positive linear relationship between recognition exposure time and goodness of view for piano.

Bench: $r = -0.057$
Because this value is extremely close to 0, there is an extremely weak negative linear relationship between recognition exposure time and goodness of view for bench.

Motorbike: $r = 0.619$
Because this value is near 0.5, there is a moderate positive linear relationship between recognition exposure time and goodness of view for motorbike.

Armchair: $r = 0.294$
Because this value is fairly close to 0, there is a weak positive linear relationship between recognition exposure time and goodness of view for armchair.

Teapot: $r = 0.949$
Because this value is very close to 1, there is a strong positive linear relationship between recognition exposure time and goodness of view for teapot.

b. **Piano:** $r^2 = 0.447^2 = 0.1998$
19.98% of the total sample variability around the sample mean recognition exposure time is explained by the linear relationship between the recognition exposure time and the goodness of view for piano.

Bench: $r^2 = (-0.057)^2 = 0.0032$
.32% of the total sample variability around the sample mean recognition exposure time is explained by the linear relationship between the recognition exposure time and the goodness of view for bench.

Motorbike: $r^2 = 0.619^2 = 0.3832$
38.32% of the total sample variability around the sample mean recognition exposure time is explained by the linear relationship between the recognition exposure time and the goodness of view for motorbike.

Armchair: $r^2 = 0.294^2 = 0.0864$
8.64% of the total sample variability around the sample mean recognition exposure time is explained by the linear relationship between the recognition exposure time and the goodness of view for armchair.

Teapot: $r^2 = 0.949^2 = 0.9006$
90.06% of the total sample variability around the sample mean recognition exposure time is explained by the linear relationship between the recognition exposure time and the goodness of view for teapot.

c. The test is:

$$H_0 : \beta_1 = 0$$
$$H_a : \beta_1 \neq 0$$

We will use $df = n - 2 = 25 - 2 = 23$ for all of the following:

Piano: $t = 2.40$
Using MINITAB, the p-value is $p = 2P(t > 2.40) = 2(1 - 0.9876) = 0.0248$.

For levels of significance greater than $\alpha = 0.0248$, H_0 can be rejected. There is sufficient evidence to indicate that there is a linear relationship between goodness of view and recognition exposure time for piano for $\alpha > 0.0248$.

Bench: $t = 0.27$
Using MINITAB, the *p*-value is $p = 2P(t > 0.27) = 2(1 - 0.6052) = 0.7896$.
For any reasonable value of α, H_0 is not rejected. There is insufficient evidence to indicate that there is a linear relationship between goodness of view and recognition exposure time for bench for any reasonable value of α.

Motorbike: $t = 3.78$
Using MINITAB, the *p*-value is $p = 2P(t > 3.78) = 2(1 - 0.9995) = 0.0010$.
H_0 can be rejected for $\alpha > 0.001$. There is sufficient evidence to indicate that there is a linear relationship between goodness of view and recognition exposure time for motorbike for $\alpha > 0.001$.

Armchair: $t = 1.47$
Using MINITAB, the *p*-value is $p = 2P(t > 1.47) = 2(1 - 0.9224) = 0.1552$.
H_0 cannot be rejected for levels of significance $\alpha < 0.1552$. There is insufficient evidence to indicate that there is a linear relationship between goodness of view and recognition exposure time for armchair for $\alpha < 0.1552$.

Teapot: $t = 14.50$
Using MINITAB, the *p*-value is $p = 2P(t > 14.50) = 2(1 - 1) = 0$.
H_0 can be rejected for $\alpha \geq 0.0001$. There is sufficient evidence to indicate that there is a linear relationship between goodness of view and recognition exposure time for teapot for $\alpha \geq 0.0001$.

11.95 a. To determine whether average payoff and punishment use are negatively correlated, we test:

$$H_0: \beta_1 = 0$$
$$H_a: \beta_1 < 0$$

b. Since the *p*-value is very small $(p = 0.0010)$, we would reject H_0 for any value of α greater than 0.001. There is sufficient evidence to indicate average payoff and punishment use are negatively correlated for $\alpha > 0.001$.

c. No. Just because two variables are significantly correlated does not mean that one variable causes the other. Something else could be causing both of the variables to react in certain ways.

11.97 Using the values computed in Exercise 11.70:

$$r = \frac{SS_{xy}}{\sqrt{SS_{xx}SS_{yy}}} = \frac{48.125}{\sqrt{379.9375(18.75)}} = 0.5702$$

Because r is moderately small, there is a rather weak positive linear relationship between blood lactate concentration and perceived recovery.

$r^2 = 0.5702^2 = 0.3251$.

32.51% of the sample variance of blood lactate concentration around the sample mean is explained by the linear relationship between blood lactate concentration and perceived recovery.

11.99 a. Using MINITAB, the results are:

Regression Analysis: Molecular versus Altitude

```
Analysis of Variance

Source        DF  Adj SS   Adj MS  F-Value  P-Value
Regression     1    2573   2573.1     9.07    0.013
  Altitude     1    2573   2573.1     9.07    0.013
Error         10    2837    283.7
Total         11    5410

Model Summary

      S     R-sq  R-sq(adj)  R-sq(pred)
16.8425   47.56%     42.32%      28.74%

Coefficients

Term          Coef  SE Coef  T-Value  P-Value   VIF
Constant      95.2     22.0     4.32    0.002
Altitude   -0.0765   0.0254    -3.01    0.013  1.00

Regression Equation

Molecular = 95.2 - 0.0765 Altitude
```

The value of r^2 is $r^2 = 0.4756$. 47.56% of the sample variation of molecular variances around the sample mean is explained by the linear relationship between molecular variance and altitude.

b. Using MINITAB, the results are:

Regression Analysis: Molecular versus PopSize

```
Analysis of Variance

Source        DF  Adj SS   Adj MS  F-Value  P-Value
Regression     1    2833   2832.6    10.99    0.008
  PopSize      1    2833   2832.6    10.99    0.008
Error         10    2577    257.7
Total         11    5410
```

```
Model Summary

      S      R-sq   R-sq(adj)   R-sq(pred)
16.0537    52.36%      47.60%       30.87%


Coefficients

Term         Coef   SE Coef   T-Value   P-Value    VIF
Constant    11.97      7.25      1.65     0.130
PopSize    0.1762    0.0532      3.32     0.008   1.00


Regression Equation

Molecular = 11.97 + 0.1762 PopSize
```

The value of r^2 is $r^2 = 0.5236$. 52.36% of the sample variation of molecular variances around the sample mean is explained by the linear relationship between molecular variance and population size.

c. From the results in parts **a** and **b**, there is a significant linear relationship between molecular variance and both altitude and population size. In both cases, approximately 50% of the sample variance is explained by the models. The value of r^2 for the model with molecular variance is smaller than the value of r^2 for the model with molecular variance and population size. Therefore, we would recommend that the researchers use population size to predict molecular variance.

11.101 For a given *x*, *y* is the actual value of the dependent variable and *E*(*y*) is the mean value of *y* at the given value of *x*.

11.103 The statement "The greater the deviation between *x* and $\bar{x}$, the wider the prediction interval for *y* will be." is true. The further *x* gets from $\bar{x}$, the less precise the interval will be.

11.105 a. $\hat{\beta}_1 = \frac{SS_{xy}}{SS_{xx}} = \frac{28}{32} = 0.875$

$\hat{\beta}_0 = \bar{y} - \hat{\beta}_1\bar{x} = 4 - 0.875(3) = 1.375$

The least squares line is $\hat{y} = 1.375 + 0.875x$.

b. The least squares line is:

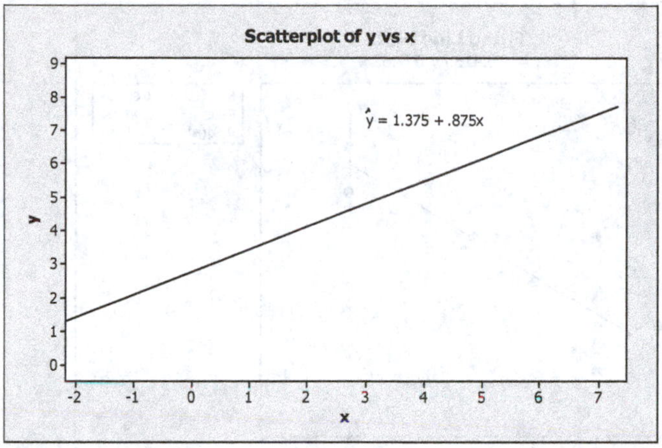

c. $\text{SSE} = \text{SS}_{yy} - \hat{\beta}_1 \text{SS}_{xy} = 26 - 0.875(28) = 1.5$

d. $s^2 = \dfrac{\text{SSE}}{n-2} = \dfrac{1.5}{10-2} = 0.1875$

e. $s = \sqrt{0.1875} = 0.4330$

The form of the confidence interval is $\hat{y} \pm t_{\alpha/2}\, s\sqrt{\dfrac{1}{n} + \dfrac{(x_p - \bar{x})^2}{\text{SS}_{xx}}}$

For $x_p = 2.5$, $\hat{y} = 1.375 + 0.875(2.5) = 3.5625$

For confidence coefficient 0.95, $\alpha = 0.05$ and $\alpha/2 = 0.05/2 = 0.025$. From Table III, Appendix B, with $df = n - 2 = 10 - 2 = 8$, $t_{0.025} = 2.306$. The confidence interval is:

$$3.5625 \pm 2.306(0.4330)\sqrt{\frac{1}{10} + \frac{(2.5-3)^2}{32}} \Rightarrow 3.5625 \pm 0.3279 \Rightarrow (3.2346,\ 3.8904)$$

f. The form of the prediction interval is $\hat{y} \pm t_{\alpha/2}\, s\sqrt{1 + \dfrac{1}{n} + \dfrac{(x_p - \bar{x})^2}{\text{SS}_{xx}}}$

For $x_p = 4$, $\hat{y} = 1.375 + 0.875(4) = 4.875$

For confidence coefficient 0.95, $\alpha = 0.05$ and $\alpha/2 = 0.05/2 = 0.025$. From Table III, Appendix B, with $df = n - 2 = 10 - 2 = 8$, $t_{0.025} = 2.306$. The prediction interval is:

$$4.875 \pm 2.306(0.4330)\sqrt{1 + \frac{1}{10} + \frac{(4-3)^2}{32}} \Rightarrow 4.875 \pm 1.062 \Rightarrow (3.813,\ 5.937)$$

11.107 a, b. The scattergram is:

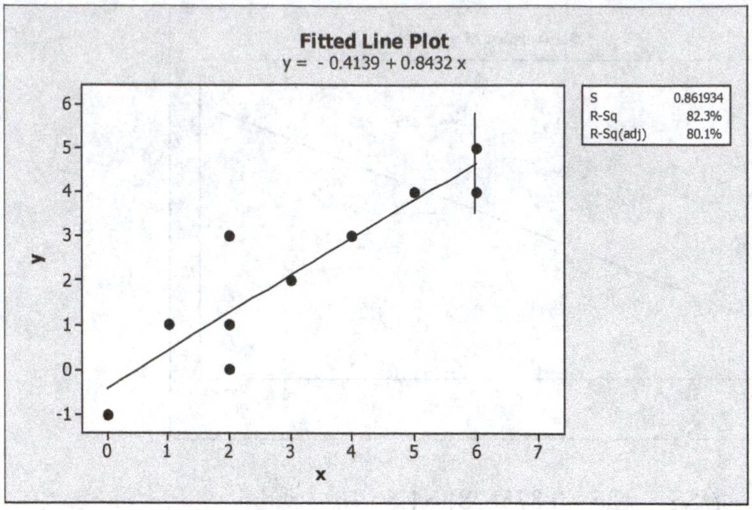

c. $\text{SSE} = \text{SS}_{yy} - \hat{\beta}_1 \text{SS}_{xy} = 33.6 - 0.843(32.8) = 5.9496$

$$s^2 = \frac{\text{SSE}}{n-2} = \frac{5.9496}{10-2} = 0.7437 \qquad s = \sqrt{0.7437} = 0.8623$$

$$\bar{x} = \frac{31}{10} = 3.1$$

The form of the confidence interval is $\hat{y} \pm t_{\alpha/2}\, s\sqrt{\frac{1}{n} + \frac{(x_p - \bar{x})^2}{\text{SS}_{xx}}}$

For $x_p = 6$, $\hat{y} = -0.414 + 0.843(6) = 4.644$

For confidence coefficient 0.95, $\alpha = 0.05$ and $\alpha/2 = 0.05/2 = 0.025$. From Table III, Appendix B, with $df = n - 2 = 10 - 2 = 8$, $t_{0.025} = 2.306$. The confidence interval is:

$$4.644 \pm 2.306(0.8623)\sqrt{\frac{1}{10} + \frac{(6-3.1)^2}{38.9}} \Rightarrow 4.644 \pm 1.118 \Rightarrow (3.526,\ 5.762)$$

d. For $x_p = 3.2$, $\hat{y} = -0.414 + 0.843(3.2) = 2.284$

The confidence interval is:

$$2.284 \pm 2.306(0.8623)\sqrt{\frac{1}{10} + \frac{(3.2-3.1)^2}{38.9}} \Rightarrow 2.284 \pm 0.630 \Rightarrow (1.654,\ 2.914)$$

For $x_p = 0$, $\hat{y} = -0.414 + 0.843(0) = -0.414$

The confidence interval is:

$$-0.414 \pm 2.306(0.8623)\sqrt{\frac{1}{10}+\frac{(0-3.1)^2}{38.9}} \Rightarrow -0.414 \pm 1.171 \Rightarrow (-1.585,\ 0.757)$$

e. The width of the confidence interval for the mean value of y depends on the distance x_p is from $\bar{x}$. The width of the interval for $x_p = 3.2$ is the smallest because 3.2 is the closest to $\bar{x} = 3$. The width of the interval for $x_p = 0$ is the widest because 0 is the farthest from $\bar{x} = 3.1$.

11.109 a. To predict the average payoff for a single player who used punishment 10 times, you would form a prediction interval for the individual payoff when punishment is 10.

b. To predict the mean of the average payoff for all players who used punishment 10 times, you would form a confidence interval for the mean payoff when punishment is 10.

11.111 For observation 7, the 95% prediction interval for the actual annual rainfall when the maximum daily temperature is 11.4 is $(92.298,\ 125.104)$. We are 95% confident that the actual rainfall in a site with a maximum daily temperature of 11.4 is between 92.298 and 125.104.

11.113 Answers may vary. One possible answer is:

The 90% confidence interval for $x = 220.00$ is $(5.64898,\ 5.83848)$. We are 90% confident that the mean sweetness index of all orange juice samples will be between 5.64898 and 5.83848 parts per million when the pectin value is 220.00.

11.115 a. The interval desired by the researcher is the 95% confidence interval for $E(y)$. In the test, it says that the researcher wants to estimate the average in-game heart rate.

b. Using MINITAB, the results are:

Prediction for HR%

```
Regression Equation

HR% = -27.2 + 0.558 VO2Max


Variable  Setting
VO2Max        150


    Fit   SE Fit         95% CI                 95% PI
56.4677  3.31256  (48.3622, 64.5733)  (41.2395, 71.6959)
```

The 95% confidence interval is $(48.3622,\ 64.5733)$.

c. We are 95% confident that the average in-game heart rate of all top-level water polo players who have a maximal oxygen uptake of 150 VO_2max is between 48.3622% and 64.5733%.

11.117 a. From Exercise 11.33 d, $\hat{y} = 39.304 + 0.493x$ and $s = 2.32153$. From Exercise 11.68 b, $SS_{xx} = 584.6528$.

For $x = 66$, $\hat{y} = 39.304 + 0.493(66) = 71.842$.

For confidence coefficient 0.95, $\alpha = 0.05$ and $\alpha/2 = 0.05/2 = 0.025$. Using MINITAB, with $df = n - 2 = 72 - 2 = 70$, $t_{0.025} = 1.994$. The 95% prediction interval is:

$$\hat{y} \pm t_{\alpha/2} s \sqrt{1 + \frac{1}{n} + \frac{(x_p - \bar{x})^2}{SS_{xx}}} \Rightarrow 71.842 \pm 1.994(2.32153)\sqrt{1 + \frac{1}{72} + \frac{(66 - 64.5694)^2}{584.6528}}$$

$$\Rightarrow 71.842 \pm 4.669 \Rightarrow (67.073,\ 76.411)$$

We are 95% confident that the actual ideal partner's height is between 67.073 and 76.411 inches when a female's height is 66 inches.

b. From Exercise 11.33 e, $\hat{y} = 23.274 + 0.596x$ and $s = 2.05688$. From Exercise 11.68 a, $SS_{xx} = 534.3467$.

For $x = 66$, $\hat{y} = 23.274 + 0.596(66) = 62.61$.

For confidence coefficient 0.95, $\alpha = 0.05$ and $\alpha/2 = 0.05/2 = 0.025$. From Table III, Appendix B, with $df = n - 2 = 75 - 2 = 73$, $t_{0.025} = 1.993$. The 95% prediction interval is:

$$\hat{y} \pm t_{\alpha/2} s \sqrt{1 + \frac{1}{n} + \frac{(x_p - \bar{x})^2}{SS_{xx}}} \Rightarrow 62.61 \pm 1.993(2.05688)\sqrt{1 + \frac{1}{75} + \frac{(66 - 71.0933)^2}{534.3467}}$$

$$\Rightarrow 62.61 \pm 4.224 \Rightarrow (58.386,\ 66.834)$$

We are 95% confident that the actual ideal partner's height is between 58.386 and 66.834 inches when a male's height is 66 inches.

c. The prediction for the ideal partner's height for males might be invalid because 66 is outside the range of observed values for male heights. The smallest observed value of male heights is 67 inches. We are not sure if the relationship between the two variables remains the same outside the observed range.

11.119 a. Using MINITAB, the output is:

```
Predicted Values for New Observations
New
Obs    Fit   SE Fit       99% CI           99% PI
 1   3.510   0.196   (2.955, 4.066)   (1.020, 6.000)

Values of Predictors for New Observations

New
Obs  Time
  1  15.0
```

From the printout, the 99% confidence interval for the mean mass of all spills with an elapsed time of 15 minutes is $(2.955,\ 4.066)$. We are 99% confident that the mean mass of all spills with an elapsed time of 15 minutes is between 2.955 and 4.066.

b. From the printout, the 99% prediction interval for the actual mass of a spill with an elapsed time of 15 minutes is $(1.020,\ 6.000)$. We are 99% confident that the actual mass of a spill with an elapsed time of 15 minutes is between 1.020 and 6.000.

c. The prediction interval in part b is larger than the confidence interval in part a. This will always be the case. The confidence interval gives a range of values for the mean of a distribution. The prediction interval gives a range of values for the actual value of a distribution. Once a mean is located the actual values can still vary around it. This is why the prediction interval is always wider than the confidence interval for the mean.

11.121 a. From Exercise 11.52, $SS_{xx} = 3000$ and $\bar{x} = 50$.

Also, for Brand A, $s = 1.211$; for Brand B, $s = 0.610$.

For Brand A, $\hat{y} = 6.62 - 0.0727(45) = 3.349$, while for Brand B, $\hat{y} = 9.31 - 0.1077(45) = 4.464$.

For both brands, $df = n - 2 = 15 - 2 = 13$. For confidence coefficient 0.90, (i.e., for all parts of this question), $\alpha = 0.10$ and $\alpha/2 = 0.10/2 = 0.05$. From Table III, Appendix B, with $df = 13$, $t_{0.05} = 1.771$.

The form of both confidence intervals is $\hat{y} \pm t_{\alpha/2} s \sqrt{\frac{1}{n} + \frac{(x_p - \bar{x})^2}{SS_{xx}}}$

For Brand A, we obtain:

$$3.349 \pm 1.771(1.211)\sqrt{\frac{1}{15} + \frac{(45-50)^2}{3000}} \Rightarrow 3.349 \pm 0.587 \Rightarrow (2.762,\ 3.936)$$

For Brand B, we obtain:

$$4.464 \pm 1.771(0.610)\sqrt{\frac{1}{15} + \frac{(45-50)^2}{3000}} \Rightarrow 4.464 \pm 0.296 \Rightarrow (4.168,\ 4.760)$$

The interval for Brand A is wider, caused by the larger value of s.

b. The form of both prediction intervals is $\hat{y} \pm t_{\alpha/2} s \sqrt{1 + \frac{1}{n} + \frac{(x_p - \bar{x})^2}{SS_{xx}}}$

For Brand A, we obtain:

$$3.349 \pm 1.771(1.211)\sqrt{1 + \frac{1}{15} + \frac{(45-50)^2}{3000}} \Rightarrow 3.349 \pm 2.224 \Rightarrow (1.125,\ 5.573)$$

For Brand B, we obtain:

$$4.464 \pm 1.771(0.610)\sqrt{1+\frac{1}{15}+\frac{(45+50)^2}{3000}} \Rightarrow 4.464 \pm 1.120 \Rightarrow (3.344,\ 5.584)$$

Again, the interval is wider for Brand A, caused by the larger value of s. Each of these intervals is wider than its counterpart from part **a**, since, for the same x, a prediction interval for an individual y is always wider than a confidence interval for the mean of y. This is due to an individual observation having a greater variance than the variance of the mean of a set of observations.

c. To obtain a confidence interval for the life of a brand A cutting tool that is operated at 100 meters per minute, we use:

$$\hat{y} \pm t_{\alpha/1} s \sqrt{1+\frac{1}{n}+\frac{\left(x_p-\bar{x}\right)^2}{\text{SS}_{xx}}}$$

For $x = 100$, $\hat{y} = 6.62 - 0.0727(100) = -0.65$.

For both brands, $df = n-2 = 15-2 = 13$. For confidence coefficient 0.95, $\alpha = 0.05$ and $\alpha/2 = 0.05/2 = 0.025$. From Table III, Appendix B, with $df = 13$, $t_{0.025} = 2.160$.

Here, we obtain:

$$-0.65 \pm 2.160(1.211)\sqrt{1+\frac{1}{15}+\frac{(100-50)^2}{3000}} \Rightarrow -0.65 \pm 3.606 \Rightarrow (-4.256,\ 2.956)$$

The additional assumption would be that the straight line model fits the data well for the x's actually observed all the way up to the value under consideration, 100. Clearly from the estimated value of −0.65, this is not true (usually, negative "useful lives" are not found).

11.123 a. Using MINITAB, the results of fitting the model are:

Regression Analysis: Efficiency versus AverageDust

```
Analysis of Variance

Source          DF   Adj SS    Adj MS   F-Value   P-Value
Regression       1   0.1255    0.1255      1.11     0.323
  AverageDust    1   0.1255    0.1255      1.11     0.323
Error            8   0.9039    0.1130
Total            9   1.0294

Model Summary

       S     R-sq   R-sq(adj)   R-sq(pred)
0.336142   12.19%       1.21%        0.00%
```

```
Coefficients

Term              Coef  SE Coef  T-Value  P-Value   VIF
Constant         1.411    0.265     5.32    0.001
AverageDust        349      331     1.05    0.323  1.00


Regression Equation

Efficiency = 1.411 + 349 AverageDust
```

The fitted regression line is $\hat{y} = 1.411 + 349x$.

b. To determine if there is a linear relationship between efficiency and average dust thickness, we test:

$$H_0 : \beta_1 = 0$$
$$H_a : \beta_1 \neq 0$$

From the printout, the test statistic is $t = 1.05$ and the p-value is $p = 0.323$. Since the p-value is not small, we would not reject H_0 for any reasonable value of α. There is insufficient evidence of a linear relationship between efficiency and average dust thickness for any reasonable value of α. In addition, the value of the coefficient of determination is $r^2 = 0.1219$. Only 12.19% of the sample variation in the efficiency values is explained by the linear relationship between efficiency and average dust thickness. We would not recommend using the model in practice.

11.125 Using MINITAB, a scatterplot of the data is:

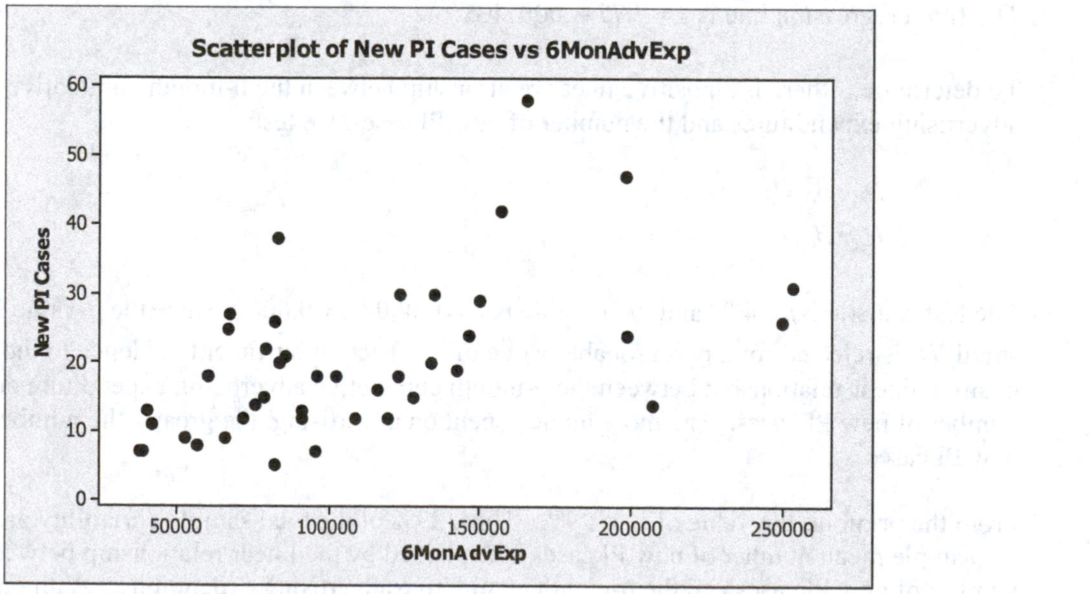

It appears that there is a positive linear relationship between the 6-month cumulative advertising expenditure and the number of new PI cases at the law firm.

The descriptive statistics for the two variables are:

Descriptive Statistics: New PI Cases, 6MonAdvExp

Variable	N	Mean	StDev	Minimum	Q1	Median	Q3	Maximum
New PI Cases	42	20.05	11.34	5.00	12.00	18.00	26.00	58.00
6MonAdvExp	42	108779	54104	37490	70293	93059	136611	253011

Using MINITAB, the results of fitting the regression line are:

Regression Analysis: New PI Cases versus 6MonAdvExp

```
The regression equation is
New PI Cases = 7.77 + 0.000113 6MonAdvExp


Predictor         Coef      SE Coef     T      P
Constant         7.767        3.385  2.29  0.027
6MonAdvExp  0.00011289   0.00002793  4.04  0.000


S = 9.67521   R-Sq = 29.0%   R-Sq(adj) = 27.2%


Analysis of Variance

Source          DF      SS      MS      F      P
Regression       1  1529.5  1529.5  16.34  0.000
Residual Error  40  3744.4    93.6
Total           41  5273.9
```

The fitted regression line is $\hat{y} = 7.77 + .000113x$.

To determine if there is a positive linear relationship between the 6-month cumulative advertising expenditures and the number of new PI cases, we test:

$$H_0 : \beta_1 = 0$$
$$H_a : \beta_1 > 0$$

The test statistic is $t = 4.04$ and the *p*-value is $p = 0.000 / 2 = 0.000$. Since the *p*-value is so small, H_0 is rejected for any reasonable value of α. There is sufficient evidence to indicate a positive linear relationship between the 6-month cumulative advertising expenditure and the number of new PI cases. The more money spent on advertising, the greater the number of new PI cases.

From the printout, the value of r^2 is 29%. Thus, 29% of the total sample variability around the sample mean number of new PI cases is explained by the linear relationship between the number of new PI cases and the 6-month cumulative advertising expenditure. With only 29% of the variability explained, we know that there might be other factors not considered that can help explain the variability in the number of new PI cases.

11.127 The general form of the straight-line model for $E(y)$ is $E(y) = \beta_0 + \beta_1 x$.

11.129 The statement "In simple linear regression, about 95% of the y-values in the sample will fall within 2s of their respective predicted values." is a true statement.

11.131 a.

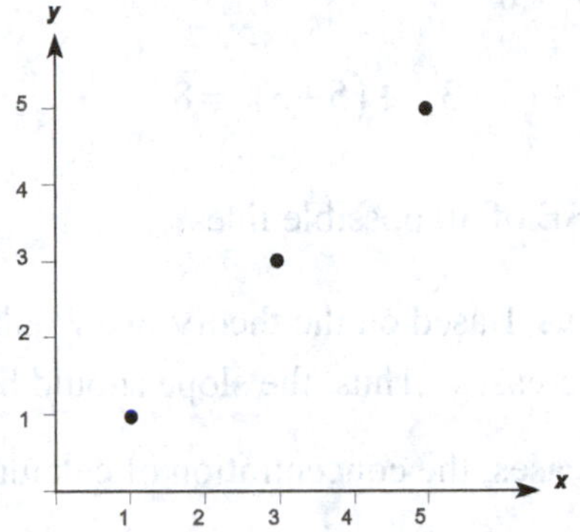

b. One possible line is $\hat{y} = x$.

x	y	$\hat{y}$	$\hat{y}-\hat{y}$
1	1	1	0
3	3	3	0
5	5	5	0
			0

For this example $\sum(y-\hat{y}) = 0$

A second possible line is $\hat{y} = 3$.

x	y	$\hat{y}$	$\hat{y}-\hat{y}$
1	1	3	−2
3	3	3	0
5	5	3	2
			0

For this example $\sum(y-\hat{y}) = 0$

c. Some preliminary calculations are:

$\sum x_i = 9 \qquad \sum x_i^2 = 35 \qquad \sum x_i y_i = 35 \qquad \sum y_i = 9 \qquad \sum y_i^2 = 35$

$$SS_{xy} = \sum x_i \sum y_i - \frac{\sum x_i \sum y_i}{n} = 35 - \frac{9(9)}{3} = 8$$

$$SS_{xx} = \sum x_i^2 - \frac{\left(\sum x_i\right)^2}{n} = 35 - \frac{9^2}{3} = 8$$

$$SS_{yy} = \sum y_i^2 - \frac{\left(\sum y_i\right)^2}{n} = 35 - \frac{9^2}{3} = 8$$

$$\hat{\beta}_1 = \frac{SS_{xy}}{SS_{xx}} = \frac{8}{8} = 1 \qquad \hat{\beta}_0 = \bar{y} - \hat{\beta}_1\bar{x} = \frac{9}{3} - 1\left(\frac{9}{3}\right) = 0$$

The least squares line is $\hat{y} = 0 + 1x = x$.

d. For $\hat{y} = x$, $SSE = SS_{yy} - \hat{\beta}_1 SS_{xy} = 8 - 1(8) = 0$

For $\hat{y} = 3$, $SSE = \sum(y_i - \hat{y}_i)^2 = (1-3)^2 + (3-3)^2 + (5-3)^2 = 8$

The least squares line has the smallest SSE of all possible lines.

11.133 a. The straight-line model is $y = \beta_0 + \beta_1 x + \varepsilon$. Based on the theory, we would expect the metal level to decrease as the distance increases. Thus, the slope should be negative.

b. Yes. As the distance from the plant increases, the concentration of calcium tends to decrease.

c. No. As the distance from the plant increases, the concentration of arsenic tends to increase.

11.135 a. $r = 0.43$. Since the value is not particularly close to 1, there is a moderately weak positive linear relationship between time allotted to sports and audience ratings.

b. $r^2 = 0.43^2 = 0.1849$. 18.49% of the total sample variability around the sample mean audience rating is explained by the linear relationship between audience rating and total time allotted to sports.

11.137 a. We would expect the slope of the line from modeling the relationship between the mean number of games won and the team's batting average to be positive. As the team's batting average increases, we would expect that the number of games won should also increase.

b. Using MINITAB, a scatterplot of the data is:

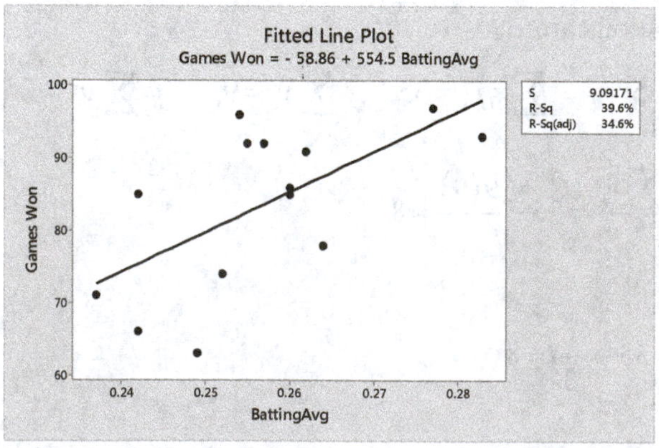

The graph indicates that batting average and games won have a positive linear relationship. As batting average increases, the number of games won tends to increase.

c. From the printout, $\hat{\beta}_0 = -58.86$ and $\hat{\beta}_1 = 554.5$. The least squares line is: $\hat{y} = -58.86 + 554.5x$.

d. The least squares line is plotted on the graph in part **a**. The data appear to fit the data somewhat well.

e. $\hat{\beta}_0 = -58.86$. Since there is no team with a batting average of 0, this has no meaning other than the y-intercept.
$\hat{\beta}_1 = 554.5$. For each additional 1 pint increase in batting average, the mean number of games won is estimated to increase by 554.5. Since batting average cannot increase by 1 full unit, it might be better to interpret $\hat{\beta}_1$ as follows: For each additional increase of 0.01 in batting average, the mean number of games won is estimated to increase by 5.545 games.

f. To determine if the mean number of games won by a major league baseball team is positively linearly related to the team's batting average, we test:

$$H_0: \beta_1 = 0$$
$$H_a: \beta_1 > 0$$

The test statistic is $t = 2.81$ and the p-value is $p = 0.016$. Since we are running only a one-tailed test, we must divide the p-value by 2. Thus, the p-value of this test is $p = 0.016/2 = 0.008$. Since the p-value is less than $\alpha = 0.05\,(p = 0.008 < 0.05)$, H_0 is rejected. There is sufficient evidence to indicate the mean number of games won by a major league baseball team is positively linearly related to the team's batting average at $\alpha = 0.05$.

g. From the printout, $r^2 = 0.396$. 39.6% of the sample variability around the sample mean number of games won is explained by the linear relationship between the number of games won and the batting average of the team.

h. Yes. There is evidence of a positive linear relationship between batting average and games won.

11.139 a. The equation for the straight-line model is $y = \beta_0 + \beta_1 x + \varepsilon$.

b. Some preliminary calculations are:

$$\bar{y} = \frac{\sum y}{n} = \frac{398}{9} = 44.2222 \qquad \bar{x} = \frac{\sum x}{n} = \frac{1,444}{9} = 160.4444$$

$$SS_{xy} = \sum xy - \frac{\sum x \sum y}{n} = 60,428 - \frac{1,444(398)}{9} = -3,428.88889$$

$$SS_{xx} = \sum x^2 - \frac{\left(\sum x\right)^2}{n} = 235,866 - \frac{1,444^2}{9} = 4,184.2222$$

$$\hat{\beta}_1 = \frac{SS_{xy}}{SS_{xx}} = \frac{-3,428.88889}{4,184.2222} = -0.819480593$$

$$\hat{\beta}_0 = \bar{y} - \hat{\beta}_1\bar{x} = \frac{398}{9} - (-0.819480593)\left(\frac{1,444}{9}\right) = 175.7033307$$

The fitted model is $\hat{y} = 175.7033 - 0.8195x$.

c. $\hat{\beta}_0 = 175.7033$. Since $x = 0$ (age of fish) is not in the observed range, $\hat{\beta}_0$ has no practical meaning.

d. $\hat{\beta}_1 = -0.8195$. For each additional day of age, the mean number of strikes is estimated to decrease by 0.8195 strikes.

e. Some additional calculations are:

$\sum y_i = 398 \qquad \sum y_i^2 = 22,078$

$$SS_{yy} = \sum y_i^2 - \frac{\left(\sum y_i\right)^2}{n} = 22,078 - \frac{398^2}{9} = 4,477.55556$$

$$SSE = SS_{yy} - \hat{\beta}_1 SS_{xy} = 4,477.55556 - (-0.819480593)(-3,428.88889) = 1,667.647659$$

$$s^2 = \frac{SSE}{n-2} = \frac{1,667.647659}{9-2} = 238.2353799$$

$$s = \sqrt{s^2} = \sqrt{238.2353799} = 15.43487544$$

$H_0: \beta_1 = 0$
$H_a: \beta_1 < 0$

The test statistic is $t = \frac{\hat{\beta}_1 - 0}{s_{\hat{\beta}_1}} = \frac{-0.8195 - 0}{15.4349 / \sqrt{4,184.2222}} = -3.434$

The rejection region requires $\alpha = 0.10$ in the lower tail of the t distribution with $df = n - 2 = 9 - 2 = 7$. From, Table III, Appendix B, $t_{0.10} = 1.415$. The rejection region is $t < -1.415$.

Since the observed value of the test statistic falls in the rejection region $(t = -3.434 < -1.415)$, H_0 is rejected. There is sufficient evidence to indicate a negative linear relationship exists between the number of strikes and the age of the fish at $\alpha = 0.10$

11.141 a. The simple linear regression model relating y to x is $y = \beta_0 + \beta_1 x + \varepsilon$.

b. $r^2 = 0.92$. 92% of the total sample variability around the sample mean metal uptake is explained by the linear relationship between metal uptake and the final concentration of metal in the solution.

11.143 a. Some preliminary calculations are:

$$\sum x_i = 62 \qquad \sum y_i = 97.8 \qquad \sum x_i y_i = 1{,}087.78$$

$$\sum x_i^2 = 720.52 \qquad \sum y_i^2 = 1{,}710.2$$

$$SS_{xy} = \sum x_i y_i - \frac{\sum x_i \sum y_i}{n} = 1{,}087.78 - \frac{62(97.8)}{6} = 77.18$$

$$SS_{xx} = \sum x_i^2 - \frac{\left(\sum x_i\right)^2}{n} = 720.52 - \frac{62^2}{6} = 79.8533333$$

$$\hat{\beta}_1 = \frac{SS_{xy}}{SS_{xx}} = \frac{77.18}{79.8533333} = 0.966521957$$

$$\hat{\beta}_o = \bar{y} - \hat{\beta}_1 \bar{x} = \frac{97.8}{6} - 0.966521957\left(\frac{62}{6}\right) = 6.312606442$$

The least squares prediction equation is $\hat{y} = 6.31 + 0.967x$

b. The y-intercept is 6.31. This value has no meaning because 0 is not in the observed range of the independent variable mean pore diameter.

c. The slope of the line is 0.967. For each unit increase in mean pore diameter, the mean porosity is estimated to increase by 0.967.

d. Using MINITAB, the results are:

Regression Analysis: Porosity versus PoreDiameter

Analysis of Variance

Source	DF	Adj SS	Adj MS	F-Value	P-Value
Regression	1	74.60	74.60	7.20	0.055
PoreDiameter	1	74.60	74.60	7.20	0.055
Error	4	41.46	10.37		
Total	5	116.06			

Model Summary

S	R-sq	R-sq(adj)	R-sq(pred)
3.21962	64.27%	55.34%	0.00%

```
Coefficients

Term           Coef  SE Coef  T-Value  P-Value   VIF
Constant       6.31     3.95     1.60    0.185
PoreDiameter  0.967    0.360     2.68    0.055  1.00

Regression Equation

Porosity = 6.31 + 0.967 PoreDiameter
```

To determine if there is a linear relationship between apparent porosity and mean pore diameter, we test:

$$H_0: \beta_1 = 0$$
$$H_a: \beta_1 \neq 0$$

The test statistic is $t = 2.68$ and the p-value is $p = 0.055$. Since the p-value is less than $\alpha = 0.10\,(p = 0.055 < 0.10)$, H_0 is rejected. There is sufficient evidence to indicate there is a linear relationship between apparent porosity and mean pore diameter at $\alpha = 0.10$.

e. From the printout, $r^2 = 0.6427$. 64.27% of the sample variability in the apparent porosity around its sample mean is explained by the linear relationship between apparent porosity and mean pore diameter.

$r = \sqrt{0.6427} = 0.8017$. There is a moderately strong positive linear relationship between apparent porosity and mean pore diameter.

f. Using MINITAB, the results are:

Prediction for Porosity

```
Regression Equation

Porosity = 6.31 + 0.967 PoreDiameter

Variable      Setting
PoreDiameter       10

    Fit    SE Fit          90% CI                    90% PI
15.9778   1.31988   (13.1640, 18.7916)   (8.55972, 23.3959)
```

The 90% prediction interval is $(8.55972, 23.3959)$. We are 90% confident that the actual apparent porosity is between 8.56 and 23.40 when the mean pore diameter is 10 micrometers.

11.145 a. Using MINITAB, the scattergram of the data is:

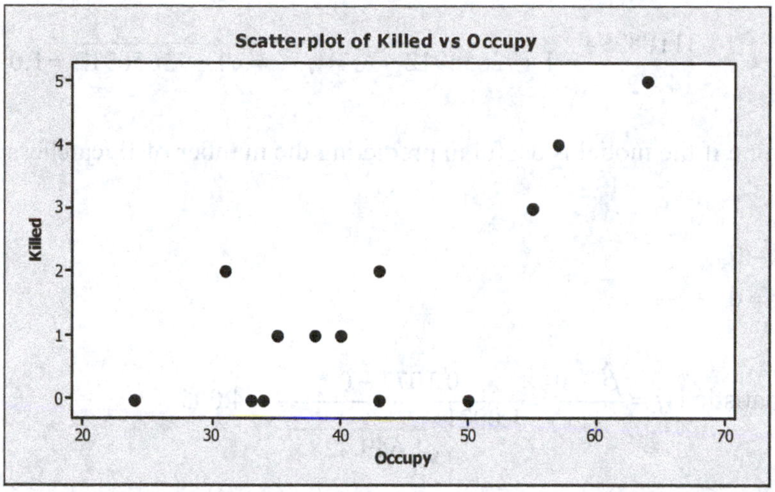

From the graph, it appears that as nest box tit occupancy increases, the number of flycatchers killed also increases.

b. Some preliminary calculations are:

$\sum x_i = 582$ $\sum y_i = 20$ $\sum x_i y_i = 1{,}009$

$\sum x_i^2 = 25{,}844$ $\sum y_i^2 = 62$

$$SS_{xy} = \sum x_i y_i - \frac{\sum x_i \sum y_i}{n} = 1{,}009 - \frac{582(20)}{14} = 177.5714286$$

$$SS_{xx} = \sum x_i^2 - \frac{\left(\sum x_i\right)^2}{n} = 25{,}844 - \frac{582^2}{14} = 1{,}649.42857$$

$$\hat{\beta}_1 = \frac{SS_{xy}}{SS_{xx}} = \frac{177.5714286}{1{,}649.42857} = 0.107656331$$

For each unit increase in the nest box tit occupancy percent, the mean number of flycatchers killed is estimated to increase by 0.1077.

$$\hat{\beta}_0 = \bar{y} - \hat{\beta}_1\bar{x} = \frac{20}{14} - (0.107656331)\left(\frac{582}{14}\right) = -3.046856046$$

Since $x = 0$ is not in the observed range, $\hat{\beta}_0$ has no meaning other than the y-intercept.

c. Some additional calculations are:

$$SS_{yy} = \sum y_i^2 - \frac{\left(\sum y_i\right)^2}{n} = 62 - \frac{20^2}{14} = 33.42857143$$

$$\text{SSE} = \text{SS}_{yy} - \hat{\beta}_1 \text{SS}_{xy} = 33.42857143 - (0.107656331)(177.5714286) = 14.31188294$$

$$s^2 = \frac{\text{SSE}}{n-2} = \frac{14.31188294}{14-2} = 1.192656912 \quad s = \sqrt{s^2} = \sqrt{1.192656912} = 1.0921$$

To determine if the model is useful in predicting the number of flycatchers killed, we test:

$$H_0: \beta_1 = 0$$
$$H_a: \beta_1 \neq 0$$

The test statistic is $t = \frac{\hat{\beta}_1 - 0}{s_{\hat{\beta}_1}} = \frac{0.1077 - 0}{1.0921 / \sqrt{1,649.42857}} = 4.005$

The rejection region requires $\alpha / 2 = 0.05 / 2 = 0.025$ in each tail of the t distribution. From Table III, Appendix B, with $df = n-2 = 14-2 = 12$, $t_{0.025} = 2.179$. The rejection region is $t < -2.179$ or $t > 2.179$.

Since the observed value of the test statistic falls in the rejection region $(t = 4.005 > 2.179)$, H_0 is rejected. There is sufficient evidence to indicate that the model is useful in predicting the number of flycatchers killed at $\alpha = 0.05$.

d. $$r = \frac{\text{SS}_{xy}}{\sqrt{\text{SS}_{xx}\text{SS}_{yy}}} = \frac{177.5714286}{\sqrt{1,649.42857(33.42857143)}} = 0.7562$$

The value of r is moderately close to 1. Thus, there is a moderately strong positive linear relationship between the number of flycatchers killed and the nest box tit occupancy.

$r^2 = 0.7562^2 = 0.572$. 57.2% of the sample variability around the sample mean number of flycatchers killed is explained by the linear relationship between the number of flycatchers killed and the nest box tit occupancy.

e. From part c, $s = 1.0921$. We would expect most of the observed number of flycatchers killed to fall within $2s = 2(1.0921) = 2.1842$ units of their predicted values.

f. Since there is a significant linear relationship between the number of flycatchers killed and the nest box tit occupancy, we would recommend using this model. In addition, the r^2 value is 0.572. Although this is not great, over half of the sample variation is explained by the model.

11.147 Using MINITAB, the output for fitting the least squares line is:

Regression Analysis: Impur versus Temp

```
The regression equation is
Impur = - 13.5 - 0.0528 Temp

Predictor        Coef    SE Coef      T      P
Constant      -13.490      2.074  -6.51  0.000
Temp        -0.052829   0.007728  -6.84  0.000


S = 0.133347   R-Sq = 85.4%    R-Sq(adj) = 83.6%


Analysis of Variance

Source          DF       SS       MS      F      P
Regression       1  0.83089  0.83089  46.73  0.000
Residual Error   8  0.14225  0.01778
Total            9  0.97315


Unusual Observations

Obs  Temp   Impur     Fit  SE Fit  Residual  St Resid
  2  -265  0.2020  0.5093  0.0492   -0.3073    -2.48R
  3  -256  0.2040  0.0339  0.1038    0.1701     2.03RX

R denotes an observation with a large standardized residual.
X denotes an observation whose X value gives it large influence.


Predicted Values for New Observations

New
Obs     Fit  SE Fit       95% CI              95% PI
  1  0.9320  0.0558  (0.8034, 1.0605)  (0.5987, 1.2652)

Values of Predictors for New Observations

New
Obs  Temp
  1  -273
```

a. $\hat{\beta}_0 = -13.490$. Since $x = 0$ is not in the observed range, $\hat{\beta}_0$ has no meaning since. It is the y-intercept.

$\hat{\beta}_1 = -0.052829$. For each additional degree, the mean proportion of impurity passing through helium is estimated to decrease by .052829 units.

b. For confidence coefficient 0.95, $\alpha = 0.05$ and $\alpha / 2 = 0.05 / 2 = 0.025$. From Table III, Appendix B, with $df = n - 2 = 10 - 2 = 8$, $t_{0.025} = 2.306$. The confidence interval is:

$$\hat{\beta}_1 \pm t_{\alpha/2} s_{\hat{\beta}_1} \Rightarrow -0.0528 \pm 2.306(0.007728) \Rightarrow -0.0528 \pm 0.0178 \Rightarrow (-0.0706, -0.0350)$$

We are 95% confident that the change in mean proportion of impurity passing through helium for each additional degree is between −0.0706 and −0.0350. Since 0 is not in the interval, there is evidence to indicate that temperature contributes information about the proportion of impurity passing through helium.

c. $r^2 = R - sq = 0.854$. 85.4% of the total sample variation around the mean proportion of impurity is explained by the linear relationship between proportion of impurity and temperature.

d. From the printout, the 95% prediction interval is $(0.5987,\ 1.2652)$. We are 95% confident that the actual percentage of impurity passing through solid helium at -273°C is between 59.87% and 126.52%. Since we cannot have more than 100% of the impurity passing through solid helium, the interval is from 59.87% to 100%.

e. We cannot be sure that the relationship between the proportion of impurity passing through helium and temperature is the same outside the observed range.

11.149 a. $r = -0.726$. Since this value is somewhat close to -1, there is a moderately strong negative linear relationship between the number of online courses taken and the weekly quiz grade.

b. To determine if a negative correlation exists, we test:

$$H_0: \rho = 0$$
$$H_a: \rho < 0$$

The test statistic is $t = \frac{r\sqrt{n-2}}{\sqrt{1-r^2}} = \frac{-0.726\sqrt{24-2}}{\sqrt{1-(-0.726)^2}} = -4.95$.

The rejection region requires $\alpha = 0.05$ in the lower tail of the t distribution with $df = n-2 = 24-2 = 22$. From Table III, Appendix B, $t_{0.05} = 1.717$. The rejection region is $t < -1.717$.

Since the observed value of the test statistic falls in the rejection region $(t = -4.95 < -1.717)$, H_o is rejected. There is sufficient evidence to indicate a negative linear relationship between the number of online courses taken and the weekly quiz grade at $\alpha = 0.05$.

11.151 Some preliminary calculations are:

$\sum x = 4305$ $\sum x^2 = 1{,}652{,}025$ $\sum xy = 76{,}652{,}695$

$\sum y = 201{,}558$ $\sum y^2 = 3{,}571{,}211{,}200$

a. $\hat{\beta}_1 = \frac{\sum xy}{\sum x^2} = \frac{76{,}652{,}695}{1{,}652{,}025} = 46.39923427 \approx 46.3992$

The least squares line is $\hat{y} = 46.3992x$.

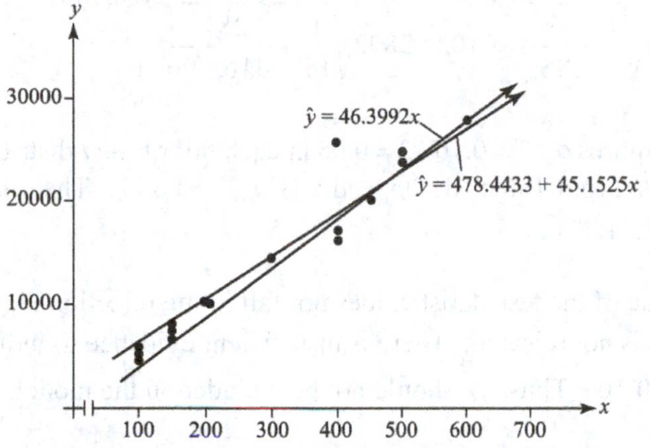

b. $$SS_{xy} = \sum xy - \frac{\sum x \sum y}{n} = 76{,}652{,}695 - \frac{4305(201{,}558)}{15} = 18{,}805{,}549$$

$$SS_{xx} = \sum x^2 - \frac{\left(\sum x\right)^2}{n} = 1{,}652{,}025 - \frac{4305^2}{15} = 416{,}490$$

$$\hat{\beta}_1 = \frac{SS_{xy}}{SS_{xx}} = \frac{18{,}805{,}549}{416{,}490} = 45.15246224 \approx 45.1525$$

$$\hat{\beta}_0 = \bar{y} - \hat{\beta}_1\bar{x} = \frac{201{,}558}{15} - 45.15246224\left(\frac{4305}{15}\right) = 478.4433$$

The least squares line is $\hat{y} = 478.4433 + 45.1525x$.

c. Because $x = 0$ is not in the observed range, we are trying to represent the data on the observed interval with the best fitting line. We are not concerned with whether the line goes through $(0, 0)$ or not.

d. Some preliminary calculations are:

$$SS_{yy} = \sum y^2 - \frac{\left(\sum y\right)^2}{n} = 3{,}571{,}211{,}200 - \frac{201{,}558^2}{15} = 862{,}836{,}042$$

$$SSE = SS_{yy} - \hat{\beta}_1 SS_{xy} = 862{,}836{,}042 - 45.15246224(18{,}805{,}549) = 13{,}719{,}200.88$$

$$s^2 = \frac{SSE}{n-2} = \frac{13{,}719{,}200.88}{15-2} = 1{,}055{,}323.145 \qquad s = \sqrt{s^2} = 1027.2892$$

$$H_0: \beta_0 = 0$$
$$H_a: \beta_0 \neq 0$$

The test statistic is $t = \dfrac{\hat{\beta}_0 - 0}{s\sqrt{\dfrac{1}{n} + \dfrac{\bar{x}^2}{SS_{xx}}}} = \dfrac{478.443}{1027.2892\sqrt{\dfrac{1}{15} + \dfrac{287^2}{416,490}}} = 0.906$

The rejection region requires $\alpha / 2 = 0.10 / 2 = 0.05$ in each tail of the t distribution with $df = n - 2 = 15 - 2 = 13$. From Table III, Appendix B, $t_{0.05} = 1.771$. The rejection region is $t < -1.771$ or $t > 1.771$.

Since the observed value of the test statistic does not fall in the rejection region $(t = 0.906 \not> 1.771)$, H_0 is not rejected. There is insufficient evidence to indicate β_0 is different from 0 at $\alpha = 0.10$. Thus, β_0 should not be included in the model.

11.153 Answers may vary. Possible answer:

The scaffold-drop survey provides the most accurate estimate of spall rate in a given wall segment. However, the drop areas were not selected at random from the entire complex; rather, drops were made at areas with high spall concentrations. Therefore, if the photo spall rates could be shown to be related to drop spall rates, then the 83 photo spall rates could be used to predict what the drop spall rates would be.

Construct a scatterplot for the data.

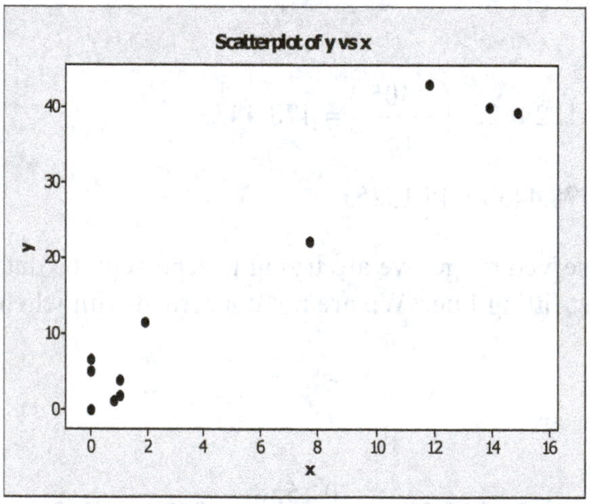

The scatterplot shows a positive linear relationship between the photo spall rate (x) and the drop spall rate (y).

Find the prediction equation for drop spall rate. The MINITAB output shows the results of the analysis.

```
The regression equation is
drop = 2.55 + 2.76 photo

Predictor        Coef       StDev        T        P
Constant        2.548       1.637     1.56    0.154
photo          2.7599      0.2180    12.66    0.000
```

```
S = 4.164          R-Sq = 94.7%        R-Sq(adj) = 94.1%
Analysis of Variance

Source              DF          SS          MS         F        P
Regression           1      2777.5      2777.5    160.23    0.000
Residual Error       9       156.0        17.3
Total               10      2933.5
```

The least squares prediction line is $\hat{y} = 2.55 + 2.76x$.

Conduct a formal statistical hypothesis test to determine if the photo spall rates contribute information for the prediction of drop spall rates.

$$H_0 : \beta_0 = 0$$
$$H_a : \beta_0 \neq 0$$

The test statistic is $t = 12.66$ and the p-value is $p = 0.000$.

Since the p-value is so small $(p = 0.000)$, H_0 is rejected for any reasonable value of α. There is sufficient evidence to indicate that photo spall rates contribute information for the prediction of drop spall rates for any reasonable value of α.

One could now use the 83 photos spall rates to predict values for 83 drop spall rates. Then use this information to estimate the true spall rate at a given wall segment and estimate to total spall damage.

Chapter 12

Multiple Regression and Model Building

12.1 a. $E(y)=\beta_0+\beta_1x_1+\beta_2x_2$

b. $E(y)=\beta_0+\beta_1x_1+\beta_2x_2+\beta_3x_3+\beta_4x_4$

c. $E(y)=\beta_0+\beta_1x_1+\beta_2x_2+\beta_3x_3+\beta_4x_4+\beta_5x_5$

12.3 The 6 steps in a multiple regression analysis are:

1. Hypothesize the deterministic component of the model. This model relates the mean, $E(y)$, to the independent variables $x_1, x_2, \ldots, x_k$. This involves the choice of the independent variables to be included in the model.

2. Use the sample data to estimate the unknown model parameters $\beta_0, \beta_1, \beta_2, \ldots, \beta_k$ in the model.

3. Specify the probability distribution of the random error term, ε, and estimate the standard deviation of the distribution, σ.

4. Check the assumptions on ε are satisfied, and make model modifications if necessary.

5. Statistically evaluate the usefulness of the model.

6. When satisfied that the model is useful, use it for prediction, estimation, and other purposes.

12.5 To test the overall adequacy of a multiple regression model, you would use a global test. The null hypothesis would be that all of the beta parameters (except β_0) are 0 and the alternative hypothesis would be that at least one beta parameter is not 0.

12.7 a. We are given $\hat{\beta}_2=2.7$, $s_{\hat{\beta}_2}=1.86$, and $n=30$.

$H_0: \beta_2=0$
$H_a: \beta_2\neq 0$

The test statistic is $t=\dfrac{\hat{\beta}_2-0}{s_{\hat{\beta}_2}}=\dfrac{2.7}{1.86}=1.45$

The rejection region requires $\alpha/2=0.05/2=0.025$ in each tail of the t distribution with $df=n-(k+1)=30-(3+1)=26$. From Table III, Appendix B, $t_{0.025}=2.056$. The rejection region is $t<-2.056$ or $t>2.056$.

Since the observed value of the test statistic does not fall in the rejection region $(t = 1.45 \not> 2.056)$, H_0 is not rejected. There is insufficient evidence to indicate $\beta_2 \neq 0$ at $\alpha = 0.05$.

b. We are given $\hat{\beta}_3 = 0.93$, $s_{\hat{\beta}_3} = 0.29$, and $n = 30$.

$$H_0: \beta_3 = 0$$
$$H_a: \beta_3 \neq 0$$

The test statistic is $t = \dfrac{\hat{\beta}_3 - 0}{s_{\hat{\beta}_3}} = \dfrac{0.93}{0.29} = 3.21$

The rejection region is the same as part **a**, $t < -2.056$ or $t > 2.056$.

Since the observed value of the test statistic falls in the rejection region $(t = 3.21 > 2.056)$, H_0 is rejected. There is sufficient evidence to indicate $\beta_3 \neq 0$ at $\alpha = 0.05$.

c. $\hat{\beta}_3$ has a smaller estimated standard error than $\hat{\beta}_2$. Therefore, the test statistic is larger for $\hat{\beta}_3$ even though $\hat{\beta}_3$ is smaller than $\hat{\beta}_2$.

12.9 The number of degrees of freedom available for estimating σ^2 is $n - (k + 1)$ where k is the number of variables in the regression model. Each additional independent variable placed in the model causes a corresponding decrease in the degrees of freedom.

12.11 a. Yes. Since $R^2 = 0.92$ is close to 1, this indicates the model provides a good fit. Also, SSE = 0.33 is fairly small which indicates the model provides a good fit.

b. $H_0: \beta_1 = \beta_2 = \cdots = \beta_5 = 0$
H_a: At least one $\beta_i \neq 0$

The test statistic is $F = \dfrac{R^2 / k}{(1 - R^2)/[n - (k + 1)]} = \dfrac{0.92/5}{(1 - 0.92)/[30 - (5 + 1)]} = 55.2$

The rejection region requires $\alpha = 0.05$ in the upper tail of the F distribution with $\nu_1 = k = 5$ and $\nu_2 = n - (k + 1) = 30 - (5 + 1) = 24$. From Table VI, Appendix B, $F_{0.05} = 2.62$. The rejection region is $F > 2.62$.

Since the observed value of the test statistic falls in the rejection region $(F = 55.2 > 2.62)$, H_0 is rejected. There is sufficient evidence to indicate the model is useful in predicting y at $\alpha = 0.05$.

12.13 a. To determine the overall adequacy of the model, we test:

$$H_0: \beta_1 = \beta_2 = \beta_3 = 0$$
$$H_a: \text{At least one } \beta_i \neq 0$$

The test statistic is $F = 31.98$ and the p-value is $p < 0.001$.

Since the p-value is less than $\alpha\left(p < 0.001 < 0.01\right)$, H_0 is rejected. There is sufficient evidence that the model is useful for predicting severity of tilting at $\alpha = 0.01$.

b. To determine if the rate of change of severity of tilting with perceived effect of experience on tilting depends on poker experience, we test:

$$H_0: \beta_3 = 0$$
$$H_a: \beta_3 \neq 0$$

The test statistic is $t = -5.61$ and the p-value is $p < 0.001$.

Since the p-value is less than $\alpha\left(p < 0.001 < 0.01\right)$, H_0 is rejected. There is sufficient evidence to indicate that the rate of change of severity of tilting with perceived effect of experience on tilting depends on poker experience at $\alpha = 0.01$.

12.15 a. Two properties of the errors of prediction that result from using the method of least squares are the sum of the errors is 0 and the sum of the squares of the errors (SSE) is minimized.

b. $\hat{\beta}_4 = 0.42$. For each unit increase in the number of shortest paths between peers, the mean lead-user rating is estimated to increase by 0.42, holding all of the other variables constant.

c. Since the p-value is less than $\alpha\left(p = 0.002 < 0.05\right)$, H_0 is rejected. There is sufficient evidence to indicate a linear relationship between the number of shortest paths between peers and lead-user rating, adjusted for the other variables in the model for $\alpha = 0.05$.

12.17 a. The model would be: $E(y) = \beta_0 + \beta_1 x_1 + \beta_2 x_2$

b. To determine if the length of an entangled whale increases linearly as the water depth increases for entanglements that are a fixed distance from land, we test:

$$H_0: \beta_2 = 0$$
$$H_a: \beta_2 > 0$$

c. Since the p-value is less than $\alpha\left(p = 0.013 < 0.05\right)$, H_0 is rejected. There is sufficient evidence to indicate the length of an entangled whale increases linearly as the water depth increases for entanglements that are a fixed distance from land for $\alpha = 0.05$.

12.19 a. From the output, the least squares prediction equation is

$$\hat{y} = 40 + 0.254x_1 + 0.407x_2 + 0.841x_3 + 1.63x_4 + 1.09x_5 + 0.05x_6 - 0.044x_7 - 0.0911x_8 - 0.122x_9.$$

b. $\hat{\beta}_0 = 40$. This value has no meaningful interpretation.

$\hat{\beta}_1 = 0.254$. We estimate the mean number of runs scored by a team in a season will increase by 0.254 for each additional walk, when all other independent variables are held constant.

$\hat{\beta}_2 = 0.407$. We estimate the mean number of runs scored by a team in a season will increase by 0.407 for each additional single, when all the other independent variables are held constant.

$\hat{\beta}_3 = 0.841$. We estimate the mean number of runs scored by a team in a season will increase by 0.841 for each additional double, when all the other independent variables are held constant.

$\hat{\beta}_4 = 1.63$. We estimate the mean number of runs scored by a team in a season will increase by 1.63 for each additional triple, when all the other independent variables are held constant.

$\hat{\beta}_5 = 1.09$. We estimate the mean number of runs scored by a team in a season will increase by 1.09 for each additional home run, when all the other independent variables are held constant.

$\hat{\beta}_6 = 0.05$. We estimate the mean number of runs scored by a team in a season will increase by 0.05 for each additional stolen base, when all the other independent variables are held constant.

$\hat{\beta}_7 = -0.044$. We estimate the mean number of runs scored by a team in a season will decrease by 0.044 for each additional player caught stealing, when all the other independent variables are held constant.

$\hat{\beta}_8 = -0.0911$. We estimate the mean number of runs scored by a team in a season will decrease by 0.0911 for each additional strikeout, when all the other independent variables are held constant.

$\hat{\beta}_9 = -0.122$. We estimate the mean number of runs scored by a team in a season will decrease by 0.122 for each additional out, when all the other independent variables are held constant.

c. $H_0: \beta_7 = 0$
$H_a: \beta_7 < 0$

The test statistic is $t = \dfrac{\hat{\beta}_7 - 0}{s_{\hat{\beta}_7}} = \dfrac{-0.0436 - 0}{0.4842} = -0.09$ and the p-value is $p = 0.929$.

Since the *p*-value is not less than $\alpha\left(p = 0.929 \not< 0.05\right)$, H_0 is not rejected. There is insufficient evidence to indicate that there is a negative linear relationship between the number of runs scored and the number of times caught stealing at $\alpha = .05$.

d. For confidence coefficient 0.95, $\alpha = 0.05$ and $\alpha/2 = 0.05/2 = 0.025$. From Table III, Appendix B, with $df = n-(k+1) = 30-(9+1) = 20$, $t_{0.025} = 2.086$. The 95% confidence interval is:

$$\hat{\beta}_5 \pm t_{0.025} s_{\hat{\beta}_5} \Rightarrow 1.0925 \pm 2.086(0.1945) \Rightarrow 1.0925 \pm 0.4057 \Rightarrow (0.687,\ 1.498)$$

We are 95% confident that for each additional home run, the mean number of runs scored will increase by anywhere from 0.687 to 1.498, holding all the other variables constant.

12.21 a. The first order model is $E(y) = \beta_0 + \beta_1 x_1 + \beta_2 x_2 + \beta_3 x_3$

b. $R^2 = 0.08$. 8% of the sample variation in the frequency of marijuana use in the past 6 months is explained by the model containing severity of inattention, severity of impulsivity-hyperactivity, and level of oppositional-defiant and conduct disorder.

c. The global *F*-test has the hypotheses:

$$H_0: \beta_1 = \beta_2 = \beta_3 = 0$$
$$H_a: \text{At least one } \beta_i \neq 0$$

The global *F*-test yielded a *p*-value less than 0.01. Since the *p*-value is so small, there is evidence to reject H_0 for $\alpha > 0.01$. There is sufficient evidence to indicate that at least one of the independent variables contributes to the prediction of the frequency of marijuana use in the past 6 months at $\alpha > 0.01$.

d. The test for β_1:

$$H_0: \beta_1 = 0$$
$$H_a: \beta_1 \neq 0$$

The *p*-value for the test is less than 0.01. Since the *p*-value is so small, there is evidence to reject H_0 for $\alpha > 0.01$. There is sufficient evidence to indicate that severity of inattention contributes to the prediction of the frequency of marijuana use in the past 6 months, adjusting for severity of impulsivity-hyperactivity and level of oppositional-defiant and conduct disorder fixed at $\alpha > 0.01$.

e. The test for β_2:

$$H_0: \beta_2 = 0$$
$$H_a: \beta_2 \neq 0$$

The p-value for the test is greater than 0.05. Since the p-value is not small, there is no evidence to reject H_0 for $\alpha \le 0.05$. There is insufficient evidence to indicate that severity of impulsivity-hyperactivity contributes to the prediction of the frequency of marijuana use in the past 6 months, adjusting for severity of inattention and level of oppositional-defiant and conduct disorder fixed at $\alpha \le 0.05$.

f. The test for β_3:

$$H_0: \beta_3 = 0$$
$$H_a: \beta_3 \ne 0$$

The p-value for the test is greater than 0.05. Since the p-value is not small, there is no evidence to reject H_0 for $\alpha \le 0.05$. There is insufficient evidence to indicate that level of oppositional-defiant and conduct disorder contributes to the prediction of the frequency of marijuana use in the past 6 months, adjusting for severity of inattention and severity of impulsivity-hyperactivity fixed at $\alpha \le 0.05$.

12.23 a. Let y = arsenic level, x_1 = latitude, x_2 = longitude, and x_3 = depth. The first-order model would be:

$$E(y) = \beta_o + \beta_1 x_1 + \beta_2 x_2 + \beta_3 x_3$$

b. Using MINITAB, the results are:

Regression Analysis: ARSENIC versus LATITUDE, LONGITUDE, DEPTH-FT

```
The regression equation is
ARSENIC = - 86868 - 2219 LATITUDE + 1542 LONGITUDE - 0.350 DEPTH-FT


327 cases used, 1 cases contain missing values


Predictor      Coef  SE Coef      T      P
Constant     -86868    31224  -2.78  0.006
LATITUDE    -2218.8    526.8  -4.21  0.000
LONGITUDE    1542.2    373.1   4.13  0.000
DEPTH-FT    -0.3496   0.1566  -2.23  0.026


S = 103.301   R-Sq = 12.8%   R-Sq(adj) = 12.0%


Analysis of Variance

Source           DF       SS      MS      F      P
Regression        3   505770  168590  15.80  0.000
Residual Error  323  3446791   10671
Total           326  3952562


Source      DF  Seq SS
LATITUDE     1  132448
LONGITUDE    1  320144
DEPTH-FT     1   53179
```

The fitted regression model is:

$\hat{y} = -86,868 - 2,218.8x_1 + 1,542.2x_2 - 0.3496x_3$

c. $\hat{\beta}_0 = -86,868$. This value has no meaningful interpretation because $x_1 = 0$, $x_2 = 0$, and $x_3 = 0$ are not in the observed range.

$\hat{\beta}_1 = -2,218.8$. We estimate the mean arsenic level will decrease by 2,218.8 for each unit increase in latitude, holding longitude and depth constant.

$\hat{\beta}_2 = 1,542.2$. We estimate the mean arsenic level will increase by 1,542.2 for each unit increase in longitude, holding latitude and depth constant.

$\hat{\beta}_3 = -0.3496$. We estimate the mean arsenic level will decrease by 0.3496 for each unit increase in depth, holding latitude and longitude constant.

d. $s = 103.301$. We would expect about 95% of the observed values of arsenic levels to fall within $2s = 2(103.301) = 206.602$ units of their least squares predicted values.

e. $R^2 = \text{R-Sq} = 0.128$. 12.8% of the sample variation in the arsenic level is explained by the model containing latitude, longitude and depth.

$R_a^2 = \text{R-Sq(adj)} = 0.120$. 12.0% of the sample variation in the arsenic level is explained by the model containing latitude, longitude and depth, adjusted for the sample size and the number of independent variables in the model.

f. To determine if the overall model is useful for predicting y, we test:

$H_0: \beta_1 = \beta_2 = \beta_3 = 0$

H_a: At least one $\beta_i \neq 0$

The test statistic is $F = 15.80$ and the p-value is $p = 0.000$.

Since the p-value is less than α $(p = 0.000 < 0.05)$, H_0 is rejected. There is sufficient evidence to indicate that at least one of the independent variables is useful in predicting arsenic levels at $\alpha = 0.05$.

g. Even though the independent variables are significant in predicting arsenic levels, the R^2 value and adjusted R^2 values are very low. Not very much of the sample variation in the arsenic levels can be explained by the model. In addition, the estimate of the standard deviation $(s = 103.301)$ is very large compared with the actual values of the data. I would not recommend using this model.

12.25 a. Using MINITAB, the results of fitting the first-order model are:

Regression Analysis: DESIRE versus GENDER, SELFESTM, BODYSAT, IMPREAL

```
The regression equation is
DESIRE = 14.0 - 2.19GENDER - 0.0479SELFESTM - 0.322BODYSAT + 0.493IMPREAL


Predictor      Coef   SE Coef      T      P
Constant    14.0107    0.7753  18.07  0.000
GENDER      -2.1865    0.6766  -3.23  0.001
SELFESTM   -0.04794   0.03669  -1.31  0.193
BODYSAT     -0.3223    0.1435  -2.25  0.026
IMPREAL      0.4931    0.1274   3.87  0.000


S = 2.25087   R-Sq = 49.8%   R-Sq(adj) = 48.5%


Analysis of Variance

Source           DF       SS      MS      F      P
Regression        4   827.83  206.96  40.85  0.000
Residual Error  165   835.95    5.07
Total           169  1663.79

Source     DF  Seq SS
GENDER      1  674.64
SELFESTM    1   57.66
BODYSAT     1   19.62
IMPREAL     1   75.91
```

The least squares prediction equation is:

$$\hat{y} = 14.0107 - 2.1865(\text{GENDER}) - 0.04794(\text{SELFESTM}) - 0.3223(\text{BODYSAT}) + 0.493(\text{IMPREAL})$$

b. $\hat{\beta}_0 = 14.0107$. This has no meaning other than the y-intercept.

$\hat{\beta}_1 = -2.1865$. The mean value of desire to have cosmetic surgery is estimated to be 2.1865 units lower for males than females, holding all other variables constant.

$\hat{\beta}_2 = -0.04794$. For each unit increase in self-esteem, the mean value of desire to have cosmetic surgery is estimated to decrease by 0.04794 units, holding all other variables constant.

$\hat{\beta}_3 = -0.3223$. For each unit increase in body satisfaction, then mean value of desire to have cosmetic surgery is estimated to decrease by 0.3223 units, holding all other variables constant.

$\hat{\beta}_4 = 0.4931$. For each unit increase in impression of reality TV, the mean value of desire to have cosmetic surgery is estimated to increase by 0.4931 units, holding all other variables constant.

c. To determine if the overall model is useful for predicting desire to have cosmetic surgery, we test:

$H_0: \beta_1 = \beta_2 = \beta_3 = \beta_4 = 0$
H_a: At least one $\beta_i \neq 0$

From the printout, the test statistic is $F = 40.85$ and the *p*-value is $p = 0.000$.

Since the *p*-value is less than $\alpha (p = 0.000 < 0.01)$, H_0 is rejected. There is sufficient evidence to indicate the overall model is useful for predicting desire to have cosmetic surgery at $\alpha = 0.01$.

d. R_a^2 is the preferred measure of model fit. From the printout, $R_a^2 = 0.485$. This indicates that 48.5% of the total sample variation in desire values is explained by the model containing gender, self-esteem, body satisfaction and impression of reality TV, adjusting for the sample size and the number of variables in the model.

e. To determine if the desire to have cosmetic surgery decreases linearly as level of body satisfaction increases, we test:

$H_0: \beta_3 = 0$
$H_a: \beta_3 < 0$

From the printout, the test statistic is $t = -2.25$ and the *p*-value is $p = 0.026/2 = 0.013$.

Since the *p*-value is less than $\alpha (p = 0.013 < 0.05)$, H_0 is rejected. There is sufficient evidence to indicate the desire to have cosmetic surgery decreases linearly as level of body satisfaction increases at $\alpha = 0.05$.

f. For confidence coefficient 0.95, $\alpha = 0.05$ and $\alpha/2 = 0.05/2 = 0.025$. From Table III, Appendix B, with $df = n-(k+1) = 170-(4+1) = 165$, $t_{0.025} \approx 1.98$. The 95% confidence interval is

$$\hat{\beta}_4 \pm t_{\alpha/2} s_{\hat{\beta}_4} \Rightarrow 0.4931 \pm 1.98(0.1274) \Rightarrow 0.4931 \pm 0.2523 \Rightarrow (0.2408,\ 0.7454)$$

We are 95% confident that the increase in mean desire for cosmetic surgery is between 0.2408 and 0.7454 for each unit increase in impression of reality TV, holding all other variables constant.

12.27 a. Let y = heat rate, x_1 = speed, x_2 = inlet temperature, x_3 = exhaust temperature, x_4 = cycle pressure ratio, and x_5 = air flow rate. The first-order model would be:

$$E(y) = \beta_o + \beta_1 x_1 + \beta_2 x_2 + \beta_3 x_3 + \beta_4 x_4 + \beta_5 x_5$$

b. Using MINITAB, the results are:

Regression Analysis: HEATRATE versus RPM, INLET-TEMP, ...

```
The regression equation is
HEATRATE = 13614 + 0.0888 RPM - 9.20 INLET-TEMP + 14.4 EXH-TEMP + 0.4
CPRATIO - 0.848 AIRFLOW

Predictor      Coef   SE Coef      T      P
Constant    13614.5     870.0  15.65  0.000
RPM         0.08879   0.01391   6.38  0.000
INLET-TEMP   -9.201     1.499  -6.14  0.000
EXH-TEMP     14.394     3.461   4.16  0.000
CPRATIO        0.35     29.56   0.01  0.991
AIRFLOW     -0.8480    0.4421  -1.92  0.060

S = 458.828   R-Sq = 92.4%   R-Sq(adj) = 91.7%

Analysis of Variance

Source          DF         SS        MS       F      P
Regression       5  155055273  31011055  147.30  0.000
Residual Error  61   12841935    210524
Total           66  167897208


Source      DF     Seq SS
RPM          1  119598530
INLET-TEMP   1   26893467
EXH-TEMP     1    7784225
CPRATIO      1       4623
AIRFLOW      1     774427
```

The fitted regression model is:

$$\hat{y} = 13,614.5 + 0.08879x_1 - 9.201x_2 + 14.394x_3 + 0.35x_4 - 0.848x_5$$

c. $\hat{\beta}_o = 13,614.5$. This value has no meaningful interpretation other than the y-intercept because $x_1 = 0$, $x_2 = 0$, $x_3 = 0$, $x_4 = 0$, and $x_5 = 0$, are not in the observed range.

$\hat{\beta}_1 = 0.08879$. We estimate the mean heat rate will increase by 0.08879 units for each unit increase in speed, holding inlet temperature, exhaust temperature, cycle pressure ratio, and air flow rate constant.

$\hat{\beta}_2 = -9.201$. We estimate the mean heat rate will decrease by 9.201 units for each unit increase in inlet temperature, holding speed, exhaust temperature, cycle pressure ratio, and air flow rate constant.

$\hat{\beta}_3 = 14.394$. We estimate the mean heat rate will increase by 14.394 units for each unit increase in exhaust temperature, holding speed, inlet temperature, cycle pressure ratio, and air flow rate constant.

$\hat{\beta}_4 = 0.35$. We estimate the mean heat rate will increase by 0.35 units for each unit increase in cycle pressure ratio, holding speed, exhaust temperature, inlet temperature, and air flow rate constant.

$\hat{\beta}_5 = -0.8480$. We estimate the mean heat rate will decrease by 0.848 units for each unit increase in air flow rate constant, holding speed, inlet temperature, exhaust temperature, and cycle pressure ratio.

d. $s = 458.828$. We would expect about 95% of the observed values of heat rate to fall within $2s = 2(458.828) = 917.656$ units of their least squares predicted values.

e. $R_a^2 = \text{R-Sq}(\text{adj}) = 0.917$. 91.7% of the sample variation in the heat rate is explained by the model containing speed, inlet temperature, exhaust temperature, cycle pressure ratio, and air flow rate, adjusted for the sample size and the number of independent variables in the model.

f. To determine if the overall model is useful for predicting y, we test:

$$H_0: \beta_1 = \beta_2 = \beta_3 = \beta_4 = \beta_5 = 0$$
$$H_a: \text{At least one } \beta_i \neq 0$$

The test statistic is $F = 147.30$ and the p-value is $p = 0.000$.

Since the p-value is less than $\alpha\ (p = 0.000 < 0.01)$, H_0 is rejected. There is sufficient evidence to indicate that at least one of the independent variables is useful in predicting heat rate at $\alpha = 0.01$.

12.29 a. The least squares prediction equation for the interstate highway model is:

$$\hat{y} = 1.81231 + 0.10875x_1 + 0.00017x_2$$

b. $\hat{\beta}_0 = 1.81231$. This value has no meaningful interpretation other than the y-intercept because $x_1 = 0$ and $x_2 = 0$ are not in the observed range.

$\hat{\beta}_1 = 0.10875$. We estimate the mean number of crashes per 3 years will increase by 0.10875 for each additional mile of roadway, holding AADT constant.

$\hat{\beta}_2 = 0.00017$. We estimate the mean number of crashes per 3 years will increase by 0.00017 for each additional vehicle per day, holding roadway mileage constant.

c. For confidence coefficient 0.99, $\alpha = 0.01$ and $\alpha/2 = 0.01/2 = 0.005$. From Table III, Appendix B, with $df = n - (k+1) = 100 - (2+1) = 97$, $t_{0.005} \approx 2.62$. The 99% confidence interval is:

$$\hat{\beta}_1 \pm t_{0.005} s_{\hat{\beta}_1} \Rightarrow 0.10875 \pm 2.62(0.03166) \Rightarrow 0.10875 \pm 0.08295 \Rightarrow (0.02580,\ \ 0.19170)$$

We are 99% confident that the increase in mean number of crashes for each additional mile of roadway is between 0.02580 and 0.19170, holding AADT constant.

d. The 99% confident interval is:

$$\hat{\beta}_2 \pm t_{0.005} s_{\hat{\beta}_2} \Rightarrow 0.00017 \pm 2.62(0.00003) \Rightarrow 0.00017 \pm 0.00008 \Rightarrow (0.00009,\ \ 0.00025)$$

We are 99% confident that the increase in mean number of crashes for each additional vehicle per day is between 0.00009 and 0.00025, holding roadway mileage constant.

e. The least squares prediction equation for the non-interstate highway model is:

$\hat{y} = 1.20785 + 0.06343x_1 + 0.00056x_2$

$\hat{\beta}_o = 1.20785$. This value has no meaningful interpretation other than the y-intercept because $x_1 = 0$ and $x_2 = 0$ are not in the observed range.

$\hat{\beta}_1 = 0.06343$. We estimate the mean number of crashes per 3 years will increase by 0.06343 for each additional mile of roadway, holding AADT constant.

$\hat{\beta}_2 = 0.00056$. We estimate the mean number of crashes per 3 years will increase by 0.00056 for each additional vehicle per day, holding roadway mileage constant.
For confidence coefficient 0.99, $\alpha = 0.01$ and $\alpha/2 = 0.01/2 = 0.005$. From Table III, Appendix B, with $df = n-(k+1) = 100-(2+1) = 97$, $t_{0.005} \approx 2.62$. The 99% confidence interval is:

$$\hat{\beta}_1 \pm t_{0.005} s_{\hat{\beta}_1} \Rightarrow 0.06343 \pm 2.62(0.01809) \Rightarrow 0.06343 \pm 0.04740 \Rightarrow (0.01603,\ \ 0.11083)$$

We are 99% confident that the increase in mean number of crashes for each additional mile of roadway is between 0.01603 and 0.11083, holding AADT constant.

The 99% confident interval is:

$$\hat{\beta}_2 \pm t_{0.005} s_{\hat{\beta}_2} \Rightarrow 0.00056 \pm 2.62(0.00012) \Rightarrow 0.00056 \pm 0.00031 \Rightarrow (0.00025,\ \ 0.00087)$$

We are 99% confident that the increase in mean number of crashes for each additional vehicle per day is between 0.00025 and 0.00087, holding roadway mileage constant.

12.319 The model we are fitting is $y = \beta_0 + \beta_1 x_1 + \beta_2 x_2 + \ldots + \beta_k x_k + \varepsilon$. The fitted model is $\hat{y} = \hat{\beta}_0 + \hat{\beta}_1 x_1 + \hat{\beta}_2 x_2 + \ldots + \hat{\beta}_k x_k$. Thus, we use the fitted model to predict the actual value of y. In addition, we could write the model as $E(y) = \beta_0 + \beta_1 x_1 + \beta_2 x_2 + \ldots + \beta_k x_k$. So, the fitted model will also estimate $E(y)$.

`12.33 a. For $x_1 = 1$, $x_2 = 10$, $x_3 = 5$, and $x_4 = 2$, the predicted lead-user rating is:

$$\hat{y} = 3.58 + 0.01(1) - 0.06(10) - 0.01(5) + 0.42(2) = 3.78$$

b. For $x_1 = 0$, $x_2 = 8$, $x_3 = 10$, and $x_4 = 4$, the predicted lead-user rating is:

$$\hat{y} = 3.58 + 0.01(0) - 0.06(8) - 0.01(10) + 0.42(4) = 4.68$$

12.35 a. The confidence interval for student 1 is $(13.4243, 14.3067)$. We are 95% confident that the true mean desire to have cosmetic surgery is between 13.4243 and 14.3067 for everyone who is female, has a self-esteem score of 24, has a body satisfaction score of 3, and has an impression of reality TV score of 4.

b. The confidence interval for student 4 is $(8.7908, 10.8910)$. We are 95% confident that the true mean desire to have cosmetic surgery is between 8.7908 and 10.8910 for everyone who is male, has a self-esteem score of 22, has a body satisfaction score of 9, and has an impression of reality TV score of 4.

12.37 a. From the printout, the 95% prediction interval is $(11{,}599.6, 13{,}665.5)$. We are 95% confident that the actual heat rate will be between 11,599.6 and 13,665.5 when the speed is 7,500 rpm, the inlet temperature is 1,000, the exhaust temperature is 525, the cycle pressure ratio is 13.5, and the air flow rate is 10.0.

b. From the printout, the 95% confidence interval for the mean is $(12{,}157.9, 13{,}107.1)$. We are 95% confident that the mean heat rate will be between 12,157.9 and 13,107.1 when the speed is 7,500 rpm, the inlet temperature is 1,000, the exhaust temperature is 525, the cycle pressure ratio is 13.5, and the air flow rate is 10.0.

c. The confidence interval for $E(y)$ will always be narrower than the prediction interval for y. The confidence interval for $E(y)$ is a confidence interval for the mean of a distribution. Once the mean is located, the actual values will vary around this mean. Thus, the width of the prediction interval includes the variation for locating the mean plus the variation of the actual values around the mean.

12.39 a. The first order model would be $E(y) = \beta_0 + \beta_1 x_1 + \beta_2 x_2 + \beta_3 x_3$.

b. Using MINITAB, the results are:

Regression Analysis: Project versus Intrapersonal, Stress, Mood

```
The regression equation is
Project = 86.9 - 0.210 Intrapersonal + 0.152 Stress + 0.0733 Mood


Predictor            Coef   SE Coef      T      P
Constant           86.902     3.198  27.17  0.000
Intrapersonal    -0.20986   0.09355  -2.24  0.037
Stress            0.15154   0.07686   1.97  0.063
Mood              0.07330   0.06714   1.09  0.289


S = 2.96544   R-Sq = 29.6%   R-Sq(adj) = 18.5%
```

```
Analysis of Variance

Source          DF       SS      MS     F      P
Regression       3   70.287  23.429  2.66  0.077
Residual Error  19  167.083   8.794
Total           22  237.370
```

The fitted model is: $\hat{y} = 86.902 - 0.20986x_1 + 0.15154x_2 + 0.07330x_3$

c. To determine if the overall model is statistically useful in predicting *y*, we test:

$H_0: \beta_1 = \beta_2 = \beta_3 = 0$
H_a: At least one $\beta_i \neq 0$

The test statistic is $F = 2.66$ and the *p*-value is $p = 0.077$.

Since the *p*-value is not less than $\alpha\left(p = 0.077 \not< 0.05\right)$, H_0 is not rejected. There is insufficient evidence to indicate that the model is adequate in predicting mean project score at $\alpha = 0.05$.

d. $R_a^2 = 0.1850$. Only 18.5% of the sample variation in the mean project scores is explained by the model containing intrapersonal scores, stress, and mood scores, adjusting for the number of parameters in the model and the sample size.

$s = 2.96544$. We would expect about 95% of the observed values of mean project scores to fall within $2s = 2(2.96544) = 5.93088$ units of their least squares predicted values.

e. Using MINITAB, the results are:

```
Predicted Values for New Observations

New
Obs     Fit  SE Fit        95% CI              95% PI
  1  89.084   0.893  (87.215, 90.952)  (82.602, 95.566)


Values of Predictors for New Observations

New
Obs  Intrapersonal  Stress  Mood
  1           20.0    30.0  25.0
```

The 95% prediction interval is $(82.602, 95.566)$. We are 95% confident that the actual mean project score is between 82.60 and 95.57 when the intrapersonal range is 20, the stress range is 30, and the mood range is 25.

12.41 Yes, we agree. The estimated coefficients for latitude and depth are both negative. Thus, to maximize arsenic levels, we would want latitude and depth to be as low as possible. The estimated coefficient for longitude is positive. To maximize arsenic levels, we would want longitude to be as large as possible.

Using MINTAB to find the prediction interval:

```
Predicted Values for New Observations

New
Obs      Fit   SE Fit          95% CI                95% PI
  1   232.33    23.23   (186.63, 278.04)   (24.03, 440.64)X

X denotes a point that is an outlier in the predictors.


Values of Predictors for New Observations

New
Obs   LATITUDE   LONGITUDE   DEPTH-FT
 1     23.755      90.662      25.0
```

The 95% prediction interval is $(24.03,\ 440.64)$. We are 95% confident that the actual value of arsenic level is between 24.03 and 440.61 when latitude is minimized at 23.803, longitude is maximized at 90.662, and depth is minimized at 25.

12.43 a. $E(y)=\beta_0+\beta_1x_1+\beta_2x_2+\beta_3x_1x_2$

b. $E(y)=\beta_0+\beta_1x_1+\beta_2x_2+\beta_3x_3+\beta_4x_1x_2+\beta_5x_1x_3+\beta_6x_2x_3$

12.45 a. $R^2=1-\dfrac{\text{SSE}}{\text{SS}_{yy}}=1-\dfrac{21}{479}=0.956$

95.6% of the total sample variability of the y values is explained by this model.

b. To test the utility of the model, we test:

$H_0:\beta_1=\beta_2=\beta_3=0$
$H_a:$ At least one $\beta_i\neq 0$, $i=1,2,3$

The test statistic is $F=\dfrac{R^2/k}{(1-R^2)/[n-(k+1)]}=\dfrac{0.956/3}{(1-0.956)/[32-(3+1)]}=202.8$

The rejection region requires $\alpha=0.05$ in the upper tail of the F distribution, with $\nu_1=k=3$ and $\nu_2=n-(k+1)=32-(3+1)=28$. From Table VI, Appendix B, $F_{0.05}=2.95$. The rejection region is $F>2.95$.

Since the observed value of the test statistic falls in the rejection region $(F=202.8>2.95)$, H_0 is rejected. There is sufficient evidence that the model is adequate for predicting y at $\alpha=0.05$.

c.

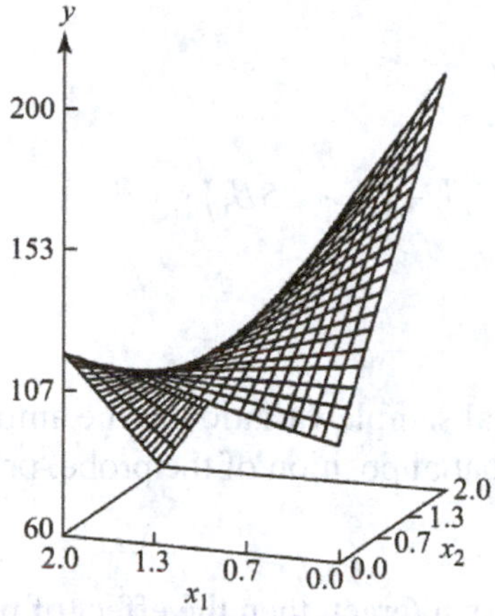

d. To determine if x_1 and x_2 interact, we test:

$$H_0: \beta_3 = 0$$
$$H_a: \beta_3 \neq 0$$

The test statistic is $t = \dfrac{\hat{\beta}_3 - 0}{s_{\hat{\beta}_3}} = \dfrac{10}{4} = 2.5$.

The rejection region requires $\alpha / 2 = 0.05/2 = 0.025$ in each tail of the t distribution with $df = n - (k+1) = 32 - (3+1) = 28$. From Table III, Appendix B, $t_{0.025} = 2.048$. The rejection region is $t < -2.048$ or $t > 2.048$.

Since the observed value of the test statistic falls in the rejection region $(t = 2.5 > 2.048)$, H_0 is rejected. There is sufficient evidence to indicate that x_1 and x_2 interact at $\alpha = 0.05$.

12.47 a. The model corresponding to Theory 1 is: $E(y) = \beta_0 + \beta_1 x_1 + \beta_2 x_2$

b. The model corresponding to Theory 2 is: $E(y) = \beta_0 + \beta_1 x_1 + \beta_2 x_2 + \beta_3 x_1 x_2$

c. Based on the graph, the second Theory would seem to fit the data better. When no compliment was given, there is basically no relationship between tip size and the size of the party. However, when compliments are given, the size of the tip tends to decrease as the party size increases. This would imply that the effect of the size of the party is greater when the server compliments the customers' menu choices than when the server does not compliment the customers' menu choices.

12.49 a. The equation for the interaction model is:

$$E(y) = \beta_0 + \beta_1 x_1 + \beta_2 x_2 + \beta_3 x_1 x_2$$

b. When $x_2 = 10$, the model becomes:

$$E(y) = \beta_0 + \beta_1 x_1 + \beta_2(10) + \beta_3 x_1(10) = \beta_0 + 10\beta_2 + (\beta_1 + 10\beta_3)x_1$$

Thus, the slope of the line is $\beta_1 + 10\beta_3$.

c. When $x_2 = 25$, the model becomes:

$$E(y) = \beta_0 + \beta_1 x_1 + \beta_2(25) + \beta_3 x_1(25) = \beta_0 + 25\beta_2 + (\beta_1 + 25\beta_3)x_1$$

Thus, the slope of the line is $\beta_1 + 25\beta_3$.

12.51 a. $R^2 = 0.994$. This means that 99.4% of the total sample variation in the amplitude values is explained by the model containing spatial position of the probe, position of the cross, and the interaction of the two variables.

b. If position of the probe and position of the cross interact, then the effect of position of the probe on the value of amplitude depends on the level of position of the cross.

c. A possible graph would look like the following:

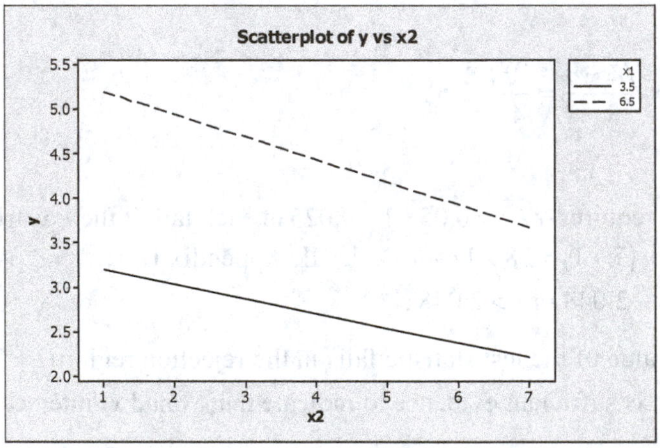

When $x_1 = 6.5$, $\hat{y} = 0.91 + 0.70(6.5) - 0.06x_2 - 0.03(6.5)x_2 = 5.46 - 0.255x_2$, the slope is -0.255. When $x_1 = 3.5$, $\hat{y} = 0.91 + 0.70(3.5) - 0.06x_2 - 0.03(3.5)x_2 = 3.36 - 0.165x_2$, the slope is -0.165. From the graph, you can see that the slope of the line corresponding to $x_1 = 6.5$ is steeper than the slope of the line corresponding to $x_1 = 3.5$.

12.53 a. The interaction model is: $E(y) = \beta_0 + \beta_1 x_1 + \beta_2 x_2 + \beta_3 x_1 x_2$

b. When the PN-ratio is 2.5, the model is:
$E(y) = \beta_0 + \beta_1 x_1 + \beta_2(2.5) + \beta_3 x_1(2.5) = \beta_0 + 2.5\beta_2 + (\beta_1 + 2.5\beta_3)x_1$.
Thus, the change in revenue for every 1-tweet increase in tweet rate is $(\beta_1 + 2.5\beta_3)$ when the PN-ratio is 2.5.

c. When the PN-ratio is 5.0, the model is:
$E(y) = \beta_0 + \beta_1 x_1 + \beta_2(5.0) + \beta_3 x_1(5.0) = \beta_0 + 5\beta_2 + (\beta_1 + 5\beta_3)x_1$.

Thus, the change in revenue for every 1-tweet increase in tweet rate is $(\beta_1 + 5\beta_3)$ when the PN-ratio is 5.0.

d. When the tweet rate is 100, the model is:
$E(y) = \beta_0 + \beta_1(100) + \beta_2 x_2 + \beta_3 x_2(100) = \beta_0 + 100\beta_1 + (\beta_2 + 100\beta_3)x_2$.
Thus, the change in revenue for every 1-unit increase in PN-ratio is $(\beta_2 + 100\beta_3)$ when the tweet rate is 100.

e. To determine if tweet rate and PN-ratio interact, we test:

$$H_0: \beta_3 = 0$$
$$H_a: \beta_3 \neq 0$$

12.55 a. The phrase "client credibility and linguistic delivery style interact" means that the affect of client credibility on likelihood depends on the level of linguistic delivery style.

b. To determine if the overall model is adequate for predicting likelihood, we test:

$$H_0: \beta_1 = \beta_2 = \beta_3 = 0$$
$$H_a: \text{At least one } \beta_i \neq 0, \quad i = 1, 2, 3$$

c. From the table, the test statistic is $F = 55.35$ and the p-value is $p < 0.0005$. Since the p value is so small, H_0 is rejected. There is sufficient evidence to indicate the model is adequate in predicting likelihood for any reasonable level of α.

d. To determine if client credibility and linguistic delivery style interact, we test:

$$H_0: \beta_3 = 0$$
$$H_a: \beta_3 \neq 0$$

e. From the table, $t = 4.008$ and the p-value is $p < 0.005$. Since the p-value is so small, H_0 is rejected. There is sufficient evidence to indicate that the client credibility and linguistic delivery style interact for any reasonable level of α.

f. The least squares prediction equation is $\hat{y} = 15.865 + 0.037x_1 - 0.678x_2 + 0.036x_1x_2$.

When client credibility is 22, the least squares line is:

$$\hat{y} = 15.865 + 0.037(22) - 0.678x_2 + 0.036(22)x_2 = 16.679 + 0.114x_2$$

When client credibility is 22, for each unit increase in linguistic delivery score, the mean likelihood score is estimated to increase by 0.114.

g. When client credibility is 46, the least squares line is:

$$\hat{y} = 15.865 + 0.037(46) - 0.678x_2 + 0.036(46)x_2 = 17.567 + 0.978x_2$$

When client credibility is 46, for each unit increase in linguistic delivery score, the mean likelihood score is estimated to increase by 0.978.

12.57 a. The model would be:

$$E(y) = \beta_o + \beta_1 x_1 + \beta_2 x_2 + \beta_3 x_3 + \beta_4 x_4 + \beta_5 x_5 + \beta_6 x_2 x_5 + \beta_7 x_3 x_5$$

b. Using MINITAB, the results are:

Regression Analysis: HEATRATE versus RPM, INLET-TEMP

```
The regression equation is
HEATRATE = 13646 + 0.0460 RPM - 12.7 INLET-TEMP + 23.0 EXH-TEMP - 3.0
CPRATIO + 1.29 AIRFLOW + 0.0161 INxAIRF - 0.0414 EXxAIRF


Predictor         Coef   SE Coef      T      P
Constant         13646      1068  12.77  0.000
RPM            0.04599   0.01602   2.87  0.006
INLET-TEMP     -12.675     1.542  -8.22  0.000
EXH-TEMP        23.003     3.768   6.11  0.000
CPRATIO          -3.02     26.42  -0.11  0.909
AIRFLOW          1.288     3.563   0.36  0.719
INxAIRF       0.016149  0.003673   4.40  0.000
EXxAIRF       -0.04143   0.01098  -3.77  0.000


S = 404.693   R-Sq = 94.2%   R-Sq(adj) = 93.6%


Analysis of Variance

Source          DF         SS        MS       F      P
Regression       7  158234406  22604915  138.02  0.000
Residual Error  59    9662802    163776
Total           66  167897208


Source      DF     Seq SS
RPM          1  119598530
INLET-TEMP   1   26893467
EXH-TEMP     1    7784225
CPRATIO      1       4623
AIRFLOW      1     774427
INxAIRF      1     849347
EXxAIRF      1    2329786
```

The least squares prediction equation is:

$$\hat{y} = 13{,}646 + 0.04599x_1 - 12.675x_2 + 23.003x_3 - 3.02x_4$$
$$+ 1.288x_5 + 0.016149x_2x_5 - 0.04143x_3x_5$$

c. To determine if inlet temperature and air flow rate interact to affect heat rate, we test:

$$H_0: \beta_6 = 0$$
$$H_a: \beta_6 \neq 0$$

From the printout, the test statistic is $t = 4.40$ and the p-value is $p = 0.000$.

Since the p-value is less than $\alpha\left(p = 0.000 < 0.05\right)$, H_0 is rejected. There is sufficient evidence to indicate that inlet temperature and air flow rate interact to affect heat rate at $\alpha = 0.05$.

d. To determine if exhaust temperature and air flow rate interact to affect heat rate, we test:

$H_0: \beta_7 = 0$
$H_a: \beta_7 \neq 0$

From the printout, the test statistic is $t = -3.77$ and the p-value is $p = 0.000$.

Since the p-value is less than $\alpha\left(p = 0.000 < 0.05\right)$, H_0 is rejected. There is sufficient evidence to indicate that exhaust temperature and air flow rate interact to affect heat rate at $\alpha = 0.05$.

e. Both of the interaction tests were significant. This means that the effect of inlet temperature on the heat rate is different at different levels of air flow and the effect of exhaust temperature on heat rate is different at different levels of air flow.

12.59 a. The interaction model would be: $E(y) = \beta_0 + \beta_1 x_1 + \beta_2 x_2 + \beta_3 x_1 x_2$

b. Using MINITAB, the results are:

Regression Analysis: Density versus MassFlux, HeatFlux, M*H

```
The regression equation is
Density = - 63238 + 18.8 MassFlux + 445486 HeatFlux - 140 M*H


Predictor      Coef  SE Coef      T      P
Constant     -63238    31150  -2.03  0.062
MassFlux      18.77    35.55   0.53  0.606
HeatFlux     445486    49589   8.98  0.000
M*H         -139.83    56.59  -2.47  0.027


S = 24360.3   R-Sq = 95.9%   R-Sq(adj) = 95.1%


Analysis of Variance

Source          DF           SS           MS       F      P
Regression       3  1.96608E+11  65535959189  110.44  0.000
Residual Error  14   8307933599    593423829
Total           17  2.04916E+11
```

The least squares prediction equation is:
$\hat{y} = -63{,}238 + 18.77x_1 + 445{,}486x_2 - 139.83x_1x_2$

c. To determine if the overall model is adequate, we test:

$H_0: \beta_1 = \beta_2 = \beta_3' = 0$
H_a: At least one $\beta_i \neq 0$

The test statistic is $F = 110.44$ and the p-value is $p = 0.000$.

Since the p-value is less than $\alpha\left(p = 0.0000 < 0.05\right)$, H_0 is rejected. There is sufficient evidence to indicate that the model is adequate at $\alpha = 0.05$.

$R_a^2 = 0.951$. Only 95.1% of the sample variation in the bubble densities is explained by the model containing mass flux, heat flux, and the interaction of mass flux and heat flux, adjusting for the number of parameters in the model and the sample size.

$s = 24{,}360.3$. We would expect about 95% of the observed values of bubble densities to fall within $2s = 2(24{,}360.3) = 48{,}720.6$ units of their least squares predicted values.

d. To determine if mass flux and heat flux interact, we test:

$H_0: \beta_3 = 0$
$H_a: \beta_3 \neq 0$

The test statistic is $t = -2.47$ and the p-value is $p = 0.027$.

Since the p-value is less than $\alpha\left(p = 0.027 < 0.05\right)$, H_0 is rejected. There is sufficient evidence to indicate that mass flux and heat flux interact at $\alpha = 0.05$.

e. When heat flux is 0.5, the prediction equation is
$\hat{y} = -63{,}238 + 18.77x_1 + 445{,}486(0.5) - 139.83x_1(0.5) = 159{,}505 - 51.145x_1$

We would expect bubble density to decrease by 51.145 liters/m^2 for each 1 kg/m^2-sec increase in mass flux when the heat flux is 0.5.

12.61 a. $E(y) = \beta_0 + \beta_1 x + \beta_2 x^2$

b. $E(y) = \beta_0 + \beta_1 x_1 + \beta_2 x_2 + \beta_3 x_1^2 + \beta_4 x_2^2 + \beta_5 x_1 x_2$

c. $E(y) = \beta_0 + \beta_1 x_1 + \beta_2 x_2 + \beta_3 x_3 + \beta_4 x_1^2 + \beta_5 x_2^2 + \beta_6 x_3^2 + \beta_7 x_1 x_2 + \beta_8 x_1 x_3 + \beta_9 x_2 x_3$

12.63 a. $H_0: \beta_2 = 0$
$H_a: \beta_2 \neq 0$

The test statistic is $t = \frac{\hat{\beta}_2 - 0}{s_{\hat{\beta}_2}} = \frac{0.47 - 0}{0.15} = 3.133$

The rejection region requires $\alpha / 2 = 0.05 / 2 = 0.025$ in each tail of the t distribution with $df = n - (k+1) = 25 - (2+1) = 22$. From Table III, Appendix B, $t_{0.025} = 2.074$. The rejection region is $t < -2.074$ or $t > 2.074$.

Since the observed value of the test statistic falls in the rejection region $(t = 3.133 > 2.074)$, H_0 is rejected. There is sufficient evidence to indicate the true relationship is given by the quadratic model at $\alpha = 0.05$.

b. $H_0: \beta_2 = 0$
$H_a: \beta_2 > 0$

The test statistic is the same as in part **a**, $t = 3.133$.

The rejection region requires $\alpha = 0.05$ in the upper tail of the t distribution with $df = 22$. From Table III, Appendix B, $t_{0.05} = 1.717$. The rejection region is $t > 1.717$.

Since the observed value of the test statistic falls in the rejection region $(t = 3.133 > 1.717)$, H_0 is rejected. There is sufficient evidence to indicate the quadratic curve opens upward at $\alpha = 0.05$.

12.65 a.

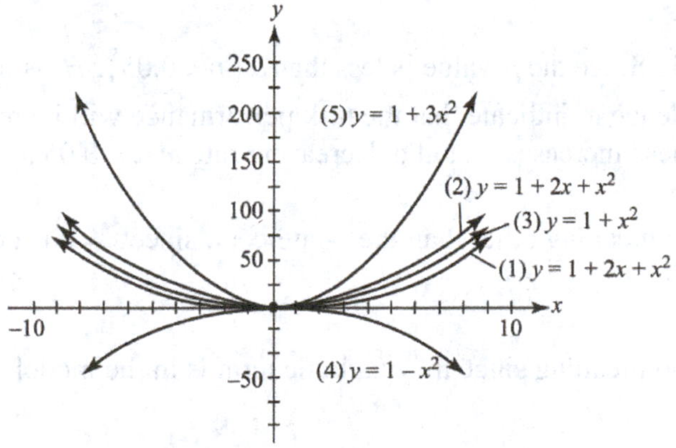

b. It moves the graph to the right $(-2x)$ or to the left $(+2x)$ compared to the graph of $y = 1 + x^2$.

c. It controls whether the graph opens up $(+x^2)$ or down $(-x^2)$. It also controls how steep the curvature is, i.e., the larger the absolute value of the coefficient of x^2, the narrower the curve is.

12.67 a. The coefficient of correlation measures the strength of the linear relationship between two variables. If the variables are curvilinearly related, then the coefficient of correlation is not appropriate.

b. The curvilinear model is: $E(y) = \beta_0 + \beta_1 x + \beta_2 x^2$

c. Answers will vary. A possible sketch would be:

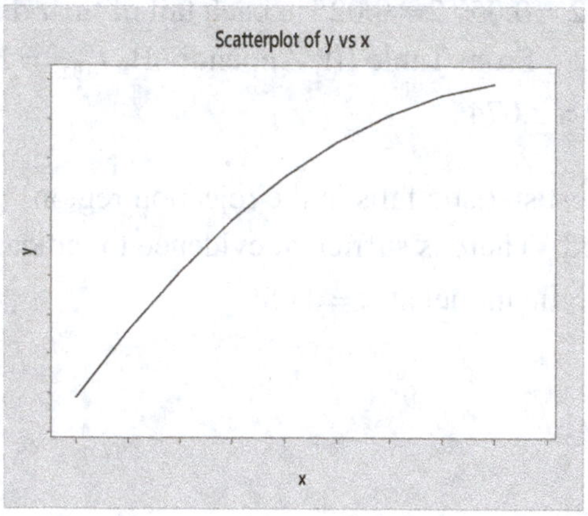

d. The sign of β_2 would be negative because the curve opens down.

e. To determine if the task performance will increase as the level of conscientiousness increases, but at a decreasing rate, we test:

$$H_0: \beta_2 = 0$$
$$H_a: \beta_2 < 0$$

The *p*-value is $p < 0.05$. Since the *p*-value is less than $\alpha\,(p < 0.05)$, H_0 is rejected. There is sufficient evidence to indicate that the task performance will increase as the level of conscientiousness increases, but at a decreasing rate at $\alpha = 0.05$.

12.69 a. $\hat{\beta}_0 = 6.13$. This has no meaning other than the *y*-intercept, since $x = 0$ is not in the observed range.

$\hat{\beta}_1 = 0.141$. This has no meaning since the quadratic term is in the model. This is a shift parameter.

$\hat{\beta}_2 = -0.0009$. Since the parameter estimate is negative, this indicates that the curve has a mound shape.

b. $R^2 = 0.226$. This means that 22.6% of the total sample variation in number of points scored is explained by the model containing number of yards from opposing goal line and number of yards from opposing goal line squared.

c. No. If you add variables to the model, the value of R^2 will either increase or stay the same. We cannot tell whether the increase in the value of R^2 is significant or not by just looking at the 2 values.

d. To determine if the quadratic model is a better fit than the linear model, we would test:

$$H_0: \beta_2 = 0$$
$$H_a: \beta_2 \neq 0$$

12.71 a.

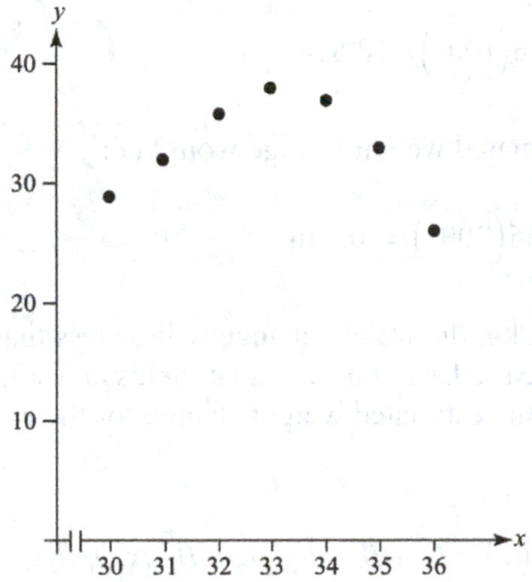

b. If information was available only for $x = 30,\ 31,\ 32$, and 33, we would suggest a first-order model where $\beta_1 > 0$. If information was available only for $x = 33,\ 34,\ 35$, and 36, we would again suggest a first-order model where $\beta_1 < 0$. If all the information was available, we would suggest a second-order model.

12.73 a. The complete second-order model is: $E(y) = \beta_o + \beta_1 x_1 + \beta_2 x_2 + \beta_3 x_1^2 + \beta_4 x_4^2 + \beta_5 x_1 x_2$

b. The terms that allow for a curvilinear relationship are $\beta_3 x_1^2$ and $\beta_4 x_4^2$.

12.75 a. Using MINITAB, the graph is:

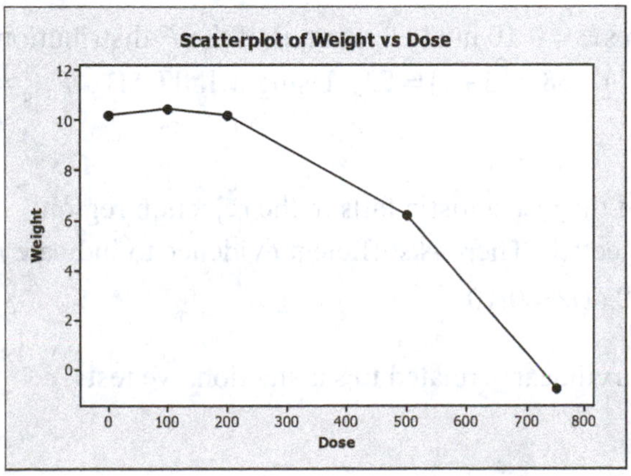

b. For a dosage of 500 mg/kg, the estimated weight change would be:

$$\hat{y} = 10.25 + 0.0053(500) - 0.0000266(500^2) = 6.25$$

c. For a dosage of 0 mg/kg, the estimated weight change would be:

$$\hat{y} = 10.25 + 0.0053(0) - 0.0000266(0^2) = 10.25$$

d. For a dosage of 100 mg/kg, the estimated weight change would be:

$$\hat{y} = 10.25 + 0.0053(100) - 0.0000266(100^2) = 10.514$$

For a dosage of 200 mg/kg, the estimated weight change would be:

$$\hat{y} = 10.25 + 0.0053(200) - 0.0000266(200^2) = 10.246$$

For any dosage greater than 200 mg/kg, the weight change will be less than 10.246. Thus, a dosage of 200 mg/kg will be the largest dosage that yields an estimated weight change that is closest to, but below the estimated weight change for the control group (10.25).

12.77 a. The complete 2nd-order model is: $E(y) = \beta_0 + \beta_1 x_1 + \beta_2 x_2 + \beta_3 x_1 x_2 + \beta_4 x_1^2 + \beta_5 x_2^2$

b. $R^2 = 0.402$. This means that 40.2% of the total sample variation of the satisfaction level values is explained by the complete 2nd-order model including needs and supplies.

c. To determine if the overall model is adequate, we test:

$H_0: \beta_1 = \beta_2 = \beta_3 = \beta_4 = \beta_5 = 0$
H_a: At least one $\beta_i \neq 0$

The test statistic is $F = \dfrac{R^2/k}{(1-R^2)/[n-(k+1)]} = \dfrac{0.402/5}{(1-0.402)/[58-(5+1)]} = 6.99$.

The rejection region requires $\alpha = 0.10$ in the upper tail of the F-distribution with $\nu_1 = k = 5$ and $\nu_2 = n-(k+1) = 58-(5+1) = 52$. Using MINITAB, $F_{0.10} = 1.96$. The rejection region is $F > 1.96$.

Since the observed value of the test statistic falls in the rejection region $(F = 6.99 > 1.96)$, H_0 is rejected. There is sufficient evidence to indicate at least one of the betas is different from 0 at $\alpha = 0.10$.

d. To determine if needs is curvilinearly related to satisfaction, we test:

$H_0: \beta_4 = 0$
$H_a: \beta_4 \neq 0$

The test statistic is $t = \dfrac{\hat{\beta}_4 - 0}{s_{\hat{\beta}_4}} = \dfrac{-0.181 - 0}{0.302} = -0.599$.

The rejection region requires $\alpha/2 = 0.10/2 = 0.05$ in each tail of the t-distribution with $df = n-(k+1) = 58-(5+1) = 52$. Using MINITAB, $t_{0.05} = 1.675$. The rejection region is $(t < -1.675 \text{ or } t > 1.675)$.

Since the observed value of the test statistic does not fall in the rejection region $(t = -0.599 \not< -1.675)$, H_0 is not rejected. There is insufficient evidence to indicate needs is curvilinearly related to satisfaction at $\alpha = 0.10$.

e. To determine if supplies is curvilinearly related to satisfaction, we test:

$H_0: \beta_5 = 0$
$H_a: \beta_5 \neq 0$

The test statistic is $t = \dfrac{\hat{\beta}_5 - 0}{s_{\hat{\beta}_5}} = \dfrac{-0.755 - 0}{0.413} = -1.83$.

The rejection region requires $\alpha/2 = 0.10/2 = 0.05$ in each tail of the t-distribution with $df = n-(k+1) = 58-(5+1) = 52$. Using MINITAB, $t_{0.05} = 1.675$. The rejection region is $(t < -1.675 \text{ or } t > 1.675)$.

Since the observed value of the test statistic falls in the rejection region $(t = -1.83 < -1.675)$, H_0 is rejected. There is sufficient evidence to indicate supplies is curvilinearly related to satisfaction at $\alpha = 0.10$.

12.79 a. Using MINITAB, the results are:

Regression Analysis: RATE versus EST, EST-SQ

```
The regression equation is
RATE = - 288 + 1.39 EST + 0.000035 EST-SQ


Predictor         Coef       SE Coef      T      P
Constant          -288          8049  -0.04  0.972
EST              1.395         3.651   0.38  0.706
EST-SQ      0.00003509    0.00009724   0.36  0.722


S = 31901.1   R-Sq = 45.9%   R-Sq(adj) = 40.8%
```

```
Analysis of Variance

Source            DF           SS          MS     F       P
Regression         2  18138955261  9069477631  8.91   0.002
Residual Error    21  21371254395  1017678781
Total             23  39510209656


Source  DF       Seq SS
EST      1  18006405335
EST-SQ   1    132549926
```

The fitted model is $\hat{y} = -288 + 1.398x + 0.0000351x^2$.

To determine if the incidence rate is curvilinearly related to estimated rate, we test:

$H_0: \beta_2 = 0$
$H_a: \beta_2 \neq 0$

From the printout, the test statistic is $t = 0.36$ and the p-value is $p = 0.722$.

Since the p-value is not less than $\alpha\left(p = 0.722 \not< 0.05\right)$, H_0 is not rejected. There is insufficient evidence that the incidence rate is curvilinearly related to estimated rate at $\alpha = 0.05$.

b. Using MINITAB, the scatterplot is:

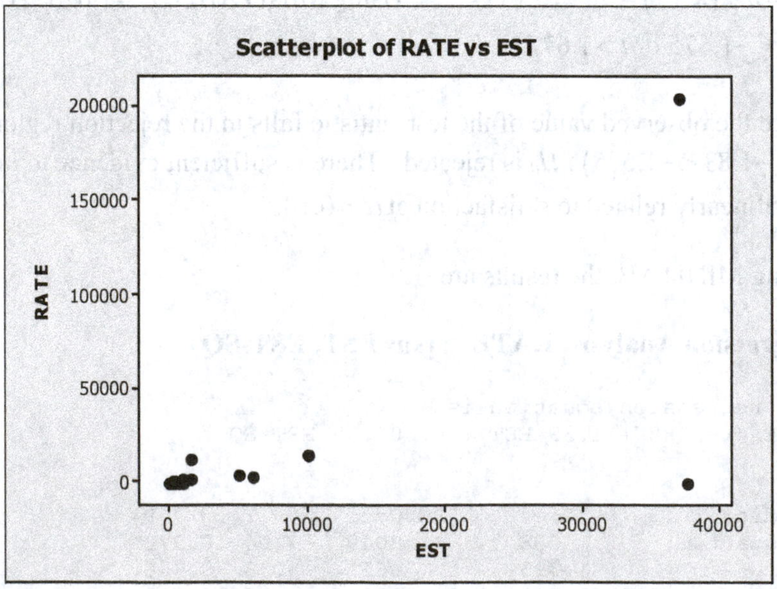

The point associated with Botulism is in the lower right corner of the graph. This point looks to be out of line with the other data points.

c. Using MINITAB the new results are:

Regression Analysis: RATE versus EST, EST-SQ

```
The regression equation is
RATE = 735 - 0.081 EST + 0.000151 EST-SQ

Predictor          Coef       SE Coef       T       P
Constant          735.0         695.9    1.06   0.303
EST             -0.0810        0.3167   -0.26   0.801
EST-SQ       0.00015052    0.00000868   17.34   0.000

S = 2756.80   R-Sq = 99.6%   R-Sq(adj) = 99.6%

Analysis of Variance

Source            DF            SS           MS        F       P
Regression         2   39251490541  19625745270  2582.35   0.000
Residual Error    20     151998825      7599941
Total             22   39403489366

Source   DF       Seq SS
EST       1  36967483627
EST-SQ    1   2284006914
```

Now, if we test to see if the incidence rate is curvilinearly related to estimated rate:

$$H_0: \beta_2 = 0$$
$$H_a: \beta_2 \neq 0$$

From the printout, the test statistic is $t = 17.34$ and the p-value is $p = 0.000$.

Since the p-value is less than $\alpha\left(p = 0.000 < 0.05\right)$, H_0 is rejected. There is sufficient evidence that the incidence rate is curvilinearly related to estimated rate at $\alpha = 0.05$. The fit of the model has improved dramatically. The R^2 value for the new model is $R^2 = 99.6\%$ compared to the R^2 value for the first model, $R^2 = 45.9\%$.

12.81 Let $x = \begin{cases} 1 & \text{if qualitative variable assumes 2nd level} \\ 0 & \text{otherwise} \end{cases}$

The model is $E(y) = \beta_0 + \beta_1 x_1$.

β_0 = mean value of y when the qualitative variable assumes the first level.

β_1 = difference in the mean values of y between levels 2 and 1 of the qualitative variable.

12.83 a. Level 1 implies $x_1 = x_2 = x_3 = 0$. $\hat{y} = 10.2$

Level 2 implies $x_1 = 1$ and $x_2 = x_3 = 0$. $\hat{y} = 10.2 - 4(1) = 6.2$

Level 3 implies $x_2 = 1$ and $x_1 = x_3 = 0$. $\hat{y} = 10.2 + 12(1) = 22.2$

Level 4 implies $x_3 = 1$ and $x_1 = x_2 = 0$. $\hat{y} = 10.2 + 2(1) = 12.2$

b. The hypotheses are:

$$\begin{array}{l} H_0: \beta_1 = \beta_2 = \beta_3 = 0 \\ H_a: \text{At least one } \beta_i \neq 0 \end{array}, \; i = 1, 2, 3$$

12.85 a. Let $x = \begin{cases} 1 \text{ if multiple birth} \\ 0 \text{ if single birth} \end{cases}$

b. The model would be: $E(y) = \beta_0 + \beta_1 x$

c. For this model, β_0 = mean number of children when the first birth was a single birth and β_1 = difference in the mean number of children between when the first birth was a multiple and when the first birth was a single.

12.87 a. For gill nets, $x_1 = 0$ and $x_2 = 0$. Thus, $E(y) = \beta_0 + \beta_1(0) + \beta_2(0) = \beta_0$.

b. β_1 = difference in mean length between whales entangled in set nets and whales entangled in gill nets.

c. To determine if the mean body lengths of entangled whales differ for the three types of fishing gear, we test:

$$H_0: \beta_1 = \beta_2 = 0$$
$$H_a: \text{At least one } \beta_i \neq 0$$

12.89 a. Let $x_1 = \begin{cases} 1 & \text{if player is white} \\ 0 & \text{otherwise} \end{cases}$ $\quad x_2 = \begin{cases} 1 & \text{if card availability is low} \\ 0 & \text{otherwise} \end{cases}$

$x_3 = \begin{cases} 1 & \text{if player position is running back} \\ 0 & \text{otherwise} \end{cases}$ $\quad x_4 = \begin{cases} 1 & \text{if player position is wide receiver} \\ 0 & \text{otherwise} \end{cases}$

$x_5 = \begin{cases} 1 & \text{if player position is tight end} \\ 0 & \text{otherwise} \end{cases}$ $\quad x_6 = \begin{cases} 1 & \text{if player position is defensive lineman} \\ 0 & \text{otherwise} \end{cases}$

$x_7 = \begin{cases} 1 & \text{if player position is linebacker} \\ 0 & \text{otherwise} \end{cases}$ $\quad x_8 = \begin{cases} 1 & \text{if player position is defensive back} \\ 0 & \text{otherwise} \end{cases}$

$x_9 = \begin{cases} 1 & \text{if player position is offensive lineman} \\ 0 & \text{otherwise} \end{cases}$

b. The model for price (y) as a function of race is: $E(y) = \beta_0 + \beta_1 x_1$.

β_0 = mean price of card of black player
β_1 = difference in mean price of card between a white player and a black player

c. The model for price (y) as a function of card availability is: $E(y) = \beta_0 + \beta_1 x_2$.

β_0 = mean price of card of high availability
β_1 = difference in mean price of card between a card of low availability and a card of high availability

d. The model for price (y) as a function of position is:

$$E(y) = \beta_0 + \beta_1 x_3 + \beta_2 x_4 + \beta_3 x_5 + \beta_4 x_6 + \beta_5 x_7 + \beta_6 x_8 + \beta_7 x_9$$

β_0 = mean price of card of quarterback
β_1 = difference in mean price of card between a running back and a quarterback
β_2 = difference in mean price of card between a wide receiver and a quarterback
β_3 = difference in mean price of card between a tight end and a quarterback
β_4 = difference in mean price of card between a defensive lineman and a quarterback
β_5 = difference in mean price of card between a linebacker and a quarterback
β_6 = difference in mean price of card between a defensive back and a quarterback
β_7 = difference in mean price of card between a offensive lineman and a quarterback

12.91 a. There would be a total of $2 \times 2 = 4$ different gene marker combinations at the two locations. They are: AA, AB, BA, BB

b. Let $x_1 = \begin{cases} 1 & \text{if gene marker is AA} \\ 0 & \text{otherwise} \end{cases}$ $x_2 = \begin{cases} 1 & \text{gene marker is AB} \\ 0 & \text{otherwise} \end{cases}$

$x_3 = \begin{cases} 1 & \text{gene marker is BA} \\ 0 & \text{otherwise} \end{cases}$

A model for $E(y)$ as a function of the gene marker combination is:

$$E(y) = \beta_0 + \beta_1 x_1 + \beta_2 x_2 + \beta_3 x_3$$

c. β_0 = mean extent of disease for gene marker combination BB
β_1 = difference in mean extent of disease between gene marker combinations AA and BB
β_2 = difference in mean extent of disease between gene marker combinations AB and BB
β_3 = difference in mean extent of disease between gene marker combinations BA and BB

d. To determine if the overall model is useful for predicting extent of the disease, we test:

$H_0: \beta_1 = \beta_2 = \beta_3 = 0$
H_a: At least one $\beta_i \neq 0$

12.93 a. Let $x_1 = \begin{cases} 1 \text{ if major depression only} \\ 0 \text{ if not} \end{cases}$ and $x_2 = \begin{cases} 1 \text{ if personality disorder only} \\ 0 \text{ if not} \end{cases}$

The model would be: $E(y) = \beta_0 + \beta_1 x_1 + \beta_2 x_2$.

b. If there are no differences in the mean number of personality disorders for the three patient groups, then $\beta_1 = \beta_2 = 0$.

c. To determine if the mean number of personality disorders for the major depression only patients is less than the corresponding mean for the patients with both major depression and personality disorder, we would test:

$H_0: \beta_1 = 0$
$H_a: \beta_1 < 0$

12.957 Let $x_1 = \begin{cases} 1 \text{ if Juror is female} \\ 0 \text{ otherwise} \end{cases}$ $x_2 = \begin{cases} 1 \text{ if no expert testimony is present} \\ 0 \text{ otherwise} \end{cases}$

To allow for the likelihood that female jurors would be more likely than male jurors to change a verdict from not guilty to guilty after deliberations if expert testimony is present, the model would have to include an interaction term between gender and expert testimony. The model would be:

$$E(y) = \beta_0 + \beta_1 x_1 + \beta_2 x_2 + \beta_3 x_1 x_2$$

A possible sketch of this relationship would be:

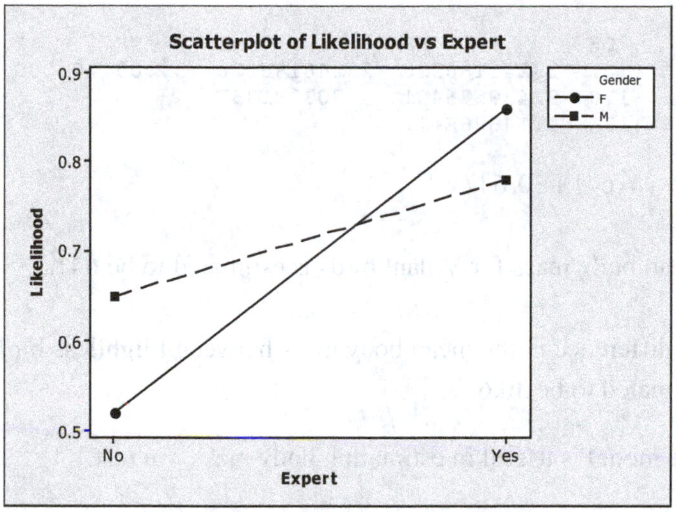

12.97 a. Let y = mean body mass and $x = \begin{cases} 1 & \text{if flightless} \\ 0 & \text{otherwise} \end{cases}$

The model is $E(y) = \beta_0 + \beta_1 x$

b. Let y = mean body mass and

$x_1 = \begin{cases} 1 & \text{if vertebrates} \\ 0 & \text{otherwise} \end{cases}$ $x_2 = \begin{cases} 1 & \text{if vegetables} \\ 0 & \text{otherwise} \end{cases}$ $x_3 = \begin{cases} 1 & \text{if invertebrates} \\ 0 & \text{otherwise} \end{cases}$

The model is $E(y) = \beta_o + \beta_1 x_1 + \beta_2 x_2 + \beta_3 x_3$

c. Let y = mean egg length and

$x_1 = \begin{cases} 1 & \text{if cavity within ground} \\ 0 & \text{otherwise} \end{cases}$ $x_2 = \begin{cases} 1 & \text{if tree} \\ 0 & \text{otherwise} \end{cases}$ $x_3 = \begin{cases} 1 & \text{if cavity above ground} \\ 0 & \text{otherwise} \end{cases}$

The model is $E(y) = \beta_0 + \beta_1 x_1 + \beta_2 x_2 + \beta_3 x_3$

d. Using MINTAB, the results are:

Regression Analysis: Body Mass versus Flt

```
The regression equation is
Body Mass = 641 + 30647 Flt


Predictor    Coef   SE Coef     T       P
Constant      641      2665  0.24   0.810
Flt         30647      5331  5.75   0.000


S = 26521.3   R-Sq = 20.3%   R-Sq(adj) = 19.7%
```

```
Analysis of Variance

Source           DF           SS           MS      F      P
Regression        1  23246186220  23246186220  33.05  0.000
Residual Error  130  91439576421    703381357
Total           131  1.14686E+11
```

The fitted model is $\hat{y} = 641 + 30,647x$

$\hat{\beta}_0 = 641$. The mean body mass for Volant birds is estimated to be 641.

$\hat{\beta}_1 = 30,647$. The difference in the mean body mass between Flightless birds and Volant birds is estimated to be 30,647.

e. To determine if the model is useful in estimating body mass, we test:

$$H_0: \beta_1 = 0$$
$$H_a: \beta_1 \neq 0$$

The test statistic is $F = 33.05$ and the *p*-value is $p = 0.000$. Since the *p*-value is less than $\alpha\left(p = 0.000 < 0.01\right)$, H_0 is rejected. There is sufficient evidence to indicate a difference in mean body mass between the two types of flight capacity birds at $\alpha = 0.01$.

f. Using MINITAB, the results are:

Regression Analysis: Body Mass versus DX1, DX2, DX3

```
The regression equation is
Body Mass = 903 + 2997 DX1 + 26206 DX2 - 660 DX3


Predictor    Coef  SE Coef      T      P
Constant      903     4171   0.22  0.829
DX1          2997    14298   0.21  0.834
DX2         26206     6090   4.30  0.000
DX3          -660     5772  -0.11  0.909


S = 27352.1   R-Sq = 16.5%   R-Sq(adj) = 14.5%
Analysis of Variance

Source           DF           SS          MS     F      P
Regression        3  18924196564  6308065521  8.43  0.000
Residual Error  128  95761566077   748137235
Total           131  1.14686E+11


Source  DF       Seq SS
X1       1     79962608
X2       1  18834456334
X3       1      9777622
```

The fitted model is $\hat{y} = 903 + 2,997x_1 + 26,206x_2 - 660x_3$

$\hat{\beta}_0 = 903$. The mean body mass for birds on a fish diet is estimated to be 903.

$\hat{\beta}_1 = 2{,}997$. The difference in the mean body mass between birds on a vertebrate diet and birds on a fish diet is estimated to be 2,997.

$\hat{\beta}_2 = 26{,}206$. The difference in the mean body mass between birds on a vegetable diet and birds on a fish diet is estimated to be 26,206.

$\hat{\beta}_3 = -660$. The difference in the mean body mass between birds on an invertebrate diet and birds on a fish diet is estimated to be −660.

g. To determine if the model is useful in estimating body mass, we test:

$$H_0: \beta_1 = \beta_2 = \beta_3 = 0$$
$$H_a: \text{At least one } \beta_i \neq 0, \quad i = 1, 2, 3$$

The test statistic is $F = 8.43$ and the *p*-value is $p = 0.000$. Since the *p*-value is less than $\alpha\left(p = 0.000 < 0.01\right)$, H_0 is rejected. There is sufficient evidence to indicate a difference in mean body mass among the four types of diets at $\alpha = 0.01$.

h. Using MINITAB, the results are:

Regression Analysis: Egg Length versus NX1, NX2, NX3

```
The regression equation is
Egg Length = 73.7 - 9.13 NX1 - 45.0 NX2 - 39.5 NX3

130 cases used, 2 cases contain missing values

Predictor      Coef  SE Coef      T      P
Constant     73.732    4.907  15.03  0.000
NX1          -9.132    9.515  -0.96  0.339
NX2          -45.01    14.45  -3.12  0.002
NX3         -39.510    9.253  -4.27  0.000

S = 40.7622   R-Sq = 16.1%   R-Sq(adj) = 14.1%
Analysis of Variance

Source           DF      SS     MS     F      P
Regression        3   40230  13410  8.07  0.000
Residual Error  126  209356   1662
Total           129  249586

Source  DF  Seq SS
NX1      1     482
NX2      1    9455
NX3      1   30293
```

The fitted model is $\hat{y} = 73.732 - 9.132x_1 - 45.01x_2 - 39.51x_3$

$\hat{\beta}_0 = 73.732$. The mean egg length for nesting on the ground is estimated to be 73.732 mm.

$\hat{\beta}_1 = -9.132$. The difference in the mean egg length between birds nesting in a cavity within the ground and birds nesting on the ground is estimated to be −9.132.

$\hat{\beta}_2 = -45.01$. The difference in the mean egg length between birds nesting in a tree and birds nesting on the ground is estimated to be −45.01.

$\hat{\beta}_3 = -39.51$. The difference in the mean egg length between birds nesting in a cavity above ground and birds nesting on the ground is estimated to be −39.51.

i. To determine if the model is useful in estimating body mass, we test:

$$H_0: \beta_1 = \beta_2 = \beta_3 = 0$$
$$H_a: \text{At least one } \beta_i \neq 0, \quad i = 1, 2, 3$$

The test statistic is $F = 8.07$ and the *p*-value is $p = 0.000$. Since the *p*-value is less than $\alpha\ (p = 0.000 < 0.01)$, H_0 is rejected. There is sufficient evidence to indicate a difference in mean egg length among the four types of nesting sites at $\alpha = 0.01$.

12.99 a. For months following the murder, $x_2 = 1$ and for Center, Texas, $x_1 = 0$. Thus, the mean violent crime rate would be:
$$E(y) = \beta_0 + \beta_1 x_1 + \beta_2 x_2 + \beta_3 x_1 x_2 = \beta_0 + \beta_1(0) + \beta_2(1) + \beta_3(0)(1) = \beta_0 + \beta_2$$

b. For months following the murder, $x_2 = 1$ and for Jasper, Texas, $x_1 = 1$. Thus, the mean violent crime rate would be:
$$E(y) = \beta_0 + \beta_1 x_1 + \beta_2 x_2 + \beta_3 x_1 x_2 = \beta_0 + \beta_1(1) + \beta_2(1) + \beta_3(1)(1) = \beta_0 + \beta_1 + \beta_2 + \beta_3$$

c. The difference between the mean violent crime rate for Jasper and Center following the murder would be: $\beta_0 + \beta_1 + \beta_2 + \beta_3 - (\beta_0 + \beta_2) = \beta_1 + \beta_3$

d. For months before the murder, $x_2 = 0$ and for Center, Texas, $x_1 = 0$. Thus, the mean violent crime rate would be:
$$E(y) = \beta_0 + \beta_1 x_1 + \beta_2 x_2 + \beta_3 x_1 x_2 = \beta_0 + \beta_1(0) + \beta_2(0) + \beta_3(0)(0) = \beta_0$$

For months before the murder, $x_2 = 0$ and for Jasper, Texas, $x_1 = 1$. Thus, the mean violent crime rate would be:
$$E(y) = \beta_0 + \beta_1 x_1 + \beta_2 x_2 + \beta_3 x_1 x_2 = \beta_0 + \beta_1(1) + \beta_2(0) + \beta_3(1)(0) = \beta_0 + \beta_1$$

The difference between the mean violent crime rate for Jasper and Center before the murder would be: $\beta_0 + \beta_1 - \beta_0 = \beta_1$

e. If x_1 and x_2 do not interact, then the effect of x_1 on the mean crime rate would be the same for each level of x_2. Here, when $x_2 = 1$ (following the murder), the difference in the mean violent crime rate between Jasper and Center is $\beta_1 + \beta_3$. When $x_2 = 0$ (before the murder), the difference in the mean violent crime rate between Jasper and Center is β_1. These are not the same, so x_1 and x_2 interact.

f. Since the observed p-value is less than $\alpha\left(p < 0.001 < 0.01\right)$, H_0 is rejected. There is sufficient evidence to indicate β_3 is not 0 for $\alpha = 0.01$. This indicates that the 2 independent variables, time and place, interact to affect violent crime rate.

g. The mean violent crime rate in Center following the murder is $\beta_0 + \beta_2$. The mean violent crime rate in Center before the murder is β_0. The difference between following and before the murder is $\beta_0 + \beta_2 - \beta_0 = \beta_2$. The estimate of this is $\hat{\beta}_2 = -169$.

The mean violent crime rate in Jasper following the murder is $\beta_0 + \beta_1 + \beta_2 + \beta_3$. The mean violent crime rate in Jasper before the murder is $\beta_0 + \beta_1$. The difference between following and before the murder is $\beta_0 + \beta_1 + \beta_2 + \beta_3 - \left(\beta_0 + \beta_1\right) = \beta_2 + \beta_3$. The estimate of this is $\hat{\beta}_2 + \hat{\beta}_3 = -169 + 255 = 86$

12.101 a. The complete second-order model is $E\left(y\right) = \beta_0 + \beta_1 x_1 + \beta_2 x_1^2$

b. The new model is $E\left(y\right) = \beta_0 + \beta_1 x_1 + \beta_2 x_1^2 + \beta_3 x_2 + \beta_4 x_3$

where $x_2 = \begin{cases} 1 \text{ if level 2 of qualitative variable} \\ 0 \text{ otherwise} \end{cases}$ $\quad x_3 = \begin{cases} 1 \text{ if level 3 of qualitative variable} \\ 0 \text{ otherwise} \end{cases}$

c. The model with the interaction terms is:

$$E\left(y\right) = \beta_0 + \beta_1 x_1 + \beta_2 x_1^2 + \beta_3 x_2 + \beta_4 x_3 + \beta_5 x_1 x_2 + \beta_6 x_1 x_3 + \beta_7 x_1^2 x_2 + \beta_8 x_1^2 x_3$$

d. The response curves will have the same shape if none of the interaction terms are present or if $\beta_5 = \beta_6 = \beta_7 = \beta_8 = 0$.

e. The response curves will be parallel lines if the interaction terms as well as the second-order terms are absent or if $\beta_2 = \beta_5 = \beta_6 = \beta_7 = \beta_8 = 0$.

f. The response curves will be identical if no terms involving the qualitative variable are present or $\beta_3 = \beta_4 = \beta_5 = \beta_6 = \beta_7 = \beta_8 = 0$.

12.103 a. When $x_2 = x_3 = 0$, $E(y) = \beta_0 + \beta_1 x_1$
When $x_2 = 1$ and $x_3 = 0$, $E(y) = \beta_0 + \beta_2 + \beta_1 x_1$
When $x_2 = 0$ and $x_3 = 1$, $E(y) = \beta_0 + \beta + \beta_1 x_{13}$

b. For level 1, $x_2 = x_3 = 0$. $\hat{y} = 44.8 + 2.2x_1 + 9.4\left(0\right) + 15.6\left(0\right) = 44.8 + 2.2x_1$

For level 2, $x_2 = 1$ and $x_3 = 0$.

$\hat{y} = 44.8 + 2.2x_1 + 9.4(1) + 15.6(0) = 44.8 + 2.2x_1 + 9.4 = 54.2 + 2.2x_1$

For level 3, $x_2 = 0$ and $x_3 = 1$.

$\hat{y} = 44.8 + 2.2x_1 + 9.4(0) + 15.6(1) = 44.8 + 2.2x_1 + 15.6 = 60.4 + 2.2x_1$

12.105 a. The least squares prediction equation is $\hat{y} = 11.779 - 1.972x_1 + 0.585x_4 - 0.553x_1x_4$.

b. For a male college student, $x_1 = 1$. The predictive level of desire is:
$\hat{y} = 11.779 - 1.972(1) + 0.585(5) - 0.553(1)(5) = 9.967$

c. To determine if the model is adequate, we test:

$H_0: \beta_1 = \beta_2 = \beta_3 = 0$
H_a: At least one $\beta_i \neq 0$

From the printout, the test statistic is $F = 45.09$ and the *p*-value is $p < 0.0001$. Since the *p*-value is less than $\alpha\ (p < 0.0001 < 0.10)$, H_0 is rejected. There is sufficient evidence to indicate the model is useful at $\alpha = 0.10$.

d. $R_a^2 = 0.4390$. This means that 43.9% of the total sample variation in level of desire scores is explained by the model containing gender, impression of reality TV and the interaction of gender and impression of reality TV, adjusting for the sample size and the number of parameters in the model.

e. $s = 2.35$. We would expect about 95% of the observed values of level of desire to fall within $2s = 2(2.35) = 4.70$ units of their least squares predicted values.

f. To determine if gender and impression of reality TV interact, we test:

$H_0: \beta_3 = 0$
$H_a: \beta_3 \neq 0$

From the printout, the test statistic is $t = -2.00$ and the *p*-value is $p = 0.0467$. Since the *p*-value is less than $\alpha\ (p = 0.0467 < 0.10)$, H_0 is rejected. There is sufficient evidence to indicate gender and impression of reality TV interact in the prediction of level of desire for cosmetic surgery at $\alpha = 0.10$.

g. For female students, $x_1 = 0$. The predictive equation for females is:
$\hat{y} = 11.779 - 1.972(0) + 0.585x_4 - 0.553(0)x_4 = 11.779 + 0.585x_4$.

For every 1-point increase in impression of reality TV show, the level of desire is estimated to increase by 0.585 units for female students.

h. For male students, $x_1 = 1$. The predictive equation for males is:

$\hat{y} = 11.779 - 1.972(1) + .585x_4 - 0.553(1)x_4 = 9.807 + 0.032x_4$.

For every 1-point increase in impression of reality TV show, the level of desire is estimated to increase by 0.032 units for male students.

12.107 a. From Exercise 12.88, let $x_1 = \begin{cases} 1 \text{ if blonde Caucasian} \\ 0 \text{ if not} \end{cases}$ and $x_2 = \begin{cases} 1 \text{ if brunette Caucasian} \\ 0 \text{ if not} \end{cases}$.

Also, let x_3 = beauty rating.

The first-order model with no interaction is $E(y) = \beta_0 + \beta_1 x_1 + \beta_2 x_2 + \beta_3 x_3$.

b. For minority females, the equation would be $E(y) = \beta_0 + \beta_1(0) + \beta_2(0) + \beta_3 x_3 = \beta_0 + \beta_3 x_3$ The change in contribution level for each 1-point increase in solicitor's beauty rating would be β_3.

For blonde Caucasian females, the equation would be $E(y) = \beta_0 + \beta_1(1) + \beta_2(0) + \beta_3 x_3 = \beta_0 + \beta_1 + \beta_3 x_3$. The change in contribution level for each 1-point increase in solicitor's beauty rating would be β_3.

For brunette Caucasian females, the equation would be $E(y) = \beta_0 + \beta_1(0) + \beta_2(1) + \beta_3 x_3 = \beta_0 + \beta_2 + \beta_3 x_3$. The change in contribution level for each 1-point increase in solicitor's beauty rating would be β_3.

c. The interaction model would be $E(y) = \beta_0 + \beta_1 x_1 + \beta_2 x_2 + \beta_3 x_3 + \beta_4 x_1 x_3 + \beta_5 x_2 x_3$.

d. For minority females, the equation would be $E(y) = \beta_0 + \beta_1(0) + \beta_2(0) + \beta_3 x_3 + \beta_4(0)x_3 + \beta_5(0)x_3 = \beta_0 + \beta_3 x_3$ The change in contribution level for each 1-point increase in solicitor's beauty rating would be β_3.

For blonde Caucasian females, the equation would be $E(y) = \beta_0 + \beta_1(1) + \beta_2(0) + \beta_3 x_3 + \beta_4(1)x_3 + \beta_5(0)x_3 = \beta_0 + \beta_1 + (\beta_3 + \beta_4)x_3$. The change in contribution level for each 1-point increase in solicitor's beauty rating would be $\beta_3 + \beta_4$.

For brunette Caucasian females, the equation would be $E(y) = \beta_0 + \beta_1(0) + \beta_2(1) + \beta_3 x_3 + \beta_4(0)x_3 + \beta_5(1)x_3 = \beta_0 + \beta_2 + (\beta_3 + \beta_5)x_3$. The change in contribution level for each 1-point increase in solicitor's beauty rating would be $\beta_3 + \beta_5$.

e. Answers will vary. A possible answer would be:

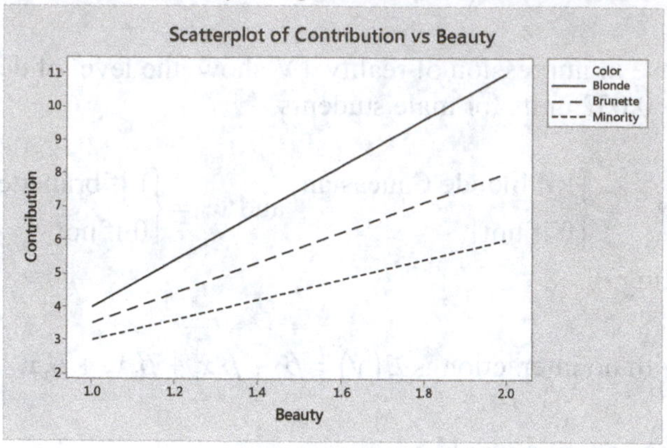

12.109 a. Let x_1 = water depth of the entanglement, $x_2 = \begin{cases} 1 \text{ if set nets} \\ 0 \text{ if not} \end{cases}$, and $x_3 = \begin{cases} 1 \text{ if pots} \\ 0 \text{ if not} \end{cases}$.

The main-effects only model is: $E(y) = \beta_0 + \beta_1 x_1 + \beta_2 x_2 + \beta_3 x_3$

b. Using MINITAB, a possible graph of the main-effects only model might look like:

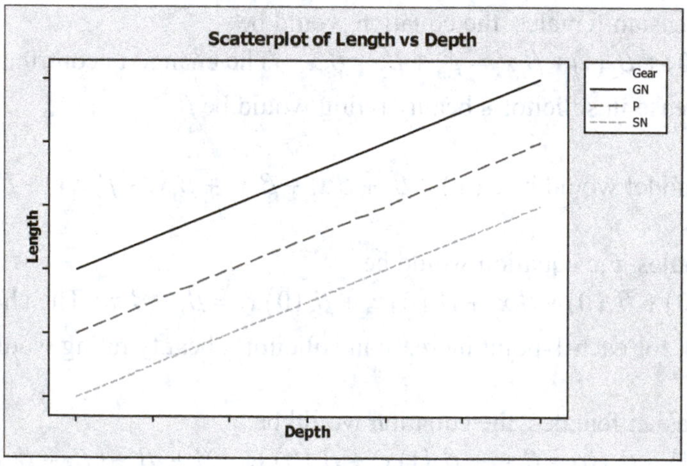

c. The interaction model would be: $E(y) = \beta_0 + \beta_1 x_1 + \beta_2 x_2 + \beta_3 x_3 + \beta_4 x_1 x_2 + \beta_5 x_1 x_3$

d. Using MINITAB, the possible graph of an interaction model might look like:

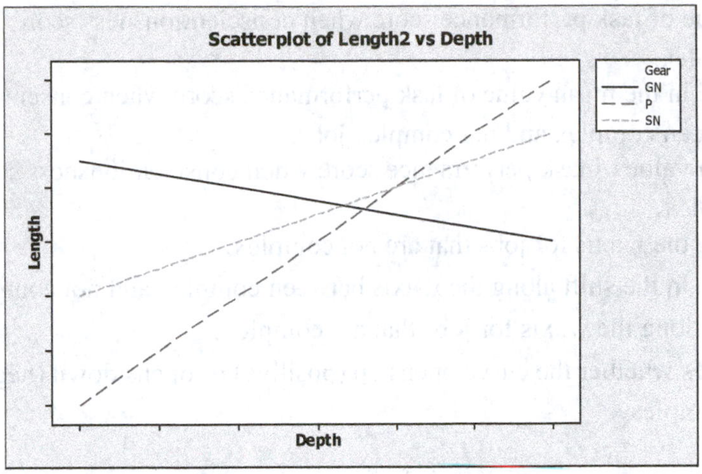

e. For set nets, $x_2 = 1$ and $x_3 = 0$. The model for set nets would be:
$E(y) = \beta_0 + \beta_1 x_1 + \beta_2(1) + \beta_3(0) + \beta_4 x_1(1) + \beta_5 x_1(0) = \beta_0 + \beta_2 + (\beta_1 + \beta_4)x_1$
The rate of change of whale length with water depth for set nets is $\beta_1 + \beta_4$.

f. For pots, $x_2 = 0$ and $x_3 = 1$. The model for set nets would be:
$E(y) = \beta_0 + \beta_1 x_1 + \beta_2(0) + \beta_3(1) + \beta_4 x_1(0) + \beta_5 x_1(1) = \beta_0 + \beta_3 + (\beta_1 + \beta_5)x_1$
The rate of change of whale length with water depth for set nets is $\beta_1 + \beta_5$.

g. For gill nets, $x_2 = 0$ and $x_3 = 0$. The model for set nets would be:
$E(y) = \beta_0 + \beta_1 x_1 + \beta_2(0) + \beta_3(0) + \beta_4 x_1(0) + \beta_5 x_1(0) = \beta_0 + \beta_1 x_1$
The rate of change of whale length with water depth for set nets is β_1.

h. To test whether the rate of change of whale length with water depth is the same for all three types of fishing gear, we test:

$H_0: \beta_4 = \beta_5 = 0$
H_a: At least one $\beta_i \neq 0$

12.111 a. For jobs that are not highly complex, the equation would be
$E(y) = \beta_0 + \beta_1 x_1 + \beta_2 (x_1)^2 + \beta_3(0) + \beta_4 x_1(0) + \beta_5 (x_1)^2(0) = \beta_0 + \beta_1 x_1 + \beta_2 (x_1)^2$

b. β_0 = mean value of task performance score when conscientiousness score is 0 and the job is not complex.
β_1 = shift along the x_1 axis.
β_2 indicates whether the curve opens up (positive) or opens down (negative).

c. For highly complex jobs, the equation would be
$E(y) = \beta_0 + \beta_1 x_1 + \beta_2 (x_1)^2 + \beta_3(1) + \beta_4 x_1(1) + \beta_5 (x_1)^2(1) = \beta_0 + \beta_3 + (\beta_1 + \beta_4)x_1 + (\beta_2 + \beta_5)(x_1)^2$

d. β_0 = mean value of task performance score when conscientiousness score is 0 and the job is not complex.
β_3 = difference in the mean value of task performance score when conscientiousness score is 0 between complex and not complex jobs.
$\beta_0 + \beta_3$ = mean value of task performance score when conscientiousness score is 0 and the job is complex.
β_1 = shift along the x_1 axis for jobs that are not complex.
β_4 = difference in the shift along the x_1 axis between complex and not complex jobs.
$\beta_1 + \beta_4$ = shift along the x_1 axis for jobs that are complex.
$\beta_2 + \beta_5$ indicates whether the curve opens up (positive) or opens down (negative) for jobs that are complex.

e. Yes. Since the interaction term, $\beta_5 (x_1)^2 x_2$, is in the model, the curvilinear relationship between task performance score and conscientiousness score depends on job complexity.

12.113 a. Let $x_2 = \begin{cases} 1 \text{ if low POS} \\ 0 \text{ if not} \end{cases}$ and $x_3 = \begin{cases} 1 \text{ if neutral POS} \\ 0 \text{ if not} \end{cases}$.

b. The model with three parallel lines would be: $E(y) = \beta_0 + \beta_1 x_1 + \beta_2 x_2 + \beta_3 x_3$

c. The model with three nonparallel lines would be:
$E(y) = \beta_0 + \beta_1 x_1 + \beta_2 x_2 + \beta_3 x_3 + \beta_4 x_1 x_2 + \beta_5 x_1 x_3$

d. The model in part ***c*** would support the findings that the intention to leave was greater at the low level of POS than at the high level of POS.

12.115 a. Let $x_1 = \begin{cases} 1 \text{ if water is from sub-surface flow} \\ 0 \text{ otherwise} \end{cases}$ $x_2 = \begin{cases} 1 \text{ if water is from overground flow} \\ 0 \text{ otherwise} \end{cases}$

and x_3 = silica concentration.

A model for $E(y)$ as a function of the water source and silica concentration is:

$E(y) = \beta_0 + \beta_1 x_1 + \beta_2 x_2 + \beta_3 x_3$

A possible graph would be:

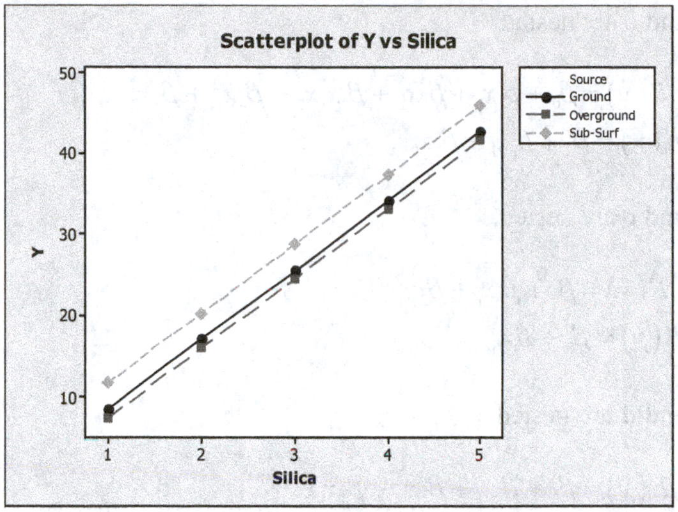

b. A model allowing for the relationship between nitrate concentrations and silica concentrations to differ for the three water sources is:

$E(y)=\beta_0+\beta_1x_1+\beta_2x_2+\beta_3x_3+\beta_4x_1x_3+\beta_5x_2x_3$

A possible sketch of the relationships is:

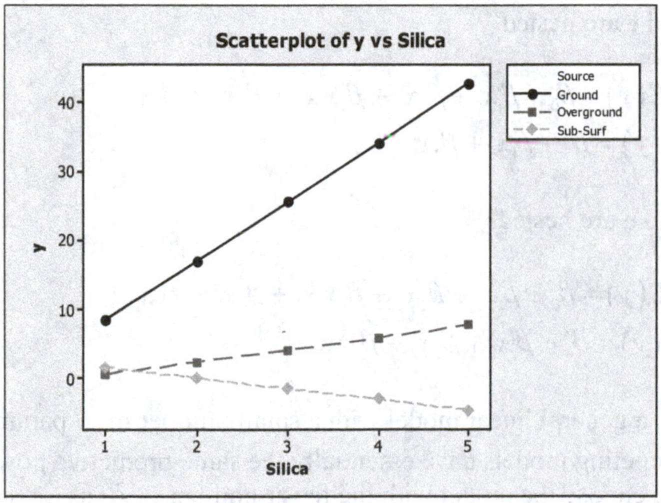

12.117 The models in parts **a** and **b** are nested:

The complete model is $E(y)=\beta_0+\beta_1x_1+\beta_2x_2$

The reduced model is $E(y)=\beta_0+\beta_1x_1$

The models in parts **a** and **d** are nested.

The complete model is $E(y)=\beta_0+\beta_1x_1+\beta_2x_2+\beta_3x_1x_2$

The reduced model is $E(y)=\beta_0+\beta_1x_1+\beta_2x_2$

The models in parts **a** and **e** are nested.

The complete model is $E(y)=\beta_0+\beta_1x_1+\beta_2x_2+\beta_3x_1x_2+\beta_4x_1^2+\beta_5x_2^2$
The reduced model is $E(y)=\beta_0+\beta_1x_1+\beta_2x_2$

The models in parts **b** and **c** are nested.

The complete model is $E(y)=\beta_0+\beta_1x_1+\beta_2x_1^2$
The reduced model is $E(y)=\beta_0+\beta_1x_1$

The models in parts **b** and **d** are nested.

The complete model is $E(y)=\beta_0+\beta_1x_1+\beta_2x_2+\beta_3x_1x_2$
The reduced model is $E(y)=\beta_0+\beta_1x_1$

The models in parts **b** and **e** are nested.

The complete model is $E(y)=\beta_0+\beta_1x_1+\beta_2x_2+\beta_3x_1x_2+\beta_4x_1^2+\beta_5x_2^2$
The reduced model is $E(y)=\beta_0+\beta_1x_1$

The models in parts **c** and **e** are nested.

The complete model is $E(y)=\beta_0+\beta_1x_1+\beta_2x_2+\beta_3x_1x_2+\beta_4x_1^2+\beta_5x_2^2$
The reduced model is $E(y)=\beta_0+\beta_1x_1+\beta_2x_1^2$

The models in parts **d** and **e** are nested.

The complete model is $E(y)=\beta_0+\beta_1x_1+\beta_2x_2+\beta_3x_1x_2+\beta_4x_1^2+\beta_5x_2^2$
The reduced model is $E(y)=\beta_0+\beta_1x_1+\beta_2x_2+\beta_3x_1x_2$

12.119 A parsimonious model is a general linear model with a small number of β parameters. In situations where two competing models have essentially the same predictive power (as determined by an F-test), choose the model with the fewer number of β 's.

12.121 a. There are five β parameters in the complete model and three in the reduced model.

b. The hypotheses are:

$$H_0: \beta_3=\beta_4=0$$
$$H_a: \text{At least one } \beta_i \neq 0, \; i = 3, 4$$

c. The test statistic is

$$F = \frac{(\text{SSE}_\text{R} - \text{SSE}_\text{C})/(k-g)}{\text{SSE}_\text{C}/\left[n-(k+1)\right]} = \frac{(160.44-152.66)/(4-2)}{152.66/\left[20-(4+1)\right]} = \frac{3.89}{10.1773} = 0.38$$

The rejection region requires $\alpha = 0.05$ in the upper tail of the F distribution with $\nu_1 = k - g = 4 - 2 = 2$ and $\nu_2 = n - (k+1) = 20 - (4+1) = 15$. From Table VI, Appendix B, $F_{0.05} = 3.68$. The rejection region is $F > 3.68$.

Since the observed value of the test statistic does not fall in the rejection region $(F = 0.38 \not> 3.68)$, H_0 is not rejected. There is insufficient evidence to indicate the complete model is better than the reduced model at $\alpha = 0.05$.

12.123 a. The complete 1st-order model is $E(y) = \beta_0 + \beta_1 x_1 + \beta_2 x_2 + \beta_3 x_3 + \beta_4 x_4 + \beta_5 x_5 + \beta_6 x_6$.

b. The reduced model would be $E(y) = \beta_0 + \beta_1 x_1 + \beta_2 x_2 + \beta_3 x_3 + \beta_4 x_4$. To determine whether the leadership score of either the purser of head flight attendant (or both) are useful for predicting team goal attainment, we test:

$H_0: \beta_5 = \beta_6 = 0$
$H_\text{a}:$ At least one $\beta_i \neq 0$

c. Because the value of R^2 increases from 0.02 to 0.25 when comparing the reduced model to the complete model for successful cabin crews, we would reject the null hypothesis and conclude that the leadership score of either the purser of head flight attendant (or both) are useful for predicting team goal attainment.

d. Since the p-value is less than $\alpha\ (p < 0.05 < 0.05)$, H_0 is rejected. There is sufficient evidence to indicate that the leadership score of either the purser of head flight attendant (or both) for successful cabin crews are useful for predicting team goal attainment at $\alpha = 0.05$.

e. Because the value of R^2 only increases from 0.14 to 0.15 when comparing the reduced model to the complete model for unsuccessful cabin crews, we would not reject the null hypothesis and conclude that the leadership score of either the purser of head flight attendant (or both) were not useful for predicting team goal attainment.

f. Since the p-value is not less than $\alpha\ (p > 0.10 \not< 0.05)$, H_0 is not rejected. There is insufficient evidence to indicate that the leadership score of either the purser of head flight attendant (or both) for unsuccessful cabin crews are useful for predicting team goal attainment at $\alpha = 0.05$.

12.125 a. The two models are nested because all of the terms in the model in part ***b*** are contained in the model in part ***c***. The complete model is

$$E(y) = \beta_0 + \beta_1 x_1 + \beta_2 x_2 + \beta_3 x_3 + \beta_4 x_1 x_2 + \beta_5 x_1 x_3$$

and the reduced model is $E(y) = \beta_0 + \beta_1 x_1 + \beta_2 x_2 + \beta_3 x_3$.

b. To compare the two models, the null hypothesis would be:

$$H_0: \beta_4 = \beta_5 = 0$$

c. If you reject H_0, then the complete model would be preferred. There is evidence that the interaction terms are significant.

d. If you fail to reject H_0, then the reduced model would be preferred. There is no evidence that the interaction terms are significant.

12.127 a. Model 1: $R^2 = 0.101$. 10.1% of the total sample variation in supervisor-directed aggression score is explained by the model including the 4 independent variables.

Model 2: $R^2 = 0.555$. 55.5% of the total sample variation in supervisor-directed aggression score is explained by the model including the 8 independent variables.

b. To determine if the additional 4 variables are useful in the prediction of supervisor-directed aggression score, we test:

$$H_0: \beta_5 = \beta_6 = \beta_7 = \beta_8 = 0$$
$$H_a: \text{At least one } \beta_i \neq 0$$

c. Yes, the models are nested. All of the variables included in Model 1 are also included in model 2.

d. Since the *p*-value is very small $(p < 0.001)$, there is evidence to reject H_0. There is sufficient evidence to indicate at least 1 of the additional 4 variables is useful in the prediction of supervisor-directed aggression score at $\alpha > 0.001$.

e. Model 3 would be:

$$\begin{aligned}
E(y) = \beta_o &+ \beta_1(\text{Age}) + \beta_2(\text{Gender}) \\
&+ \beta_3(\text{Interactional injustice at 2nd job}) + \beta_4(\text{Abusive supervisor at 2nd job}) \\
&+ \beta_5(\text{Self-esteem}) + \beta_6(\text{History of agression}) \\
&+ \beta_7(\text{Interactional injustice at primary job}) \\
&+ \beta_8(\text{Abusive supervisor at primary job}) \\
&+ \beta_9(\text{Self-esteem})(\text{History of agression}) \\
&+ \beta_{10}(\text{Self-esteem})(\text{Interactional injustice at primary job}) \\
&+ \beta_{11}(\text{Self-esteem})(\text{Abusive supervisor at primary job}) \\
&+ \beta_{12}(\text{History of agression})(\text{Interactional injustice at primary job}) \\
&+ \beta_{13}(\text{History of agression})(\text{Abusive supervisor at primary job}) \\
&+ \beta_{14}(\text{Interactional injustice at primary job})(\text{Abusive supervisor at primary job})
\end{aligned}$$

f. Since the *p*-value is $p > 0.10$, there is no evidence to indicate that any of the interaction terms are useful in the prediction of supervisor-directed aggression score for $\alpha \le 0.10$.

12.129 a. To determine the overall adequacy of the model, we test:

$$H_0: \beta_1 = \beta_2 = \beta_3 = \beta_4 = \beta_5 = 0$$
$$H_a: \text{At least one } \beta_i \ne 0$$

b. To determine whether task performance score and conscientiousness score are curvilinearly related, we test:

$$H_0: \beta_2 = \beta_5 = 0$$
$$H_a: \text{At least one } \beta_i \ne 0$$

c. To determine whether the curvilinear relationship between task performance score and conscientiousness score depends on job complexity, we test:

$$H_0: \beta_5 = 0$$
$$H_a: \beta_5 \ne 0$$

d. For the test in part a, the reduced model is $E(y) = \beta_0$. The test statistics is $F = \dfrac{\text{MSR}}{\text{MSE}}$.

For the test in part b, the reduced model is $E(y) = \beta_0 + \beta_1 x_1 + \beta_3 x_2 + \beta_4 x_1 x_2$. The test statistic is $F = \dfrac{(\text{SSE}_\text{R} - \text{SSE}_\text{C})/(k-g)}{\text{SSE}_\text{C}/[n-(k+1)]}$.

For the test in part c, the reduced model is $E(y) = \beta_0 + \beta_1 x_1 + \beta_2 (x_1)^2 + \beta_3 x_2 + \beta_4 x_1 x_2$. The test statistic is $F = \dfrac{(\text{SSE}_\text{R} - \text{SSE}_\text{C})/(k-g)}{\text{SSE}_\text{C}/[n-(k+1)]}$.

12.131 a. To determine if there is any difference in the mean lengths of entangled whales for the three gear types, we test:

$$H_0: \beta_2 = \beta_3 = \beta_4 = \beta_5 = 0$$
$$H_a: \text{At least one } \beta_i \ne 0$$

b. The reduced model would be: $E(y) = \beta_0 + \beta_1 x_1$

c. If we reject H_0 above, we would conclude that the mean lengths of entangled whales for the three gear types differ.

d. To determine if the rate of change of whale length with water depth is the same for all three types of fishing gear, we test:

$H_0: \beta_4 = \beta_5 = 0$
H_a: At least one $\beta_i \neq 0$

e. The reduced model would be: $E(y) = \beta_0 + \beta_1 x_1 + \beta_2 x_2 + \beta_3 x_3$

f. If we fail to reject H_0 above, we would conclude that the rate of change of whale length with water depth does not differ for the three types of fishing gear.

12.133 a. Let x_1 = speed, x_2 = rate, and x_3 = percent weight. The interaction model would be:
$E(y) = \beta_0 + \beta_1 x_1 + \beta_2 x_2 + \beta_3 x_3 + \beta_4 x_1 x_2 + \beta_5 x_1 x_3 + \beta_6 x_2 x_3$

b. For $x_1 = 1000$ and $x_2 = 50$, the model is:
$$E(y) = \beta_0 + \beta_1(1000) + \beta_2(50) + \beta_3 x_3 + \beta_4(1000)(50) + \beta_5(1000)x_3 + \beta_6(50)x_3$$
$$= \beta_0 + 1000\beta_1 + 50\beta_2 + 50{,}000\beta_4 + (\beta_3 + 1000\beta_5 + 50\beta_6)x_3$$
Thus, the slope of the line is $\beta_3 + 1000\beta_5 + 50\beta_6$.

c. For $x_1 = 1000$ and $x_2 = 150$, the model is:
$$E(y) = \beta_0 + \beta_1(1000) + \beta_2(150) + \beta_3 x_3 + \beta_4(1000)(150) + \beta_5(1000)x_3 + \beta_6(150)x_3$$
$$= \beta_0 + 1000\beta_1 + 150\beta_2 + 150{,}000\beta_4 + (\beta_3 + 1000\beta_5 + 150\beta_6)x_3$$
Thus, the slope of the line is $\beta_3 + 1000\beta_5 + 150\beta_6$.

d. Using MINITAB, the results are:

Regression Analysis: FORCE versus SPEED, RATE, PCTWT, S_R, S_P, R_P

```
The regression equation is
FORCE = 343 + 0.0181 SPEED + 1.99 RATE + 9.37 PCTWT - 0.000013 S_R
        - 0.000375 S_P + 0.0125 R_P


Predictor          Coef      SE Coef       T      P
Constant         343.13        19.05   18.02  0.000
SPEED          0.018125     0.007315    2.48  0.035
RATE             1.9875       0.1463   13.58  0.000
PCTWT             9.375        1.463    6.41  0.000
S_R         -0.00001250   0.00004877   -0.26  0.803
S_P          -0.0003750    0.0004877   -0.77  0.462
R_P            0.012500     0.009754    1.28  0.232

S = 9.75392   R-Sq = 99.6%   R-Sq(adj) = 99.3%

Analysis of Variance

Source          DF      SS     MS       F      P
Regression       6  216287  36048  378.90  0.000
Residual Error   9     856     95
Total           15  217144
```

The least squares prediction equation is:
$\hat{y} = 343 + 0.0181x_1 + 1.99x_2 + 9.37x_3 - 0.000013x_1x_2 - 0.000375x_1x_3 + 0.0125x_2x_3$

e. We must fit the reduced model without the interaction terms. Using MINITAB, the results are:

Regression Analysis: FORCE versus SPEED, RATE, PCTWT

```
The regression equation is
FORCE = 341 + 0.0131 SPEED + 2.09 RATE + 9.87 PCTWT


Predictor       Coef    SE Coef      T      P
Constant     340.625      8.531  39.93  0.000
SPEED       0.013125   0.002366   5.55  0.000
RATE         2.08750    0.04732  44.11  0.000
PCTWT         9.8750     0.4732  20.87  0.000

S = 9.46485    R-Sq = 99.5%    R-Sq(adj) = 99.4%

Analysis of Variance

Source          DF      SS     MS       F      P
Regression       3  216069  72023  803.98  0.000
Residual Error  12    1075     90
Total           15  217144
```

To determine if any of the interaction terms are statistically significant, we test:

$H_0: \beta_4 = \beta_5 = \beta_6 = 0$
H_a: At least one $\beta_i \neq 0$

The test statistic is

$$F = \frac{(\text{SSE}_R - \text{SSE}_C)/(k-g)}{\text{SSE}_C/[n-(k+1)]} = \frac{(1075-856)/(6-3)}{856/[16-(6+1)]} = \frac{73}{95.1111} = 0.768 .$$

The rejection region requires $\alpha = 0.05$ in the upper tail of the F distribution with $\nu_1 = k - g = 6 - 3 = 3$ and $\nu_2 = n - (k+1) = 16 - (6+1) = 9$. From Table VI, Appendix B, $F_{0.05} = 3.86$. The rejection region is $F > 3.86$.

Since the test statistic does not fall in the rejection region $(F = 0.768 \not> 3.86)$, H_0 is not rejected. There is insufficient evidence to indicate that any of the interaction terms are statistically significant at $\alpha = 0.05$.

12.135 In a stepwise model, several steps are involved in selecting the model. First, all one-variable models are compared. The one variable model with the highest r^2 is selected. Next, all 2 variable models that include the previously selected variable are compared. The model with the highest R^2 is selected. This process is continued until no additional variables can be added to the model (no more contribute significantly to the prediction of y). In a standard regression model, the model is specified first and then all variables thought to be important are included in the model.

12.137 a. The best one-variable predictor of y is the one whose t statistic has the largest absolute value. The t statistics for each of the variables are:

Independent Variable	$t = \dfrac{\hat{\beta}_i}{s_{\hat{\beta}_i}}$
x_1	$t = 1.6 / 0.42 = 3.81$
x_2	$t = -0.9 / 0.01 = -90$
x_3	$t = 3.4 / 1.14 = 2.98$
x_4	$t = 2.5 / 2.06 = 1.21$
x_5	$t = -4.4 / 0.73 = -6.03$
x_6	$t = 0.3 / 0.35 = 0.86$

The variable x_2 is the best one-variable predictor of y. The absolute value of the corresponding t score is 90. This is larger than any of the others.

b. Yes. In the stepwise procedure, the first variable entered is the one which has the largest absolute value of t, provided the absolute value of the t falls in the rejection region.

c. Once x_2 is entered, the next variable that is entered is the one that, adjusted for x_2, has the largest absolute t value associated with it.

12.139 Using stepwise regression would probably not be the best approach to finding the best model to predict the total number of convictions because two of the independent variables are qualitative (type of crime and race). Race will probably have several levels and it would be somewhat difficult to include those in a stepwise regression. Maybe a better approach would be to fit a first order model with the 3 independent variables (type of crime, race, and age at first conviction) in the model. Then we could fit a model with all the two-way interaction terms in the model and test to see if any of the two-way interactions are significant. If none of the two-way interaction terms are significant, then go back to the first order model and keep only those independent variables that are significant. If any of the interaction terms are significant, you must determine which of the interaction terms are significant and keep only those that are significant.

12.141 a. The hypothesized equation of the final stepwise regression model is:

$E(y) = \beta_0 + \beta_1 x_1 + \beta_2 x_2$, where x_1 = leg length and x_2 = pressure.

b. $R^2 = 0.771$. This means that 77.1% of the sample variation in heel depth is explained by the model containing leg length and pressure.

c. To determine if the overall model is useful in predicting heel depth, we test:

$$H_0: \beta_1 = \beta_2 = 0$$
$$H_a: \text{At least one } \beta_i \neq 0$$

Since the p-value is so small $(p < 0.001)$, we would reject H_0 for any reasonable value of α. There is sufficient evidence to indicate the model is useful for predicting heel depth.

d. In the first step of the stepwise regression, there were 6 individual t-tests run. In the second step, there are another $6-1=5$ individual t-tests run. In the third step, there are an additional $6-2=4$ individual t-tests run. Thus, at a minimum, there are a total of $6+5+4=15$ individual t-tests run.

e. Since there were so many individual t-tests run, the probability of making at least one Type I error during the stepwise analysis is very high. The probability of making at least one Type I error is 1 – probability of not making any Type I error in 15 tests. Thus, P(at least 1 Type I error) = 1 – P(no Type I errors) = $1-(1-\alpha)^{15}$.

12.143 a. In the first step of stepwise regression, 11 models are fit. These would be all the one-variable models.

b. In step two, all two variable models that include the variable selected in step one are fit. There will be a total of 10 models fit in this step.

c. In step 11, there is one model fit – the model containing all the independent variables.

d. The equation for $E(y)$ is:

$$E(y)=\beta_0+\beta_1x_{11}+\beta_2x_4+\beta_3x_2+\beta_4x_7+\beta_5x_{10}+\beta_6x_1+\beta_7x_9+\beta_8x_3$$

e. $R^2=0.677$. 67.7% of the sample variation in the overall satisfaction with BRT scores is explained by the model containing the 8 independent variables.

f. In any stepwise regression, many t-tests are performed. This inflates the probability of committing a Type I error. In addition, there are certain combinations of variables that may never be reached because of the first few variables put in the model. Finally, you should consider including interaction and higher order terms in the model.

12.145 a. Since there were 11 potential independent variables, there were 11 t-tests performed in Step 1.

b. In step 2, there were $11-1=10$ t-tests performed.

c. The global test for testing whether at least one of the two variables is significant in predicting TME has a p-value of 0.001. Since this p-value is so small, there is evidence to indicate the model is useful in predicting TME. The value of R^2 is 0.988. This means that 98.8% of the sample variation in TME values is explained by the model containing both AMAP and NDF.

$\hat{\beta}_1=2.14$. For each unit increase in AMAP (amylopectin/amylase), the mean value of TME is estimated to increase by 2.14, holding NDF (neutral detergent fiber) constant.

$\hat{\beta}_2 = 0.16$. For each unit increase in NDF (neutral detergent fiber), the mean value of TME is estimated to increase by 0.16, holding AMAP (amylopectin/amylase) constant.

d. One should always be wary of using a model selected by stepwise regression. This stepwise regression model selected the variables AMAP and NDF. The best model for describing the dependent variable may not include either of these 2 variables. The stepwise procedure may prevent one from ever getting to the best model. Another reason to be wary is that a large number of *t*-tests have been conducted, leading to a high probability of making one or more Type I or Type II errors.

e. The complete 2nd-order model for TME would be:

$$E(y) = \beta_0 + \beta_1 AMAP + \beta_2 NDF + \beta_3 AMAP \bullet NDF + \beta_4 AMAP^2 + \beta_5 NDF^2$$

f. To determine if the terms in the model in part e that allow for curvature are statistically useful for predicting TME, we would test:

$$H_0: \beta_4 = \beta_5 = 0$$
$$H_a: \text{At least one } \beta_i \neq 0$$

12.147 A residual that is larger than 3*s* (in absolute value) is considered to be an outlier.

12.149 The statement "Regression models fit to time series data typically result in uncorrelated errors." is false. Time series data tend to be correlated.

12.151 Three indicators of a multicollinearity problem are:

1. Significant correlations between pairs of independent variables.
2. Nonsignificant *t*-tests for all (or nearly all) of the individual β parameters when the *F*-test for overall model adequacy is significant.
3. Signs opposite from what is expected in the estimated β parameters.

12.153 Yes. x_2 and x_4 are highly correlated $(r_{24} = 0.93)$, as well as x_4 and x_5 $(r_{45} = 0.86)$. When highly correlated independent variables are present in a regression model, the results can be confusing. The researcher may want to include only one of the correlated variables.

12.155 If the highest correlation)in absolute value) for any pair of independent variables is $r = -0.16$, the regression analysis would probably not exhibit multicollinearity problems because this shows that the independent variables are not highly correlated with each other.

12.157 For this problem, there are 6 correlation coefficients with absolute values in the range from 0.2 to 0.8. This indicates there would be moderate multicollinearity. The variables x_1 and x_3 have a correlation coefficient of 0.62. We would suggest maybe using only one of the 2 variables in the model. Likewise, the variables x_4 and x_5 have a correlation coefficient of -0.62. We would again suggest maybe using only one of the 2 variables in the model.

12.159 a. No, there is no evidence of extreme multicollinearity. The independent variables would be irritability, trait anger, and narcissism. The largest pairwise correlation (in absolute value) between independent variables is $r = 0.57$. This is not extremely large and would probably not lead to extreme multicollinearity.

b. Yes, there would be some evidence of multicollinearity. The independent variables would be aggressive behavior, irritability, and trait anger. The largest pairwise correlation (in absolute value) between independent variables is $r = 0.80$. This is fairly large and would probably lead to a problem with multicollinearity.

12.161 One possible reason for the sign of the estimated parameter multiplied by x_1 to be the opposite of what is expected is that company role of estimator and previous accuracy could be highly correlated to each other. If two independent variables are highly correlated to each other and appear in the regression equation together, multicollinearity could be present.

12.163 a. Using MINITAB, the plots of the residuals versus each of the independent variables are:

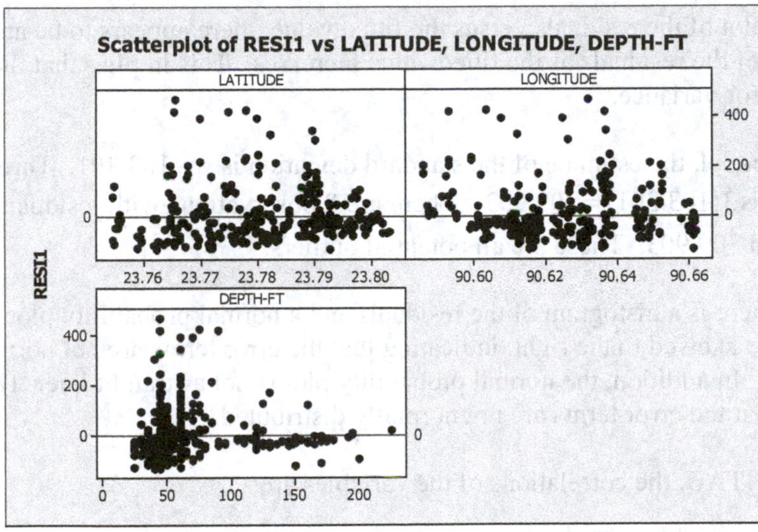

From each of the plots, there is no apparent U-shape to the residuals. This indicates that the model was probably not misspecified. Thus, the assumption of mean error = 0 is probably valid.

b. Using MINITAB, the residual plots are:

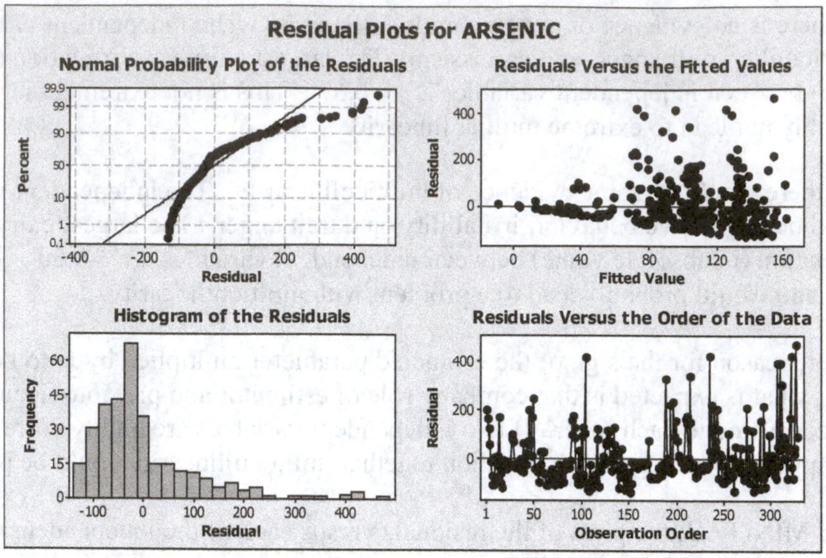

Using the plot of the residuals versus the fitted value, there appears to be an increase in the spread of the residuals as the fitted value increases. This implies that there is not constant error variance.

c. From the model, the estimate of the standard deviation is $s = 103.301$. Three standard deviations is $3(103.301) = 309.903$. There are 8 observations with residuals that are greater than 309.903. These are all potential outliers.

d. In part b, there is a histogram of the residuals and a normal probability plot. The histogram is skewed to the right, indicating that the error terms are not normally distributed. In addition, the normal probability plot is not a straight line. This also indicates that the error terms are not normally distributed.

e. Using MINITAB, the correlations of the variables are:

Correlations: ARSENIC, LATITUDE, LONGITUDE, DEPTH-FT

```
             ARSENIC   LATITUDE  LONGITUDE
LATITUDE      -0.186
               0.001

LONGITUDE      0.192      0.311
               0.000      0.000

DEPTH-FT      -0.245      0.151     -0.328
               0.000      0.006      0.000

Cell Contents: Pearson correlation
               P-Value
```

All of the pairs of independent variables have correlation coefficients that are less than 0.33 in absolute value. These correlations are very low. This indicates that there probably is not a problem with multicollinearity.

12.165 Using MINITAB, the plots of the residuals versus each of the independent variables are:

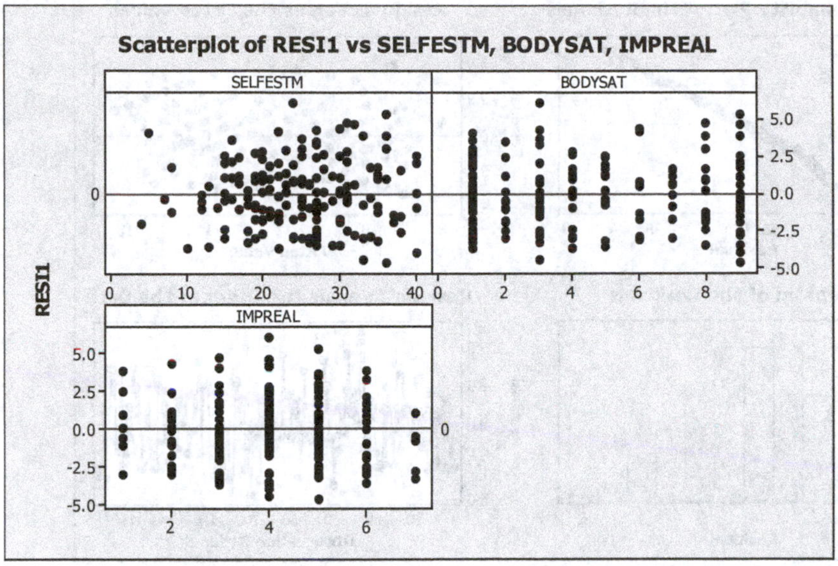

From the plot of the residuals versus self-esteem, there is no U-shaped pattern to the residuals. This implies that the relationship between desire and self-esteem has not been misspecified. Thus, the assumption of mean error = 0 is appropriate. From the analysis, the estimate of the standard deviation is $s = 2.25$. There are no points that are more than 3 standard deviations from the mean. Thus, there does not appear to be any outliers.

From the plot of the residuals versus body satisfaction, there is no U-shaped pattern to the residuals. This implies that the relationship between desire and body satisfaction has not been misspecified. Thus, the assumption of mean error = 0 is appropriate. In addition, there are no points that appear to be outliers.

From the plot of the residuals versus impression of reality TV, there is no U-shaped pattern to the residuals. This implies that the relationship between desire and impression of reality TV has not been misspecified. Thus, the assumption of mean error = 0 is appropriate. In addition, there are no points that appear to be outliers.

Using MINITAB, additional plots of the residuals are:

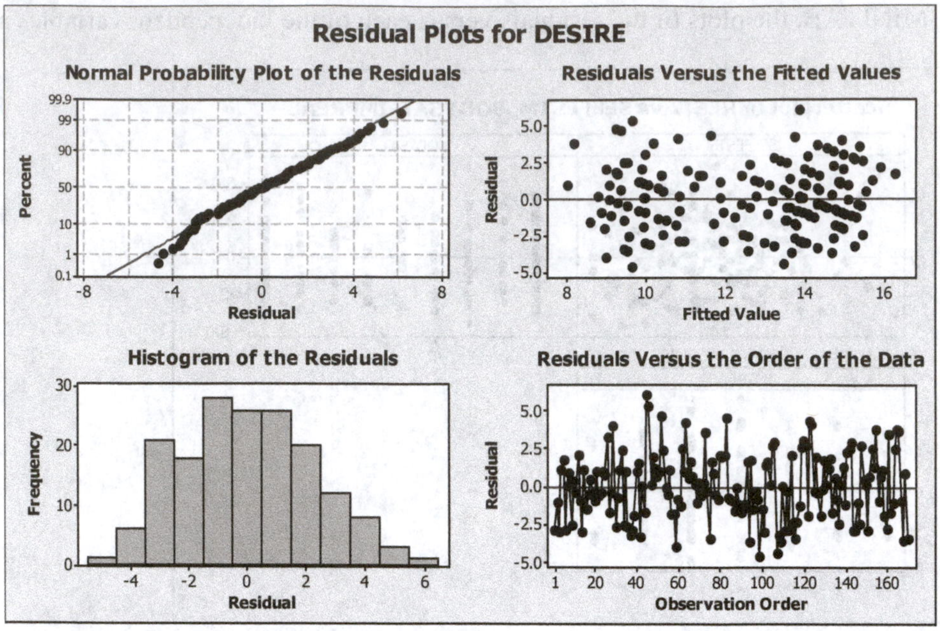

The plot of the residuals versus the fitted values does not appear to be funnel-shaped or football-shaped. Thus, the assumption of equal variance appears to be met.

The histogram of the residuals appears to be mound-shaped, which implies that the assumption of normal error terms is satisfied. The normal probability plot is very close to a straight line, also indicating that the error terms are normally distributed.

Using MINITAB, the correlations are:

Correlations: DESIRE, SELFESTM, BODYSAT, IMPREAL, GENDER

```
           DESIRE  SELFESTM    BODYSAT   IMPREAL
SELFESTM   -0.485
            0.000

BODYSAT    -0.644     0.757
            0.000     0.000

IMPREAL     0.132     0.167     0.143
            0.087     0.030     0.062

GENDER     -0.637     0.511     0.828     0.065
            0.000     0.000     0.000     0.398


Cell Contents: Pearson correlation
               P-Value
```

The correlation between body satisfaction and self-esteem is 0.757. This indicates a possible moderate problem with multicollinearity. Self-esteem and body satisfaction should probably not both appear in the model together. The correlations between impression of reality TV and both body satisfaction and self-esteem are below 0.2, indicating there is not a problem with multicollinearity.

12.167 Using MINITAB, the correlations among the variables are:

Correlation: Intrapersonal, Stress, Mood, Project

	Intrapersonal	Stress	Mood
Stress	0.083		
Mood	0.201	-0.198	
Project	-0.368	0.310	0.054

Cell Contents: Pearson correlation

None of the correlation coefficients among the independent variables are high in absolute values. This indicates there is not a problem with multicollinearity. In addition, the signs of the coefficients in the model are what would be expected based on the signs of the correlation coefficients between the project scores and the independent variables. Again, this indicates that multicollinearity is not an issue.

12.169 $E(y)=\beta_0+\beta_1x_1+\beta_2x_2+\beta_3x_3$

where $x_1=\begin{cases}1, & \text{if level 2}\\ 0, & \text{otherwise}\end{cases}$ $\quad x_2=\begin{cases}1, & \text{if level 3}\\ 0, & \text{otherwise}\end{cases}$ $\quad x_3=\begin{cases}1, & \text{if level 4}\\ 0, & \text{otherwise}\end{cases}$

12.171 a. $E(y)=\beta_0+\beta_1x_1+\beta_2x_2+\beta_3x_3$

where $x_2=\begin{cases}1, & \text{if level 2}\\ 0, & \text{otherwise}\end{cases}$ $\quad x_3=\begin{cases}1, & \text{if level 3}\\ 0, & \text{otherwise}\end{cases}$

b. $E(y)=\beta_0+\beta_1x_1+\beta_2x_1^2+\beta_3x_2+\beta_4x_3+\beta_5x_1x_2+\beta_6x_1x_3+\beta_7x_1^2x_2+\beta_8x_1^2x_3$

where x_1, x_2, and x_3 are as in part **a**.

12.173 a. To determine if at least one of the β parameters is not zero, we test:

$H_0: \beta_1=\beta_2=\beta_3=\beta_4=0$
H_a: At least one $\beta_i \neq 0$

The test statistic is $F=\dfrac{R^2/k}{(1-R^2)/[n-(k+1)]}=\dfrac{0.83/4}{(1-0.83)/[25-(4+1)]}=24.41$

The rejection region requires $\alpha=0.05$ in the upper tail of the F distribution with $\nu_1=k=4$ and $\nu_2=n-(k=1)=25-(4+1)=20$. From Table VI, Appendix B, $F_{0.05}=2.87$. The rejection region is $F>2.87$.

Since the observed value of the test statistic falls in the rejection region $(F=24.41>2.87)$, H_0 is rejected. There is sufficient evidence to indicate at least one of the β parameters is nonzero at $\alpha=0.05$.

b. $H_0: \beta_1 = 0$
$H_a: \beta_1 < 0$

The test statistic is $t = \frac{\hat{\beta}_1 - 0}{s_{\hat{\beta}_1}} = \frac{-2.43 - 0}{1.21} = -2.01$

The rejection region requires $\alpha = 0.05$ in the lower tail of the t distribution with $df = n-(k+1) = 25-(4+1) = 20$. From Table III, Appendix B, $t_{0.05} = 1.725$. The rejection region is $t < -1.725$.

Since the observed value of the test statistic falls in the rejection region $(t = -2.01 < -1.725)$, H_0 is rejected. There is sufficient evidence to indicate β_1 is less than 0 at $\alpha = 0.05$.

c. $H_0: \beta_2 = 0$
$H_a: \beta_2 > 0$

The test statistic is $t = \frac{\hat{\beta}_2 - 0}{s_{\hat{\beta}_2}} = \frac{0.05 - 0}{0.16} = 0.31$

The rejection region requires $\alpha = 0.05$ in the upper tail of the t distribution. From part **b** above, the rejection region is $t > 1.725$.

Since the observed value of the test statistic does not fall in the rejection region $(t = 0.31 > 1.725)$, H_0 is not rejected. There is insufficient evidence to indicate β_2 is greater than 0 at $\alpha = 0.05$.

d. $H_0: \beta_3 = 0$
$H_a: \beta_3 \neq 0$

The test statistic is $t = \frac{\hat{\beta}_3 - 0}{s_{\hat{\beta}_3}} = \frac{0.62 - 0}{0.26} = 2.38$

The rejection region requires $\alpha / 2 = 0.05 / 2 = 0.025$ in each tail of the t distribution with $df = 20$. From Table III, Appendix B, $t_{0.025} = 2.086$. The rejection region is $t < -2.086$ or $t > 2.086$.

Since the observed value of the test statistic falls in the rejection region $(t = 2.38 > 2.086)$, H_0 is rejected. There is sufficient evidence to indicate β_3 is different from 0 at $\alpha = 0.05$.

12.175 The error of prediction is smallest when the values of x_1, x_2, and x_3 are equal to their sample means. The further x_1, x_2, and x_3 are from their means, the larger the error. When $x_1 = 60$, $x_2 = 0.4$, and $x_3 = 900$, the observed values are outside the observed ranges of the x values. When $x_1 = 30$, $x_2 = 0.6$, and $x_3 = 1300$, the observed values are within the observed ranges and consequently the x values are closer to their means. Thus, when $x_1 = 30$, $x_2 = 0.6$, and $x_3 = 1300$, the error of prediction is smaller.

12.177 Even though $SSE = 0$, we cannot estimate σ^2 because there are no degrees of freedom corresponding to error. With three data points, there are only two degrees of freedom available. The degrees of freedom corresponding to the model is $k = 2$ and the degrees of freedom corresponding to error is $n-(k+1)=3-(2+1)=0$. Without an estimate for σ^2, no inferences can be made.

12.179 a. $R^2 = 0.31$. 31% of the sample variation in the ln(level of CO_2 emissions in current year) is explained by the model containing ln(foreign investments 16 years earlier), gross domestic investment 16 years earlier, trade exports 16 years earlier, ln(GNP 16 years earlier), agricultural production 16 years earlier, African country, and ln(level of CO_2 emissions 16 years earlier).

b. $H_0: \beta_1 = \beta_2 = \cdots = \beta_7 = 0$
H_a: At least one $\beta_i \neq 0$

The test statistic is $F = \dfrac{R^2/k}{(1-R^2)/[n-(k+1)]} = \dfrac{0.31/7}{(1-0.31)/[66-(7+1)]} = 3.723$

The rejection region requires $\alpha = 0.01$ in the upper tail of the F distribution. From Table XI, Appendix B, with $\nu_1 = k = 7$ and $\nu_2 = n-(k+1) = 66-(7+1) = 58$, $F_{0.01} \approx 2.95$. The rejection region is $F > 2.95$.

Since the observed value of the test statistic falls in the rejection region $(F = 3.723 > 2.95)$, H_0 is rejected. There is sufficient evidence to indicate that at least one of the 7 independent variables contributes to the prediction of ln (level of CO_2 emissions in current year) at $\alpha = 0.01$.

c. We would not advise using individual t-tests on each of the independent variables to test for overall model adequacy. If we did, we would be very likely to make one or more errors in deciding which terms to retain in the model and which to exclude.

d. To determine if foreign investments 16 years earlier is a statistically useful predictor of CO_2 emissions in current year, we test:

$H_0: \beta_1 = 0$
$H_a: \beta_1 \neq 0$

e. The test statistic is $t = 2.52$ and the p-value is $p < 0.05$. Since the p-value is less than $\alpha = 0.05$, H_0 is rejected. There is sufficient evidence to indicate that foreign investment 16 years earlier is a statistically useful predictor of CO_2 emissions in current year at $\alpha = 0.05$.

f. The independent variables that are highly correlated are x_4 (ln(GNP 16 years earlier)) and x_5 (agricultural production 16 years earlier). The correlation coefficient is $r_{45} = -0.84$. Several other variables have moderate correlation: x_1 and x_3 with $r_{13} = 0.57$, x_4 and x_6 with $r_{46} = -0.53$, and x_5 and x_7 with $r_{57} = -0.50$. If these highly correlated variables are included in the model at the same time, signs of the estimates of the beta parameters may be different than what is expected. Also, the global test could indicate that at least one of the variables contributes to the prediction of *y*, but none of the individual variable tests are significant.

12.181 a. The first order model for *E*(*y*) as a function of the first five independent variables is:

$$E(y) = \beta_0 + \beta_1 x_1 + \beta_2 x_2 + \beta_3 x_3 + \beta_4 x_4 + \beta_5 x_5$$

b. To test the utility of the model, we test:

$$H_0: \beta_1 = \beta_2 = \beta_3 = \beta_4 = \beta_5 = 0$$
$$H_a: \text{At least one } \beta_i \neq 0, \quad i = 1, 2, 3, 4, 5$$

The test statistic is $F = 34.47$ and the *p*-value is $p < 0.001$.

Since the *p*-value is so small, there is sufficient evidence to indicate the model is useful for predicting GSI at $\alpha > 0.001$.

$R^2 = 0.469$. 46.9% of the variability in the GSI scores is explained by the model including the first five independent variables.

c. The first order model for *E*(*y*) as a function of the first seven independent variables is:

$$E(y) = \beta_0 + \beta_1 x_1 + \beta_2 x_2 + \beta_3 x_3 + \beta_4 x_4 + \beta_5 x_5 + \beta_6 x_6 + \beta_7 x_7$$

d. $R^2 = 0.603$. 60.3% of the variability in the GSI scores is explained by the model including the first seven independent variables.

e. Since the *p*-values associated with the variables DES and PDEQ-SR are both less than 0.001, there is evidence that both variables contribute to the prediction of GSI, adjusted for all the other variables already in the model for $\alpha > 0.001$.

12.183 a. A regression model containing the interaction between blade position and turntable speed would be:

$$E(y) = \beta_o + \beta_1 x_1 + \beta_2 x_2 + \beta_3 x_1 x_2$$

b. "Blade position and turntable speed interact" means that the effect of turntable speed on the number of defects (y) depends on the blade position.

c. This means that the interaction term in the model in part **a** would be significant. The slope of the line relating the number of defects and the turntable speed is different for different levels of blade position. Since the slope is steeper for lower values of cutting blade position, this implies that $\beta_3 < 0$.

12.185 a. $R^2_{adj} = 0.76$. 76% of the sample variability in the SAT-Math scores about their means is explained by the linear relationship between the SAT-Math scores and the independent variables score on PSAT and whether the student was coached or not, adjusted for the sample size and the number of parameters in the model.

b. For confidence coefficient 0.95, $\alpha = 0.05$ and $\alpha/2 = 0.05/2 = 0.025$. From Table III, Appendix B, with $df = n-(k+1) = 3{,}492-(2+1) = 3{,}489$, $t_{0.025} = 1.96$. The 95% confidence interval is:

$$\beta_2 \pm t_{0.025} s_{\hat{\beta}_2} \Rightarrow 19 \pm 1.96(3) \Rightarrow 19 \pm 5.88 \Rightarrow (13.12,\ 24.88)$$

We are 95% confident that the difference in the mean SAT-Math scores between students who were coached and those who were not is between 13.12 and 24.88 points, holding PSAT scores constant.

c. Since the interval did not contain 0, there is evidence to indicate that coaching had an effect on the final SAT-Math scores at $\alpha = 0.05$.

d. The hypothesized alternative model is

$$E(y) = \beta_0 + \beta_1 x_1 + \beta_2 x_2 + \beta_3 x_3 + \beta_4 x_4 + \beta_5 x_5 + \beta_6 x_6 + \beta_7 x_7 + \beta_8 x_8 + \beta_9 x_9 + \beta_{10} x_{10}$$

e. The null hypothesis would be $H_0: \beta_3 = \beta_4 = \beta_5 = \beta_6 = \beta_7 = \beta_8 = \beta_9 = \beta_{10} = 0$

f. The test was statistically significant. Thus, H_0 was rejected. There is sufficient evidence to indicate that at least one of the "control" variables contributes to the prediction of SAT-Math scores.

g. $R^2_a = 0.79$. 79% of the sample variability of the SAT-Math scores around their means is explained by the proposed model relating SAT-Math scores to the 10 independent variables, adjusting for the sample size and the number of β parameters in the model.

h. For confidence coefficient 0.95, $\alpha = 0.05$ and $\alpha/2 = 0.05/2 = 0.025$. From Table III, Appendix B, with $df = n-(k+1) = 3{,}492-(10+1) = 3{,}481$, $t_{0.025} = 1.96$. The 95% confidence interval is:

$$\hat{\beta}_2 \pm t_{0.025} s_{\hat{\beta}_2} \Rightarrow 14 \pm 1.96(3) \Rightarrow 14 \pm 5.88 \Rightarrow (8.12,\ 19.88)$$

We are 95% confident that the difference in the mean SAT-Math scores between students who were coached and those who were not is between 8.12 and 19.88 points, holding all the other variables constant.

i. Yes. From part *b*, the confidence interval for β_2 was $(13.12,\ 24.88)$. In part **h**, the confidence interval for β_2 was $(8.12,\ 19.88)$. Even though coaching is significant in both models, the change in the mean SAT-Math scores is not as great if the control variables are added to the model.

j. The complete model would be:

$$E(y)=\beta_0+\beta_1x_1+\beta_2x_2+\beta_3x_3+\beta_4x_4+\beta_5x_5+\beta_6x_6+\beta_7x_7+\beta_8x_8+\beta_9x_9+\beta_{10}x_{10}+\beta_{11}x_1x_2$$
$$+\beta_{12}x_3x_2+\beta_{13}x_4x_2+\beta_{14}x_5x_2+\beta_{15}x_6x_2+\beta_{16}x_7x_2+\beta_{17}x_8x_2+\beta_{18}x_9x_2+\beta_{19}x_{10}x_2$$

k. The null hypothesis would be H_0: $\beta_{11}=\beta_{12}=\beta_{13}=\beta_{14}=\beta_{15}=\beta_{16}=\beta_{17}=\beta_{18}=\beta_{19}=0$. To perform this test, you would fit the complete model specified in part **j**. You would also fit the reduced model specified in part **d**. Then, you would perform the test comparing the complete and reduced models.

12.187 a. The first order model is $E(y)=\beta_0+\beta_1x_1+\beta_2x_2$.

b. $\hat{\beta}_1=-0.002$. We estimate the mean average size of female beetles will decrease 0.002 millimeters for each degree increase in temperature, when Julian date is held fixed.

c. To determine if the average size of female Japanese beetles decrease linearly as temperature increases, we test:

$$H_0:\beta_1=0$$
$$H_a:\beta_1<0$$

The test statistic is $t=-0.72$ and the *p*-value is $p=0.49$.

Since the *p*-value is greater than $\alpha\ (p=0.49>0.05)$, H_0 is not rejected. There is insufficient evidence to indicate the average size of female Japanese beetles decrease linearly as temperature increases, when Julian date is held constant at $\alpha=0.05$.

12.189 a. Using MINITAB, the results are:

Regression Analysis: WeightChg versus Digest, Fiber

```
The regression equation is
WeightChg = 12.2 - 0.0265 Digest - 0.458 Fiber

Predictor        Coef     SE Coef        T       P
Constant       12.180       4.402     2.77   0.009
Digest       -0.02654     0.05349    -0.50   0.623
Fiber         -0.4578      0.1283    -3.57   0.001

S = 3.519        R-Sq = 52.9%      R-Sq(adj) = 50.5%

Analysis of Variance

Source           DF         SS          MS        F       P
Regression        2     542.03      271.02    21.88   0.000
Residual Error   39     483.08       12.39
Total            41    1025.12
```

$\hat{y} = 12.180 - .02654x_1 - .4578x_2$

b. $\hat{\beta}_0 = 12.180$. This is the estimate of the y-intercept. It has no other interpretation.

$\hat{\beta}_1 = -0.0265$. We estimate that the mean weight change will decrease by 0.0265% for each additional increase of 1% in digestion efficiency, with acid-detergent fibre held constant.

$\hat{\beta}_2 = -0.458$. We estimate that the mean weight change will decrease by 0.458% for each additional increase of 1% in acid-detergent fibre, with digestion efficiency held constant.

c. To determine if digestion efficiency is a useful predictor of weight change, we test:

$$H_0: \beta_1 = 0$$
$$H_a: \beta_1 \neq 0$$

The test statistic is $t = -0.50$ and the p-value is $p = 0.623$. Since the p-value is greater than $\alpha\left(p = 0.623 \not< 0.01\right)$, H_0 is not rejected. There is insufficient evidence to indicate that digestion efficiency is a useful linear predictor of weight change at $\alpha = 0.01$.

d. For confidence coefficient 0.99, $\alpha = 0.01$ and $\alpha/2 = 0.01/2 = 0.005$. From Table III, Appendix B, with $df = n-(k+1) = 42-(2+1) = 39$, $t_{0.005} \approx 2.704$. The 99% confidence interval is:

$$\hat{\beta}_2 \pm t_{0.005} s_{\hat{\beta}_2} \Rightarrow -0.4578 \pm 2.704(0.1283) \Rightarrow -0.4578 \pm 0.3469$$
$$\Rightarrow (-0.8047, -0.1109)$$

We are 99% confident that the change in mean weight change for each unit change in acid-detergent fiber, holding digestion efficiency constant is between −0.8047 and −0.1109.

e. $R^2 = \text{R-Sq} = 0.529$. 52.9% of the total sample variation in the weight change scores is explained by the model containing the independent variables digestion efficiency and acid-detergent fiber. $R_a^2 = \text{R-Sq}(\text{adj}) = 0.505$. 50.5% of the total sample variation in the weight change scores is explained by the model containing the independent variables digestion efficiency and acid-detergent fiber, adjusting for the sample size and the number of independent variables in the model. R_a^2 is the preferred measure of model fit because as more variables are added to the model, R_a^2 can actually decrease if the variables do not improve the model, where R^2 will either stay the same or increase as variables are added to the model.

f. To determine if the overall model is useful in the prediction of weight change, we test:

$H_0: \beta_1 = \beta_2 = 0$
H_a: At least one $\beta_i \neq 0$, $i = 1, 2$

The test statistic is $F = 21.88$ and the p-value is $p = 0.000$. Since the p-value is less than $\alpha\left(p = 0.000 < 0.05\right)$, H_0 is rejected. There is sufficient evidence to indicate at least one of the variables digestion efficiency and acid-detergent fibre is useful in the prediction weight change at $\alpha = 0.05$.

g. The first-order model that allows for different slopes is $E(y) = \beta_0 + \beta_1 x_1 + \beta_2 x_2 + \beta_3 x_1 x_2$

where x_1 = digestive efficiency and $x_2 = \begin{cases} 1 & \text{if plants} \\ 0 & \text{otherwise} \end{cases}$

h. Using MINITAB, the results are:

Regression Analysis: WeightChg versus X1, X2, X1X2

```
The regression equation is
WeightChg = 8.14 - 0.016 X1 - 10.4 X2 + 0.095 X1X2


Predictor       Coef   SE Coef      T      P
Constant       8.144     8.446   0.96  0.341
X1           -0.0165    0.1329  -0.12  0.902
X2           -10.354     8.538  -1.21  0.233
X1X2          0.0948    0.1418   0.67  0.508


S = 3.88175   R-Sq = 44.1%   R-Sq(adj) = 39.7%


Analysis of Variance

Source          DF       SS      MS      F      P
Regression       3   452.54  150.85  10.01  0.000
Residual Error  38   572.58   15.07
Total           41  1025.12


Source  DF  Seq SS
X1       1  384.24
X2       1   61.57
X1X2     1    6.73
```

The least squares prediction equation is $\hat{y} = 8.14 - .016x_1 - 10.4x_2 + 0.095x_1x_2$.

i. For diet = plants, $x_2 = 1$. The prediction equation is

$\hat{y} = 8.14 - 0.016x_1 - 10.4(1) + 0.095x_1(1) = -2.26 + 0.079x_1$. The slope of the line is 0.079. For each unit increase is diet efficiency, the mean gosling weight change is estimated to increase by 0.079 for goslings fed plants.

j. For diet = duck chow, $x_2 = 0$. The prediction equation is

$\hat{y} = 8.14 - 0.016x_1 - 10.4(1) + 0.095x_1(1) = 8.14 - 0.016x_1$. The slope of the line is $-.016$.

For each unit increase is diet efficiency, the mean gosling weight change is estimated to decrease by 0.016 for goslings fed duck chow.

k. To determine if the slopes associated with the two diets are different, we test:

$H_0: \beta_3 = 0$
$H_a: \beta_3 \neq 0$

From the printout, the test statistic is $t = 0.67$ and the p-value is $p = 0.508$. Since the p-value is greater than $\alpha \left(p = 0.508 \not< 0.05\right)$, H_0 is not rejected. There is insufficient evidence to indicate the slopes associated with the two diets differ at $\alpha = 0.05$.

12.191 a. The correlation between the number of females in managerial positions and the number of females with a college degree is $r = 0.983$. The number of females in managerial positions is the dependent variable and the number of females with a college degree is an independent variable. Multicollinearity is a problem if there is correlation among the independent variables. Thus, this correlation does not result in a potential multicollinearity problem.

b. The correlation between the number of females in managerial positions and the number of female high school graduates with no college degree is $r = 0.074$. Again, the number of females in managerial positions is the dependent variable and the number of female high school graduates with no college degree is an independent variable. Multicollinearity is a problem if there is correlation among the independent variables. Thus, this correlation does not result in a potential multicollinearity problem.

c. The correlation between the number of males in managerial positions and the number of males with a college degree is $r = 0.772$. Both of these variables are independent variables. Since this value is fairly close to 1, this would indicate a potential moderate multicollinearity problem.

d. The correlation between the number of males in managerial positions and the number of male high school graduates with no college degree is $r = 0.528$. Both of these variables are independent variables. Since this value is not close to 1 or 0, this would indicate a potential moderate multicollinearity problem.

12.193 a. Since in the proposed model the funniness of the joke should increase and then decrease, the value of β_2 is expected to be negative.

b. To determine if the quadratic model relating pain to funniness rating is useful, we test:

$H_0: \beta_1 = \beta_2 = 0$
$H_a:$ At least one $\beta_i \neq 0$, $i = 1, 2$

The test statistic is $F = 1.60$.

The rejection region requires $\alpha = 0.05$ in the upper tail of the F distribution with $\nu_1 = k = 2$ and $\nu_2 = n-(k+1) = 32-(2+1) = 29$. From Table VI, Appendix B, $F_{0.05} = 3.33$. The rejection region is $F > 3.33$.

Since the observed value of the test statistic does not fall in the rejection region $(F = 1.60 \not> 3.33)$, H_0 is not rejected. There is insufficient evidence to indicate that the quadratic model relating pain to funniness rating is useful at $\alpha = 0.05$.

c. To determine if the quadratic model relating aggression/hostility to funniness rating is useful, we test:

$$H_0: \beta_1 = \beta_2 = 0$$
$$H_a: \text{At least one } \beta_i \neq 0, \quad i = 1, 2$$

The test statistic is $F = 1.61$.

The rejection region requires $\alpha = 0.05$ in the upper tail of the F distribution with $\nu_1 = k = 2$ and $\nu_2 = n-(k+1) = 32-(2+1) = 29$. From Table VI, Appendix B, $F_{0.05} = 3.33$. The rejection region is $F > 3.33$.

Since the observed value of the test statistic does not fall in the rejection region $(F = 1.61 \not> 3.33)$, H_0 is not rejected. There is insufficient evidence to indicate that the quadratic model relating aggression/hostility to funniness rating is useful at $\alpha = 0.05$.

12.195 a. The first order model is $E(y) = \beta_0 + \beta_1 x$.

b. The second order model is $E(y) = \beta_0 + \beta_1 x + \beta_2 x^2$.

c. Using MINITAB, the scatterplot of the data is:

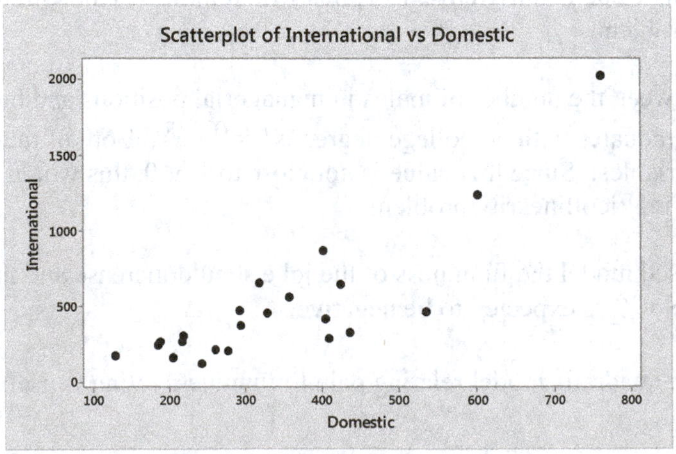

It appears that the second-order model might explain the relationship between International gross revenues and Domestic gross revenues somewhat better than the first-order model. The relationship looks like there might be an upward curve to the data.

d. Using MINITAB, the results are:

Regression Analysis: International versus Domestic, Dom-sq

```
The regression equation is
International = 384 - 1.66 Domestic + 0.00485 Dom-sq


Predictor        Coef   SE Coef      T      P
Constant        384.3     226.4   1.70  0.106
Domestic       -1.661     1.180  -1.41  0.175
Dom-sq       0.004853  0.001366   3.55  0.002


S = 187.669   R-Sq = 82.9%   R-Sq(adj) = 81.1%


Analysis of Variance

Source          DF       SS       MS      F      P
Regression       2  3238896  1619448  45.98  0.000
Residual Error  19   669171    35220
Total           21  3908067


Source     DF   Seq SS
Domestic    1  2794144
Dom-sq      1   444752
```

To determine if there is a curvilinear relationship between domestic and international gross revenues, we test:

$$H_0: \beta_2 = 0$$
$$H_a: \beta_2 \neq 0$$

The test statistic is $t = 3.55$ and the p-value is $p = 0.002$. Since the p-value is smaller than $\alpha\left(p = 0.002 < 0.05\right)$, H_0 is rejected. There is sufficient evidence to indicate a curvilinear relationship between domestic and international gross revenues at $\alpha = 0.05$.

e. Since we rejected H_0 is part d, there is evidence of a curvilinear relationship between the 2 variables. Thus, we would say that the better model is the second-order model.

12.197 a. $E(y) = \beta_0 + \beta_1 x_1 + \beta_2 x_6 + \beta_3 x_7$

where $x_6 = \begin{cases} 1 & \text{if condition is good} \\ 0 & \text{otherwise} \end{cases}$ $\quad x_7 = \begin{cases} 1 & \text{if condition is fair} \\ 0 & \text{otherwise} \end{cases}$

b. Using MINITAB, the plot is:

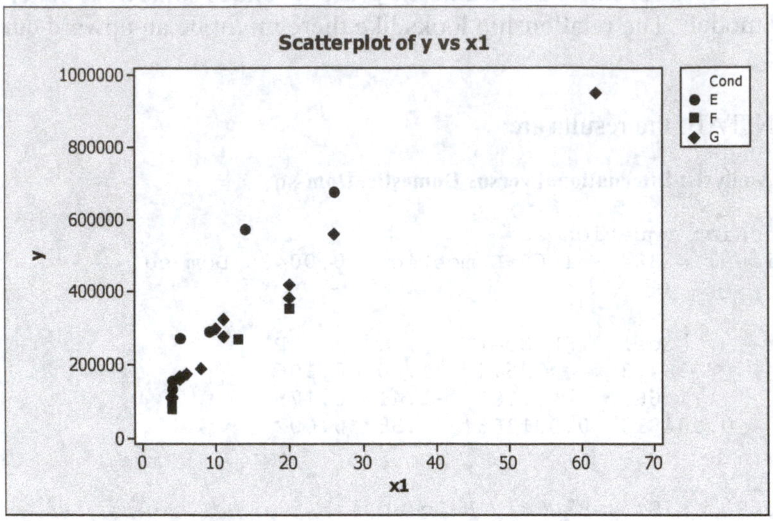

The model specified in part **a** seems appropriate. The points for E, F, and G cluster around three parallel lines.

c. Using MINITAB, the output is

Regression Analysis: y versus x1, x6, x7

```
The regression equation is
y = 188875 + 15617 x1 - 103046 x6 - 152487 x7


Predictor      Coef  SE Coef      T      P
Constant     188875    28588   6.61  0.000
x1            15617     1066  14.66  0.000
x6          -103046    31784  -3.24  0.004
x7          -152487    39157  -3.89  0.001


S = 64623.6   R-Sq = 91.8%   R-Sq(adj) = 90.7%


Analysis of Variance

Source          DF           SS           MS      F      P
Regression       3  9.86170E+11  3.28723E+11  78.71  0.000
Residual Error  21  87700442851   4176211564
Total           24  1.07387E+12


Source  DF       Seq SS
x1       1  9.15776E+11
x6       1   7061463149
x7       1  63332198206
```

The fitted model is $\hat{y} = 188,875 + 15,617x_1 - 103,046x_6 - 152,487x_7$

For excellent condition, $x_6 = 0$ and $x_7 = 0$:
$\hat{y} = 188{,}875 + 15{,}617x_1 - 103{,}046(0) - 152{,}487(0) = 188{,}875 + 15{,}617x_1$

For good condition, $x_6 = 1$ and $x_7 = 0$:
$\hat{y} = 188{,}875 + 15{,}617x_1 - 103{,}046(1) - 152{,}487(0) = 85{,}829 + 15{,}617x_1$

For fair condition, $x_6 = 0$ and $x_7 = 1$:
$\hat{y} = 188{,}875 + 15{,}617x_1 - 103{,}046(0) - 152{,}487(1) = 36{,}388 + 15{,}617x_1$

d. Using MINITAB the plot is:

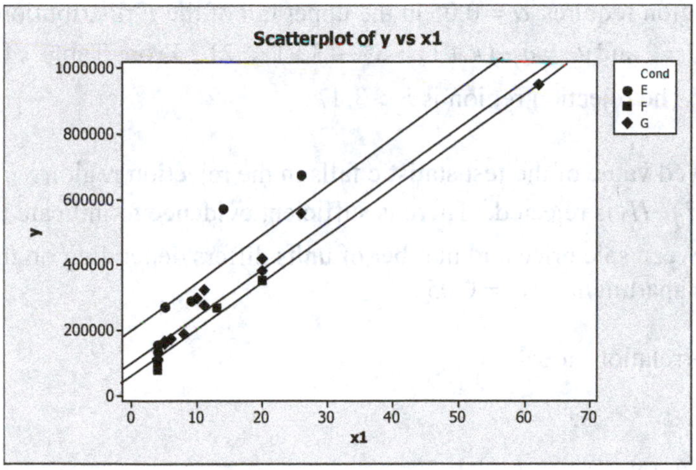

e. We must first fit a reduced model with just x_1, number of apartments. Using MINITAB, the output is:

Regression Analysis: y versus x1

```
The regression equation is
y = 101786 + 15525 x1


Predictor      Coef  SE Coef      T      P
Constant     101786    23291   4.37  0.000
x1            15525     1345  11.54  0.000


S = 82907.5   R-Sq = 85.3%   R-Sq(adj) = 84.6%


Analysis of Variance

Source          DF          SS          MS       F      P
Regression       1  9.15776E+11  9.15776E+11  133.23  0.000
Residual Error  23  1.58094E+11   6873656705
Total           24  1.07387E+12
```

To determine if the relationship between sale price and number of units differs depending on the physical condition of the apartments, we test:

$$H_0: \beta_6 = \beta_7 = 0$$
$$H_a: \text{At least one } \beta_i \neq 0, \quad i = 6, 7$$

The test statistic is:

$$F = \frac{(\text{SSE}_R - \text{SSE}_C)/(k-g)}{\text{SSE}_C/[n-(k+1)]} = \frac{(1.58094 \times 10^{11} - 87700442851)/2}{4176211564} = 8.43$$

The rejection region requires $\alpha = 0.05$ in the upper tail of the F distribution with $\nu_1 = k - g = 3 - 1 = 2$ and $\nu_2 = n - (k+1) = 25 - (3+1) = 21$. From Table VI, Appendix B, $F_{0.05} = 3.47$. The rejection region is $F > 3.47$.

Since the observed value of the test statistic falls in the rejection region $(F = 8.43 > 3.47)$, H_0 is rejected. There is sufficient evidence to indicate that the relationship between sale price and number of units differs depending on the physical condition of the apartments at $\alpha = 0.05$.

f. The pairwise correlations are:

	x1	x2	x3	x4	x5	x6
x2	-0.014					
x3	0.800	-0.188				
x4	0.224	-0.363	0.166			
x5	0.878	0.027	0.673	0.089		
x6	0.175	-0.447	0.271	0.112	0.020	
x7	-0.128	0.392	-0.118	0.050	-0.238	-0.564

When highly correlated independent variables are present in a regression model, the results are confusing. The researchers may only want to include one of the variables. This may be the case for the variables: x_1 and x_3 $(r_{13} = 0.800)$, x_1 and x_5 $(r_{15} = 0.878)$, and x_3 and x_5 $(r_{35} = 0.673)$.

g. Use the following plots to check the assumptions on ε.

standardized residuals vs x_1
standardized residuals vs x_2
standardized residuals vs x_3
standardized residuals vs x_4
standardized residuals vs x_5
standardized residuals vs predicted values
histogram of the standardized residuals
normal probability plot

From the plots of the residuals, there do not appear to be any outliers - no standardized residuals are larger than 2.38 in magnitude. In all the plots of the residuals vs x_i, there is no trend that would indicate non-constant variance (no funnel shape). In addition, there is no U or upside-down U shape that would indicate that any of the variables should be squared. In the histogram of the residuals, the plot is fairly mound-shaped, that would indicate the residuals are approximately normally distributed. The normal probability plot is a fairly straight line. This indicates that the assumption of normality is valid. All of the assumptions appear to be met.

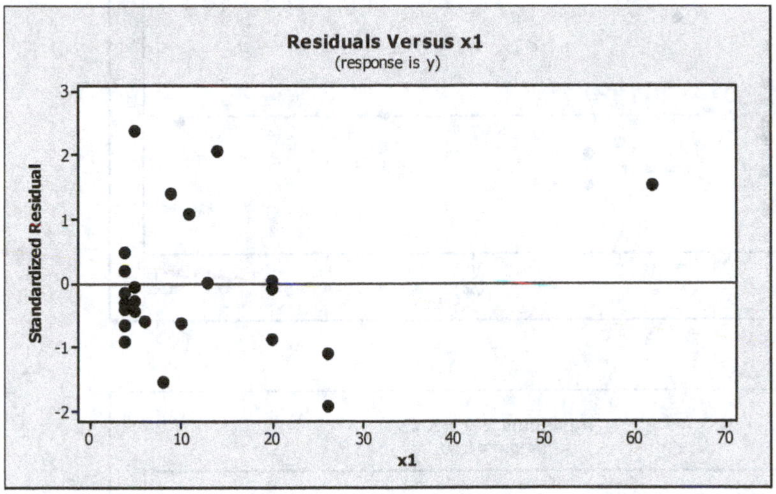

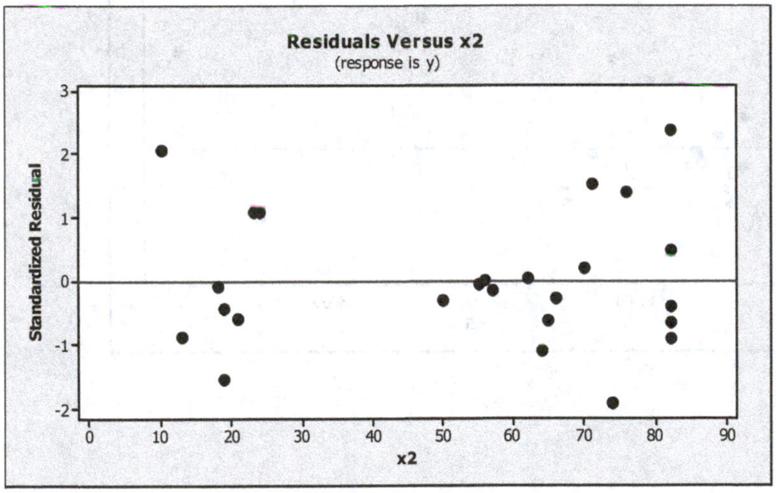

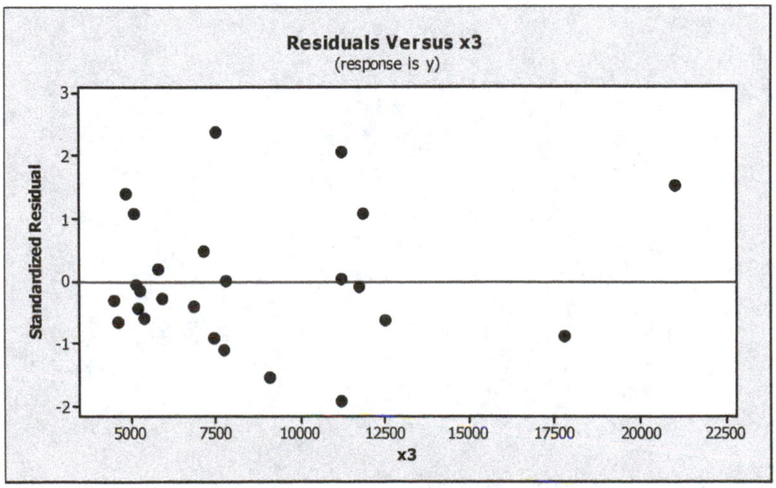

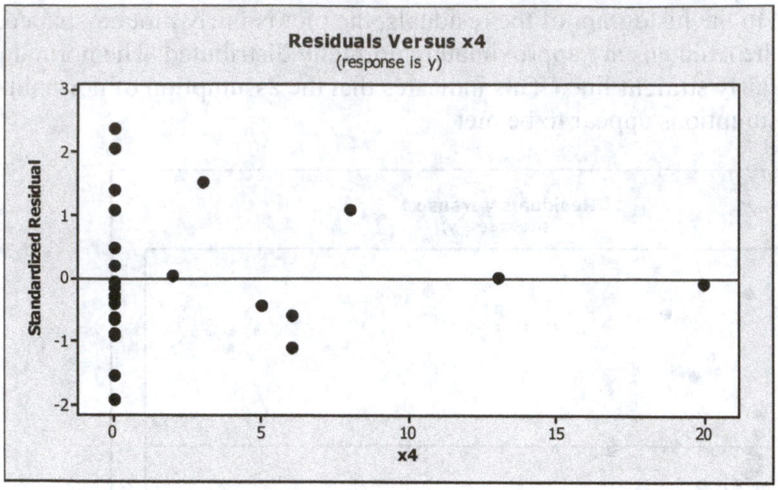
Residuals Versus x4
(response is y)
Standardized Residual
x4
3
2
1
0
-1
-2
0
5
10
15
20

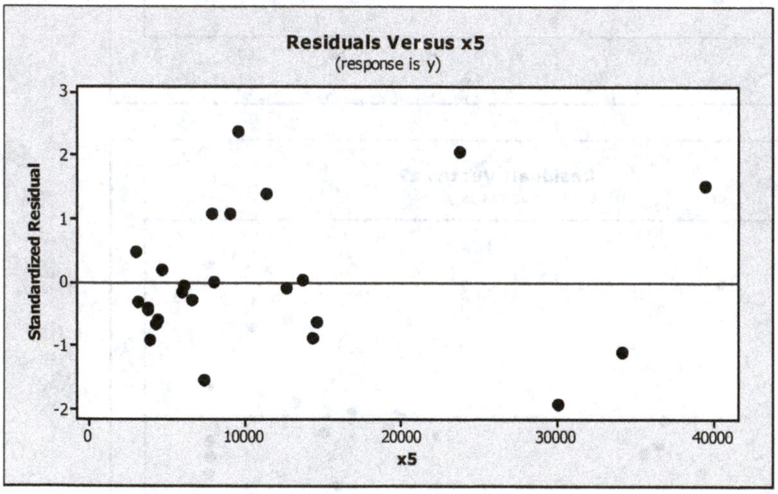
Residuals Versus x5
(response is y)
Standardized Residual
x5
3
2
1
0
-1
-2
0
10000
20000
30000
40000

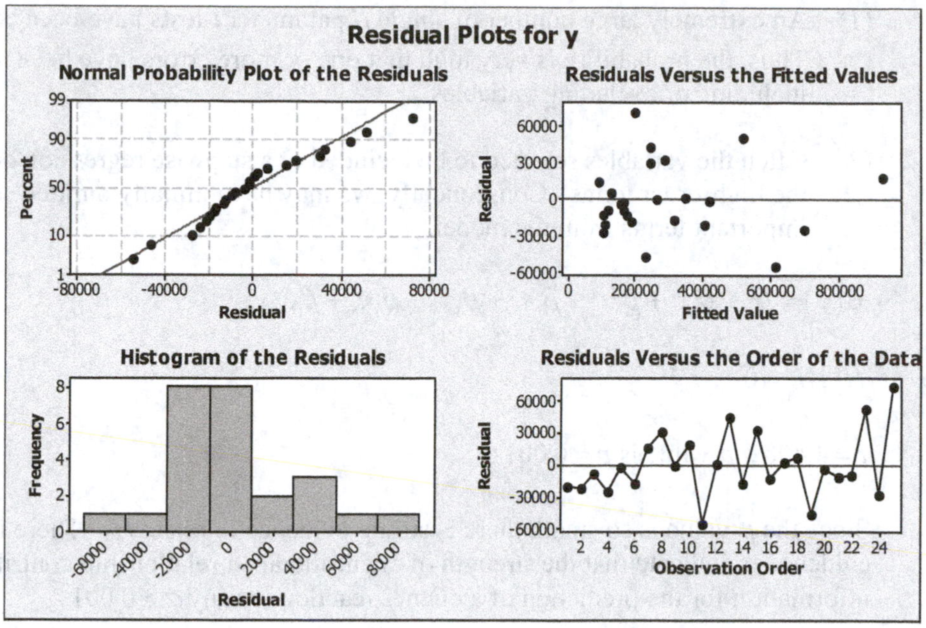

12.199 a. In Step 1, all one-variable models are fit to the data. These models are of the form:

$$E(y) = \beta_0 + \beta_1 x_i$$

Since there are 7 independent variables, 7 models are fit. (Note: There are actually only 6 independent variables. One of the qualitative variables has three levels and thus two dummy variables. Some statistical packages will allow one to bunch these two variables together so that they are either both in or both out. In this answer, we are assuming that each x_i stands by itself.

b. In Step 2, all two-variable models are fit to the data, where the variable selected in Step 1 is one of the variables. These models are of the form:

$$E(y) = \beta_0 + \beta_1 x_1 + \beta_2 x_i$$

Since there are 6 independent variables remaining, 6 models are fit.

c. In Step 3, all three-variable models are fit to the data, where the variables selected in Step 2 are two of the variables. These models are of the form:

$$E(y) = \beta_0 + \beta_1 x_1 + \beta_2 x_2 + \beta_3 x_i$$

Since there are 5 independent variables remaining, 5 models are fit.

d. The procedure stops adding independent variables when none of the remaining variables, when added to the model, have a p-value less than some predetermined value. This predetermined value is usually $\alpha = 0.05$.

e. Two major drawbacks to using the final stepwise model as the "best" model are:

(1) An extremely large number of single β parameter t-tests have been conducted. Thus, the probability is very high that one or more errors have been made in including or excluding variables.

(2) Often the variables selected to be included in a stepwise regression do not include the high-order terms. Consequently, we may have initially omitted several important terms from the model.

12.201 a. $E(y)=\beta_0+\beta_1x_1+\beta_2x_2+\beta_3x_3+\beta_4x_4+\beta_5x_5+\beta_6x_1x_2$

b. $H_0:\beta_4=0$

c. $t=4.408$, p-value is $p=0.001$.

Since the p-value is so small, there is strong evidence to reject H_0. There is sufficient evidence to indicate that the strength of client-therapist relationship contributes information for the prediction of a client's reaction for any $\alpha>0.001$.

d. Yes. For female therapists, the coefficient associated with age is β_1. For male therapists, the coefficient associated with age is $\beta_1+\beta_6$. If β_6 is positive and highly significant, then the slope of the age line is more positive for males than for females.

e. $R^2=0.2946$. 29.46% of the variability in the client's reaction scores can be explained by this model.

12.203 a. (1) Transect location is qualitative.
(2) Land use is qualitative
(3) Total number of trees in transect is quantitative.

b. $E(y)=\beta_0+\beta_1x_1$ where x_1 = total number of trees in a transect

c. $E(y)=\beta_0+\beta_1x_1+\beta_2x_2+\beta_3x_3$

where $x_2=\begin{cases}1 & \text{if small pasture field}\\ 0 & \text{otherwise}\end{cases}$ $x_3=\begin{cases}1 & \text{if small arable field}\\ 0 & \text{otherwise}\end{cases}$

A sketch of the response curve might be:

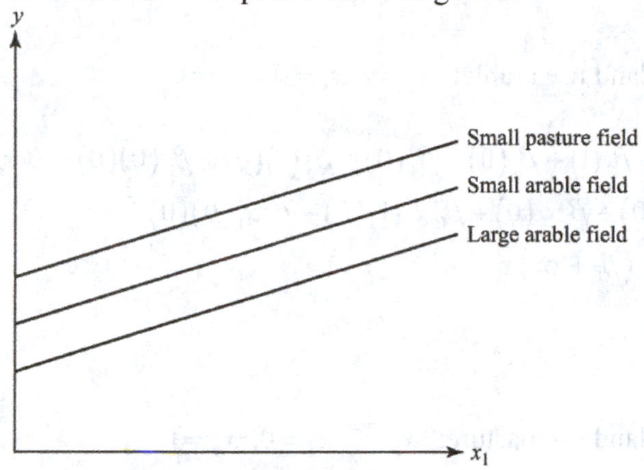

d. $E(y) = \beta_0 + \beta_1 x_1 + \beta_2 x_2 + \beta_3 x_3 + \beta_4 x_4$

where $x_4 = \begin{cases} 1 & \text{if land use is pasture} \\ 0 & \text{if land use is arable} \end{cases}$

The slopes are the same for any combination of transect location and land use. The slopes are all β_1.

e. $E(y) = \beta_0 + \beta_1 x_1 + \beta_2 x_2 + \beta_3 x_3 + \beta_4 x_4 + \beta_5 x_2 x_4 + \beta_6 x_3 x_4$

No, the slope is not affected because all the values of β_2, β_3, β_4, β_5 and β_6 are multiplied by values of 0 or 1 only.

f.
$$E(y) = \beta_0 + \beta_1 x_1 + \beta_2 x_2 + \beta_3 x_3 + \beta_4 x_4 + \beta_5 x_2 x_4 + \beta_6 x_3 x_4 + \beta_7 x_1 x_2 + \beta_8 x_1 x_3 + \beta_9 x_1 x_4 + \beta_{10} x_1 x_2 x_4 + \beta_{11} x_1 x_3 x_4$$

Large arable field and land use arable: $x_2 = x_3 = x_4 = 0$.

$$\begin{aligned} E(y) &= \beta_0 + \beta_1 x_1 + \beta_2(0) + \beta_3(0) + \beta_4(0) + \beta_5(0)(0) + \beta_6(0)(0) + \beta_7 x_1(0) \\ &\quad + \beta_8 x_1(0) + \beta_9 x_1(0) + \beta_{10} x_1(0)(0) + \beta_{11} x_1(0)(0) \\ &= \beta_0 + \beta_1 x_1 \end{aligned}$$

The slope is β_1.

Large arable field and land use pasture: $x_2 = x_3 = 0$, $x_4 = 1$

$$\begin{aligned} E(y) &= \beta_0 + \beta_1 x_1 + \beta_2(0) + \beta_3(0) + \beta_4(1) + \beta_5(0)(1) + \beta_6(0)(1) + \beta_7 x_1(0) \\ &\quad + \beta_8 x_1(0) + \beta_9 x_1(1) + \beta_{10} x_1(0)(1) + \beta_{11} x_1(0)(1) \\ &= (\beta_0 + \beta_4) + (\beta_1 + \beta_9) x_1 \end{aligned}$$

The slope is $(\beta_1 + \beta_9)$.

Small pasture field and land use arable: $x_2 = 1,\ x_3 = 0 = x_4 = 0$

$$\begin{aligned} E(y) &= \beta_0 + \beta_1 x_1 + \beta_2(1) + \beta_3(0) + \beta_4(0) + \beta_5(1)(0) + \beta_6(0)(0) + \beta_7 x_1(1) \\ &\quad + \beta_8 x_1(0) + \beta_9 x_1(0) + \beta_{10} x_1(1)(0) + \beta_{11} x_1(0)(0) \\ &= (\beta_0 + \beta_2) + (\beta_1 + \beta_7) x_1 \end{aligned}$$

The slope is $(\beta_1 + \beta_7)$.

Small pasture field and land use pasture: $x_2 = 1,\ x_3 = 0,\ x_4 = 1$

$$\begin{aligned} E(y) &= \beta_0 + \beta_1 x_1 + \beta_2(1) + \beta_3(0) + \beta_4(1) + \beta_5(1)(1) + \beta_6(0)(1) + \beta_7 x_1(1) \\ &\quad + \beta_8 x_1(0) + \beta_9 x_1(1) + \beta_{10} x_1(1)(1) + \beta_{11} x_1(0)(1) \\ &= (\beta_0 + \beta_2 + \beta_4) + (\beta_1 + \beta_7 + \beta_9 + \beta_{10}) x_1 \end{aligned}$$

The slope is $(\beta_1 + \beta_7 + \beta_9 + \beta_{10})$.

Small arable field and land use arable: $x_2 = 0,\ x_3 = 1,\ x_4 = 0$

$$\begin{aligned} E(y) &= \beta_0 + \beta_1 x_1 + \beta_2(0) + \beta_3(1) + \beta_4(0) + \beta_5(0)(0) + \beta_6(1)(0) + \beta_7 x_1(0) \\ &\quad + \beta_8 x_1(1) + \beta_9 x_1(0) + \beta_{10} x_1(0)(0) + \beta_{11} x_1(1)(0) \\ &= (\beta_0 + \beta_3) + (\beta_1 + \beta_8) x_1 \end{aligned}$$

The slope is $(\beta_1 + \beta_8)$.

Small arable field and land use pasture: $x_2 = 0,\ x_3 = 1,\ x_4 = 1$

$$\begin{aligned} E(y) &= \beta_0 + \beta_1 x_1 + \beta_2(0) + \beta_3(1) + \beta_4(1) + \beta_5(0)(1) + \beta_6(1)(1) + \beta_7 x_1(0) \\ &\quad + \beta_8 x_1(1) + \beta_9 x_1(1) + \beta_{10} x_1(0)(1) + \beta_{11} x_1(1)(1) \\ &= (\beta_0 + \beta_3 + \beta_4 + \beta_6) + (\beta_1 + \beta_8 + \beta_9 + \beta_{11}) x_1 \end{aligned}$$

The slope is $(\beta_1 + \beta_8 + \beta_9 + \beta_{11})$.

12.205 All of the independent variables are continuous except District. Four dummy variables were created for District before using the stepwise regression. Also, note that several observations had missing values. Thus, only 217 observations were used in the analysis. Since Ratio was computed from the ratio of Price and DOT estimate, it was not used as a predictor variable. Using SAS and $\alpha = 0.10$ for keeping variables in the model, the Stepwise regression results are:

```
                          The REG Procedure
                            Model: MODEL1
                      Dependent Variable: PRICE

                      Stepwise Selection: Step 1

        Variable DOTEST Entered: R-Square = 0.9763 and C(p) = 14.2985

                         Analysis of Variance

                                    Sum of           Mean
Source                   DF        Squares         Square    F Value    Pr > F

Model                     1      687136244      687136244    8856.79    <.0001
Error                   215       16680334          77583
Corrected Total         216      703816578

                  Parameter     Standard
      Variable     Estimate        Error   Type II SS  F Value  Pr > F

      Intercept    33.24768     22.19263       174129     2.24  0.1356
      DOTEST        0.90637      0.00963    687136244  8856.79  <.0001

                    Bounds on condition number: 1, 1
------------------------------------------------------------------------------

                      Stepwise Selection: Step 2

         Variable STATUS Entered: R-Square = 0.9774 and C(p) = 5.9631

                         Analysis of Variance

                                    Sum of           Mean
Source                   DF        Squares         Square    F Value    Pr > F

Model                     2      687894714      343947357    4622.87    <.0001
Error                   214       15921865          74401
Corrected Total         216      703816578

                          The REG Procedure
                            Model: MODEL1
                      Dependent Variable: PRICE

                      Stepwise Selection: Step 2

                  Parameter     Standard
      Variable     Estimate        Error   Type II SS  F Value  Pr > F

      Intercept   -12.92808     26.10499        18247     0.25  0.6209
      DOTEST        0.91233      0.00961    669924194  9004.21  <.0001
      STATUS      132.16624     41.39440       758469    10.19  0.0016
               Bounds on condition number: 1.0392, 4.1569
```

Stepwise Selection: Step 3

Variable DAYS Entered: R-Square = 0.9779 and C(p) = 3.4115

Analysis of Variance

Source	DF	Sum of Squares	Mean Square	F Value	Pr > F
Model	3	688228730	229409577	3134.76	<.0001
Error	213	15587848	73182		
Corrected Total	216	703816578			

Variable	Parameter Estimate	Standard Error	Type II SS	F Value	Pr > F
Intercept	-59.79987	33.93608	227239	3.11	0.0795
DOTEST	0.88616	0.01552	238455655	3258.37	<.0001
STATUS	139.40952	41.19370	838166	11.45	0.0008
DAYS	0.35761	0.16739	334016	4.56	0.0338

Bounds on condition number: 2.7653, 19.698

All variables left in the model are significant at the 0.1000 level.

The REG Procedure
Model: MODEL1
Dependent Variable: PRICE

Stepwise Selection: Step 3

No other variable met the 0.1500 significance level for entry into the model.

Summary of Stepwise Selection

Step	Variable Entered	Variable Removed	Number Vars In	Partial R-Square	Model R-Square	C(p)	F Value	Pr > F
1	DOTEST		1	0.9763	0.9763	14.2985	8856.79	<.0001
2	STATUS		2	0.0011	0.9774	5.9631	10.19	0.0016
3	DAYS		3	0.0005	0.9779	3.4115	4.56	0.0338

From the results, only three independent variables are selected to predict price using the stepwise regression and $\alpha = 0.05$. These variables are DOT estimate, Status, and Days.

A new model was fit with just the variables DOT estimate, Status, and Days using MINITAB. The results are:

Regression Analysis: Price versus Dotest, Status, Days

The regression equation is
Price = - 59.8 + 0.886 Dotest + 139 Status + 0.358 Days

Predictor	Coef	SE Coef	T	P
Constant	-59.80	33.94	-1.76	0.079
Dotest	0.88616	0.01552	57.08	0.000
Status	139.41	41.19	3.38	0.001

```
Days          0.3576    0.1674    2.14  0.034

S = 270.522   R-Sq = 97.8%    R-Sq(adj) = 97.8%


Analysis of Variance

Source           DF          SS         MS        F      P
Regression        3   688228730  229409577  3134.76  0.000
Residual Error  213    15587848      73182
Total           216   703816578
```

We see that this is a fairly good model. The R-square value is 0.9779. 97.79% of the variation in the PRICE values is explained by the model containing DODEST, STATUS, and DAYS.

We note that the parameter estimate for the parameter associated Bid Status is 139.40952. Since Bid Status is either 1 (fixed) or 0 (competitive), the PRICE for the fixed bid is estimated to be 139.40952 units higher than the competitive bid, all other variables held constant.

To check the model assumptions, we will look at the plots. The plots of the residuals versus the independent variables are:

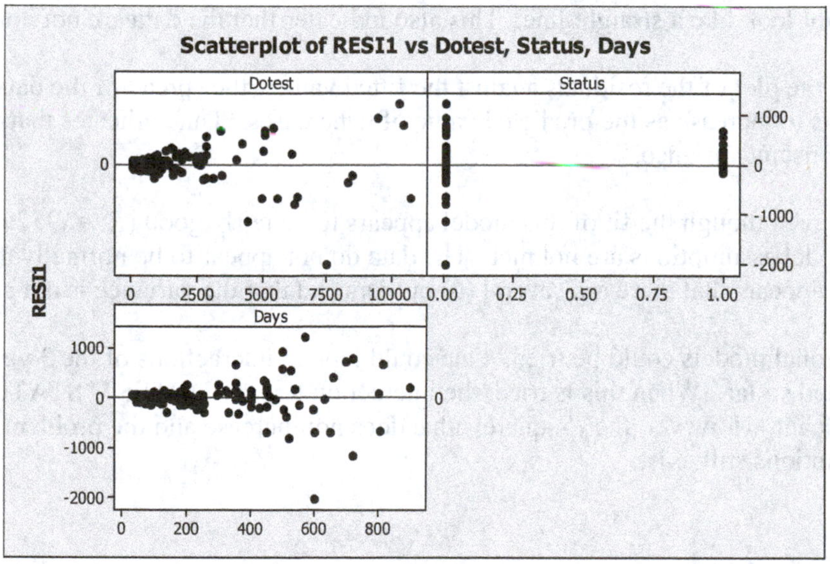

There are no 'U' or 'upside-down U' shapes to these plots. Thus, the relationship between PRICE and the independent variables appear to be linear. There are a total of 6 observations that are more than 3 standard deviations from the mean. These 6 points could be outliers.

Additional residual plots are:

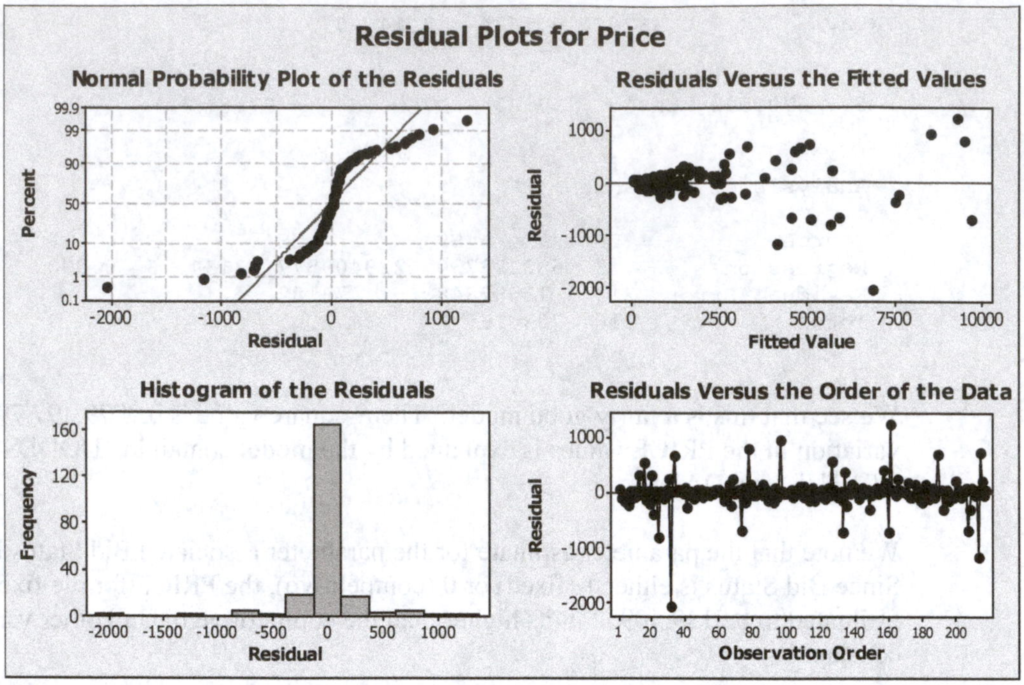

The histogram of the residuals does not look to be mound-shaped, so the assumption that the data are normal appears to be violated. In addition, the normal probability plot does not look like a straight line. This also indicates that the data are not normal.

Using the plot of the residuals against the fitted values, the spread of the data points appears to increase as the predicted value of *y* increases. This indicates that there is non-constant variance.

Thus, even though the fit of this model appears to be fairly good ($R^2 = .9779$), some of the model assumptions are not met. The data do not appear to be normally distributed. There appears that there are several (6) outliers and that the variance is not constant.

Additional models could be tried. One could look at interactions of the 3 variables included so far. When this is tried, the interaction term of DOTEST*STATUS is significant. However, the *R*-square value does not increase and the problems with the assumptions still exist.

Chapter 13

Categorical Data Analysis

13.1 The characteristics of the multinomial experiment are:

1. The experiment consists of n identical trials.
2. There are k possible outcomes to each trial.
3. The probabilities of the k outcomes, denoted $p_1, p_2, \ldots, p_k$, remain the same from trial to trial, where $p_1 + p_2 + \cdots + p_k = 1$.
4. The trials are independent.
5. The random variables of interest are the counts $n_1, n_2, \ldots, n_k$ in each of the k cells.

The characteristics of the binomial are the same as those for the multinomial with $k = 2$.

13.3 a. With $df = 10$, $\chi^2_{0.05} = 18.3070$

b. With $df = 50$, $\chi^2_{0.990} = 29.7067$

c. With $df = 16$, $\chi^2_{0.10} = 23.5418$

d. With $df = 50$, $\chi^2_{0.005} = 79.4900$

13.5 a. The rejection region requires $\alpha = 0.05$ in the upper tail of the χ^2 distribution with $df = k - 1 = 3 - 1 = 2$. From Table IV, Appendix B, $\chi^2_{0.05} = 5.99147$. The rejection region is $\chi^2 > 5.99147$.

b. The rejection region requires $\alpha = 0.10$ in the upper tail of the χ^2 distribution with $df = k - 1 = 5 - 1 = 4$ 4. From Table IV, Appendix B, $\chi^2_{0.10} = 7.77944$. The rejection region is $\chi^2 > 7.77944$.

c. The rejection region requires $\alpha = 0.01$ in the upper tail of the χ^2 distribution with $df = k - 1 = 4 - 1 = 3$. From Table IV, Appendix B, $\chi^2_{0.01} = 11.3449$. The rejection region is $\chi^2 > 11.3449$.

13.7 Some preliminary calculations are:

If the probabilities are the same, $p_{1,0} = p_{2,0} = p_{3,0} = p_{4,0} = 0.25$

$$E_1 = E_2 = E_3 = E_4 = np_{1,0} = 205(0.25) = 51.25$$

a. To determine if the multinomial probabilities differ, we test:

$$H_0: p_1 = p_2 = p_3 = p_4 = 0.25 \qquad H_a: \text{At least one } p_i \neq 0.25, \quad i = 1, 2, 3, 4$$

The test statistic is

$$\chi^2 = \sum \frac{[n_i - E_i]^2}{E_i} = \frac{(43-51.25)^2}{51.25} + \frac{(56-51.25)^2}{51.25} + \frac{(59-51.25)^2}{51.25} + \frac{(47-51.25)^2}{51.25} = 3.293$$

The rejection region requires $\alpha = 0.05$ in the upper tail of the χ^2 distribution with $df = k - 1 = 4 - 1 = 3$. From Table IV, Appendix B, $\chi^2_{0.05} = 7.81473$. The rejection region is $\chi^2 > 7.81473$.

Since the observed value of the test statistic does not fall in the rejection region $(\chi^2 = 3.293 \not> 7.81473)$, H_0 is not rejected. There is insufficient evidence to indicate the multinomial probabilities differ at $\alpha = 0.05$.

b. The Type I error is concluding the multinomial probabilities differ when, in fact, they do not.

The Type II error is concluding the multinomial probabilities are equal, when, in fact, they are not.

c. For confidence coefficient 0.95, $\alpha = 0.05$ and $\alpha / 2 = 0.05 / 2 = 0.025$. From Table II, Appendix B, $z_{0.025} = 1.96$.

$\hat{p}_3 = 59 / 205 = 0.288$

The confidence interval is:

$$\hat{p} \pm z_{0.025}\sqrt{\frac{\hat{p}\hat{q}}{n}} \Rightarrow 0.288 \pm 1.96\sqrt{\frac{0.288(0.712)}{205}} \Rightarrow 0.288 \pm 0.062 \Rightarrow (0.226, 0.350)$$

13.9 a. The qualitative variable of interest is nonfunctional jaw habits. It has 4 levels: bruxism, teeth clenching, both bruxism and clenching, and neither.

b. A one-way table of the data is:

Cell

	Bruxism	Clench	Bruxism & Clench	Neither
n_i	3	11	30	16

c. To determine whether the percentages associated with the admitted habits differ, we test:

$$H_0: p_1 = p_2 = p_3 = p_4 = 0.25 \qquad H_a: \text{At least one } p_i \neq 0.25, \quad i = 1, 2, 3, 4$$

d. $E_1 = E_2 = E_3 = E_4 = np_{1,0} = 60(0.25) = 15$.

e. The test statistic is

$$\chi^2 = \sum \frac{[n_i - E_i]^2}{E_i} = \frac{[3-15]^2}{15} + \frac{[11-15]^2}{15} + \frac{[30-15]^2}{15} + \frac{[16-15]^2}{15} = 25.733$$

f. The rejection region requires $\alpha = 0.05$ in the upper tail of the χ^2 distribution with $df = k - 1 = 4 - 1 = 3$. From Table IV, Appendix B, $\chi^2_{0.05} = 7.81473$. The rejection region is $\chi^2 > 7.81473$.

g. Since the observed value of the test statistic falls in the rejection region $(\chi^2 = 25.733 > 7.81473)$, H_0 is rejected. There is sufficient evidence to indicate the percentages associated with the admitted habits differ at $\alpha = 0.05$.

h. $\hat{p}_3 = \dfrac{30}{60} = 0.5$

For confidence coefficient 0.95, $\alpha = 0.05$ and $\alpha / 2 = 0.05 / 2 = 0.025$. From Table II, Appendix B, $z_{0.025} = 1.96$. The 95% confidence interval is:

$$\hat{p}_i \pm z_{0.025}\sqrt{\frac{\hat{p}_i \hat{q}_i}{n}} \Rightarrow 0.5 \pm 1.96\sqrt{\frac{0.5(0.5)}{60}} \Rightarrow 0.5 \pm .127 \Rightarrow (0.373,\ 0.627)$$

We are 95% confident that the true proportion of dental patients who admit to both habits is between 0.373 and 0.627.

13.11 a. The variable measured on each robot is the type of legs/wheels that each has. There were 4 possible answers: legs only, wheels only, legs and wheels, and neither legs nor wheels.

b. To test the design engineer's claim, we test:

$H_0: p_1 = 0.50,\ p_2 = 0.30,\ p_3 = 0.10,\ p_4 = 0.10$

H_a: At least one p_i differs from its hypothesized value

c. $E_1 = np_{1,0} = 106(0.50) = 53;\quad E_2 = np_{2,0} = 106(0.30) = 31.8$

$E_3 = np_{3,0} = 106(0.10) = 10.6;\quad E_4 = np_{4,0} = 106(0.10) = 10.6$

d. $$\chi^2 = \sum \frac{[n_i - E_i]^2}{E_i} = \frac{[63-53]^2}{53} + \frac{[20-31.8]^2}{31.8} + \frac{[8-10.6]^2}{10.6} + \frac{[15-10.6]^2}{10.6} = 8.73$$

e. The rejection region requires $\alpha = 0.05$ in the upper tail of the χ^2 distribution with $df = k - 1 = 4 - 1 = 3$. From Table IV, Appendix B, $\chi^2_{0.05} = 7.81473$. The rejection region is
$\chi^2 > 7.81473$.

Since the observed value of the test statistic falls in the rejection region $\left(\chi^2 = 8.73 > 7.81473\right)$, H_0 is rejected. There is sufficient evidence to indicate that at least one probability of type of legs/wheels differs from its hypothesized value at $\alpha = 0.05$.

13.13 a. If it is just as likely to have a boy as a girl, then for any child, $P(G) = P(B) = \frac{1}{2}$. Since the gender of one child is independent of the next, $P(GG) = \frac{1}{2}\left(\frac{1}{2}\right) = 0.25$,

$P(GB) = \frac{1}{2}\left(\frac{1}{2}\right) = 0.25$, $P(BG) = \frac{1}{2}\left(\frac{1}{2}\right) = 0.25$, and $P(BB) = \frac{1}{2}\left(\frac{1}{2}\right) = 0.25$.

b. Since each probability is 0.25, the expected number of families of each combination is $np_i = 42,888(0.25) = 10,722$.

c. The test statistics is

$$\chi^2 = \sum \frac{[n_i - E_i]^2}{E_i} = \frac{[9,523 - 10,722]^2}{10,722} + \frac{[11,118 - 10,722]^2}{10,722} + \frac{[10,913 - 10,722]^2}{10,722} + \frac{[11,334 - 10,722]^2}{10,722} = 187.04$$

d. The rejection region requires $\alpha = 0.10$ in the upper tail of the χ^2 distribution with $df = k - 1 = 4 - 1 = 3$. From Table IV, Appendix B, $\chi^2_{0.10} = 6.25139$. The rejection region is $\chi^2 > 6.25139$.

Since the observed value of the test statistic falls in the rejection region $\left(\chi^2 = 187.04 > 6.25139\right)$, H_0 is rejected. There is sufficient evidence to indicate that at least one probability of gender configuration for a two-child family is not 0.25 at $\alpha = 0.10$.

e. The expected numbers of families for each gender configuration are:

$$GG: np_1 = 42,888(0.23795) = 10,205.2$$
$$GB: np_1 = 42,888(0.23795) = 10,205.2$$
$$BG: np_1 = 42,888(0.23795) = 10,205.2$$
$$BB: np_1 = 42,888(0.23795) = 10,205.2$$

The test statistics is

$$\chi^2 = \sum \frac{[n_i - E_i]^2}{E_i} = \frac{[9{,}523 - 10{,}205.2]^2}{10{,}205.2} + \frac{[11{,}118 - 10{,}715.6]^2}{10{,}715.6} + \frac{[10{,}913 - 10{,}715.6]^2}{10{,}715.6}$$

$$+ \frac{[11{,}334 - 11{,}251.7]^2}{11{,}251.7} = 64.95$$

Since the observed value of the test statistic falls in the rejection region $(\chi^2 = 64.95 > 6.25139)$, H_0 is rejected. There is sufficient evidence to indicate that at least one probability of gender configuration for a two-child family is not the proposed probability at $\alpha = 0.10$.

13.15 Some preliminary calculations are:

$$E_1 = np_{1,0} = 80(0.40) = 32; \quad E_2 = np_{2,0} = 80(0.40) = 32$$
$$E_3 = np_{3,0} = 80(0.10) = 8; \quad E_4 = np_{4,0} = 80(0.05) = 4 \quad E_5 = np_{5,0} = 80(0.05) = 4$$

To determine if the proportions of gang shootings over the 5-year period have the following values, we test:

$H_0: p_1 = 0.40,\ p_2 = 0.40,\ p_3 = 0.10,\ p_4 = 0.05,\ p_5 = 0.05$
H_a: At least one p_i differs from its hypothesized value

The test statistic is

$$\chi^2 = \sum \frac{[n_i - E_i]^2}{E_i} = \frac{[37-32]^2}{32} + \frac{[30-32]^2}{32} + \frac{[4-8]^2}{8} + \frac{[5-4]^2}{4} + \frac{[4-4]^2}{4} = 3.156$$

Since no α is given, we will use $\alpha = 0.05$. The rejection region requires $\alpha = 0.05$ in the upper tail of the χ^2 distribution with $df = k - 1 = 5 - 1 = 4$. From Table IV, Appendix A, $\chi^2_{0.05} = 9.48773$. The rejection region is $\chi^2 > 9.48773$.

Since the observed value of the test statistic does not fall in the rejection region $(\chi^2 = 3.156 \not> 9.48773)$, H_0 is not rejected. There is insufficient evidence to indicate the proportions of gang shootings over the 5-year period have the hypothesized values at $\alpha = 0.05$.

13.17 Some preliminary calculations are:

$$E_1 = np_{1,0} = 2{,}097(0.02) = 41.94; \quad E_2 = np_{2,0} = 2{,}097(0.25) = 524.25;$$
$$E_3 = np_{3,0} = 2{,}097(0.73) = 1{,}530.81$$

To determine if the distribution of the E4/E4 genotypes for the population of young adults differs from the norm, we test:

$H_0: p_1 = 0.02,\ p_2 = 0.25,\ p_3 = 0.73$
H_a: At least one p_i differs from its hypothesized value

The test statistic is

$$\chi^2 = \sum \frac{[n_i - E_i]^2}{E_i} = \frac{[56-41.94]^2}{41.94} + \frac{[517-524.25]^2}{524.25} + \frac{[1524-1530.81]^2}{1530.81} = 4.84$$

The rejection region requires $\alpha = 0.05$ in the upper tail of the χ^2 distribution with $df = k-1 = 3-1 = 2$. From Table IV, Appendix B, $\chi^2_{0.05} = 5.99146$. The rejection region is $\chi^2 > 5.99146$.

Since the observed value of the test statistic does not fall in the rejection region $\left(\chi^2 = 4.84 \not> 5.99146\right)$, H_0 is not rejected. There is insufficient evidence to indicate the distribution of the E4/E4 genotypes for the population of young adults differs from the norm at $\alpha = 0.05$.

13.19 a. Some preliminary calculations are: $E_1 = E_2 = E_3 = E_4 = E_5 = np_{i,0} = 81(0.20) = 16.2$

For child-led interactions, to determine if differences exist in the proportions associated with the types of interactions, we test:

$H_0: p_1 = p_2 = p_3 = p_4 = p_5 = 0.20$
H_a: At least one p_i differs from its hypothesized value

The test statistic is

$$\chi^2 = \sum \frac{[n_i - E_i]^2}{E_i} = \frac{[26-16.2]^2}{16.2} + \frac{[21-16.2]^2}{16.2} + \frac{[21-16.2]^2}{16.2} + \frac{[12-16.2]^2}{16.2} + \frac{[1-16.2]^2}{16.2} = 24.123$$

The rejection region requires $\alpha = 0.01$ in the upper tail of the χ^2 distribution with $df = k-1 = 5-1 = 4$. From Table IV, Appendix B, $\chi^2_{0.01} = 13.2767$. The rejection region is $\chi^2 > 13.2767$.

Since the observed value of the test statistic fall in the rejection region $\left(\chi^2 = 24.123 > 13.2767\right)$, H_0 is rejected. There is sufficient evidence to indicate differences exist in the proportions associated with the types of interactions for child-led interactions at $\alpha = 0.01$.

b. Some preliminary calculations are: $E_1 = E_2 = E_3 = E_4 = E_5 = np_{i,0} = 89(0.20) = 17.8$

For adult-led interactions, to determine if differences exist in the proportions associated with the types of interactions, we test:

$H_0: p_1 = p_2 = p_3 = p_4 = p_5 = 0.20$

H_a : At least one p_i differs from its hypothesized value

The test statistic is

$$\chi^2 = \sum \frac{[n_i - E_i]^2}{E_i} = \frac{[0-17.8]^2}{17.8} + \frac{[64-17.8]^2}{17.8} + \frac{[10-17.8]^2}{17.8} + \frac{[9-17.8]^2}{17.8} + \frac{[6-17.8]^2}{17.8} = 153.303$$

The rejection region requires $\alpha = 0.01$ in the upper tail of the χ^2 distribution with $df = k-1 = 5-1 = 4$. From Table IV, Appendix B, $\chi^2_{0.01} = 13.2767$. The rejection region is $\chi^2 > 13.2767$.

Since the observed value of the test statistic fall in the rejection region $(\chi^2 = 153.303 > 13.2767)$, H_0 is rejected. There is sufficient evidence to indicate differences exist in the proportions associated with the types of interactions for adult-led interactions at $\alpha = 0.01$.

13.21 a. Some preliminary calculations are:

$E_A = np_{A,0} = 350(0.09) = 31.5$ $\quad E_B = np_{B,0} = 350(0.02) = 7$

$E_C = np_{C,0} = 350(0.02) = 7$ $\quad E_D = np_{D,0} = 350(0.04) = 14$

$E_E = np_{E,0} = 350(0.12) = 42$ $\quad E_F = np_{F,0} = 350(0.02) = 7$

$E_G = np_{G,0} = 350(0.03) = 10.5$ $\quad E_H = np_{H,0} = 350(0.02) = 7$

$E_I = np_{I,0} = 350(0.09) = 31.5$ $\quad E_J = np_{J,0} = 350(0.01) = 3.5$

$E_K = np_{K,0} = 350(0.01) = 3.5$ $\quad E_L = np_{L,0} = 350(0.04) = 14$

$E_M = np_{M,0} = 350(0.02) = 7$ $\quad E_N = np_{N,0} = 350(0.06) = 21$

$E_O = np_{O,0} = 350(0.08) = 28$ $\quad E_P = np_{P,0} = 350(0.02) = 7$

$E_Q = np_{Q,0} = 350(0.01) = 3.5$ $\quad E_R = np_{R,0} = 350(0.06) = 21$

$E_S = np_{S,0} = 350(0.04) = 14$ $\quad E_T = np_{T,0} = 350(0.06) = 21$

$E_U = np_{U,0} = 350(0.04) = 14$ $\quad E_V = np_{V,0} = 350(0.02) = 7$

$E_W = np_{W,0} = 350(0.02) = 7$ $\quad E_X = np_{X,0} = 350(0.01) = 3.5$

$E_Y = np_{Y,0} = 350(0.02) = 7$ $\quad E_Z = np_{Z,0} = 350(0.01) = 3.5$

$E_{BLANK} = np_{BL,0} = 350(0.02) = 7$

To determine if the ScrabbleExpress™ presents the player with "unfair word selection opportunities" that are different from the Scrabble™ board game, we test:

$H_0: p_A = 0.09, p_B = 0.02, p_C = 0.02, p_D = 0.04, p_E = 0.12, p_F = 0.02, p_G = 0.03, p_H = 0.02,$
$p_I = 0.09, p_J = 0.01, p_K = 0.01, p_L = 0.04, p_M = 0.02, p_N = 0.06, p_O = 0.08, p_P = 0.02,$
$p_Q = 0.01, p_R = 0.06, p_S = 0.04, p_T = 0.06, p_U = 0.04, p_V = 0.02, p_W = 0.02, p_X = 0.01,$
$p_Y = 0.02, p_Z = .01, p_{BLANK} = 0.02$

H_a: At least one p_i differs from its hypothesized value

The test statistic is

$$\chi^2 = \sum \frac{[n_i - E_i]^2}{E_i} = \frac{[20-31.5]^2}{31.5} + \frac{[8-7]^2}{7} + \frac{[15-7]^2}{7} + \cdots + \frac{[19-7]^2}{7} = 190.91$$

The rejection region requires $\alpha = 0.05$ in the upper tail of the χ^2 distribution with $df = k - 1 = 27 - 1 = 26$. From Table IV, Appendix B, $\chi^2_{0.05} = 38.8852$. The rejection region is $\chi^2 > 38.8852$.

Since the observed value of the test statistic falls in the rejection region $(\chi^2 = 190.91 > 38.8852)$, H_0 is rejected. There is sufficient evidence to indicate the ScrabbleExpress™ presents the player with "unfair word selection opportunities" that are different from the Scrabble™ board game at $\alpha = 0.05$.

b. The true proportion of letters drawn from the board game that are vowels is

$$p_A + p_E + p_I + p_O + p_U = 0.09 + 0.12 + 0.09 + 0.08 + 0.04 = 0.42$$

The number of vowels drawn by the electronic game is $20 + 16 + 12 + 10 + 11 = 69$. The proportion of vowels drawn is $69/350 = 0.197$.

For confidence coefficient 0.95, $\alpha = 0.05$ and $\alpha/2 = 0.05/2 = 0.025$. From Table II, Appendix B, $z_{0.025} = 1.96$. The 95% confidence interval is:

$$\hat{p}_i \pm z_{0.025}\sqrt{\frac{\hat{p}_i\hat{q}_i}{n}} \Rightarrow 0.197 \pm 1.96\sqrt{\frac{0.197(0.803)}{350}} \Rightarrow 0.197 \pm 0.042 \Rightarrow (0.155,\ 0.239)$$

We are 95% confident that the true proportion of vowels drawn by the electronic game is between 0.155 and 0.239. The actual proportion of vowels drawn in the board game is 0.42. Since this value is not contained in the 95% confidence interval, one can conclude that the ScrabbleExpress™ game produces too few vowels in the 7-letter draws.

13.23 A contingency table with fixed marginals is a table where the column totals for one qualitative variable are known in advance. For example, we could ask a sample of 150 males and a sample of 150 females their preference to a particular brand. The column totals for gender are known in advance.

13.25 The conditions that are required for a valid chi-square test of data from a contingency table are:

1. The n observed counts are a random sample from the population of interest.
2. The sample size, n, will be large enough so that, for every cell, the expected count, E_{ij}, will be equal to 5 or more.

13.27 a. H_0: The row and column classifications are independent
H_a: The row and column classifications are dependent

b. The test statistic is $\chi^2 = \sum\sum \frac{\left[n_{ij} - \hat{E}_{ij}\right]^2}{\hat{E}_{ij}}$

The rejection region requires $\alpha = 0.01$ in the upper tail of the χ^2 distribution with $df = (r-1)(c-1) = (2-1)(3-1) = 2$. From Table IV, Appendix B, $\chi^2_{0.01} = 9.21034$. The rejection region is $\chi^2 > 9.21034$.

c. The expected cell counts are:

$$\hat{E}_{11} = \frac{R_1C_1}{n} = \frac{96(25)}{167} = 14.37 \qquad \hat{E}_{21} = \frac{R_2C_1}{n} = \frac{71(25)}{167} = 10.63$$

$$\hat{E}_{12} = \frac{R_1C_2}{n} = \frac{96(64)}{167} = 36.79 \qquad \hat{E}_{22} = \frac{R_2C_2}{n} = \frac{71(64)}{167} = 27.21$$

$$\hat{E}_{13} = \frac{R_1C_3}{n} = \frac{96(78)}{167} = 44.84 \qquad \hat{E}_{23} = \frac{R_2C_3}{n} = \frac{71(78)}{167} = 33.16$$

d. The test statistic is

$$\chi^2 = \sum\sum \frac{\left[n_{ij} - \hat{E}_{ij}\right]^2}{\hat{E}_{ij}} = \frac{(9-14.37)^2}{14.37} + \frac{(34-36.79)^2}{36.79} + \frac{(53-44.84)^2}{44.84} + \frac{(16-10.63)^2}{10.63}$$

$$+ \frac{(30-27.21)^2}{27.21} + \frac{(25-33.16)^2}{33.16} = 8.71$$

Since the observed value of the test statistic does not fall in the rejection region $\left(\chi^2 = 8.71 \not> 9.21034\right)$, H_0 is not rejected. There is insufficient evidence to indicate the row and column classifications are dependent at $\alpha = 0.01$.

13.29 Some preliminary calculations are:

$$\hat{E}_{11} = \frac{R_1C_1}{n} = \frac{154(134)}{439} = 47.007 \qquad \hat{E}_{21} = \frac{186(134)}{439} = 56.774$$

$$\hat{E}_{12} = \frac{154(163)}{439} = 57.180 \qquad \hat{E}_{22} = \frac{186(163)}{439} = 69.062$$

$$\hat{E}_{13} = \frac{154(142)}{439} = 49.813 \qquad \hat{E}_{23} = \frac{186(142)}{439} = 60.164$$

$$\hat{E}_{31} = \frac{99(134)}{439} = 30.219 \qquad \hat{E}_{33} = \frac{99(142)}{439} = 32.023$$

$$\hat{E}_{32} = \frac{99(163)}{439} = 36.759$$

To determine if the row and column classifications are dependent, we test:

H_0: The row and column classifications are independent
H_a: The row and column classifications are dependent

The test statistic is

$$\chi^2 = \sum\sum \frac{\left[n_{ij} - \hat{E}_{ij}\right]^2}{\hat{E}_{ij}} = \frac{(40-47.007)^2}{47.007} + \frac{(72-57.180)^2}{57.180} + \frac{(42-49.813)^2}{49.813} + \frac{(63-56.774)^2}{56.774}$$
$$+ \frac{(53-69.062)^2}{69.062} + \frac{(70-60.164)^2}{60.164} + \frac{(31-32.023)^2}{32.023} + \frac{(38-36.759)^2}{36.759} + \frac{(30-32.023)^2}{32.023}$$
$$= 12.33$$

The rejection region requires $\alpha = 0.05$ in the upper tail of the χ^2 distribution with $df = (r-1)(c-1) = (3-1)(3-1) = 4$. From Table IV, Appendix B, $\chi^2_{0.05} = 9.48773$. The rejection region is $\chi^2 > 9.48773$.

Since the observed value of the test statistic falls in the rejection region $\left(\chi^2 = 12.33 > 9.48773\right)$, H_0 is rejected. There is sufficient evidence to indicate the row and column classifications are dependent at $\alpha = 0.05$.

13.31. a. $\hat{p}_1 = 6/77 = 0.078$

b. $\hat{p}_2 = 11/70 = 0.157$

c. The proportion of girls drawing a dog is almost 2 times the proportion of boys drawing a dog. It is possible that the likelihood of drawing a dog depends on gender.

d. To determine if the likelihood of drawing a dog depends on gender, we test:

H_0: Presence of a dog and gender are independent
H_a: Presence of a dog and gender are dependent

e. From the printout, the test statistic is $\chi^2 = 2.250$ and the p-value is $p = 0.134$.

Since the p-value is not less than $\alpha\left(p = 0.134 \nless 0.05\right)$, H_0 is not rejected. There is insufficient evidence to indicate the likelihood of drawing a dog depends on gender at $\alpha = 0.05$.

f. Using MINITAB, the results are:

Tabulated statistics: TV, GENDER

```
Using frequencies in NUMBER

Rows: TV    Columns: GENDER

          Boy    Girl     All

No         66      61     127
        66.52   60.48  127.00

Yes        11       9      20
        10.48    9.52   20.00

All        77      70     147
        77.00   70.00  147.00

Cell Contents:        Count
                      Expected count


Pearson Chi-Square = 0.064, DF = 1, P-Value = 0.801
Likelihood Ratio Chi-Square = 0.064, DF = 1, P-Value = 0.801
```

To determine if the likelihood of drawing a TV in the bedroom depends on gender, we test:

H_0: Presence of TV and gender are independent
H_a: Presence of TV and gender are dependent

From the printout, the test statistic is $\chi^2 = 0.064$ and the *p*-value is $p = 0.801$.

Since the *p*-value is not less than $\alpha\left(p = 0.801 \nless 0.05\right)$, H_0 is not rejected. There is insufficient evidence to indicate the likelihood of drawing TV depends on gender at $\alpha = 0.05$.

13.33 a. $\hat{p}_1 = \frac{111}{170} = 0.653$

b. $\hat{p}_2 = \frac{61}{110} = 0.555$

c. $\hat{p}_3 = \frac{11}{30} = 0.355$

d. The sample proportions range from 0.367 to 0.653. It appears that the proportion of news stories that are deceptive depends on story tone.

e. To determine whether the authenticity of a news story depends on tone, we test:

H_0: Authenticity of news story and tone are independent
H_a: Authenticity of news story and tone are dependent

f. From the printout, the test statistic is $\chi^2 = 10.427$ and the p-value is $p = 0.005$. Since the p-value is less than $\alpha\left(p = 0.005 < 0.05\right)$, H_0 is rejected. There is sufficient evidence to indicate that the authenticity of a news story depends on tone at $\alpha = 0.05$.

13.35 a. To determine if package design and sound pitch combination influences the consumer's opinion on product taste, we test:

H_0: Package design/sound pitch and opinion on product taste are independent
H_a: Package design/sound pitch and opinion on product taste are dependent

b. $\hat{E}_{11} = \dfrac{R_1C_1}{n} = \dfrac{42(40)}{80} = 21$ $\quad\hat{E}_{12} = \dfrac{R_1C_2}{n} = \dfrac{42(40)}{80} = 21$

$\hat{E}_{21} = \dfrac{R_2C_1}{n} = \dfrac{38(40)}{80} = 19$ $\quad\hat{E}_{22} = \dfrac{R_2C_2}{n} = \dfrac{38(40)}{80} = 19$

c. The test statistic is

$$\chi^2 = \sum \frac{\left[n_{ij} - \hat{E}_{ij}\right]^2}{\hat{E}_{ij}} = \frac{[35-21]^2}{21} + \frac{[7-21]^2}{21} + \frac{[5-19]^2}{19} + \frac{[33-19]^2}{19} = 39.298$$

d. Since the p-value is so small, H_0 is rejected. Three is sufficient evidence to indicate package design and sound pitch combination influences the consumer's opinion on product taste for any reasonable value of α.

13.37 The contingency table is:

	Responses			
Presentation	**4,100**	**5,000**	**Other**	*Totals*
Group 1 (2 sec/black)	17	33	10	60
Group 2 (2 sec/red)	8	42	10	60
Group 3 (15 sec/red)	12	36	12	60
Group 4 (1 sec/black)	13	42	5	60
Group 5 (15 sec/black)	20	31	9	60
Totals	70	184	46	300

Some preliminary calculations are:

$\hat{E}_{11} = \dfrac{R_1C_1}{n} = \dfrac{60(70)}{300} = 14$ $\quad\hat{E}_{12} = \dfrac{R_1C_2}{n} = \dfrac{60(184)}{300} = 36.8$ $\quad\hat{E}_{13} = \dfrac{R_1C_3}{n} = \dfrac{60(46)}{300} = 9.2$

$\hat{E}_{21} = \dfrac{R_2C_1}{n} = \dfrac{60(70)}{300} = 14$ $\quad\hat{E}_{22} = \dfrac{R_2C_2}{n} = \dfrac{60(184)}{300} = 36.8$ $\quad\hat{E}_{23} = \dfrac{R_2C_3}{n} = \dfrac{60(46)}{300} = 9.2$

$$\hat{E}_{31}=\frac{R_3C_1}{n}=\frac{60(70)}{300}=14 \quad \hat{E}_{32}=\frac{R_3C_2}{n}=\frac{60(184)}{300}=36.8 \quad \hat{E}_{33}=\frac{R_3C_3}{n}=\frac{60(46)}{300}=9.2$$

$$\hat{E}_{41}=\frac{R_4C_1}{n}=\frac{60(70)}{300}=14 \quad \hat{E}_{42}=\frac{R_4C_2}{n}=\frac{60(184)}{300}=36.8 \quad \hat{E}_{43}=\frac{R_4C_3}{n}=\frac{60(46)}{300}=9.2$$

$$\hat{E}_{51}=\frac{R_5C_1}{n}=\frac{60(70)}{300}=14 \quad \hat{E}_{52}=\frac{R_5C_2}{n}=\frac{60(184)}{300}=36.8 \quad \hat{E}_{53}=\frac{R_5C_3}{n}=\frac{60(46)}{300}=9.2$$

To determine if presentation of the numbers to be added influence performance, we test:

H_0: Presentation and performance are independent
H_a: Presentation and performance are dependent

The test statistic is

$$\chi^2=\sum\frac{\left[n_{ij}-\hat{E}_{ij}\right]^2}{\hat{E}_{ij}}=\frac{[17-14]^2}{14}+\frac{[8-14]^2}{14}+\frac{[12-14]^2}{14}+\frac{[13-14]^2}{14}+\frac{[20-14]^2}{14}$$

$$+\frac{[33-36.8]^2}{36.8}+\frac{[42-36.8]^2}{36.8}+\frac{[36-36.8]^2}{36.8}+\frac{[42-36.8]^2}{36.8}+\frac{[31-36.8]^2}{36.8}$$

$$+\frac{[10-9.2]^2}{9.2}+\frac{[10-9.2]^2}{9.2}+\frac{[12-9.2]^2}{9.2}+\frac{[5-9.2]^2}{9.2}+\frac{[9-9.2]^2}{9.2}=11.849$$

The rejection region requires $\alpha = 0.01$ in the upper tail of the χ^2 distribution with $df=(r-1)(c-1)=(5-1)(3-1)=8$. From Table IV, Appendix B, $\chi^2_{0.01}=20.0902$. The rejection region is $\chi^2>20.0902$.

Since the observed value of the test statistics does not fall in the rejection region $\left(\chi^2=11.849 \not> 20.0902\right)$, H_0 is not rejected. There is insufficient evidence to indicate presentation of the numbers to be added influence performance at $\alpha = 0.01$.

13.39 Some preliminary calculations are:

$$\hat{E}_{11}=\frac{R_1C_1}{n}=\frac{2097(149)}{6560}=47.630 \qquad \hat{E}_{12}=\frac{R_1C_2}{n}=\frac{2097(1647)}{6560}=526.488$$

$$\hat{E}_{13}=\frac{R_1C_3}{n}=\frac{2097(4764)}{6560}=1522.882 \qquad \hat{E}_{21}=\frac{R_2C_1}{n}=\frac{2182(149)}{6560}=49.561$$

$$\hat{E}_{22}=\frac{R_2C_2}{n}=\frac{2182(1647)}{6560}=547.828 \qquad \hat{E}_{23}=\frac{R_2C_3}{n}=\frac{2182(4764)}{6560}=1584.611$$

$$\hat{E}_{31}=\frac{R_3C_1}{n}=\frac{2281(149)}{6560}=51.809 \qquad \hat{E}_{32}=\frac{R_3C_2}{n}=\frac{2281(1647)}{6560}=572.684$$

$$\hat{E}_{33} = \frac{R_3C_3}{n} = \frac{2281(4764)}{6560} = 1656.507$$

To determine if there are significant genotype differences across the three age groups, we test:

H_0: Genotype and age group are independent
H_a: Genotype and age group are dependent

The test statistic is

$$\chi^2 = \sum \frac{\left[n_{ij} - \hat{E}_{ij}\right]^2}{\hat{E}_{ij}} = \frac{[56-47.630]^2}{47.630} + \frac{[517-526.488]^2}{526.488} + \frac{[1524-1522.882]^2}{1522.882}$$
$$+ \frac{[45-49.561]^2}{49.561} + \frac{[566-547.828]^2}{547.828} + \frac{[1571-1584.611]^2}{1584.611}$$
$$+ \frac{[48-51.809]^2}{51.809} + \frac{[564-572.684]^2}{572.684} + \frac{[1669-1656.507]^2}{1656.507} = 3.288$$

The rejection region requires $\alpha = 0.05$ in the upper tail of the χ^2 distribution with $df = (r-1)(c-1) = (3-1)(3-1) = 4$. From Table IV, Appendix B, $\chi^2_{0.05} = 9.48773$. The rejection region is $\chi^2 > 9.48773$.

Since the observed value of the test statistics does not fall in the rejection region $\left(\chi^2 = 3.288 \not> 9.48773\right)$, H_0 is not rejected. There is insufficient evidence to indicate there are significant genotype differences across the three age groups at $\alpha = 0.05$.

13.41 Using MINITAB, the results of the analyses are:

Results for: AIRTHR~1.MTP

Tabulated statistics: Instruction, Strategy

```
Rows: Instruction    Columns: Strategy

              Guess   Other    TTBC      All

Cue               5       6      13       24
              7.000   8.500   8.500   24.000

Pattern           9      11       4       24
              7.000   8.500   8.500   24.000

All              14      17      17       48
             14.000  17.000  17.000   48.000

Cell Contents:        Count
                      Expected count

Pearson Chi-Square = 7.378, DF = 2, P-Value = 0.025
Likelihood Ratio Chi-Square = 7.668, DF = 2, P-Value = 0.022
```

To determine if the choice of heuristic strategy depends on type of instruction provided, we test:

H_0: Choice of heuristic strategy and instruction provided are independent
H_0: Choice of heuristic strategy and instruction provided are dependent

From the printout, the test statistic is $\chi^2 = 7.378$ and the p-value is $p = 0.025$.

Since the p-value is less than $\alpha\left(p = 0.025 < 0.05\right)$, H_0 is rejected. There is sufficient evidence to indicate the choice of heuristic strategy depends on type of instruction provided at $\alpha = 0.05$.

If $\alpha = 0.01$, then the p-value is not less than $\alpha\left(p = 0.025 \nless 0.01\right)$. H_0 would not be rejected at $\alpha = 0.01$

13.43 Some preliminary calculations are:

$$\hat{E}_{11} = \frac{R_1C_1}{n} = \frac{50(40)}{160} = 12.5 \quad \hat{E}_{12} = \frac{R_1C_2}{n} = \frac{50(40)}{160} = 12.5 \quad \hat{E}_{13} = \frac{R_1C_3}{n} = \frac{50(40)}{160} = 12.5$$

$$\hat{E}_{14} = \frac{R_1C_4}{n} = \frac{50(40)}{160} = 12.5 \quad \hat{E}_{21} = \frac{R_2C_1}{n} = \frac{26(40)}{160} = 6.5 \quad \hat{E}_{22} = \frac{R_2C_2}{n} = \frac{26(40)}{160} = 6.5$$

$$\hat{E}_{23} = \frac{R_2C_3}{n} = \frac{26(40)}{160} = 6.5 \quad \hat{E}_{24} = \frac{R_2C_4}{n} = \frac{26(40)}{160} = 6.5 \quad \hat{E}_{31} = \frac{R_3C_1}{n} = \frac{52(40)}{160} = 13$$

$$\hat{E}_{32} = \frac{R_3C_2}{n} = \frac{52(40)}{160} = 13 \quad \hat{E}_{33} = \frac{R_3C_3}{n} = \frac{52(40)}{160} = 13 \quad \hat{E}_{34} = \frac{R_3C_4}{n} = \frac{52(40)}{160} = 13$$

$$\hat{E}_{41} = \frac{R_4C_1}{n} = \frac{32(40)}{160} = 8 \quad \hat{E}_{42} = \frac{R_4C_2}{n} = \frac{32(40)}{160} = 8 \quad \hat{E}_{43} = \frac{R_4C_3}{n} = \frac{32(40)}{160} = 8$$

$$\hat{E}_{44} = \frac{R_4C_4}{n} = \frac{32(40)}{160} = 8$$

To determine if one circuit appears to be more difficult to analyze than any other, we test:

H_0: Circuits and difficulty are independent
H_a: Circuits and difficulty are dependent

The test statistic is

$$\chi^2 = \sum \frac{\left[n_{ij} - \hat{E}_{ij}\right]^2}{\hat{E}_{ij}} = \frac{[31-12.5]^2}{12.5} + \frac{[10-12.5]^2}{12.5} + \frac{[5-12.5]^2}{12.5} + \cdots + \frac{[10-8]^2}{8} = 64.237$$

Since no α is given, we will use $\alpha = 0.05$. The rejection region requires $\alpha = 0.05$ in the upper tail of the χ^2 distribution with $df = (r-1)(c-1) = (4-1)(4-1) = 9$. From Table IV, Appendix B, $\chi^2_{0.05} = 16.9190$. The rejection region is $\chi^2 > 16.9190$.

Since the observed value of the test statistics falls in the rejection region $(\chi^2 = 64.237 > 16.9190)$, H_0 is rejected. There is sufficient evidence to indicate that one circuit appears to be more difficult to analyze than any other at $\alpha = 0.05$.

13.45 a. The contingency table is:

	Test Results		
Patient Group	Positive	Negative	Totals
Unvaccinated	22	9	31
Vaccinated	2	5	7
Totals	24	14	38

Some preliminary calculations are:

$$\hat{E}_{11} = \frac{R_1C_1}{n} = \frac{31(24)}{38} = 19.58 \qquad \hat{E}_{12} = \frac{R_1C_2}{n} = \frac{31(14)}{38} = 11.42$$

$$\hat{E}_{21} = \frac{R_2C_1}{n} = \frac{7(24)}{38} = 4.42 \qquad \hat{E}_{22} = \frac{R_2C_2}{n} = \frac{7(14)}{38} = 2.58$$

To determine whether the vaccine is effective in treating the particular strain of HIV, we test:

H_0: Vaccine and particular strain are independent
H_a: Vaccine and particular strain are dependent

The test statistic is

$$\chi^2 = \sum\sum \frac{\left[n_{ij} - \hat{E}_{ij}\right]^2}{\hat{E}_{ij}} = \frac{(22-19.58)^2}{19.58} + \frac{(9-11.42)^2}{11.42} + \frac{(2-4.42)^2}{4.42} + \frac{(5-2.58)^2}{2.58} = 4.407$$

The rejection region requires $\alpha = 0.05$ in the upper tail of the χ^2 distribution with $df = (r-1)(c-1) = (2-1)(2-1) = 1$. From Table IV, Appendix B, $\chi^2_{0.05} = 3.84146$. The rejection region is $\chi^2 > 3.84146$.

Since the observed value of the test statistic falls in the rejection region $(\chi^2 = 4.407 > 3.84146)$, H_0 is rejected. There is sufficient evidence to indicate that the vaccine is effective in treating the particular strain of HIV at $\alpha = 0.05$.

b. The necessary assumptions are:

1. The n observed counts are a random sample from the population of interest.

2. The sample size, n, will be large enough so that, for every cell, the expected count, $E(n_{ij})$, will be equal to 5 or more.

For this example, the second assumption is violated. Two of the expected cell counts are less than 5. What we have computed for the test statistic may not have a χ^2 distribution.

c. For this contingency table:

$$p = \frac{\binom{7}{2}\binom{31}{22}}{\binom{38}{24}} = \frac{\frac{7!}{2!5!}\,\frac{31!}{22!9!}}{\frac{38!}{24!14!}} = 0.0438$$

d. The two new contingency tables are:

	Test Results		
Patient Group	Positive	Negative	Totals
Unvaccinated	23	8	31
Vaccinated	1	6	7
	24	14	38

	Test Results		
Patient Group	Positive	Negative	Totals
Unvaccinated	24	7	31
Vaccinated	0	7	7
	24	14	38

If vaccine and the particular strain are independent, then the proportion of positive results should be relatively the same for both patient groups. In the two tables presented, the proportion of positive results for the vaccinated group is smaller than the proportion for the original table.

For the first table,

$$p = \frac{\binom{7}{1}\binom{31}{23}}{\binom{38}{24}} = \frac{\frac{7!}{1!6!}\,\frac{31!}{23!8!}}{\frac{38!}{24!14!}} = 0.0057$$

For the second table:

$$p = \frac{\binom{7}{0}\binom{31}{24}}{\binom{38}{24}} = \frac{\frac{7!}{0!7!}\,\frac{31!}{24!7!}}{\frac{38!}{24!14!}} = 0.0003$$

e. The p-value for Fisher's exact test is $p = 0.0438 + 0.0057 + 0.0003 = 0.0498$. Since the p-value is small, there is evidence to reject H_0. There is sufficient evidence to indicate that the vaccine is effective in treating the particular strain of HIV at $\alpha > 0.0498$.

13.47 The statement "Rejecting the null hypothesis in a chi-square test for independence implies that a causal relationship between the two categorical variables exists." is false. If the variables are dependent (reject H_0), it simply implies that they are related. One cannot infer causal relationships from contingency table analysis.

13.49 a. Some preliminary calculations are:

$$\hat{E}_{11} = \frac{R_1C_1}{n} = \frac{50(50)}{250} = 10 \qquad \hat{E}_{21} = \frac{R_2C_1}{n} = \frac{100(50)}{250} = 20$$

$$\hat{E}_{12} = \frac{R_1C_2}{n} = \frac{50(90)}{250} = 18 \qquad \hat{E}_{22} = \frac{R_2C_2}{n} = \frac{100(90)}{250} = 36$$

$$\hat{E}_{13} = \frac{R_1C_3}{n} = \frac{50(110)}{250} = 22 \qquad \hat{E}_{23} = \frac{R_2C_3}{n} = \frac{100(110)}{250} = 44$$

$$\hat{E}_{31} = \frac{R_3C_1}{n} = \frac{100(50)}{250} = 20 \qquad \hat{E}_{32} = \frac{R_3C_2}{n} = \frac{100(90)}{250} = 36$$

$$\hat{E}_{33} = \frac{R_3C_3}{n} = \frac{100(110)}{250} = 44$$

To determine if the rows and columns are dependent, we test:

H_0: Rows and columns are independent
H_a: Rows and columns are dependent

The test statistic is $\chi^2 = \sum\sum \frac{\left[n_{ij} - E_{ij}\right]^2}{\hat{E}_{ij}} = \frac{(20-10)^2}{10} + \cdots + \frac{(30-44)^2}{44} = 54.14$

The rejection region requires $\alpha = 0.05$ in the upper tail of the χ^2 distribution with $df = (r-1)(c-1) = (3-1)(3-1) = 2$. From Table IV, Appendix B, $\chi^2_{0.05} = 9.48773$. The rejection region is $\chi^2 > 9.48773$.

Since the observed value of the test statistic falls in the rejection region $\left(\chi^2 = 54.14 > 9.48773\right)$, H_0 is rejected. There is sufficient evidence to indicate the rows and columns are dependent at $\alpha = 0.05$.

b. No, the analysis remains identical.

c. Yes, the assumptions on the sampling differ.

d. The percentages are in the table below.

		Column			
		1	**2**	**3**	**Totals**
	1	$\frac{20}{50}\times100\% = 40\%$	$\frac{20}{90}\times100\% = 22.2\%$	$\frac{10}{110}\times100\% = 9.1\%$	$\frac{50}{250}\times100\% = 20\%$
Row	**2**	$\frac{10}{50}\times100\% = 20\%$	$\frac{20}{90}\times100\% == 22.2\%$	$\frac{70}{110}\times100\% = 63.6\%$	$\frac{100}{250}\times100\% = 40\%$
	3	$\frac{20}{50}\times100\% = 40\%$	$\frac{50}{90}\times100\% = 55.6\%$	$\frac{30}{110}\times100\% = 37.3\%$	$\frac{100}{250}\times100\% = 40\%$

e. The graph is:

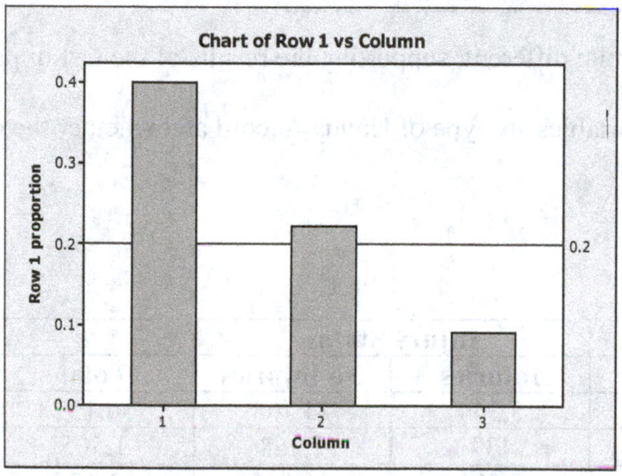

If the rows and columns are independent, then the proportion of observations for each column should be relatively the same. In this graph, the percentages are quite different, supporting the results of the test in part **a**.

f. The graph for row 2 is:

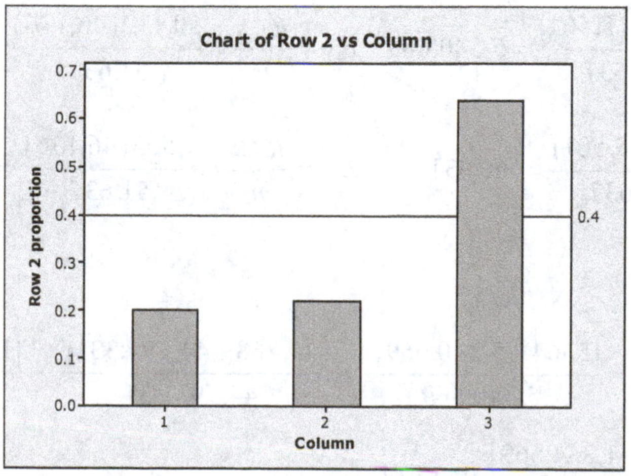

The percentages are quite different, supporting the results of the test in part **a**.

g. The graph for row 3 is:

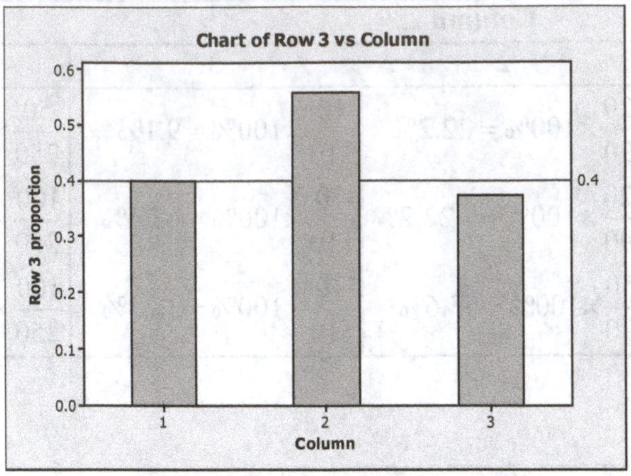

The percentages are quite different, supporting the results of the test in part **a**.

13.51 a. The two qualitative variables are type of Honda Accord and whether the collision involved injuries.

b. The contingency table is:

	Injury Status		
Automobile Type	**Injuries**	**No Injuries**	**Total**
Accord	5,364	44,768	50,132
Accord Hybrid	137	1,368	1,505
Total	5,501	46,136	51,637

c. To determine if injury rate for collision claims depends on Accord model, we test:

H_0: Injury rate and Accord model are independent
H_a: Injury rate and Accord model are dependent

d. $\hat{E}_{11} = \frac{R_1C_1}{n} = \frac{50{,}132(5{,}501)}{51{,}637} = 5{,}340.669$ $\hat{E}_{12} = \frac{R_1C_2}{n} = \frac{50{,}132(46{,}136)}{51{,}637} = 44{,}791.331$

$\hat{E}_{21} = \frac{R_2C_1}{n} = \frac{1{,}505(5{,}501)}{51{,}637} = 160.331$ $\hat{E}_{22} = \frac{R_2C_2}{n} = \frac{1{,}505(46{,}136)}{51{,}637} = 1{,}344.669$

e. The test statistic is

$$\chi^2 = \sum \frac{\left[n_{ij} - \hat{E}_{ij}\right]^2}{\hat{E}_{ij}} = \frac{[5364 - 5{,}340.669]^2}{5{,}340.669} + \frac{[44{,}768 - 44{,}791.331]^2}{44{,}791.331} + \frac{[137 - 160.331]^2}{160.331}$$

$$+ \frac{[1{,}368 - 1{,}344.669]^2}{1{,}344.669} = 3.914$$

The answer compares to that in the SAS printout.

f. The rejection region requires $\alpha = 0.05$ in the upper tail of the χ^2 distribution with $df = (r-1)(c-1) = (2-1)(2-1) = 1$. From Table IV, Appendix B, $\chi^2_{0.05} = 3.84146$. The rejection region is $\chi^2 > 3.84146$. The test statistic is just larger than the critical value and the p-value is just smaller than $\alpha = 0.05$.

g. Since the observed value of the test statistics falls in the rejection region $\left(\chi^2 = 3.914 > 3.84146\right)$, H_0 is rejected. There is sufficient evidence to indicate that injury rate for collision claims depends on Accord model at $\alpha = 0.05$.

Since the p-value is less than $\alpha \left(p = 0.043 < 0.05\right)$, H_0 is rejected. There is sufficient evidence to indicate that injury rate for collision claims depends on Accord model at $\alpha = 0.05$.

h. $\hat{p}_1 = \dfrac{x_1}{n_1} = \dfrac{5{,}364}{50{,}132} = 0.107$ and $\hat{p}_2 = \dfrac{x_2}{n_2} = \dfrac{137}{1{,}505} = 0.091$

For confidence coefficient 0.95, $\alpha = 0.05$ and $\alpha/2 = 0.05/2 = 0.025$. From Table II, Appendix B, $z_{0.025} = 1.96$. The 95% confidence interval is:

$$(\hat{p}_1 - \hat{p}_2) \pm z_{0.025}\sqrt{\frac{\hat{p}_1\hat{q}_1}{n_1} + \frac{\hat{p}_2\hat{q}_2}{n_2}} \Rightarrow (0.107 - 0.091) \pm 1.96\sqrt{\frac{(0.107)(0.893)}{50{,}132} + \frac{(0.091)(0.909)}{1{,}505}}$$

$$\Rightarrow 0.016 \pm 0.015 \Rightarrow (0.001,\ 0.031)$$

Because the interval constructed contains only positive numbers, there is evidence that the injury rate for hybrid Accords is less than the injury rate for conventional Accords with 95% confidence.

13.53 a. The 2×2 contingency table is:

	LLD		
Genetic Trait	**Yes**	**No**	**Total**
Yes	21	15	36
No	150	306	456
Total	171	321	492

b. Some preliminary calculations are:

$$\hat{E}_{11} = \frac{R_1C_1}{n} = \frac{36(171)}{492} = 12.51 \qquad \hat{E}_{12} = \frac{R_1C_2}{n} = \frac{36(321)}{492} = 23.49$$

$$\hat{E}_{21} = \frac{R_2C_1}{n} = \frac{456(171)}{492} = 158.49 \qquad \hat{E}_{22} = \frac{R_2C_2}{n} = \frac{456(321)}{492} = 297.51$$

To determine if the genetic trait occurs at a higher rate in LDD patients than in the controls, we test:

H_0: Genetic trait and LDD group are independent
H_a: Genetic trait and LDD group are dependent

The test statistic is

$$\chi^2 = \sum\sum\frac{\left[n_{ij} - \hat{E}_{ij}\right]^2}{\hat{E}_{ij}} = \frac{(21-12.51)^2}{12.51} + \frac{(15-23.49)^2}{23.49} + \frac{(150-158.49)^2}{158.49} + \frac{(306-297.51)^2}{297.51} = 9.52$$

The rejection region requires $\alpha = 0.01$ in the upper tail of the χ^2 distribution with $df = (r-1)(c-1) = (2-1)(2-1) = 1$. From Table IV, Appendix B, $\chi^2_{0.01} = 6.63490$. The rejection region is $\chi^2 > 6.63490$.

Since the observed value of the test statistic falls in the rejection region $\left(\chi^2 = 9.52 > 6.63490\right)$, H_0 is rejected. There is sufficient evidence to indicate that the genetic trait occurs at a higher rate in LDD patients than in the controls at $\alpha = 0.01$.

c. To construct a bar graph, we will first compute the proportion of LDD/Control patients that have the genetic trait. Of the LLD patients, $21/171 = 0.123$ have the trait. For the Controls, $15/321 = 0.043$. For the row totals, $36/492 = 0.073$. The bar graph is:

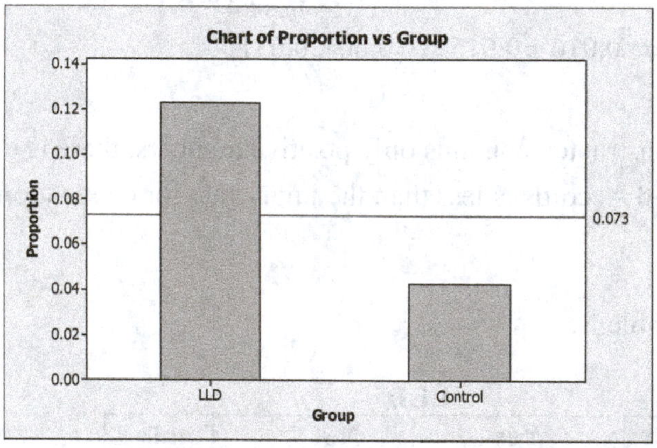

Because the bars are not close to being the same height, it indicates that the genetic trait appears at a higher rate among LLD patients than among the controls. This supports the conclusion of the test in part **b**.

13.55 Some preliminary calculations are:

$$\hat{E}_{11} = \frac{R_1C_1}{n} = \frac{62(106)}{275} = 23.90 \qquad \hat{E}_{12} = \frac{R_1C_2}{n} = \frac{62(112)}{275} = 25.25$$

$$\hat{E}_{13} = \frac{R_1C_3}{n} = \frac{62(57)}{275} = 12.85 \qquad \hat{E}_{21} = \frac{R_2C_1}{n} = \frac{59(106)}{275} = 22.74$$

$$\hat{E}_{22} = \frac{R_2C_2}{n} = \frac{59(112)}{275} = 24.03 \qquad \hat{E}_{23} = \frac{R_2C_3}{n} = \frac{59(57)}{275} = 12.23$$

$$\hat{E}_{31} = \frac{R_3C_1}{n} = \frac{59(106)}{275} = 22.74 \qquad \hat{E}_{32} = \frac{R_3C_2}{n} = \frac{59(112)}{275} = 24.03$$

$$\hat{E}_{33} = \frac{R_3C_3}{n} = \frac{59(57)}{275} = 12.23 \qquad \hat{E}_{41} = \frac{R_4C_1}{n} = \frac{95(106)}{275} = 36.62$$

$$\hat{E}_{42} = \frac{R_4C_2}{n} = \frac{95(112)}{275} = 38.69 \qquad \hat{E}_{43} = \frac{R_4C_3}{n} = \frac{95(57)}{275} = 19.69$$

To determine if political strategy of ethnic groups depends on world region, we test:

H_0: Political strategy and world region are independent
H_a: Political strategy and world region are dependent

The test statistic is

$$\chi^2 = \sum\sum \frac{\left[n_{ij} - \hat{E}_{ij}\right]^2}{\hat{E}_{ij}} = \frac{(24-23.90)^2}{23.90} + \frac{(31-25.25)^2}{25.25} + \frac{(7-12.85)^2}{12.85} + \frac{(32-22.74)^2}{22.74}$$

$$+ \frac{(23-24.03)^2}{24.03} + \frac{(4-12.23)^2}{12.23} + \frac{(11-22.74)^2}{22.74} + \frac{(22-24.03)^2}{24.03}$$

$$+ \frac{(26-12.23)^2}{12.23} + \frac{(39-36.62)^2}{36.62} + \frac{(36-38.69)^2}{38.69} + \frac{(20-19.69)^2}{19.69} = 35.409$$

The rejection region requires $\alpha = 0.10$ in the upper tail of the χ^2 distribution with $df = (r-1)(c-1) = (4-1)(3-1) = 6$. From Table IV, Appendix B, $\chi^2_{0.10} = 10.6446$. The rejection region is $\chi^2 > 10.6446$.

Since the observed value of the test statistic falls in the rejection region $\left(\chi^2 = 35.409 > 10.6446\right)$, H_0 is rejected. There is sufficient evidence to indicate that the political strategy of the ethnic groups depends on world region at $\alpha = 0.10$.

To graph the data, we will first compute the percent of observations in each category of Political Strategy for each World Region. To do this, we divide each cell frequency by the row total and then multiply by 100%. Using MINITAB, a graph of the data is:

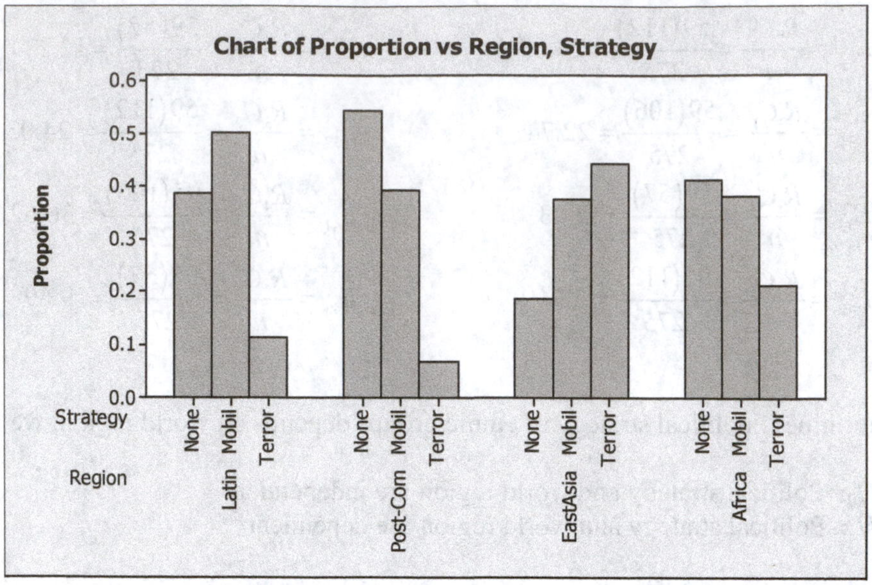

The graph supports the results of the test above. The percents in each level of Political Strategy differ for the different values of World Region.

13.57 To determine if the proportions of college students in the four drinking groups differ, we test:

$$H_0: p_1 = p_2 = p_3 = p_4 = 0.25$$
$$H_a: \text{At least one } p_i \text{ differs from its hypothesized value}$$

From the printout, the test statistic is $\chi^2 = 127.493$ and the *p*-value is $p = 0.000$.

Since the *p*-value is so small, H_0 is rejected. There is sufficient evidence to indicate that the proportions of college students in the four drinking groups differ for any reasonable value of α.

13.59 a. Based on the percentages in the table, it does appear that male and female tourists differ in their responses to purchasing photographs, postcards, and paintings. It appears that females $(28\% + 31\% = 59\%)$ tend to buy these items more frequently than males $(16\% + 27\% = 43\%)$. However, these percentages alone cannot be used to draw conclusions. The actual sample sizes are necessary to make these decisions.

b. To find the counts in each of the cells, we multiply the proportions in the "Male Tourist" column by 1,500 and the proportions in the "Female Tourist" column by 1,700. The results are:

Photos, postcards, paintings	Tourist		
	Male	**Female**	**Totals**
Always	240	476	716
Often	405	527	932
Occasionally	525	493	1018
Rarely or Never	330	204	534
Totals	1500	1700	3200

c. To determine whether male and female tourists differ in their responses to purchasing photographs, postcards, or paintings, we test:

H_0: Gender of tourists and purchasing photos/postcards/paintings are independent
H_a: Gender of tourists and purchasing photos/postcards/paintings are dependent

d. The test statistic is $\chi^2 = 112.433$ and the p-value is $p = 0.000$.

e. Since the p-value is less than $\alpha\left(p = 0.000 < 0.01\right)$, H_0 is rejected. There is sufficient evidence to indicate that male and female tourists differ in their responses to purchasing photographs, postcards, or paintings at $\alpha = 0.01$.

13.61 a. Some preliminary calculations are:

$E_1 = np_{1,0} = 85(0.26) = 22.1$ $\quad E_2 = np_{2,0} = 85(0.30) = 25.5$

$E_1 = np_{3,0} = 85(0.11) = 9.35$ $\quad E_4 = np_{4,0} = 85(0.14) = 11.9$

$E_5 = np_{5,0} = 85(0.19) = 16.15$

To determine of probabilities differ from the hypothesized values, we test:

$H_0: p_1 = 0.26,\ p_2 = 0.30,\ p_3 = 0.11,\ p_4 = 0.14,\ p_5 = 0.19$
H_a: At least one p_i differs from its hypothesized value

The test statistic is

$$\chi^2 = \sum\frac{\left[n_i - E_i\right]^2}{E_i} = \frac{(32-22.1)^2}{22.1} + \frac{(26-25.5)^2}{25.5} + \frac{(15-9.35)^2}{9.35} + \frac{(6-11.9)^2}{11.9} + \frac{(6-16.15)^2}{16.15} = 17.16$$

The rejection region requires $\alpha = 0.05$ in the upper tail of the χ^2 distribution with $df = k - 1 = 5 - 1 = 4$. From Table IV, Appendix B, $\chi^2_{0.05} = 9.48773$. The rejection region is $\chi^2 > 9.48773$.

Since the observed value of the test statistic falls in the rejection region $\left(\chi^2 = 17.16 > 9.48773\right)$, reject H_0. There is sufficient evidence to indicate the probabilities differ from their hypothesized values at $\alpha = 0.05$.

b. $\hat{p}_1 = \frac{32}{85} = 0.376$

For confidence coefficient 0.95, $\alpha = 0.05$ and $\alpha/2 = 0.05/2 = 0.025$. From Table II, Appendix B, $z_{0.025} = 1.96$. The 95% confidence interval is:

$$\hat{p}_i \pm z_{0.025}\sqrt{\frac{\hat{p}_i\hat{q}_i}{n}} \Rightarrow 0.376 \pm 1.96\sqrt{\frac{0.376(0.624)}{85}} \Rightarrow 0.376 \pm 0.103 \Rightarrow (0.273,\ \ 0.479)$$

We are 95% confident that the true proportion of Avonex MS patients who are exacerbation-free during a 2-year period is between 0.273 and 0.479.

c. From previous studies, it is known that 26% of the MS patients on a placebo experienced no exacerbations in a 2-year period. Since 0.26 does not fall in the 95% confidence interval, there is evidence that Avonex patients are more likely to have no exacerbations than the placebo patients at $\alpha = .05$.

13.63 a. Some preliminary calculations are:

$$\hat{E}_{11} = \frac{R_1C_1}{n} = \frac{23(320)}{1914} = 3.845 \qquad \hat{E}_{12} = \frac{R_1C_2}{n} = \frac{23(1594)}{1914} = 19.155$$

$$\hat{E}_{21} = \frac{R_2C_1}{n} = \frac{1891(320)}{1914} = 316.155 \qquad \hat{E}_{22} = \frac{R_2C_2}{n} = \frac{1891(1594)}{1914} = 1574.845$$

To determine if defensibility status is dependent on site type, we test:

H_0: Defensibility status and site type are independent
H_a: Defensibility status and site type are dependent

The test statistic is

$$\chi^2 = \sum\frac{\left[n_{ij} - \hat{E}_{ij}\right]^2}{\hat{E}_{ij}} = \frac{[9-3.845]^2}{3.845} + \frac{[14-19.155]^2}{19.155} + \frac{[311-316.155]^2}{316.155}$$
$$+ \frac{[1580-1574.845]^2}{1574.845} = 8.398$$

Since no α was given, we will use $\alpha = 0.05$. The rejection region requires $\alpha = 0.05$ in the upper tail of the χ^2 distribution with $df = (r-1)(c-1) = (2-1)(2-1) = 1$. From Table IV, Appendix B, $\chi^2_{0.05} = 3.84146$. The rejection region is $\chi^2 > 3.84146$.

Since the observed value of the test statistics falls in the rejection region $(\chi^2 = 8.398 > 3.84146)$, H_0 is rejected. There is sufficient evidence to indicate that defensibility status is dependent on site type at $\alpha = 0.05$.

b. Since the p-value is so small $(p = 0.008)$, H_0 is rejected. There is sufficient evidence to indicate that defensibility status is dependent on site type for any reasonable value of α. Fisher's exact test is preferred over the χ^2 because one of the assumptions necessary for the χ^2 test to be valid is that all the expected cell sizes must be greater than or equal to 5. For one cell in this table, the expected cell size is 3.845.

13.65. Some preliminary calculations are:

$$\hat{E}_{11} = \frac{R_1C_1}{n} = \frac{192(115)}{526} = 41.98 \qquad \hat{E}_{12} = \frac{R_1C_2}{n} = \frac{192(411)}{526} = 150.02$$

$$\hat{E}_{21} = \frac{R_2C_1}{n} = \frac{157(115)}{526} = 34.33 \qquad \hat{E}_{22} = \frac{R_2C_2}{n} = \frac{157(411)}{526} = 122.67$$

$$\hat{E}_{31} = \frac{R_3C_1}{n} = \frac{177(115)}{526} = 38.70 \qquad \hat{E}_{32} = \frac{R_3C_2}{n} = \frac{177(411)}{526} = 138.30$$

To determine if the proportion of students diagnosed with MR depends on the IQ test/retest method, we test:

H_0: MR diagnosis and Test/Retest are independent
H_a: MR diagnosis and Test/Retest are dependent

The test statistic is

$$\chi^2 = \sum\sum\frac{\left[n_{ij} - \hat{E}_{ij}\right]^2}{E_{ij}} = \frac{[25-41.98]^2}{41.98} + \frac{[54-34.33]^2}{34.33} + \frac{[36-38.70]^2}{38.70} + \frac{[167-150.02]^2}{150.02}$$

$$+\frac{[103-122.67]^2}{122.67} + \frac{[141-138.30]^2}{138.30} = 23.46$$

The rejection region requires $\alpha = 0.01$ in the upper tail of the χ^2 distribution with $df = (r-1)(c-1) = (3-1)(2-1) = 2$. From Table IV, Appendix B, $\chi^2_{0.01} = 9.21034$. The rejection region is $\chi^2 > 9.21034$.

Since the observed value of the test statistic falls in the rejection region $(\chi^2 = 23.46 > 9.21034)$, H_0 is rejected. There is sufficient evidence to indicate the proportion of students diagnosed with MR depends on the IQ test/retest method at $\alpha = 0.01$.

13.67 a. The observed frequencies for the 3 groups are:

Mail-only – 262
Internet-only – 43
Both – 135

Some preliminary calculations are: $E_1 = E_2 = E_3 = np_{i,0} = 440(1/3) = 146.67$

To determine if the proportions of mail-only, internet-only and both users are different, we test:

$$H_0: p_1 = p_2 = p_3 = p_4 = 1/3$$
$$H_a: \text{At least one } p_i \neq 1/3$$

The test statistic is

$$\chi^2 = \sum \frac{[n_i - E_i]^2}{E_i} = \frac{[262 - 146.67]^2}{146.67} + \frac{[43 - 146.67]^2}{146.67} + \frac{[135 - 146.67]^2}{146.67} = 164.89$$

The rejection region requires $\alpha = 0.05$ in the upper tail of the χ^2 distribution with $df = k - 1 = 3 - 1 = 2$. From Table IV, Appendix B, $\chi^2_{0.05} = 5.99147$. The rejection region is $\chi^2 > 5.99147$.

Since the observed value of the test statistic falls in the rejection region $(\chi^2 = 164.89 > 5.99147)$, H_0 is rejected. There is sufficient evidence to indicate the proportions of mail-only, internet-only and both users are different at $\alpha = 0.05$.

Using MINITAB, a graph of the categories is:

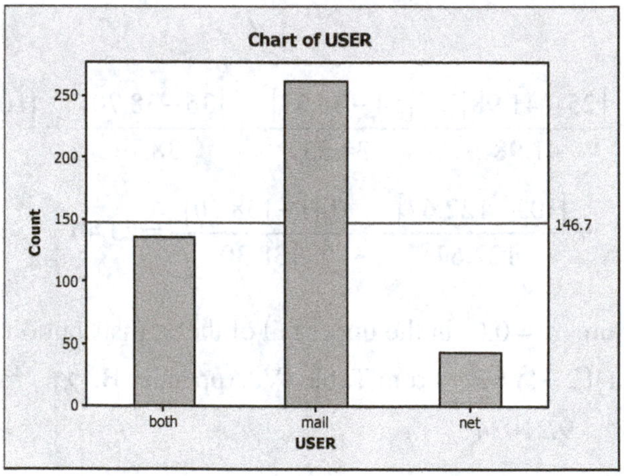

b. **Type of User vs Gender**:

Using MINITAB, the contingency table is:

Tabulated statistics: USER, GENDER

```
Rows: USER   Columns: GENDER

         Female    Male     All

both        104      31     135
          77.04   22.96  100.00
          32.70   25.41   30.68

mail        178      84     262
          67.94   32.06  100.00
          55.97   68.85   59.55

net          36       7      43
          83.72   16.28  100.00
          11.32    5.74    9.77

All         318     122     440
          72.27   27.73  100.00
         100.00  100.00  100.00

Cell Contents:      Count
                    % of Row
                    % of Column

Pearson Chi-Square = 6.797, DF = 2, P-Value = 0.033
Likelihood Ratio Chi-Square = 7.105, DF = 2, P-Value = 0.029
```

To determine if coupon user type is related to gender, we test:

H_0: Type of coupon user and gender are independent
H_0: Type of coupon user and gender are dependent

From the printout, the test statistic is $\chi^2 = 6.797$ and the p-value is $p = 0.033$.

Since the observed p-value is less than $\alpha\left(p = 0.033 < 0.05\right)$, H_0 is rejected. There is sufficient evidence to indicate the type of coupon user is related to gender at $\alpha = 0.05$.

Using MINITAB, a graph of the results is:

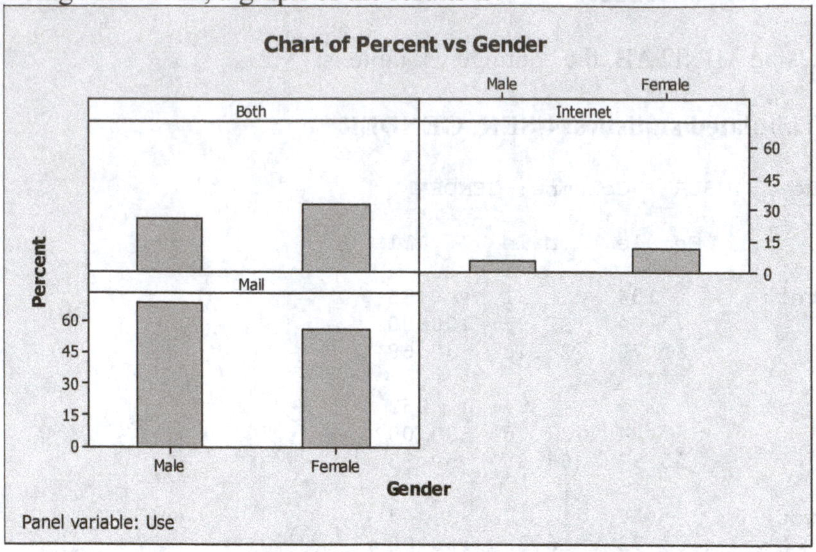

The majority of male and female users use coupons distributed through the mail. However, the proportion of male users who use coupons distributed through the mail is larger than the proportion of females. Very few users use coupons distributed via the Internet, but a higher proportion of females use coupons distributed via the Internet than males.

Type of User vs Education:

Using MINITAB, the contingency table is:

Tabulated statistics: USER, EDUC

```
Rows: USER   Columns: EDUC

          COLL     GRAD      HS     PROF      All

both        62       38      19       16      135
         45.93    28.15   14.07    11.85   100.00
         34.83    28.79   31.67    22.86    30.68

mail        96       85      34       47      262
         36.64    32.44   12.98    17.94   100.00
         53.93    64.39   56.67    67.14    59.55

net         20        9       7        7       43
         46.51    20.93   16.28    16.28   100.00
         11.24     6.82   11.67    10.00     9.77

All        178      132      60       70      440
         40.45    30.00   13.64    15.91   100.00
        100.00   100.00  100.00   100.00   100.00

Cell Contents:      Count
                    % of Row
                    % of Column

Pearson Chi-Square = 6.587, DF = 6, P-Value = 0.361
Likelihood Ratio Chi-Square = 6.786, DF = 6, P-Value = 0.341
```

To determine if coupon user type is related to education, we test:

H_0: Type of coupon user and education are independent
H_0: Type of coupon user and education are dependent

From the printout, the test statistic is $\chi^2 = 6.587$ and the p-value is $p = 0.361$.

Since the observed p-value is not less than $\alpha\left(p = 0.361 \not< 0.05\right)$, H_0 is not rejected. There is insufficient evidence to indicate the type of coupon user is related to education at $\alpha = 0.05$.

Using MINITAB, a graph of the results is:

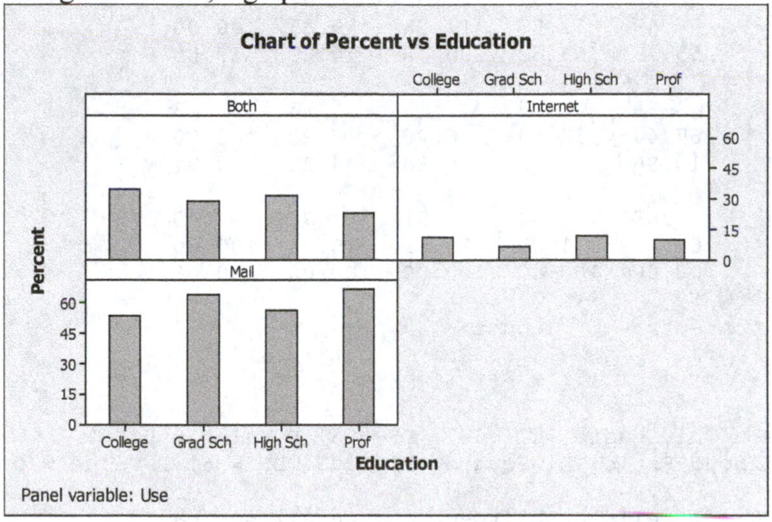

For each category of use, the proportions of each education level are about the same. This corresponds to the conclusion above that coupon use is not related to education level.

Type of User vs Work Status:

Using MINITAB, the contingency table is:

Tabulated statistics: USER, WORK

```
Rows: USER    Columns: WORK

              1        2        3        4       All

both         90       13       17       15       135
          66.67     9.63    12.59    11.11    100.00
          33.71    25.00    33.33    21.43     30.68

mail        148       31       31       52       262
          56.49    11.83    11.83    19.85    100.00
          55.43    59.62    60.78    74.29     59.55

net          29        8        3        3        43
          67.44    18.60     6.98     6.98    100.00
          10.86    15.38     5.88     4.29      9.77

All         267       52       51       70       440
          60.68    11.82    11.59    15.91    100.00
         100.00   100.00   100.00   100.00    100.00

Cell Contents:          Count
                        % of Row
                        % of Column

Pearson Chi-Square = 11.687, DF = 6, P-Value = 0.069
Likelihood Ratio Chi-Square = 12.208, DF = 6, P-Value = 0.057

* NOTE * 1 cells with expected counts less than 5
```

To determine if coupon user type is related to work status, we test:

H_0: Type of coupon user and work status are independent
H_a: Type of coupon user and work status are dependent

From the printout, the test statistic is $\chi^2 = 11.687$ and the *p*-value is $p = 0.069$.

Since the observed *p*-value is not less than $\alpha\left(p = 0.069 \not< 0.05\right)$, H_0 is not rejected. There is insufficient evidence to indicate the type of coupon user is related to work status at $\alpha = 0.05$.

Using MINITAB, a graph of the results is:

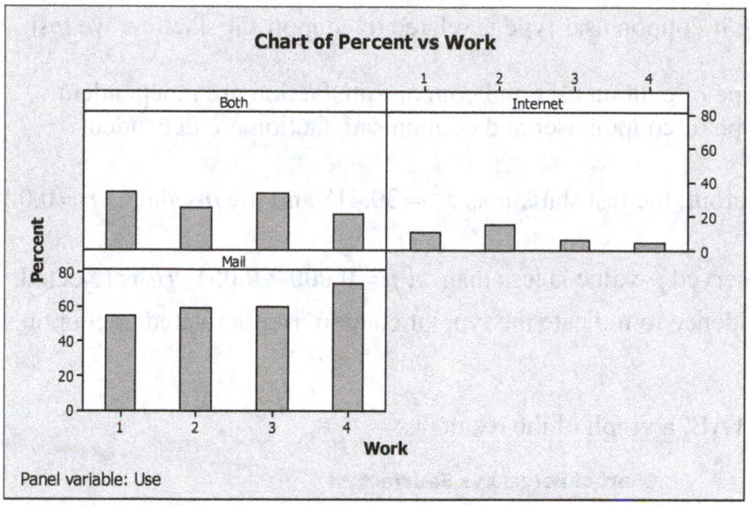

Again, for each category of use, the proportions of each work level are about the same. This corresponds to the conclusion above that coupon use is not related to work level.

Type of User vs Coupon satisfaction:

Using MINITAB, the contingency table is:

Tabulated statistics: USER, SATISF

```
Rows: USER    Columns: SATISF

            No    Some     Yes     All

both         3       9     123     135
          2.22    6.67   91.11  100.00
          8.57   11.25   37.85   30.68

mail        28      62     172     262
         10.69   23.66   65.65  100.00
         80.00   77.50   52.92   59.55

net          4       9      30      43
          9.30   20.93   69.77  100.00
         11.43   11.25    9.23    9.77

All         35      80     325     440
          7.95   18.18   73.86  100.00
        100.00  100.00  100.00  100.00

Cell Contents:      Count
                    % of Row
                    % of Column


Pearson Chi-Square = 30.418, DF = 4, P-Value = 0.000
Likelihood Ratio Chi-Square = 34.934, DF = 4, P-Value = 0.000

* NOTE * 1 cells with expected counts less than 5
```

To determine if coupon user type is related to coupon satisfaction, we test:

H_0: Type of coupon user and coupon satisfaction are independent
H_0: Type of coupon user and coupon satisfaction are dependent

From the printout, the test statistic is $\chi^2 = 30.418$ and the p-value is $p = 0.000$.

Since the observed p-value is less than $\alpha\left(p = 0.000 < 0.05\right)$, H_0 is rejected. There is sufficient evidence to indicate the type of coupon user is related to coupon satisfaction at $\alpha = 0.05$.

Using MINITAB, a graph of the results is:

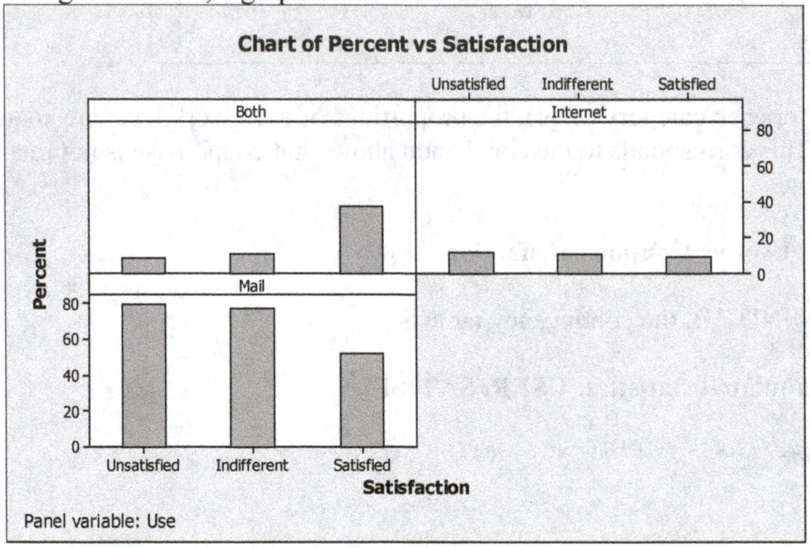

Again, very few users use coupons distributed via the Internet. The proportions of the different levels of satisfaction are about the same for this category of coupon use. For coupons distributed through the mail, the proportions of users who were unsatisfied or indifferent are larger than the proportion satisfied. However, for users of coupons distributed through both the mail and the Internet, the proportion of users who were satisfied is greater than the proportions of users unsatisfied or indifferent.

13.69 a. Using MINITAB, the results of the analyses are:

Tabulated statistics: SHSS, CAHS

```
Using frequencies in Fr

Rows: SHSS   Columns: CAHS

            L       M       H      V     All

L          32      14       2      0      48
        18.09   16.25   11.45   2.22   48.00

M          11      14       6      0      31
        11.68   10.49    7.39   1.43   31.00

H           6      14      19      3      42
        15.83   14.22   10.02   1.94   42.00

V           0       2       4      3       9
         3.39    3.05    2.15   0.42    9.00

All        49      44      31      6     130
        49.00   44.00   31.00   6.00  130.00

Cell Contents:       Count
                     Expected count

Pearson Chi-Square = 60.103, DF = 9
Likelihood Ratio Chi-Square = 59.638, DF = 9

* WARNING * 1 cells with expected counts less than 1
* WARNING * Chi-Square approximation probably invalid

* NOTE * 7 cells with expected counts less than 5
```

From the printout, 7 cells have expected counts less than 5. In order for the test to be valid, all of the cells should have expected counts greater than 5. Thus, we should not proceed with the analysis.

b. Combining the High and Very High categories, the new table is:

SHSS: C LEVEL / **CAHS LEVEL**

	Low	**Medium**	**High/Very High**
Low	32	14	2
Medium	11	14	6
High/Very High	6	16	29

c. The expected cell counts are:

$$\hat{E}_{11} = \frac{R_1C_1}{n} = \frac{48(49)}{130} = 18.0923 \qquad \hat{E}_{21} = \frac{R_2C_1}{n} = \frac{31(49)}{130} = 11.6846$$

$$\hat{E}_{12} = \frac{R_1C_2}{n} = \frac{48(44)}{130} = 16.2462 \qquad \hat{E}_{22} = \frac{R_2C_2}{n} = \frac{31(44)}{130} = 10.4923$$

$$\hat{E}_{13} = \frac{R_1C_3}{n} = \frac{48(37)}{130} = 13.6615 \qquad \hat{E}_{23} = \frac{R_2C_3}{n} = \frac{31(37)}{130} = 8.8231$$

$$\hat{E}_{31} = \frac{R_3C_1}{n} = \frac{51(49)}{130} = 19.2231 \qquad \hat{E}_{32} = \frac{R_3C_2}{n} = \frac{51(44)}{130} = 17.2615$$

$$\hat{E}_{33} = \frac{R_3C_3}{n} = \frac{51(37)}{130} = 14.5154$$

Since all of the expected cell counts are now 5 or greater, the assumption is met.

d. To determine whether CAHS Levels and SHSS:C Levels are dependent, we test:

H_0: CAHS Levels and SHSS:C Levels are independent
H_a: CAHS Levels and SHSS:C Levels are dependent

The test statistic is

$$\chi^2 = \sum\sum \frac{\left[n_{ij} - \hat{E}_{ij}\right]^2}{\hat{E}_{ij}} = \frac{(32-18.0923)^2}{18.0923} + \frac{(14-16.2462)^2}{16.2462} + \frac{(2-13.6615)^2}{13.6615} + \frac{(11-11.6846)^2}{11.6846}$$

$$+ \frac{(14-10.4923)^2}{10.4923} + \frac{(6-8.8231)^2}{8.8231} + \frac{(6-19.2231)^2}{19.2231} + \frac{(16-17.2615)^2}{17.2615} + \frac{(29-14.5154)^2}{14.5154} = 46.71$$

The rejection region requires $\alpha = 0.05$ in the upper tail of the χ^2 distribution with $df = (r-1)(c-1) = (3-1)(3-1) = 4$. From Table IV, Appendix B, $\chi^2_{0.05} = 9.48773$. The rejection region is $\chi^2 > 9.48773$.

Since the observed value of the test statistic falls in the rejection region $\left(\chi^2 = 46.70 > 9.48773\right)$, H_0 is rejected. There is sufficient evidence to indicate that CAHS Levels and SHSS:C Levels are dependent at $\alpha = 0.05$.

13.71 a.

$$\chi^2 = \sum \frac{[n_i - E_i]^2}{E_i} = \frac{(26-23)^2}{23} + \frac{(146-136)^2}{136} + \frac{(361-341)^2}{341} + \frac{(143-136)^2}{136} + \frac{(13-23)^2}{23}$$

$$= 9.647$$

b. From Table IV, Appendix B, with $df = 5$, $\chi^2_{0.05} = 11.0705$.

c. No. Since the observed value of the test statistic does not fall in the rejection region $\left(\chi^2 = 9.647 \not> 11.0705\right)$, H_0 is not rejected. There is insufficient evidence to indicate the salary distribution is nonnormal for $\alpha = 0.05$.

d. The p-value is $p = P\left(\chi^2 \geq 9.647\right)$.

Using Table IV, Appendix B, with $df = 5$, $0.05 < P\left(\chi^2 \geq 9.647\right) < 0.10$.

13.73 Using MINITAB, the results of the analyses are:

Tabulated statistics: Candidate, Time

```
Using frequencies in Fr

Rows: Candidate    Columns: Time

                1       2       3       4       5       6     All

Coppin         55      51     109      98      88     104     505
            13.89   13.56   13.37   13.98   14.08   14.04   13.82
             54.7    52.0   112.6    96.9    86.4   102.4   505.0

Montes        133     117     255     211     186     227    1129
            33.59   31.12   31.29   30.10   29.76   30.63   30.90
            122.4   116.2   251.8   216.6   193.1   229.0  1129.0

Smith         208     208     451     392     351     410    2020
            52.53   55.32   55.34   55.92   56.16   55.33   55.28
            218.9   207.9   450.5   387.5   345.5   409.6  2020.0

All           396     376     815     701     625     741    3654
           100.00  100.00  100.00  100.00  100.00  100.00  100.00
            396.0   376.0   815.0   701.0   625.0   741.0  3654.0

Cell Contents:      Count
                    % of Column
                     Expected count

Pearson Chi-Square = 2.284, DF = 10, P-Value = 0.994
Likelihood Ratio Chi-Square = 2.272, DF = 10, P-Value = 0.994
```

To determine if time period and votes received by the candidates are independent, we test:

H_0: Votes received by candidates and time period are independent
H_a: Votes received by candidates and time period are dependent

From the printout, the test statistic is $\chi^2 = 2.284$ and the p-value is $p = 0.994$.

Since the p-value is so large, H_0 would not be rejected for any reasonable value of α. There is insufficient evidence to indicate that the votes received by the candidates are dependent on the time period. In other words, there is no evidence that the percentages of votes received by the candidates change over time. If one looks at the column percentages (the second number in each cell above), the values are very similar for each time period.

A graph of the column percents is:

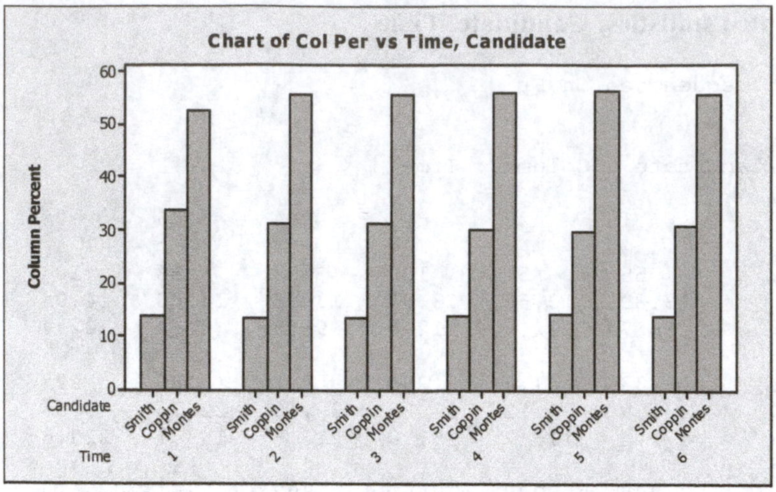

As one can see from the graph, the percent of votes received by the candidates are essentially the same for each time period.

All of the above indicates that the probability of a candidate receiving votes is independent of the time period. However, this does not necessarily imply a rigged election. If each time period is considered a random sample of voters, then the percentage of votes received by each candidate should be similar to the actual percentage of voters who favor each candidate.

Chapter 14

Nonparametric Statistics

14.1 The sign test is preferred to the t-test when the population from which the sample is selected is not normal.

14.3 a. $P(x \geq 6) = 1 - P(x \leq 5) = 1 - 0.937 = 0.063$

b. $P(x \geq 5) = 1 - P(x \leq 4) = 1 - 0.500 = 0.500$

c. $P(x \geq 8) = 1 - P(x \leq 7) = 1 - 0.996 = 0.004$

d. $P(x \geq 10) = 1 - P(x \leq 9) = 1 - 0.849 = 0.151$

$\mu = np = 15(0.5) = 7.5$ and $\sigma = \sqrt{npq} = \sqrt{15(0.5)(0.5)} = 1.9365$

$$P(x \geq 10) \approx P\left(z \geq \frac{(10-0.5)-7.5}{1.9365}\right) = P(z \geq 1.03) = 0.5 - 0.3485 = 0.1515$$

e. $P(x \geq 15) = 1 - P(x \leq 14) = 1 - 0.788 = 0.212$

$\mu = np = 25(0.5) = 12.5$ and $\sigma = \sqrt{npq} = \sqrt{25(0.5)(0.5)} = 2.5$

$$P(x \geq 15) \approx P\left(z \geq \frac{(15-0.5)-12.5}{2.5}\right) = P(z \geq 0.80) = 0.5 - 0.2881 = 0.2119$$

14.5 To determine if the median is greater than 80, we test:

$H_0: \eta = 80$
$H_a: \eta > 80$

The test statistic is $S = S_A =$ {number of measurements greater than 80} = 16.

The $p-$value $= P(x \geq 16)$ where x is a binomial random variable with $n = 25$ and $p = 0.5$. From Table I,

$p-$value $= P(x \geq 16) = 1 - P(x \leq 15) = 1 - 0.885 = 0.115$

Since the $p-$value $= 0.115 > \alpha = 0.10$, H_0 is not rejected. There is insufficient evidence to indicate the median is greater than 80 at $\alpha = 0.10$.

We must assume the sample was randomly selected from a continuous probability distribution.
Note: Since $n \geq 10$, we could use the large-sample approximation.

14.7 a. To determine if the median molar depth of all cheek teeth differs from 15, we test:

$$H_0: \eta = 15$$
$$H_a: \eta \neq 15$$

b. The sign test is appropriate because the data do not appear to be normally distributed and the sample size is only $n = 18$.

c. The test statistic is $S = S_A = 14$.

d. From the printout, the *p*-value is $p = 0.0309$. Since the $p-\text{value} = 0.0309 < \alpha = 0.05$, H_0 is rejected. There is sufficient evidence to indicate the median molar depth of all cheek teeth differs from 15 at $\alpha = 0.05$.

14.9 a. To determine if female college students have a median emotional scale score higher than 2.8, we test:

$$H_0: \eta = 2.8$$
$$H_a: \eta > 2.8$$

b. The test statistics is $S = S_A = \{\text{number of observations greater than } 2.8\} = 21$. Because $n = 30$, we will use the large sample approximation:

$$z = \frac{(S-0.5)-0.5n}{0.5\sqrt{n}} = \frac{(21-0.5)-0.5(30)}{0.5\sqrt{30}} = 2.01$$

c. The $p-\text{value} = P(z \geq 2.01) = 0.5 - 0.4778 = 0.222$ using Table I, Appendix B.

d. Since the $p-\text{value} = 0.222 \nless \alpha = 0.01$, H_0 is not rejected. There is insufficient evidence to indicate that female college students have a median emotional scale score higher than 2.8 at $\alpha = 0.01$.

14.11 To determine if the median level of MTBE in New Hampshire groundwater wells is less than 0.5 micrograms per liter, we test:

$$H_0: \eta = 0.5$$
$$H_a: \eta < 0.5$$

The test statistic is $S = S_B = \{\text{number of observations less than } 0.5\} = 180$ and the *p*-value is $p-\text{value} = 0.0000$.

Since the *p*-value is so small, H_0 is rejected. There is sufficient evidence to indicate the median level of MTBE in New Hampshire groundwater wells is less than 0.5 for any reasonable value of α.

14.13 a. If the data do not come from a normal distribution, then the test performed in Exercise 8.59 is not valid.

b. If the data are not normal, then we could use the sign test for the median with

$$H_0: \eta = 95$$
$$H_a: \eta \neq 95$$

c. S_B = number of observations less than 95 = 5 and
S_A = number of observations greater than 95 = 2.
The test statistic is S = larger of S_A and S_B = 5.

d. The $p-\text{value} = 2P(x \geq 5)$ where x is a binomial random variable with $n = 7$ and $p = 0.5$. From Table I,

$$p-\text{value} = 2P(x \geq 5) = 2(1 - P(x \leq 4)) = 2(1 - 0.773) = 0.454$$

e. In Exercise 8.59, we used $\alpha = 0.10$. Since the $p-\text{value} = 0.454 \nless \alpha = 0.10$, H_0 is not rejected. There is insufficient evidence to indicate the median trap spacing measurement differs from 95 at $\alpha = 0.10$.

14.15 To determine if the median ratio of repetition differs from 0.5, we test:

$$H_0: \eta = 0.5$$
$$H_a: \eta \neq 0.5$$

S_B = number of observations less than 0.5 = 6 and
S_A = number of observations greater than 0.5 = 2.

The test statistic is S = larger of S_A and S_B = 6.

The $p-\text{value} = P(x \geq 6)$ where x is a binomial random variable with $n = 8$ and $p = 0.5$. From Table I,

$$p-\text{value} = P(x \geq 6) = 1 - P(x \leq 5) = 1 - 0.855 = 0.145$$

We selected $\alpha = 0.05$ for Exercise 8.67, so we will use $\alpha = 0.05$ for this Exercise. Since the $p-\text{value} = 0.145 > \alpha = 0.05$, H_0 is not rejected. There is insufficient evidence to indicate the median ratio of repetition differs from 0.5at $\alpha = 0.05$.

14.17 To determine if the population median skidding distance is more than 400 meters, we test:

$$H_0: \eta = 400$$
$$H_a: \eta > 400$$

The test statistic is $S = S_A$ = {number of measurements greater than 400} = 8.

The $p-\text{value} = P(x \geq 8)$ where x is a binomial random variable with $n = 19$ and $p = 0.5$. (One observation is eliminated because it is equal to 400.) Using MINITAB,

$$p-\text{value} = P(x \geq 8) = 1 - P(x \leq 7) = 1 - 0.180 = 0.820$$

Since the $p-\text{value} = 0.820 \not< \alpha = 0.10$, H_0 is not rejected. There is insufficient evidence to indicate the population median skidding distance is more than 400 meters at $\alpha = 0.10$.

14.19 The statement "If the rank sum for sample 1 is much larger than the rank sum for sample 2 when $n_1 = n_2$, then the distribution of population1 is likely to be shifted to the right of the distribution of population 2." is true.

14.21 a. The test statistic is T_1, the rank sum of population A (because $n_1 < n_2$).

The rejection region is $T_1 \leq 41$ or $T_1 \geq 71$, from Table IX, Appendix B, with $n_1 = 7$, $n_2 = 8$, and $\alpha = 0.10$.

b. The test statistic is T_1, the rank sum of population A (because $n_1 = n_2$).

The rejection region is $T_1 \geq 50$, from Table IX, Appendix B, with $n_1 = 6$, $n_2 = 6$, and $\alpha = 0.05$.

c. The test statistic is T_1, the rank sum of population A (because $n_1 < n_2$).

The rejection region is $T_1 \leq 43$, from Table IX, Appendix B, with $n_1 = 7$, $n_2 = 10$, and $\alpha = 0.025$.

d. Since $n_1 = n_2 = 20$, the test statistic is $z_c = \dfrac{T_1 - \dfrac{n_1(n_1+n_2+1)}{2}}{\sqrt{\dfrac{n_1 n_2(n_1+n_2+1)}{12}}}$

The rejection region is $z < -z_{\alpha/2}$ or $z > z_{\alpha/2}$. For $\alpha = 0.05$ and $\alpha/2 = 0.05/2 = 0.025$, $z_{0.025} = 1.96$ from Table II, Appendix B. The rejection region is $z < -1.96$ or $z > 1.96$.

14.23 To determine if distribution A is shifted to the left of distribution B, we test:

H_0: The probability distributions of treatments A and B are identical
H_a: The probability distribution of treatment A lies to the left of that for treatment B

The test statistic is

$$z_c = \frac{T_1 - \dfrac{n_1(n_1+n_2+1)}{2}}{\sqrt{\dfrac{n_1 n_2(n_1+n_2+1)}{12}}} = \frac{173 - \dfrac{15(15+15+1)}{2}}{\sqrt{\dfrac{15(15)(15+15+1)}{12}}} = \frac{-59.5}{24.1091} = -2.47$$

The rejection region requires $\alpha = 0.05$ in the lower tail of the z distribution. From Table II, Appendix B, $z_{0.05} = 1.645$. The rejection region is $z < -1.645$.

Since the observed value of the test statistic falls in the rejection region $(z = -2.47 < -1.645)$, H_0 is rejected. There is sufficient evidence to conclude that distribution A is shifted to the left of distribution B for $\alpha = 0.05$.

14.25

Sample from Population 1 (A)	Rank	Sample from Population 2 (B)	Rank
15	13	5	2.5
10	8.5	12	10.5
12	10.5	9	6.5
16	14	9	6.5
13	12	8	4.5
8	4.5	4	1
	$T_1 = 62.5$	5	2.5
		10	8.5
			$T_2 = 42.5$

a. To determine if there is a shift in location for the two distributions, we test:

H_0: The two sampled populations have identical probability distributions
H_a: The probability distribution for population A is shifted to the left or to the right of that for B

The test statistic is $T_1 = 62.5$ since sample A has the smallest number of measurements.

The null hypothesis will be rejected if $T_1 \le T_L$ or $T_1 \ge T_U$ where T_L and T_U correspond to $\alpha = 0.05$ (two-tailed), $n_1 = 6$ and $n_2 = 8$. From Table IX, Appendix B, $T_L = 29$ and $T_U = 61$.

Reject H_0 if $T_1 \le 29$ or $T_1 \ge 61$.

Since the observed value of the test statistic falls in the rejection region $(T_1 = 62.5 \ge 61)$, H_0 is rejected. There is sufficient evidence to indicate population A is shifted to the left or right of population B at $\alpha = 0.05$.

b. H_0: The two sampled populations have identical probability distributions
H_a: The probability distribution for population A is shifted to the right of population B

The test statistic remains $T_1 = 62.5$.

The null hypothesis will be rejected if $T_1 \geq T_U$ where T_U corresponds to $\alpha = 0.05$ (one-tailed), $n_1 = 6$ and $n_2 = 8$. From Table IX, Appendix B, $T_U = 58$.

Reject H_0 if $T_1 \geq 58$.

Since the observed value of the test statistic falls in the rejection region $(T_1 = 62.5 \geq 58)$, H_0 is rejected. There is sufficient evidence to indicate population A is shifted to the right of population B at $\alpha = 0.05$.

14.27 a. To determine if the median response for the two groups of students differs, we test:

H_0: The distributions of the responses for the two groups of students are identical

H_a: The distribution of the responses for the students in the texting section is shifted to the right of that for the students in the non-texting section

b. Since the *p*-value is so small $(p = 0.004)$, H_0 is rejected. There is sufficient evidence to indicate that the median response of the students in the texting section is shifted to the right or left of that for the students in the non-texting section for any reasonable value of α. Since the median for the students in the texting section is greater than the median for the students in the nob-texting section, it indicates that students in the texting section had more of a preference for face-to-face meetings with their professors.

14.29 a. The appropriate test would be the Wilcoxon Rank Sum Test.

b. To determine if the low-handicapped golfers tend to have higher X-factors than high-handicapped golfers, we test:

H_0: The distributions of the X-factors for the two groups of golfers are identical

H_a: The distribution of the X-factors for the low-handicapped golfers is shifted to the right of that for the high-handicapped golfers

c. Let T_1 = rank sum of low-handicapped golfers and T_2 = rank sum of high-handicapped golfers. Since $n_1 = 8$ and $n_2 = 7$, the test statistic is T_2. The rejection region is $T_2 \leq T_L$ where $\alpha = 0.05$ (one-tailed). From Table IX, Appendix B, $T_L = 41$. The rejection region is $T_2 \leq 41$.

d. Since the *p*-value is not less than $\alpha\left(p = 0.487 \not< 0.05\right)$, H_0 is not rejected. There is insufficient evidence to indicate the low-handicapped golfers tend to have higher X-factors than high-handicapped golfers at $\alpha = 0.05$.

14.31 a. To determine if the distribution of FNE scores for bulimic students is shifted above the corresponding distribution for female students with normal eating habits, we test:

H_0: The distribution of FNE scores for bulimic students is the same as the corresponding distribution for female students with normal eating habits

H_a: The distribution of FNE scores for bulimic students is shifted above the corresponding distribution for female students with normal eating habits

b. The data ranked are:

Bulimic Students	Rank	Normal Students	Rank
21	21.5	13	8.5
13	8.5	6	1
10	4.5	16	13.5
20	19.5	13	8.5
25	25	8	3
19	17	19	17
16	13.5	23	23
21	21.5	18	15
24	24	11	6
13	8.5	19	17
14	11	7	2
	$T_1 = 174.5$	10	4.5
		15	12
		20	19.5
			$T_2 = 150.5$

c. The sum of the ranks of the 11 FNE scores for bulimic students is $T_1 = 174.5$.

d. The sum of the ranks of the 14 FNE scores for normal students is $T_2 = 150.5$.

e. The test statistic is $z_c = \dfrac{T_1 - \dfrac{n_1(n_1+n_2+1)}{2}}{\sqrt{\dfrac{n_1 n_2(n_1+n_2+1)}{12}}} = \dfrac{174.5_1 - \dfrac{11(11+14+1)}{2}}{\sqrt{\dfrac{11(14)(11+14+1)}{12}}} = 1.72$

The rejection region requires $\alpha = 0.10$ in the upper tail of the z distribution. From Table II, Appendix B, $z_{0.10} = 1.28$. The rejection region is $z > 1.28$.

f. Since the observed value of the test statistic falls in the rejection region $(z_c = 1.72 > 1.28)$, H_0 is rejected. There is sufficient evidence to indicate that the distribution of FNE scores for bulimic students is shifted above the corresponding distribution for female students with normal eating habits at $\alpha = 0.10$.

14.33 To determine if the distribution of MTBE levels in public wells is shifted above or below that of private wells, we test:

H_0: The distributions of MTBE levels in public and private wells are identical

H_a: The distribution of MTBE levels in public wells is shifted above or below that of private wells

Since the *p*-value is not small $(p = 0.1109)$, H_0 is not rejected. There is insufficient evidence to indicate that distribution of MTBE levels in public wells is shifted above or below that of private wells for any value of $\alpha < 0.1109$.

To determine if the distribution of MTBE levels in bedrock aquifers is shifted above or below that of unconsolidated aquifers, we test:

H_0: The distributions of MTBE levels in bedrock and unconsolidated aquifers are identical

H_a: The distribution of MTBE levels in bedrock aquifers is shifted above or below that of unconsolidated aquifers

Since the *p*-value is small $(p = 0.0337)$, H_0 is rejected. There is sufficient evidence to indicate that distribution of MTBE levels in bedrock aquifers is shifted above or below that of unconsolidated aquifers for any value of $\alpha > 0.0337$.

14.35 First, we rank the data:

Honey Dosage				DM Dosage			
Honey	Rank	Honey	Rank	DM	Rank	DM	Rank
12	52	12	52	4	5	6	11.5
11	44	8	21	6	11.5	8	21
15	65	12	52	9	28	12	52
11	44	9	28	4	5	12	52
10	37	11	44	7	16	4	5
13	59.5	15	65	7	16	12	52
10	37	10	37	7	16	13	59.5
4	5	15	65	9	28	7	16
15	65	9	28	12	52	10	37
16	68	13	59.5	10	37	13	59.5
9	28	8	21	11	44	9	28
14	62	12	52	6	11.5	4	5
10	37	10	37	3	1	4	5
6	11.5	8	21	4	5	10	37
10	37	9	28	9	28	15	65
8	21	5	9	12	52	9	28
11	44	12	52	7	16		
12	52						
			$T_1 = 1440.5$				$T_2 = 905.5$

To determine if honey is a preferable treatment for the cough and sleep difficulty, we test:

H_0: The distributions of the cough improvement scores for the two groups of children are identical

H_a: The distribution of the cough improvement scores for the honey dosage children is shifted to the right of that for the DM dosage children

The test statistic is

$$z_c = \frac{T_1 - \dfrac{n_1(n_1+n_2+1)}{2}}{\sqrt{\dfrac{n_1n_2(n_1+n_2+1)}{12}}} = \frac{1440.5 - \dfrac{35(35+33+1)}{2}}{\sqrt{\dfrac{35(33)(35+33+1)}{12}}} = \frac{233}{81.4939} = 2.86$$

The rejection region requires $\alpha = 0.05$ in the upper tail of the z distribution. From Table II, Appendix B, $z_{0.05} = 1.645$. The rejection region is $z > 1.645$.

Since the observed value of the test statistic falls in the rejection region $(z_c = 2.86 > 1.645)$, H_0 is rejected. There is sufficient evidence to indicate that honey is a preferable treatment for the cough and sleep difficulty over DM dosage at $\alpha = 0.05$.

14.37 a. Using MINITAB, the results are:

Mann-Whitney Test and CI: Nurses-4, Doctors-4

```
             N  Median
Nurses-4    31   3.000
Doctors-4   46   5.000


Point estimate for η1 - η2 is -1.000
95.1 Percent CI for η1 - η2 is (-2.001,-0.000)
W = 1007.5
Test of η1 = η2 vs η1 ≠ η2 is significant at 0.0368
The test is significant at 0.0334 (adjusted for ties)
```

To determine if the response distributions differ for nurses and doctors on question 4, we test:

H_0: The distributions of responses on Question 4 for nurses and doctors are identical

H_a: The distribution of the responses on Question 4 for nurses is shifted to the right or left of that for doctors

Since the p-value is less than $\alpha (p = 0.0368 < 0.05)$, H_0 is rejected. There is sufficient evidence to indicate that the response distributions differ for nurses and doctors on question 4 at $\alpha = 0.05$.

b. Using MINITAB, the results are:

Mann-Whitney Test and CI: Nurses-5, Doctors-5

```
             N  Median
Nurses-5    31   4.000
Doctors-5   46   5.000

Point estimate for η1 - η2 is -1.000
95.1 Percent CI for η1 - η2 is (-2.000,-1.000)
W = 872.0
```

```
Test of η1 = η2 vs η1 ≠ η2 is significant at 0.0005
The test is significant at 0.0003 (adjusted for ties)
```

To determine if the response distributions differ for nurses and doctors on question 5, we test:

H_0: The distributions of responses on Question 5 for nurses and doctors are identical

H_a: The distribution of the responses on Question 5 for nurses is shifted to the right or left of that for doctors

Since the *p*-value is less than $\alpha\left(p = 0.0005 < 0.05\right)$, H_0 is rejected. There is sufficient evidence to indicate that the response distributions differ for nurses and doctors on question 5 at $\alpha = 0.05$.

14.39 a. Since the data can take on a very limited number of values, the data are probably not from a normal distribution.

b. Using SAS, the output for the Wilcoxon test is:

```
                     The NPAR1WAY Procedure

           Wilcoxon Scores (Rank Sums) for Variable GSHWS
                     Classified by Variable COND

                        Sum of      Expected       Std Dev          Mean
COND          N         Scores      Under H0      Under H0         Score
--------------------------------------------------------------------------
ATIPS        98         8039.0       11123.0    476.284142     82.030612
TIPS        128        17612.0       14528.0    476.284142    137.593750

                 Average scores were used for ties.


                      Wilcoxon Two-Sample Test

                    Statistic                8039.0000

                    Normal Approximation
                    Z                          -6.4741
                    One-Sided Pr <  Z           <.0001
                    Two-Sided Pr > |Z|          <.0001

                    t Approximation
                    One-Sided Pr <  Z           <.0001
                    Two-Sided Pr > |Z|          <.0001

            Z includes a continuity correction of 0.5.


                        Kruskal-Wallis Test

                    Chi-Square                 41.9273
                    DF                               1
                    Pr > Chi-Square             <.0001
```

To determine if there was a difference in the level of involvement in science homework assignments between TIPS and ATIPS students, we test:

H_0: The distributions of the involvement scores for science homework for the two groups have the same location

H_a: The distributions of the involvement scores for science homework for the TIPS students is shifted to the right or left of that for the ATIPS students

From the printout, the test statistic is $z = -6.4741$ and the p-value is $p < 0.0001$. Since the p-value is less than $\alpha\left(p < 0.0001 < 0.05\right)$, H_0 is rejected. There is sufficient evidence to indicate that there is a difference in the level of involvement in science homework assignments between TIPS and ATIPS students at $\alpha = 0.05$.

c. Using SAS, the output for the Wilcoxon test is:

```
                    The NPAR1WAY Procedure

          Wilcoxon Scores (Rank Sums) for Variable MTHHWS
                    Classified by Variable COND

                     Sum of      Expected       Std Dev          Mean
COND        N        Scores      Under H0       Under H0         Score
--------------------------------------------------------------------------
ATIPS      98       10936.0      11123.0     472.485707     111.591837
TIPS      128       14715.0      14528.0     472.485707     114.960938

                 Average scores were used for ties.


                     Wilcoxon Two-Sample Test

                  Statistic                10936.0000

                  Normal Approximation
                  Z                           -0.3947
                  One-Sided Pr <  Z            0.3465
                  Two-Sided Pr > |Z|           0.6930

                  t Approximation
                  One-Sided Pr <  Z            0.3467
                  Two-Sided Pr > |Z|           0.6934
             Z includes a continuity correction of 0.5.


                       Kruskal-Wallis Test

                  Chi-Square                   0.1566
                  DF                                1
                  Pr > Chi-Square              0.6923
```

To determine if there was a difference in the level of involvement in mathematics homework assignments between TIPS and ATIPS students, we test:

H_0: The distributions of the involvement scores for mathematics homework for the two groups have the same location

H_a: The distributions of the involvement scores for mathematics homework for the TIPS students is shifted to the right or left of that for the ATIPS students

From the printout, the test statistic is $z = -0.3947$ and the p-value is $p = 0.6930$. Since the p-value is not less than $\alpha\left(p = 0.6930 \not< 0.05\right)$, H_0 is not rejected. There is insufficient evidence to indicate that there is a difference in the level of involvement in mathematics homework assignments between TIPS and ATIPS students at $\alpha = 0.05$.

d. Using SAS, the output for the Wilcoxon test is:

```
                    The NPAR1WAY Procedure

          Wilcoxon Scores (Rank Sums) for Variable LAHWS
                    Classified by Variable COND

                      Sum of       Expected       Std Dev          Mean
COND          N       Scores       Under H0       Under H0         Score
--------------------------------------------------------------------------
ATIPS        98      10496.0        11123.0     464.844149    107.102041
TIPS        128      15155.0        14528.0     464.844149    118.398438

                Average scores were used for ties.


                    Wilcoxon Two-Sample Test

               Statistic                  10496.0000

               Normal Approximation
               Z                             -1.3478
               One-Sided Pr <  Z              0.0889
               Two-Sided Pr > |Z|             0.1777

               t Approximation
               One-Sided Pr <  Z              0.0895
               Two-Sided Pr > |Z|             0.1791

          Z includes a continuity correction of 0.5.


                      Kruskal-Wallis Test

               Chi-Square                     1.8194
               DF                                  1
               Pr > Chi-Square                0.1774
```

To determine if there was a difference in the level of involvement in language arts homework assignments between TIPS and ATIPS students, we test:

H_0: The distributions of the involvement scores for language arts homework for the two groups have the same location
H_a: The distributions of the involvement scores for language arts homework for the TIPS students is shifted to the right or left of that for the ATIPS students

From the printout, the test statistic is $z = -1.3478$ and the *p*-value is $p = 0.1777$. Since the *p*-value is not less than $\alpha\left(p = 0.1777 \not< 0.05\right)$, H_0 is not rejected. There is insufficient evidence to indicate that there is a difference in the level of involvement in language arts homework assignments between TIPS and ATIPS students at $\alpha = 0.05$.

14.41 We assume that the probability distribution of differences is continuous so that the absolute differences will have unique ranks. Although tied (absolute) differences can be assigned average ranks, the number of ties should be small relative to the number of observations to assure validity.

14.43 a. The hypotheses are:

H_0: The two sampled populations have identical probability distributions
H_a: The probability distributions for population A is shifted to the right of that for population B

b. Some preliminary calculations are:

Treatment A	Treatment B	Difference A − B	Rank of Absolute Difference
54	45	9	5
60	45	15	10
98	87	11	7
43	31	12	9
82	71	11	7
77	75	2	2.5
74	63	11	7
29	30	−1	1
63	59	4	4
80	82	−2	2.5
			$T_- = 3.5$

The test statistic is $T_- = 3.5$.

The rejection region is $T_- \le 8$, from Table X, Appendix B, with $n = 10$ and $\alpha = 0.025$.

Since the observed value of the test statistic falls in the rejection region $(T_- = 3.5 \le 8)$, H_0 is rejected. There is sufficient evidence to indicate the responses for A tend to be larger than those for B at $\alpha = 0.025$.

14.45 a. H_0: The two sampled populations have identical probability distributions
H_a: The probability distribution for population A is located to the right of that for population B

b. The test statistic is:

$$z_c = \frac{T_+ - \frac{n(n+1)}{4}}{\sqrt{\frac{n(n+1)(2n+1)}{24}}} = \frac{354 - \frac{30(30+1)}{4}}{\sqrt{\frac{30(30+1)(60+1)}{24}}} = \frac{121.5}{48.6184} = 2.499$$

The rejection region requires $\alpha = 0.05$ in the upper tail of the z distribution. From Table II, Appendix B, $z_{0.05} = 1.645$. The rejection region is $z > 1.645$.

Since the observed value of the test statistic falls in the rejection region $(z_c = 2.499 > 1.645)$, H_0 is rejected. There is sufficient evidence to indicate population A is located to the right of that for population B at $\alpha = 0.05$.

c. The p-value is $p = P(z \ge 2.499) = 0.5 - 0.4938 = 0.0062$ (using Table II, Appendix B).

d. The necessary assumptions are:

1. The sample of differences is randomly selected from the population of differences.
2. The probability distribution from which the sample of paired differences is drawn is continuous.

14.47 a. Since two measurements were obtained from each patient (before and after), the measurements are not independent of each other. Each "before" measurement is paired with an "after" measurement.

b. The distributions for both before and after treatment are skewed to the right and the spread of the distribution before treatment is much larger than the spread after treatment. The distributions are not normal.

c. Since the *p*-value is so small $(p < 0.0001)$, H_0 would be rejected. There is sufficient evidence to indicate the ichthyotherapy was effective in treating psoriasis for any reasonable value of α.

14.49 a. From the printout, the test statistic is $z = -4.638$.

b. To determine if handling a museum object has a positive impact on a sick patient's wellbeing, we test:

H_0: The distributions of patient's health status are identical before and after handling museum objects
H_a: The distribution of patient's health status after handling museum objects is shifted to the right of that before

The test statistic is $z = -4.638$ and the *p*-value is $p = 0.000 / 2 = 0.000$.

Since the *p*-value is less than $\alpha\left(p < 0.000 < 0.01\right)$, H_0 is rejected. There is sufficient evidence to indicate handling a museum object has a positive impact on a sick patient's well-being at $\alpha = 0.01$.

14.51 a. To determine if the distributions of the FSI scores for good and average readers differ in location, we test:

H_0: The distribution of the FSI scores of good readers is identical to the distribution of the FSI scores of average readers

H_a: The distribution of the of the FSI scores of good readers is shifted to the right or left of the distribution of the FSI scores of average readers

b, c.

	FSI Scores			
STRATEGY	**Good Readers**	**Average Readers**	**Difference**	**Rank of Absolute Difference**
Word meaning	0.38	0.32	0.06	2
Words in context	0.29	0.25	0.04	1
Literal comprehension	0.42	0.25	0.17	5
Draw inference from single string	0.60	0.26	0.34	8
Draw inference from multiple string	0.45	0.31	0.14	4
Interpretation of metaphor	0.32	0.14	0.18	6.5
Find salient or main idea	0.21	0.03	0.18	6.5
Form judgment	0.73	0.80	–0.07	3

d. The positive rank sum is $T_+ = 33$. The negative rank sum is $T_- = 3$. The test statistic is the smaller of T_+ and T_- and is $T_- = 3$.

e. The rejection region is $T_- \leq 4$, from Table X, Appendix B, with $n = 8$ and $\alpha = 0.05$ (two-tails).

f. Since the observed value of the test statistic falls in the rejection region $(T_- = 3 \leq 4)$, H_0 is rejected. There is sufficient evidence to indicate the distributions of the FSI scores for good and average readers differ in location at $\alpha = 0.05$.

14.53 Some preliminary calculations:

Teacher	Pre-test	Post-test	Difference *Pre-Post*	Rank of Absolute Differences
1	53	74	-21	9
2	73	80	-7	8
3	70	94	-24	10
4	72	78	-6	6.5
5	77	78	-1	1.5
6	81	84	-3	5
7	73	71	2	3.5
8	87	88	-1	1.5
9	61	63	-2	3.5
10	76	83	-6	6.5
			Positive rank sum $T_+ = 3.5$	

To determine if the program was effective, we test:

H_0: The distribution of racial tolerance before the program has the same location as the distribution of racial tolerance after program

H_a: The distribution of racial tolerance before the program is shifted to the left of that for racial tolerance after the program

The test statistic is $T_+ = 3.5$.

Reject H_0 if $T_+ \leq T_0$ where T_0 is based on $\alpha = 0.01$ and $n = 10$ (one-tailed):

Reject H_0 if $T_+ \leq 5$ (from Table X, Appendix B)

Since the observed value of the test statistic falls in the rejection region $(T_+ = 3.5 \leq 5)$, H_0 is rejected. There is sufficient evidence to indicate that the program was effective at $\alpha = 0.01$.

14.55 To determine if the distribution of healthy food item percentages using the inventory method is shifted to the right or left of the distribution of healthy food item percentages using the checklist method, we test:

H_0: The distributions of healthy food item percentages using the inventory method and the checklist method are identical

H_a: The distribution of healthy food item percentages using the inventory method is shifted to the right or left of that for the distribution of healthy food item percentages using the checklist method

Since the p-value is less than α $(p = 0.003 < 0.05)$, H_0 is rejected. There is sufficient evidence to indicate that the distribution of healthy food item percentages using the inventory method is shifted to the right or left of the distribution of healthy food item percentages using the checklist method at $\alpha = 0.05$. Because the test statistic is large, it indicates that the percentages using the inventory method is larger than the percentage using the checklist method, we recommend using the inventory method.

14.57 Some preliminary calculations are:

Patient	Before Thickness	After Thickness	Difference	Rank of Absolute difference
1	11.0	11.5	−0.5	6
2	4.0	6.4	−2.4	20
3	6.3	6.1	0.2	3
4	12.0	10.0	2.0	15.5
5	18.2	14.7	3.5	24
6	9.2	7.3	1.9	14
7	7.5	6.1	1.4	12
8	7.1	6.4	0.7	8.5
9	7.2	5.7	1.5	13
10	6.7	6.5	0.2	3
11	14.2	13.2	1.0	10
12	7.3	7.5	−0.2	3
13	9.7	7.4	2.3	18.5
14	9.5	7.2	2.3	18.5
15	5.6	6.3	−0.7	8.5
16	8.7	6.0	2.7	21
17	6.7	7.3	−0.6	7
18	10.2	7.0	3.2	23
19	6.6	5.3	1.3	11
20	11.2	9.0	2.2	17

21	8.6	6.6	2.0	15.5
22	6.1	6.3	−0.2	3
23	10.3	7.2	3.1	22
24	7.0	7.2	−0.2	3
25	12.0	8.0	4.0	25
			Negative rank sum $T_{-} = 50.5$	
			Positive rank sum $T_{+} = 274.5$	

To determine if the treatment for tendonitis tends to reduce the thickness of tendons, we test:

H_0: The distribution of the thickness before treatment has the same location as the distribution of the thickness after the treatment

H_a: The distribution of the thickness before the treatment is shifted to the right of that for the thickness after the treatment

The test statistics is T = smaller of T_{-} and T_{+} which is $T_{-} = 50.5$.

Reject H_0 if $T_{-} \leq T_0$ where T_0 is based on $\alpha = 0.10$ and $n = 25$ (one-tailed).

Reject H_0 if $T_{-} \leq 101$ (From Table X, Appendix B)

(Note: There is no value for $\alpha = 0.10$ for a one-tailed test. However, if we reject H_0 for $\alpha = 0.05$, we would also reject H_0 for $\alpha = 0.10$.)

Since the observed value of the test statistic falls in the rejection region $(T_{-} = 50.5 \leq 101)$, H_0 is rejected. There is sufficient evidence to indicate the treatment for tendonitis tends to reduce the thickness of tendons at $\alpha = 0.10$.

14.59 The distribution of H is approximately a χ^2 if the null hypothesis is true and if the sample sizes, n_j, for each of the k distributions is more than 5.

14.61 a. A completely randomized design was used.

b. The hypotheses are:

H_0: The three probability distributions are identical
H_a: At least two of the three probability distributions differ in location

c. The rejection region requires $\alpha = 0.01$ in the upper tail of the χ^2 distribution with $df = k - 1 = 3 - 1 = 2$. From Table IV, Appendix B, $\chi^2_{0.01} = 9.21034$. The rejection region is $H > 9.21034$.

d. Some preliminary calculations are:

I		II		III	
Observation	**Rank**	**Observation**	**Rank**	**Observation**	**Rank**
34	5.5	24	2	72	14
56	10	18	1	101	20
65	12	27	3	91	19
59	11	41	7	76	16
82	18	34	5.5	80	17
70	13	42	8	75	15
45	9	33	4		
	$R_A = 78.5$		$R_B = 30.5$		$R_C = 101$

$$\bar{R}_A = \frac{R_A}{n_A} = \frac{78.5}{7} = 11.214 \qquad \bar{R}_B = \frac{R_B}{n_B} = \frac{30.5}{7} = 4.357$$

$$\bar{R}_C = \frac{R_C}{n_C} = \frac{101}{6} = 16.833 \qquad \bar{R} = \frac{1}{2}(n+1) = \frac{1}{2}(20+1) = 10.5$$

The test statistic is

$$H = \frac{12}{n(n+1)} \sum n_j \left(\bar{R}_j - \bar{R}\right)^2$$

$$= \frac{12}{20(20+1)}\left[7(11.214-10.5)^2 + 7(4.357-10.5)^2 + 6(16.833-10.5)^2\right] = 14.525$$

Since the observed value of the test statistic falls in the rejection region $(H = 14.5252 > 9.21034)$, H_0 is rejected. There is sufficient evidence to indicate at least two of the three probability distributions differ in location at $\alpha = 0.01$.

14.63 a. Using MINITAB. Boxplots of the data are:

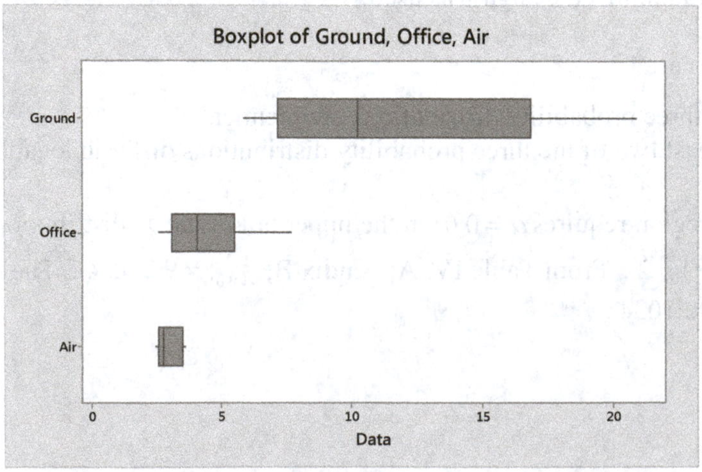

From the plots, it is very evident that the variances of the 3 groups are not the same, a requirement for the ANOVA F test.

b. The appropriate test will be the Kruskal-Wallis test. The hypotheses are:

H_0: The three probability distributions are identical
H_a: At least two of the three probability distributions differ in location

c. Some preliminary calculations are:

I		II		III	
Ground	**Rank**	**Office**	**Rank**	**Air**	**Rank**
7.6	16	5.4	13	2.5	1
10.8	19	2.8	4.5	3.4	8
20.9	21	3.9	10	2.7	3
15.5	20	5.9	14.5	2.8	4.5
9.7	18	4.3	11	3.6	9
5.9	14.5	4.6	12		
		2.6	2		
		3.3	7		
		3.2	6		
		7.7	17		
	$R_A = 108.5$		$R_B = 97$		$R_C = 25.5$

d. $\bar{R}_A = \frac{R_A}{n_A} = \frac{108.5}{6} = 18.0833$ $\quad \bar{R}_B = \frac{R_B}{n_B} = \frac{97}{10} = 9.7$

$\bar{R}_C = \frac{R_C}{n_C} = \frac{25.5}{5} = 5.1$ $\quad \bar{R} = \frac{1}{2}(n+1) = \frac{1}{2}(21+1) = 11$

The test statistic is

$$H = \frac{12}{n(n+1)}\sum n_j(\bar{R}_j - \bar{R})^2$$

$$= \frac{12}{21(21+1)}\left[6(18.0833-11)^2 + 10(9.7-11)^2 + 5(5.1-11)^2\right] = 12.779$$

e. The rejection region requires $\alpha = 0.10$ in the upper tail of the χ^2 distribution with $df = k-1 = 3-1 = 2$. From Table IV, Appendix B, $\chi^2_{0.10} = 4.60517$. The rejection region is $H > 4.60517$.

f. Since the test statistic falls in the rejection region $(H = 12.779 > 4.60517)$, H_0 is rejected. There is sufficient evidence to indicate that at least two of the three distributions of estimated containment times differ in location at $\alpha = 0.10$.

14.65 a. To determine if the probability distributions of interaction lengths differ among the three groups, we test:

H_0: The probability distributions of interaction lengths for the three groups are identical
H_a: At least two of the three probability distributions differ in location

b. The rejection region requires $\alpha = 0.05$ in the upper tail of the χ^2 distribution with $df = k - 1 = 3 - 1 = 2$. From Table IV, Appendix B, $\chi^2_{0.05} = 5.99147$. The rejection region is $H > 5.99147$.

c. The test statistic is $H = 1.1$ and the *p*-value is $p = 0.60$. Since the observed value of the test statistic does not fall in the rejection region $(H = 1.1 \not> 5.99147)$, H_0 is not rejected. There is insufficient evidence to indicate the probability distributions of interaction lengths differ among the three groups at $\alpha = 0.05$.

14.67 a – d. The ranks and sums of ranks appear in the table:

Scopolamine	Rank	Placebo	Rank	No Drug	Rank
5	2.5	8	13	8	13
8	13	10	20.5	9	17
8	13	12	26.5	11	23.5
6	6.5	10	20.5	12	26.5
6	6.5	9	17	11	23.5
6	6.5	7	10	10	20.5
6	6.5	9	17	12	26.5
8	13	10	20.5	12	26.5
6	6.5				
4	1				
5	2.5				
6	6.5				
	$R_1 = 84$		$R_2 = 145$		$R_3 = 177$

e. $\bar{R}_1 = \frac{R_1}{n_1} = \frac{84}{12} = 7$ $\quad \bar{R}_2 = \frac{R_2}{n_2} = \frac{145}{8} = 18.125$ $\quad \bar{R}_3 = \frac{R_3}{n_3} = \frac{177}{8} = 22.125$

$$\bar{R} = \frac{1}{2}(n+1) = \frac{1}{2}(28+1) = 14.5$$

$$H = \frac{12}{n(n+1)} \sum n_j \left(\bar{R}_j - \bar{R}\right)^2$$

$$= \frac{12}{28(28+1)}\left[12(7-14.5)^2 + 8(18.125-14.5)^2 + 8(22.125-14.5)^2\right] = 18.403$$

f. To determine if the distributions of the number of word pairs recalled for the three groups have different locations, we test:

H_0: The three probability distributions are identical
H_a: At least two of the three probability distributions differ in location

The test statistic is $H = 18.403$.

The rejection region requires $\alpha = 0.05$ in the upper tail of the χ^2 distribution with $df = k-1 = 3-1 = 2$. From Table IV, Appendix B, $\chi^2_{0.05} = 5.99147$. The rejection region is $H > 5.99147$.

Since the observed value of the test statistic falls in the rejection region $(H = 18.403 > 5.99147)$, H_0 is rejected. There is sufficient evidence to indicate the distributions of the number of word pairs recalled for the three groups have different locations at $\alpha = 0.05$.

g. Re-ranking the data for only groups 1 and 2, we get:

Scopolamine	Rank	Placebo	Rank
5	2.5	8	12.5
8	12.5	10	18
8	12.5	12	20
6	6.5	10	18
6	6.5	9	15.5
6	6.5	7	10
6	6.5	9	15.5
8	12.5	10	18
6	6.5		
4	1		
5	2.5		
6	6.5		
	$T_1 = 82.5$		$T_2 = 127.5$

To determine if the distributions of the number of word pairs recalled for the Scopolamine group is shifted to the left of that for the Placebo group, we test:

H_0: The two probability distributions are identical
H_a: The probability distribution of the Scopolamine group is shifted to the left of that for the Placebo group

The test statistic is the T_2 because $n_1 < n_2$.

There is no table value for this problem since $n_1 = 12$ and the largest n in Table XII is 10. We will use the normal approximation even though both sample sizes are not greater than or equal to $10\left(n_2 = 8\right)$.

The test statistic is $z_c = \dfrac{T_1 - \dfrac{n_1(n_1+n_2+1)}{2}}{\sqrt{\dfrac{n_1 n_2(n_1+n_2+1)}{12}}} = \dfrac{82.5 - \dfrac{12(12+8+1)}{2}}{\sqrt{\dfrac{12(8)(12+8+1)}{12}}} = -3.36$

The rejection region requires $\alpha = 0.05$ in the lower tail of the *z* distribution. From Table II, Appendix B, $z_{0.05} = 1.645$. The rejection region is $z < -1.645$.

Since the observed value of the test statistic falls in the rejection region $(z_c = -3.36 < -1.645)$, H_0 is rejected. There is sufficient evidence to indicate the distributions of the number of word pairs recalled for the Scopolamine group is shifted to the left of that for the Placebo group at $\alpha = 0.05$.

14.69 a. Some preliminary calculations are:

VHR	Rank	VRD	Rank	Control	Rank
-20	4	-12	6	51	20
-56	1	63	21	21	17
-34	3	12	15	8	14
0	11	-7	8.5	0	11
16	16	29	18	4	13
0	11				
-14	5				
-7	8.5				
-44	2				
43	19				
-11	7				
	$R_1 = 87.5$		$R_2 = 68.5$		$R_3 = 75$

$$\bar{R}_1 = \frac{R_1}{n_1} = \frac{87.5}{11} = 7.955, \quad \bar{R}_2 = \frac{R_2}{n_2} = \frac{68.5}{5} = 13.7, \quad \bar{R}_3 = \frac{R_3}{n_3} = \frac{75}{5} = 15,$$

$$\bar{R} = \frac{1}{2}(n+1) = \frac{1}{2}(21+1) = 11$$

To determine if the probability distributions of differences in pain intensity levels differ in location among the three treatments, we test:

H_0: The probability distributions of differences in pain intensity levels for the three groups are identical
H_a: At least two of the three probability distributions differ in location

The test statistic is

$$H = \frac{12}{n(n+1)}\sum n_j\left(\bar{R}_j - \bar{R}\right)^2$$

$$= \frac{12}{21(21+1)}\left[11(7.955-11)^2 + 5(13.7-11)^2 + 5(15-11)^2\right] = 5.674$$

The rejection region requires $\alpha = 0.05$ in the upper tail of the χ^2 distribution with $df = k - 1 = 3 - 1 = 2$. From Table IV, Appendix B, $\chi^2_{0.05} = 5.99147$. The rejection region is $H > 5.99147$.

Since the observed value of the test statistic does not fall in the rejection region $(H = 5.674 \not> 5.99147)$, H_0 is not rejected. There is insufficient evidence to indicate the probability distributions of differences in pain intensity levels differ in location among the three treatments at $\alpha = 0.05$.

b. Some preliminary calculations are:

VHR	Rank	VRD+ Control	Rank
-20	4	-12	6
-56	1	63	21
-34	3	12	15
0	11	-7	8.5
16	16	29	18
0	11	51	20
-14	5	21	17
-7	8.5	8	14
-44	2	0	11
43	19	4	13
-11	7		
	$T_1 = 87.5$		$T_2 = 143.5$

To determine if the probability distributions of differences in pain intensity levels differ in location between the two treatments, we test:

H_0: The probability distributions of differences in pain intensity levels for the two groups are identical

H_a: The probability distribution of differences in pain intensity levels for the VRH group is shifted to the right or left of that for the VRD and Control groups combined

The test statistic is

$$z_c = \frac{T_1 - \frac{n_1(n_1+n_2+1)}{2}}{\sqrt{\frac{n_1 n_2(n_1+n_2+1)}{12}}} = \frac{87.5 - \frac{11(11+10+1)}{2}}{\sqrt{\frac{11(10)(11+10+1)}{12}}} = \frac{-33.5}{14.2009} = -2.36$$

The rejection region requires $\alpha/2 = 0.05/2 = 0.025$ in each tail of the z distribution. From Table II, Appendix B, $z_{0.025} = 1.96$. The rejection region is $z < -1.96$ or $z > 1.96$.

Since the observed value of the test statistic falls in the rejection region $(z_c = -2.36 < -1.96)$, H_0 is rejected. There is sufficient evidence to indicate the probability distributions of differences in pain intensity levels differ in location between the VHR group and the combination of the VRD and control group at $\alpha = 0.05$.

14.71 Some preliminary calculations are:

AR	Rank	AC	Rank	A	Rank	P	Rank
0.51	38	0.50	36	0.16	11.5	0.58	41.5
0.58	41.5	0.30	21	0.10	7.5	0.12	9
0.52	39	0.47	33.5	0.20	15	0.62	44
0.47	33.5	0.36	25	0.29	20	0.43	29.5
0.61	43	0.39	26	-0.14	3	0.26	19
0.00	4	0.22	18	0.18	13	0.50	36
0.32	23	0.20	15	-0.35	1	0.44	31
0.53	40	0.21	17	0.31	22	0.20	15
0.50	36	0.15	10	0.16	11.5	0.42	28
0.46	32	0.10	7.5	0.04	6	0.43	29.5
0.34	24	0.02	5	-0.25	2	0.40	27
	$R_1 = 354$		$R_2 = 214$		$R_3 = 112.5$		$R_4 = 309.5$

$$\bar{R}_1 = \frac{R_1}{n_1} = \frac{354}{11} = 32.182 \qquad \bar{R}_2 = \frac{R_2}{n_2} = \frac{214}{11} = 19.455 \qquad \bar{R}_3 = \frac{R_3}{n_3} = \frac{112.5}{11} = 10.227$$

$$\bar{R}_4 = \frac{R_4}{n_4} = \frac{309.5}{11} = 28.136 \qquad \bar{R} = \frac{1}{2}(n+1) = \frac{1}{2}(44+1) = 22.5$$

To determine if the distributions of the task scores differ in location among the 4 groups, we test:

H_0: The distributions of the task scores for the 4 groups are identical
H_a: At least two of the four distributions differ in location

The test statistic is

$$H = \frac{12}{n(n+1)} \sum n_j \left(\bar{R}_j - \bar{R}\right)^2$$

$$= \frac{12}{44(44+1)} \left[\begin{array}{c} 11(32.182-22.5)^2 + 11(19.455-22.5)^2 + 11(10.227-22.5)^2 \\ +11(28.136-22.5)^2 \end{array} \right] = 19.027$$

The rejection region requires $\alpha = 0.05$ in the upper tail of the χ^2 distribution with $df = k-1 = 4-1 = 3$. From Table IV, Appendix B, $\chi^2_{0.05} = 7.81473$. The rejection region is $H > 7.81473$.

Since the observed value of the test statistic falls in the rejection region $(H = 19.027 > 7.81473)$, H_0 is rejected. There is sufficient evidence to indicate the distributions of the task scores differ in location among the 4 groups at $\alpha = 0.05$. We can infer that the task scores are affected by the amount of alcohol and reward presented.

14.73 a. Using MINITAB, the histograms of the three data sets are:

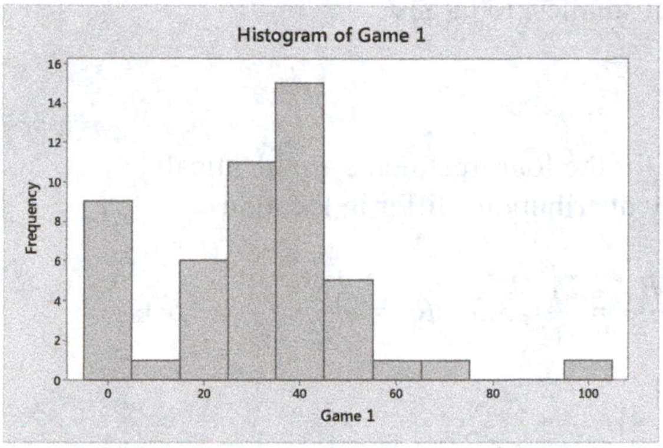

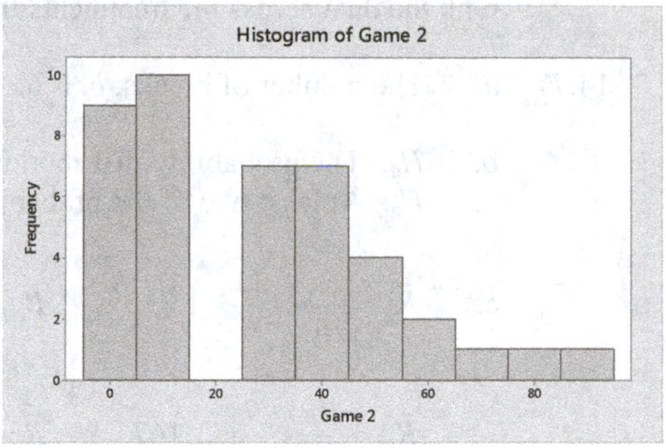

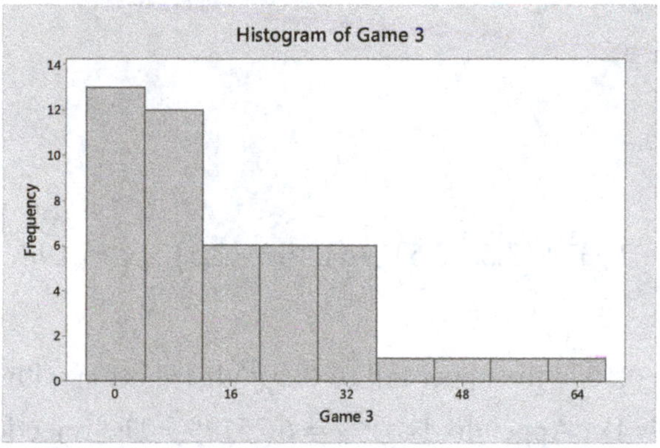

None of these three graphs look approximately normal. Thus, the assumption of normality necessary for ANOVA is likely violated.

b. Using MINITAB, the results of the Kruskal-Wallis analysis are:

```
Kruskal-Wallis Test on Percent

Game       N    Median    Ave Rank       Z
1         50    32.000        83.1    2.88
2         42    26.000        73.0    0.59
3         47     9.000        53.3   -3.49
Overall  139                  70.0

H = 13.61  DF = 2  P = 0.001
H = 13.66  DF = 2  P = 0.001 (adjusted for ties)
```

To determine if the percentages of names recalled are different for the three retrieval methods, we test:

H_0: The distributions of the three retrieval methods are identical
H_a: At least two of the three retrieval methods have probability distributions that differ in location

From the printout, the test statistic is $H = 13.66$ and the p-value is $p = 0.001$. Since the p-value is less than $\alpha\left(p = 0.001 < 0.05\right)$, H_0 is rejected. There is sufficient evidence to indicate that the percentages of names recalled are different for the three retrieval methods at $\alpha = 0.05$.

14.75 To correctly rank the data in a randomized block design, we would use b - for each block, rank the data across the treatments from smallest to largest.

14.77 a. The number of blocks, b, is 6.

b. H_0: The probability distributions for the four treatments are identical
H_a: At least two of the probability distributions differ in location

c. $\bar{R}_A = \frac{R_A}{b} = \frac{11}{6} = 1.833$ $\quad \bar{R}_B = \frac{R_B}{b} = \frac{21}{6} = 3.5$ $\quad \bar{R}_C = \frac{R_C}{b} = \frac{21}{6} = 3.5$

$\bar{R}_D = \frac{R_D}{b} = \frac{7}{6} = 1.167$ $\quad \bar{R} = \frac{k+1}{2} = \frac{4+1}{2} = 2.5$

The test statistic is

$$F_r = \frac{12b}{k(k+1)}\sum\left(\bar{R}_j - \bar{R}\right)^2$$

$$= \frac{12(6)}{4(4+1)}\left[(1.833-2.5)^2 + (3.5-2.5)^2 + (3.5-2.5)^2 + (1.167-2.5)^2\right] = 15.198$$

The rejection region requires $\alpha = 0.10$ in the upper tail of the χ^2 distribution with $df = k - 1 = 4 - 1 = 3$. From Table IV, Appendix B, $\chi^2_{0.10} = 6.25139$. The rejection region is $F_r > 6.25139$.

Since the observed value of the test statistic falls in the rejection region $(F_r = 15.198 > 6.25139)$, reject H_0. There is sufficient evidence to indicate a difference in the location for at least two of the four treatments at $\alpha = 0.10$.

d. The p-value is $p = P(F_r \geq 15.2) = P\left(\chi^2 \geq 15.198\right)$. With 3 degrees of freedom, $F_r = 15.198$ falls above $\chi^2_{0.005}$; therefore, $p < 0.005$.

14.79 $R_1 = 16$ $\quad R_2 = 7$ $\quad R_3 = 23$ $\quad R_4 = 14$

$\bar{R}_1 = \frac{R_1}{b} = \frac{16}{6} = 2.667$ $\quad \bar{R}_2 = \frac{R_2}{b} = \frac{7}{6} = 1.167$ $\quad \bar{R}_3 = \frac{R_3}{b} = \frac{23}{6} = 3.833$

$\bar{R}_4 = \frac{R_4}{b} = \frac{14}{6} = 2.333$ $\quad \bar{R} = \frac{1}{2}(k+1) = \frac{1}{2}(4+1) = 2.5$

To determine if at least two of the treatment probability distributions differ in location, we test:

H_0: The probability distributions of the four treatments are identical
H_a: At least two of the probability distributions differ in location

The test statistic is

$$F_r = \frac{12b}{k(k+1)}\sum\left(\bar{R}_j - \bar{R}\right)^2$$

$$= \frac{12(6)}{(4)(4+1)}\left[(2.667-2.5)^2 + (1.167-2.5)^2 + (3.833-2.5)^2 + (2.333-2.5)^2\right] = 12.994$$

The rejection region requires $\alpha = 0.05$ in the upper tail of the χ^2 distribution with $df = k-1 = 4-1 = 3$. From Table IV, Appendix B, $\chi^2_{0.05} = 7.84173$. The rejection region is $F_r > 7.84173$.

Since the observed value of the test statistic falls in the rejection region $(F_r = 12.994 > 7.84173)$, reject H_0. There is sufficient evidence to indicate a difference in the location for at least two of the probability distributions at $\alpha = 0.05$.

14.81 a. From the printout, the rank sums are $R_1 = 25.5$, $R_2 = 11$, $R_3 = 18.5$, and $R_4 = 25$.

b. Some preliminary calculations are:

$$\bar{R}_1 = \frac{R_1}{b} = \frac{25.5}{8} = 3.1875, \quad \bar{R}_2 = \frac{R_2}{b} = \frac{11.0}{8} = 1.375, \quad \bar{R}_3 = \frac{R_3}{b} = \frac{18.5}{8} = 2.3125,$$

$$\bar{R}_4 = \frac{R_4}{b} = \frac{25}{8} = 3.125, \quad \bar{R} = \frac{k+1}{2} = \frac{4+1}{2} = 2.5$$

$$F_r = \frac{12b}{k(k+1)}\sum\left(\bar{R}_j - \bar{R}\right)^2$$

$$= \frac{12(8)}{4(4+1)}\left[(3.1875-2.5)^2 + (1.375-2.5)^2 + (2.3125-2.5)^2 + (3.125-2.5)^2\right]$$

$$= 10.3875$$

From the printout, the value of the test statistic is $S = 10.39$. This agrees with the value computed above.

c. From the printout, the *p*-value is $p = 0.016$.

d. Since the *p*-value is less than $\alpha\,(p = 0.016 < 0.05)$, H_0 is rejected. There is sufficient evidence to indicate at least two of the distributions of heart rates differ in location among the 4 pre-slaughter phases at $\alpha = 0.05$.

14.83 a. The data were collected using a randomized block design – each subject had a response for each treatment. IN an ANOVA *F*-test, we have to assume that the distributions in all $10 \times 4 = 40$ block-treatment combinations are normally distributed. This may not be the case with this data.

b. The test statistic is $H = 14.37$ and the *p*-value is $p = 0.002$.

c. To determine if distributions of walking times differ for the four time segments, we test:

H_0: The distributions of walking times for the time segments are identical
H_a: At least two of the distributions of walking times differ in location

Since the *p*-value is less than $\alpha\left(p = 0.0020 < 0.01\right)$, H_0 is rejected. There is sufficient evidence to indicate that at least two of the distributions of walking times differ in location at $\alpha = 0.01$.

14.85 a. The rejection region requires $\alpha = 0.01$ in the upper tail of the χ^2 distribution with $df = k - 1 = 3 - 1 = 2$. From Table IV, Appendix B, $\chi^2_{0.01} = 9.21034$. The rejection region is $F_r > 9.21034$.

b. Since the observed value of the test statistic falls in the rejection region $\left(F_r = 19.16 > 9.21034\right)$, H_0 is rejected. There is sufficient evidence to indicate the distributions of proportions of eye fixations on the interior mirror differ in location among the 3 treatments at $\alpha = 0.01$.

c. Since the observed value of the test statistic does not fall in the rejection region $\left(F_r = 7.80 \not> 9.21034\right)$, H_0 is not rejected. There is insufficient evidence to indicate the distributions of proportions of eye fixations on the side mirror differ in location among the 3 treatments at $\alpha = 0.01$.

d. Since the observed value of the test statistic falls in the rejection region $\left(F_r = 20.67 > 9.21034\right)$, H_0 is rejected. There is sufficient evidence to indicate the distributions of proportions of eye fixations on the speedometer differ in location among the 3 treatments at $\alpha = 0.01$.

14.87 a. If the data were not normally distributed, then the results of the ANOVA will not be valid. One of the conditions for the ANOVA to be valid is that the data are normally distributed.

b. To determine if the distributions of the TT indices for the 15 weeks differ in location, we test:

H_0: The distributions of the TT indices for the 15 weeks are identical
H_a: At least two of the probability distributions differ in location

c. The rejection region requires $\alpha = 0.01$ in the upper tail of the χ^2 distribution with $df = k - 1 = 15 - 1 = 14$. From Table IV, Appendix B, $\chi^2_{0.01} = 29.1413$. The rejection region is $F_r > 29.1413$.

d. The TT index scores for each of the 15 weeks within one year are ranked from 1 to 15. This is repeated for each of the 6 years. The sum of the ranks for each of the 15 weeks is then computed and the average rank for each week is found by dividing the sum by 6. These values are then substituted into the formula for F_r.

e. Any p-value greater than 0.01 would lead us to conclude that no week is any more "topsy-turvey" than any other.

14.89 Some preliminary calculations are:

Boxer	M1	Rank	R1	Rank	M5	Rank	R5	Rank
1	1243	1	1244	2	1291	4	1262	3
2	1147	2	1053	1	1169	3	1177	4
3	1247	1	1375	4	1309	2	1321	3
4	1274	2	1235	1	1290	4	1285	3
5	1177	2	1139	1	1233	3	1238	4
6	1336	2	1313	1	1366	4	1362	3
7	1238	1	1279	4	1275	3	1261	2
8	1261	2	1152	1	1289	4	1266	3
		$R_1 = 13$		$R_2 = 15$		$R_3 = 27$		$R_4 = 25$

$$\bar{R}_1 = \frac{R_1}{b} = \frac{13}{8} = 1.625, \quad \bar{R}_2 = \frac{R_2}{b} = \frac{15}{8} = 1.875, \quad \bar{R}_3 = \frac{R_3}{b} = \frac{27}{8} = 3.375,$$

$$\bar{R}_4 = \frac{R_4}{b} = \frac{25}{8} = 3.125, \quad \bar{R} = \frac{k+1}{2} = \frac{4+1}{2} = 2.5$$

To compare the punching power means of the four interventions, we test:

H_0: The four probability distributions of punching power are the same
H_a: At least two of the probability distributions of punching power differ in location

The test statistic is

$$F_r = \frac{12b}{k(k+1)} \sum \left(\bar{R}_j - \bar{R}\right)^2$$

$$= \frac{12(8)}{4(4+1)} \left[(1.625-2.5)^2 + (1.875-2.5)^2 + (3.375-2.5)^2 + (3.125-2.5)^2\right] = 11.1$$

Since no α was given, we will use $\alpha = 0.05$. The rejection region requires $\alpha = 0.05$ in the upper tail of the χ^2 distribution with $df = k - 1 = 4 - 1 = 3$. From Table IV, Appendix B, $\chi^2_{0.05} = 7.81473$. The rejection region is $F_r > 7.81473$.

Since the observed value of the test statistic falls in the rejection region $(F_r = 11.1 > 7.81473)$, H_0 is rejected. There is sufficient evidence to indicate that the punching power means of the four interventions differ at $\alpha = 0.05$.

This is the same result that we obtained in Exercise 10.82.

14.91 The value of r_s when there is perfect negative rank correlation between two variables is $r_s = -1$. The value of r_s when there is perfect positive rank correlation between two variables is $r_s = 1$.

14.93 a. For $n = 22$, $P(r_s > 0.508) = 0.01$

b. For $n = 28$, $P(r_s > 0.448) = 0.01$

c. For $n = 10$, $P(r_s \le 0.648) = 1 - 0.025 = 0.975$

d. For $n = 8$, $P(r_s < -0.738 \text{ or } r_s > 0.738) = 2(0.025) = 0.05$

14.95 Since there are no ties, we will use the shortcut formula.

a. Some preliminary calculations are:

x Rank (u_i)	y Rank (v_i)	$d_i = u_i - v_i$	d_i^2
3	2	1	1
5	4	1	1
2	5	–3	9
1	1	0	0
4	3	1	1
			$\sum d_i^2 = 12$

$$r_s = 1 - \frac{6\sum d_i^2}{n(n^2-1)} = 1 - \frac{6(12)}{5(5^2-1)} = 1 - 0.6 = 0.4$$

b.

x Rank (u_i)	y Rank (v_i)	$d_i = u_i - v_i$	d_i^2
2	3	–1	1
3	4	–1	1
4	2	2	4
5	1	4	16
1	5	–4	16
			$\sum d_i^2 = 38$

$$r_s = 1 - \frac{6\sum d_i^2}{n(n^2-1)} = 1 - \frac{6(38)}{5(5^2-1)} = 1 - 1.9 = -0.9$$

c.

x Rank (u_i)	y Rank (v_i)	$d_i = u_i - v_i$	d_i^2
1	2	–1	1
4	1	3	9
2	3	–1	1
3	4	–1	1
			$\sum d_i^2 = 12$

$$r_s = 1 - \frac{6\sum d_i^2}{n(n^2-1)} = 1 - \frac{6(12)}{4(4^2-1)} = 1 - 1.2 = -0.2$$

d.

x Rank (u_i)	y Rank (v_i)	$d_i = u_i - v_i$	d_i^2
2	1	1	1
5	3	2	4
4	5	–1	1
3	2	1	1
1	4	–3	9
			$\sum d_i^2 = 16$

$$r_s = 1 - \frac{6\sum d_i^2}{n(n^2-1)} = 1 - \frac{6(16)}{5(5^2-1)} = 1 - 0.8 = 0.2$$

14.97 a, b. The ranks of the angular sizes and heights are listed in the table:

Angle	Rank, u	*Height*	Rank, v	u^2	v^2	uv
321.9	1	17	1	1	1	1
322.3	2	18	2	4	4	4
322.4	3	26	3	9	9	9
323.2	4	32	4	16	16	16
323.4	5	38	5	25	25	25
324.4	6	42	6	36	36	36
325.0	7.5	49	7	56.25	49	52.5
325.7	9	52	8	81	64	72
325.8	10.5	57	9	110.25	81	94.5
325.0	7.5	60	10	56.25	100	75
326.9	13	63	11	169	121	143
326.0	12	67	12	144	144	144
325.8	10.5	73	13	110.25	169	136.5
	$\sum u = 91$		$\sum v = 91$	$\sum u^2 = 818$	$\sum v^2 = 819$	$\sum uv = 808.5$

c. Using the additional calculations in the table above:

$$SS_{uv} = \sum uv - \frac{(\sum u)(\sum v)}{n} = 808.5 - \frac{91(91)}{13} = 171.5$$

$$SS_{uu} = \sum u^2 - \frac{(\sum u)^2}{n} = 818 - \frac{(91)^2}{13} = 181$$

$$SS_{vv} = \sum v^2 - \frac{(\sum v)^2}{n} = 819 - \frac{(91)^2}{13} = 182$$

$$r_s = \frac{SS_{uv}}{\sqrt{SS_{uu}SS_{vv}}} = \frac{171.5}{\sqrt{181(182)}} = 0.9449$$

d. To determine if angular size and height above the horizon have a positive rank correlation, we test:

$H_0: \rho_s = 0$
$H_a: \rho_s > 0$

Reject H_0 if $r_s > r_{s,\alpha}$ where $\alpha = 0.05$ and $n = 13$:

Reject H_0 if $r_s > 0.475$ (from Table XI, Appendix B).

Since the observed value of the test statistic falls in the rejection region, $(r_s = 0.9449 > 0.475)$, H_0 is rejected. There is sufficient evidence to indicate angular size and height above the horizon have a positive rank correlation at $\alpha = 0.05$.

14.99 a. Form the printout, $r_s = 0.93415$.

b.

HR	Rank, u	VO2Max	Rank, v	u^2	v^2	uv
55	2	148	1	4	1	2
54	1	157	2	1	4	2
70	4	160	3	16	9	12
67	3	179	4.5	9	20.25	13.5
74	5	179	4.5	25	20.25	22.5
77	6	180	6	36	36	36
78	7	194	7	49	49	49
85	8	197	8	64	64	64
	$\sum u = 36$		$\sum v = 36$	$\sum u^2 = 204$	$\sum v^2 = 203.5$	$\sum uv = 201$

$$SS_{uv} = \sum uv - \frac{(\sum u)(\sum v)}{n} = 201 - \frac{36(36)}{8} = 39$$

$$SS_{uu} = \sum u^2 - \frac{(\sum u)^2}{n} = 204 - \frac{(36)^2}{8} = 42$$

$$SS_{vv} = \sum v^2 - \frac{(\sum v)^2}{n} = 203.5 - \frac{(36)^2}{8} = 41.5$$

$$r_s = \frac{SS_{uv}}{\sqrt{SS_{uu}SS_{vv}}} = \frac{39}{\sqrt{42(41.5)}} = 0.93415$$

c. To determine if there is positive rank correlation between mean heart rate and maximal oxygen uptake, we test:

$H_0: \rho_s = 0$
$H_a: \rho_s > 0$

The test statistic is $r_s = 0.93415$ and the p-value is $p = 0.0007/2 = 0.00035$.

Since the p-value is less than $\alpha\left(p=0.00035<0.01\right)$, H_0 is rejected. There is sufficient evidence to indicate there is positive rank correlation between mean heart rate and maximal oxygen uptake at $\alpha=0.01$.

14.101 a. Some preliminary calculations are:

Blood Lactate Level	Rank u	Perceived Recovery	Rank v	u^2	v^2	uv
3.8	3	7	1.5	9	2.25	4.5
4.2	5.5	7	1.5	30.25	2.25	8.25
4.8	7	11	3	49	9	21
4.1	4	12	5	16	25	20
5.0	8	12	5	64	25	40
5.3	9.5	12	5	90.25	25	47.5
4.2	5.5	13	7	30.25	49	38.5
2.4	1	17	9	1	81	9
3.7	2	17	9	4	81	18
5.3	9.5	17	9	90.25	81	85.5
5.8	12	18	11.5	144	132.25	138
6.0	14	18	11.5	196	132.25	161
5.9	13	21	14.5	169	210.25	188.5
6.3	15	21	14.5	225	210.25	217.5
5.5	12	20	13	121	169	143
6.5	16	24	16	256	256	256
	$\sum u=136$		$\sum v=136$	$\sum u^2=1495$	$\sum v^2=1490.5$	$\sum uv=1396.25$

$$SS_{uv}=\sum uv-\frac{\left(\sum u\right)\left(\sum v\right)}{n}=1396.25-\frac{136(136)}{16}=240.25$$

$$SS_{uu}=\sum u^2-\frac{\left(\sum u\right)^2}{n}=1495-\frac{136^2}{16}=339$$

$$SS_{vv}=\sum v^2-\frac{\left(\sum v\right)^2}{n}=1490.5-\frac{136^2}{16}=334.5$$

a. The ranks of the blood lactate levels are stored in the "u" column above.

b. The ranks of the perceived recovery values are stored in the "v" column above.

c. $r_s=\dfrac{SS_{uv}}{\sqrt{SS_{uu}SS_{vv}}}=\dfrac{240.25}{\sqrt{339(334.5)}}=0.713$

Since r_s is relatively close to 1, blood lactate level and perceived recovery have a fairly strong, positive relationship.

d. Reject H_0 if $r_s<-r_{s,\alpha/2}$ or if $r_s>r_{s,\alpha/2}$ where $\alpha/2=0.05$ and $n=16$;
Reject H_0 if $r_s<-0.425$ or $r_s>0.425$ (from Table XI, Appendix B)

e. To determine if blood lactate level and perceived recovery are rank correlated, we test:

H_0: Blood lactate level and Perceived recovery are not rank correlated
H_a: Blood lactate level and Perceived recovery are rank correlated

The test statistic is $r_s = 0.713$.

Since the observed value of the test statistic falls in the rejection region, $(r_s = 0.713 > 0.425)$, H_0 is rejected. There is sufficient evidence to indicate that blood lactate level and perceived recovery are rank correlated at $\alpha = 0.10$.

14.103 **For Private wells**:

To determine if there is rank correlation MTBE levels and depth of well, we test:

$$H_0: \rho_s = 0$$
$$H_a: \rho_s \neq 0$$

The test statistic is $r_s = -0.410$ and the *p*-value is $p = 0.103$.

Since no value of α is given, we will use $\alpha = 0.05$. Since the *p*-value is not less than $\alpha\left(p = 0.103 \not< 0.05\right)$, H_0 is not rejected. There is insufficient evidence to indicate there is rank correlation MTBE levels and depth of well for private wells at $\alpha = 0.05$. Thus, there does not appear to be a relationship between the levels of MTBE and the depth of the wells for private wells.

For Public wells:

To determine if there is rank correlation MTBE levels and depth of well, we test:

$$H_0: \rho_s = 0$$
$$H_a: \rho_s \neq 0$$

The test statistic is $r_s = 0.444$ and the *p*-value is $p = 0.002$.

Since no value of α is given, we will use $\alpha = 0.05$. Since the *p*-value is less than $\alpha\left(p = 0.002 < 0.05\right)$, H_0 is rejected. There is sufficient evidence to indicate there is rank correlation MTBE levels and depth of well for public wells at $\alpha = 0.05$. Thus, there appears to be a relationship between the levels of MTBE and the depth of the wells for public wells. As the depth of the well increases, the levels of MTBE tend to increase.

14.105 Yes. The 'supertasters' have the greatest taste intensity. The rank correlation between the perceived sensory intensity (PSI) and the perceived hedonic intensity for favorite foods (PHI-F) is $r_s = 0.374$ and the *p*-value is $p = 0.000$. This indicates that there is a significant positive rank correlation between PSI and PHI-F. This means that as the PSI increases, the PHI-F also tends to increase. The correlation between the perceived sensory intensity (PSI) and the perceived hedonic intensity for least favorite foods (PHI-L) is $r_s = -0.403$ and the *p*-

value is $p = 0.000$. This indicates that there is a significant negative linear relationship between PSI and PHI-L. This means that as the PSI increases, the PHI-L tends to decrease. This confirms that the 'supertasters' tend to experience more extreme food likes and dislikes.

14.107 Using MINITAB, the results are:

Spearman Rho: Inventory, Checklist

```
Spearman rho for Inventory and Checklist = 0.341
P-Value = 0.039
```

To determine the detailed inventory approach and the checklist approach have a positive rank correlation, we test:

$$H_0: \rho_s = 0$$
$$H_a: \rho_s > 0$$

The test statistic is $r_s = 0.341$ and the *p*-value is $p = 0.039/2 = 0.0145$.

Since the *p*-value is less than $\alpha\left(p = 0.0145 < 0.05\right)$, H_0 is rejected. There is sufficient evidence to indicate that the detailed inventory approach and the checklist approach have a positive rank correlation at $\alpha = 0.05$.

14.109 a. Using MINITAB, histograms of the two data sets are:

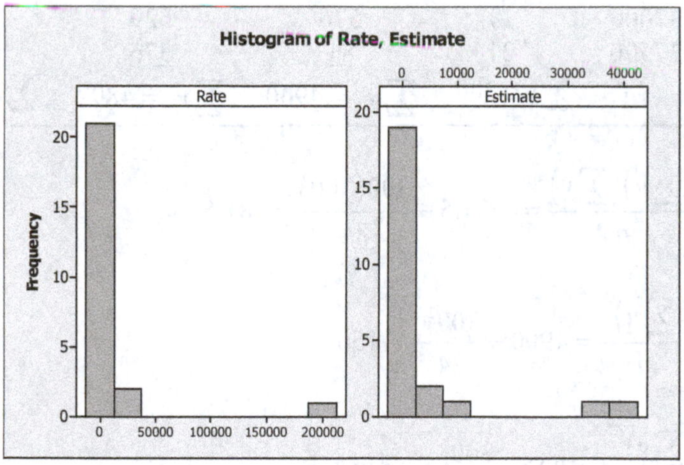

From the graphs, both data sets are skewed to the right and are not normal.

b. The analysis will not be valid because the data are not normal.

c. Some preliminary calculations are:

Incidence rate	Rank, u	Estimate	Rank, v	u^2	v^2	uv
.25	1	300	9	1	81	9
1	2	1000	16.5	4	272.25	33
1.75	3	691	14	9	196	42
2	4	20	6	16	36	24
3	5	17.5	2	25	4	10
5	6	.8	1	36	1	6
9	7	1000	16.5	49	272.25	115.5
10	8	150	4	64	16	32
22	10	326.5	10	100	100	100
23	11	146.5	3	121	9	33
39	12	370	11	144	121	132
98	13	400	12.5	169	156.25	162.5
119	14	225	8	196	64	112
152	15	200	6	225	36	90
179	16	200	6	256	36	96
936	17	400	12.5	289	156.25	212.5
1514	18	1500	19.5	324	380.25	351
1627	19	1000	16.5	361	272.25	313.5
2926	20	6000	22	400	484	440
4019	21	5000	21	441	441	441
12619	22	1500	19.5	484	380.25	429
14889	23	1000	16.5	529	272.25	379.5
203864	24	37000	23	576	529	552
15	9	37500	24	81	576	216
	$\sum u = 300$		$\sum v = 300$	$\sum u^2 = 4900$	$\sum v^2 = 4892$	$\sum uv = 4331.5$

$$SS_{uv} = \sum uv - \frac{\left(\sum u\right)\left(\sum v\right)}{n} = 4331.5 - \frac{300(300)}{24} = 581.5$$

$$SS_{uu} = \sum u^2 - \frac{\left(\sum u\right)^2}{n} = 4900 - \frac{300^2}{24} = 1150$$

$$SS_{vv} = \sum v^2 - \frac{\left(\sum v\right)^2}{n} = 4892 - \frac{300^2}{24} = 1142$$

$$r_s = \frac{SS_{uv}}{\sqrt{SS_{uu}SS_{vv}}} = \frac{581.5}{\sqrt{1150(1142)}} = .5074$$

Since the value of r_s is about half way between 0 and 1, there is a moderate positive relationship between incidence rate and estimate.

d. To determine if there is a positive association between actual incidence and estimated incidence, we test:

$H_0 : \rho_s = 0$
$H_a : \rho_s > 0$

The test statistic is $r_s = 0.5074$.

From Table XI, with $n = 24$ and $\alpha = 0.01$, the rejection region is $r_s > 0.485$.

Since the observed value of r_s falls in the rejection region $(r_s = 0.5074 > 0.485)$, H_0 is rejected. There is sufficient evidence to indicate a positive association between incidence rate and estimated rate at $\alpha = 0.01$.

14.111 a. To compare two populations with independent samples, one would use the Wilcoxon Rank Sum test.

b. To make an inference about a population median, one would use the sign test.

c. To compare three or more populations with independent samples, one would use the Kruskal-Wallis H test.

d. To make inferences about a rank correlation, one would use the Spearman's rank correlation coefficient test.

e. To compare two populations with matched pairs, one would use the Wilcoxon Signed Rank Test.

f. To compare three or more populations with a block design, one would use the Friedman's F_r test.

14.113 a. Some preliminary calculations are:

Pair	X	Rank u_i	Y	Rank v_i	u_i^2	v_i^2	$u_i v_i$
1	19	5	12	5	25	25	25
2	27	7	19	8	49	64	56
3	15	2	7	1	4	1	2
4	35	9	25	9	81	81	81
5	13	1	11	4	1	16	4
6	29	8	10	2.5	64	6.25	20
7	16	3.5	16	6	12.25	36	21
8	22	6	10	2.5	36	6.25	15
9	16	3.5	18	7	12.25	49	24.5
		$\sum u_i = 45$		$\sum v_i = 45$	$\sum u_i^2 = 284.5$	$\sum v_i^2 = 284.5$	$\sum u_i v_i = 248.5$

$$SS_{uv} = \sum u_i v_i - \frac{\sum u_i v_i}{n} = 248.5 - \frac{45(45)}{9} = 23.5$$

$$SS_{uu} = \sum u_i^2 - \frac{\left(\sum u_i\right)^2}{n} = 284.5 - \frac{45^2}{9} = 59.5$$

$$SS_{vv} = \sum v_i^2 - \frac{\left(\sum v_i\right)^2}{n} = 284.5 - \frac{45^2}{9} = 59.5$$

To determine if the Spearman rank correlation differs from 0, we test:

$H_0: \rho_s = 0$
$H_a: \rho_s \neq 0$

The test statistic is $r_s = \frac{SS_{uv}}{\sqrt{SS_{uu}SS_{vv}}} = \frac{23.5}{\sqrt{59.5(59.5)}} = 0.395$

Reject H_0 if $r_s < -r_{s,\alpha/2}$ or if $r_s > r_{s,\alpha/2}$ where $\alpha/2 = 0.025$ and $n = 9$:

Reject H_0 if $r_s < -0.683$ or $r_s > 0.683$ (from Table XI, Appendix B)

Since the observed value of the test statistic does not fall in the rejection region $(r_s = 0.395 \not> 0.683)$, H_0 is not rejected. There is insufficient evidence to indicate that Spearman's rank correlation between *x* and *y* is significantly different from 0 at $\alpha = 0.05$.

b. Use the Wilcoxon signed rank test. Some preliminary calculations are:

Pair	X	Y	Difference	Rank of Absolute Difference
1	19	12	7	3
2	27	19	8	4.5
3	15	7	8	4.5
4	35	25	10	6
5	13	11	2	1.5
6	29	10	19	8
7	16	16	0	(eliminated)
8	22	10	12	7
9	16	18	−2	1.5
				$T_- = 1.5$

To determine if the probability distribution of *x* is shifted to the right of that for *y*, we test:

H_0: The probability distributions are identical for the two variables
H_a: The probability distribution of *x* is shifted to the right of the probability distribution of *y*

The test statistic is $T = T_- = 1.5$.

Reject H_0 if $T \leq T_0$ where T_0 is based on $\alpha = 0.05$ and $n = 8$ (one-tailed):

Reject H_0 if $T \leq 6$ (from Table X, Appendix B).

Since the observed value of the test statistic falls in the rejection region $(T = 1.5 \leq 6)$, H_0 is rejected. There is sufficient evidence to conclude that the probability distribution of *x* is shifted to the right of that for *y* at $\alpha = 0.05$.

14.115 Some preliminary calculations are:

Block	1	Rank	2	Rank	3	Rank	4	Rank	5	Rank
1	75	4	65	1	74	3	80	5	69	2
2	77	3	69	1	78	4	80	5	72	2
3	70	4	63	1.5	69	3	75	5	63	1.5
4	80	3.5	69	1	80	3.5	86	5	77	2
		$R_1 = 14.5$		$R_2 = 4.5$		$R_3 = 13.5$		$R_4 = 20$		$R_5 = 7.5$

$$\bar{R}_1 = \frac{R_1}{b} = \frac{14.5}{4} = 3.625 \qquad \bar{R}_2 = \frac{R_2}{b} = \frac{4.5}{4} = 1.125 \qquad \bar{R}_3 = \frac{R_3}{b} = \frac{13.5}{4} = 3.375$$

$$\bar{R}_4 = \frac{R_4}{b} = \frac{20}{4} = 5 \qquad \bar{R}_5 = \frac{R_5}{b} = \frac{7.5}{4} = 1.875 \qquad \bar{R} = \frac{k+1}{2} = \frac{5+1}{2} = 3$$

To determine whether at least two of the treatment probability distributions differ in location, use Friedman F_r test.

H_0: The five treatments have identical probability distributions
H_a: At least two of the populations have probability distributions that differ in location

The test statistic is

$$F_r = \frac{12b}{k(k+1)}\sum\left(\bar{R}_j - \bar{R}\right)^2$$

$$= \frac{12(4)}{5(5+1)}\left[(3.625-3)^2 + (1.125-3)^2 + (3.375-3)^2 + (5-3)^2 + (1.875-3)^2\right] = 14.9$$

The rejection region requires $\alpha = 0.05$ in the upper tail of the χ^2 distribution with $df = k-1 = 5-1 = 4$. From Table IV, Appendix B, $\chi^2_{0.05} = 9.48773$. The rejection region is $F_r > 9.48773$.

Since the observed value of the test statistic falls in the rejection region $(F_r = 14.9 > 9.48773)$, H_0 is rejected. There is sufficient evidence to indicate that at least two of the treatment means differ in location at $\alpha = 0.05$.

14.117 To determine if the median cesium amount of lichen differs from 0.003, we test:

$H_0: \eta = 0.003$
$H_a: \eta \neq 0.003$

The test statistic is $S = S_A$ = {number of observations greater than 0.003} =8. (from printout). The p-value is $p = 0.0391$. Since the p-value is less than $\alpha\,(p = 0.0391 < 0.10)$, H_0 is rejected. There is sufficient evidence to indicate the median cesium amount of lichen differs from 0.003 at $\alpha = 0.10$. This result agrees with the t-test in Exercise 8.163.

14.119 Some preliminary calculations:

Theme	High School Teachers	Geography Alumni	Difference $T-A$	Rank of Absolute Differences
Tourism	10	2	8	11
Physical	2	1	1	1.5
Transportation	7	3	4	8
People	1	6	−5	9
History	2	5	−3	6
Climate	6	4	2	3.5
Forestry	5	8	−3	6
Agriculture	7	10	−3	6
Fishing	9	7	2	3.5
Energy	2	8	−6	10
Mining	10	11	−1	1.5
Manufacturing	12	12	0	(eliminated)
			Positive rank sum $T_+ = 27.5$	

To determine if the distributions of theme rankings for the two groups differ, we test:

H_0: The probability distributions for the two populations are identical
H_a: The probability distribution of the high school teachers is shifted to the right or left of the probability distribution of the geography alumni

The test statistic is $T_+ = 27.5$.

Reject H_0 if $T_+ \le T_0$ where T_0 is based on $\alpha = 0.05$ and $n = 11$ (two-tailed):

Reject H_0 if $T_+ \le 11$ (from Table X, Appendix B)

Since the observed value of the test statistic does not fall in the rejection region $(T_+ = 27.5 \not\le 11)$, H_0 is not rejected. There is insufficient evidence to indicate that the distributions of these rankings for the two groups differ at $\alpha = 0.05$. Practically, this means that the thematic content of a new atlas could be based on the views of either educators or geography alumni.

14.121 a. Some preliminary calculations are:

Week	Number of Strikes	Rank u_i	Age of fish	Rank v_i	$d_i=u_i-v_i$	d_i^2
1	85	9	120	1	−8	64
2	63	8	136	2	−6	36
3	34	3	150	3	0	0
4	39	5	155	4	1	1
5	58	7	162	5	2	4
6	35	4	169	6	−2	4
7	57	6	178	7	−1	1
8	12	1	184	8	−7	49
9	15	2	190	9	−7	49
						$\sum d_i^2 = 208$

$$r_s = 1 - \frac{6\sum d_i^2}{n(n^2-1)} = 1 - \frac{6(208)}{9(9^2-1)} = 1 - 1.733 = -0.733$$

b. To determine whether the number of strikes and age are negatively correlated, we test:

$H_0: \rho_s = 0$
$H_a: \rho_s < 0$

The test statistic is $r_s = -0.733$.

Reject H_0 if $r_s < -r_{s,\alpha}$ where $\alpha = 0.01$ and $n = 9$.

Reject H_0 if $r_s < -0.783$ (From Table XI, Appendix B)

Since the observed value of the test statistic does not fall in the rejection region $(r_s = -0.733 \not< -0.783)$, H_0 is not rejected. There is insufficient evidence to indicate that the number of strikes and age are negatively correlated at $\alpha = 0.01$.

14.123 a. To determine if the distributions of chest injury ratings for the driver and passenger differ, we test:

H_0: The distribution of driver chest injury ratings has the same location as the distribution of passenger chest injury ratings
H_a: The distribution of driver chest injury ratings is shifted to the right or left of that for passenger chest injury ratings

b. Using MIINITAB, the results are:

Wilcoxon Signed Rank Test: Diff

```
Test of median = 0.000000 versus median not = 0.000000

              N
             for   Wilcoxon            Estimated
         N  Test  Statistic        P      Median
Diff    18    16       23.0    0.021      -4.000
```

The test statistic is $T_- \leq 19$.

c. Reject H_0 if $T_- \leq T_0$ where T_0 is based on $\alpha = 0.01$ and $n = 16$ pairs (two-tailed)

Reject H_0 if $T_- \leq 19$ (From Table X, Appendix B)

d. Since the observed value of the test statistic does not fall in the rejection region $(T_- = 23 \not\leq 19)$, H_0 is not rejected. There is insufficient evidence to indicate the distributions of chest injury ratings for the driver and passenger differ at $\alpha = 0.01$.

From the printout, the p-value is $p = 0.021$.

14.125 a. The Students' t procedure requires that the populations sampled from be normal. The distributions of the zinc measurements in the 3 locations may not be normal.

b. The ranks of the observations and the sums for the two groups are:

Text line	Rank	Intersection	Rank
0.335	4	0.393	7
0.374	6	0.353	5
0.440	8	0.285	1
		0.295	2
		0.319	3
	$T_1 = 18$		$T_2 = 18$

To determine if the distributions of zinc measurements for the text line and the intersection have different centers of location, we test:

H_0: The distributions of zinc measurements for the text line and the intersection are identical

H_a: The distribution of zinc measurements for the text line is shifted to the right or left of that for the intersection

The test statistic is $T_1 = 18$.

The null hypothesis will be rejected if $T_1 \le T_L$ or $T_1 \ge T_U$ where $\alpha = 0.05$ (two-tailed), $n_1 = 3$ and $n_2 = 5$. From Table IX, Appendix B, $T_L = 6$ and $T_U = 21$. The rejection region is $T_1 \le 6$ or $T_1 \ge 21$.

Since the observed value of the test statistic does not fall in the rejection region $(T_1 = 18 \not\le 6 \text{ or } T_1 = 18 \not\ge 21)$, H_0 is not rejected. There is insufficient evidence to indicate the distributions of zinc measurements for the text line and the intersection have different centers of location at $\alpha = 0.05$.

c. The ranks of the observations and the sums for the two groups are:

Witness line	Rank	Intersection	Rank
0.210	2	0.393	9
0.262	3	0.353	8
0.188	1	0.285	4
0.329	7	0.295	5
0.439	11	0.319	6
0.397	10		
	$T_1 = 34$		$T_2 = 32$

To determine if the distributions of zinc measurements for the witness line and the intersection have different centers of location, we test:

H_0: The distributions of zinc measurements for the witness line and the intersection are identical

H_a: The distribution of zinc measurements for the witness line is shifted to the right or left of that for the intersection

The test statistic is $T_2 = 32$.

The null hypothesis will be rejected if $T_2 \le T_L$ or $T_2 \ge T_U$ where $\alpha = 0.05$ (two-tailed), $n_1 = 6$ and $n_2 = 5$. From Table IX, Appendix B, $T_L = 19$ and $T_U = 41$. The rejection region is $T_2 \le 19$ or $T_2 \ge 41$.

Since the observed value of the test statistic does not fall in the rejection region $(T_2 = 32 \not\le 19 \text{ and } T_2 = 32 \not\ge 41)$, H_0 is not rejected. There is insufficient evidence to indicate the distributions of zinc measurements for the witness line and the intersection have different centers of location at $\alpha = 0.05$.

d. Some preliminary calculations are:

Text line	Rank	Witness line	Rank	Inter-section	Rank
0.335	8	0.210	2	0.393	11
0.374	10	0.262	3	0.353	9
0.440	14	0.188	1	0.285	4
		0.329	7	0.295	5
		0.439	13	0.319	6
		0.397	12		
	$R_1 = 32$		$R_2 = 38$		$R_3 = 35$

$$\bar{R}_1 = \frac{R_1}{n_1} = \frac{32}{3} = 10.667, \quad \bar{R}_2 = \frac{R_2}{n_2} = \frac{38}{6} = 6.333, \quad \bar{R}_3 = \frac{R_3}{n_3} = \frac{35}{5} = 7,$$

$$\bar{R} = \frac{n+1}{2} = \frac{14+1}{2} = 7.5$$

To determine if the probability distributions of zinc measurements differ in location among the notebook locations, we test:

H_0: The probability distributions of zinc measurements for the three notebook locations are identical

H_a: At least two of the three probability distributions differ in location

The test statistic is

$$H = \frac{12}{n(n+1)}\sum n_j\left(\bar{R}_j - \bar{R}\right)^2$$

$$= \frac{12}{14(14+1)}\left[3(10.667-7.5)^2 + 6(6.333-7.5)^2 + 5(7-7.5)^2\right] = 2.258$$

The rejection region requires $\alpha = 0.05$ in the upper tail of the χ^2 distribution with $df = k - 1 = 3 - 1 = 2$. From Table IV, Appendix B, $\chi^2_{0.05} = 5.99147$. The rejection region is $H > 5.99147$.

Since the observed value of the test statistic does not fall in the rejection region $(H = 2.258 \not> 5.99147)$, H_0 is not rejected. There is insufficient evidence to indicate the probability distributions of zinc measurements for the three notebook locations differ in location at $\alpha = 0.05$.

e. The median zinc score for the text line is 0.374. The median zinc score for the witness line is 0.296. The median zinc score for the intersection is 0.319. We found no difference in the distributions of zinc scores among the three notebook locations in part d. We found no difference in the distributions of zinc scores between the text line and the intersection in part b. We also found no difference in the distributions of zinc scores between the witness line and the intersection in part c. Although we did not compare the text line and the witness line zinc scores, we can still conclude these two locations are not different. These two distributions appear to be the furthest apart and there was no difference among all three notebook locations from part d.

14.127 Some preliminary calculations are:

Metal	I	Rank	II	Rank	III	Rank
1	4.6	2	4.2	1	4.9	3
2	7.2	3	6.4	1	7.0	2
3	3.4	1.5	3.5	3	3.4	1.5
4	6.2	3	5.3	1	5.9	2
5	8.4	3	6.8	1	7.8	2
6	5.6	2	4.8	1	5.7	3
7	3.7	1.5	3.7	1.5	4.1	3
8	6.1	1	6.2	2	6.4	3
9	4.9	3	4.1	1	4.2	2
10	5.2	3	5.0	1	5.1	2
		$R_1 = 23$		$R_2 = 13.5$		$R_3 = 23.5$

$\bar{R}_1 = \frac{R_1}{b} = \frac{23}{10} = 2.3$ $\quad \bar{R}_2 = \frac{R_2}{b} = \frac{13.5}{10} = 1.35$ $\quad \bar{R}_3 = \frac{R_3}{b} = \frac{23.5}{10} = 2.35$ $\quad \bar{R} = \frac{k+1}{2} = \frac{3+1}{2} = 2$

To determine if there is a difference in the probability distributions of the amounts of corrosion among the three types of sealers, we test:

H_0: The probability distributions of corrosion amounts are identical for the three types of sealers

H_a: At least two of the probability distributions differ in location

The test statistic is

$$F_r = \frac{12b}{k(k+1)}\sum(\bar{R}_j - \bar{R})^2 = \frac{12(10)}{3(3+1)}\left[(2.3-2)^2 + (1.35-2)^2 + (2.35-2)^2\right] = 6.35$$

The rejection region requires $\alpha = 0.05$ in the upper tail of the χ^2 distribution with $df = k - 1 = 3 - 1 = 2$. From Table IV, Appendix B, $\chi^2_{0.05} = 5.99147$. The rejection region is $F_r > 5.99147$.

Since the observed value of the test statistic falls in the rejection region $(F_r = 6.35 > 5.99147)$, reject H_0. There is sufficient evidence to indicate a difference in the probability distributions among the three types of sealers at $\alpha = 0.05$.

14.129 Some preliminary calculations are:

UMRB-1	Rank	UMRB-2	Rank	UMBR-3	Rank	SD	Rank	SWRA	Rank
3.75	16	3.32	12	4.06	22	2.73	3.5	2.73	3.5
4.05	21	4.09	23.5	4.56	25	2.95	5	2.55	2
3.81	17	3.90	20	3.60	15	2.25	1	3.06	6
3.23	9	5.06	26	3.27	10				
3.13	7	3.85	18	4.09	23.5				
3.30	11	3.88	19	3.38	14				
3.21	8			3.37	13				
	$R_1 = 89$		$R_2 = 118.5$		$R_3 = 122.5$		$R_4 = 9.5$		$R_5 = 11.5$

$$\bar{R}_1 = \frac{R_1}{n_1} = \frac{89}{7} = 12.714, \quad \bar{R}_2 = \frac{R_2}{n_2} = \frac{118.5}{6} = 19.75, \quad \bar{R}_3 = \frac{R_3}{n_3} = \frac{122.5}{7} = 17.5,$$

$$\bar{R}_4 = \frac{R_4}{n_4} = \frac{9.5}{3} = 3.167, \quad \bar{R}_5 = \frac{R_5}{n_5} = \frac{11.5}{3} = 3.833 \quad \bar{R} = \frac{n+1}{2} = \frac{26+1}{2} = 13.5$$

To determine if the distributions of ratios differ in location among the 5 groups, we test:

H_0: The distributions ratios for the 5 groups are identical
H_a: At least two of the five distributions differ in location

The test statistic is

$$H = \frac{12}{n(n+1)}\sum n_j\left(\bar{R}_j - \bar{R}\right)^2$$

$$= \frac{12}{26(26+1)}\left[\begin{array}{c}7(12.714-13.5)^2+6(19.75-13.5)^2+7(17.5-13.5)^2\\ +3(3.167-13.5)^2+3(3.833-13.5)^2\end{array}\right] = 16.263$$

The rejection region requires $\alpha = 0.10$ in the upper tail of the χ^2 distribution with $df = k-1 = 5-1 = 4$. From Table IV, Appendix B, $\chi^2_{0.10} = 7.77944$. The rejection region is $H > 7.77944$.

Since the observed value of the test statistic falls in the rejection region $(H = 16.263 > 7.77944)$, H_0 is rejected. There is sufficient evidence to indicate the distributions of ratios differ in location among the 5 groups at $\alpha = 0.10$.

14.131 a. Some preliminary calculations are:

Change KSADS-MRS	Rank, *u*	Improve CGI-BP	Rank, *v*	u^2	v^2	*uv*
80	18	6	18	324	324	324
65	17	5	17	289	289	289
20	15.5	4	15	240.25	225	232.5
−15	13	4	15	169	225	195
−50	9	4	15	81	225	135
20	15.5	3	12	240.25	144	186
−30	11	3	12	121	144	132
−70	5.5	3	12	30.25	144	66
−10	14	2	6	196	36	84
−25	12	2	6	144	36	72
−35	10	2	6	100	36	60
−65	7.5	2	6	56.25	36	45
−65	7.5	2	6	26.25	36	45
−70	5.5	2	6	30.25	36	33
−80	4	2	6	16	36	24
−90	2.5	2	6	6.25	36	15
−95	1	2	6	1	36	6
−90	2.5	1	1	6.25	1	2.5
	$\sum u = 171$		$\sum v = 171$	$\sum u^2 = 2107$	$\sum v^2 = 2045$	$\sum uv = 1946$

$$SS_{uv} = \sum uv - \frac{\sum u \sum v}{n} = 1946 - \frac{171(171)}{18} = 321.5$$

$$SS_{uu} = \sum u^2 - \frac{\left(\sum u\right)^2}{n} = 2107 - \frac{171^2}{18} = 482.5$$

$$SS_{vv} = \sum v^2 - \frac{\left(\sum v\right)^2}{n} = 2045 - \frac{171^2}{18} = 420.5$$

$$r_s = \frac{SS_{uv}}{\sqrt{SS_{uu}SS_{vv}}} = \frac{321.5}{\sqrt{482.5(420.5)}} = .714$$

b. To determine if there is a positive rank correlation between the two test score changes in the population of all pediatric patients with manic symptoms, we test:

$H_0: \rho_s = 0$
$H_a: \rho_s > 0$

The test statistic is $r_s = 0.714$.

Reject H_0 if $r_s > r_{0.05}$ with $n = 18$

Reject H_0 if $r_s > 0.399$ (From Table XI, Appendix B)

Since the observed value of the test statistic falls in the rejection region $(r_s = 0.714 > 0.399)$, H_0 is rejected. There is sufficient evidence to indicate there is a positive rank correlation between the two test score changes in the population of all pediatric patients with manic symptoms at $\alpha = 0.05$.

14.133 a. Answers may vary. Possible answer may include:

The nonparametric test requires no assumptions concerning the population probability distributions. Since each court possesses different requirements and restrictions, the DEF values for the three populations may not be normally distributed with a common variance.

b. The Kruskal-Wallis *H*-test is appropriate for this analysis because the comparison involves three populations.

H_0: The probability distributions of the DEF values are the same for all three tax courts.
H_a: At least two of the three tax courts have probability distributions of DEF values that differ in location.

c. $\bar{R}_1 = \frac{R_1}{n_1} = \frac{5{,}335}{67} = 79.627$, $\bar{R}_2 = \frac{R_2}{n_2} = \frac{3{,}937}{57} = 69.070$, $\bar{R}_3 = \frac{R_3}{n_3} = \frac{3{,}769}{37} = 101.865$,

$$\bar{R} = \frac{1}{2}(n+1) = \frac{1}{2}(161+1) = 81$$

$$H = \frac{12}{n(n+1)}\sum n_j\left(\bar{R}_j - \bar{R}\right)^2$$
$$= \frac{12}{161(161+1)}\left[67(79.627-81)^2 + 57(69.070-81)^2 + 37(101.865-81)^2\right] = 11.20$$

d. The *p*-value is $p = 0.0037$. Since the *p*-value is so small, H_0 would be rejected for any $\alpha > 0.0037$. There is sufficient evidence to indicate that the mean DEF values for the three tax courts differ for $\alpha > 0.0037$.

14.135 To determine if 50% of the ingots have a freckle index of 10 or higher, we test:

$$H_0: \eta = 10$$
$$H_a: \eta > 10$$

$S = S_A =$ {number of measurements greater than 10} = 10.

The $p-\text{value} = P(x \geq 10)$ where x is a binomial random variable with $n = 17$ and $p = 0.5$. (One observation is eliminated because it is equal to 10.) Using MINITAB, the $p-\text{value} = P(x \geq 10) = 1 - P(x \leq 9) = 1 - 0.6855 = 0.3145$.

Since the $p-\text{value} = 0.3145 \not< \alpha = 0.01$, H_0 is not rejected. There is insufficient evidence to indicate that 50% of the ingots have a freckle index of 10 or higher at $\alpha = 0.01$.

14.137 a. The largest rank correlation is $r_s = 0.592$ and is between Methods I and III. There is a moderate positive relationship between Method I and Method III.

b. The smallest rank correlation is $r_s = 0.189$ and is between Methods I and II. There is a weak positive relationship between Method I and Method II.